D1626053

Nuclear Energy
Sixth Edition

Nuclear Energy

An Introduction to the Concepts, Systems, and Applications of Nuclear Processes

Sixth Edition

Raymond L. Murray

Nuclear Engineering Department

North Carolina State University

AMSTERDAM • BOSTON • HEIDELBERG • LONDON
NEW YORK • OXFORD • PARIS • SAN DIEGO
SAN FRANCISCO • SINGAPORE • SYDNEY • TOKYO

Butterworth-Heinemann is an imprint of Elsevier

Butterworth-Heinemann is an imprint of Elsevier
30 Corporate Drive, Suite 400, Burlington, MA 01803, USA
Linacre House, Jordan Hill, Oxford OX2 8DP, UK

Library of Congress Cataloging-in-Publication Data
Murray, Raymond LeRoy, 1920-
Nuclear energy : an introduction to the concepts, systems, and applications of nuclear
processes / author, Raymond L. Murray. – 6th ed.
p. cm.
1. Nuclear engineering. 2. Nuclear energy. I. Title.
TK9145.M87 2008
621.48–dc22
2008039209

British Library Cataloguing-in-Publication Data
A catalogue record for this book is available from the British Library.

ISBN: 978-0-12-370547-1

For information on all Butterworth–Heinemann publications
visit our Web site at www.elsevierdirect.com

Printed in the United States of America
    09 10 11 12 10 9 8 7 6 5 4 3 2

For Elizabeth, Steve and Carol, Ilah and Jim,
Marshall and Tonia, Tucker and Lin, Michael, Nancy and Rob,
Jim and Susan, Nick and Katy, Andy and Nicole, Ashley and Jesse,
Alfia and Art, Joshua, Kathryn and Doug, Ian and Barb,
Ryan and Catharine, Kerri, Amber, Saura and Kevin, Amos,
Hannah, Ben, Kelsey, Alfred, Andrew, Makena, Madeleine,
Merritt, Jamie, Sidney, Dominique, Ethan, Lauren, Jillian,
Evan, Adrian, and Natalie

Contents

Preface xv
About the Author xvii

Part I Basic Concepts

Chapter 1 Energy **3**
  1.1 Forces and Energy 3
  1.2 Thermal Energy 5
  1.3 Radiant Energy 7
  1.4 The Equivalence of Matter and Energy 8
  1.5 Energy and the World 9
  1.6 Summary 10
  1.7 Exercises 10
  1.8 General References 12
  1.9 References 13

Chapter 2 Atoms and Nuclei **15**
  2.1 Atomic Theory 15
  2.2 Gases 16
  2.3 The Atom and Light 16
  2.4 Laser Beams 20
  2.5 Nuclear Structure 21
  2.6 Sizes and Masses of Nuclei 23
  2.7 Binding Energy 23
  2.8 Summary 25
  2.9 Exercises 25
  2.10 References 27

Chapter 3 Radioactivity **29**
  3.1 Radioactive Decay 29
  3.2 The Decay Law 30
  3.3 Radioactive Chains 33
  3.4 Measurement of Half-Life 35
  3.5 Summary 36
  3.6 Exercises 37
  3.7 References 39

Chapter 4 Nuclear Processes **41**
  4.1 Transmutation of Elements 41

4.2 Energy and Momentum Conservation 43
4.3 Reaction Rates 44
4.4 Particle Attenuation 48
4.5 Neutron Cross Sections 49
4.6 Neutron Migration 51
4.7 Summary 56
4.8 Exercises 56
4.9 References 60

Chapter 5 Radiation and Materials 63
5.1 Excitation and Ionization by Electrons 63
5.2 Heavy Charged Particle Stopping by Matter 64
5.3 Gamma Ray Interactions with Matter 66
5.4 Neutron Reactions 70
5.5 Summary 70
5.6 Exercises 71
5.7 References 72

Chapter 6 Fission 73
6.1 The Fission Process 73
6.2 Energy Considerations 74
6.3 Byproducts of Fission 76
6.4 Energy from Nuclear Fuels 79
6.5 Summary 80
6.6 Exercises 80
6.7 References 81

Chapter 7 Fusion 83
7.1 Fusion Reactions 83
7.2 Electrostatic and Nuclear Forces 84
7.3 Thermonuclear Reactions in a Plasma 86
7.4 Summary 88
7.5 Exercises 88
7.6 References 89

Part II Nuclear Systems

Chapter 8 Particle Accelerators 93
8.1 Electric and Magnetic Forces 93
8.2 High-Voltage Machines 95
8.3 Linear Accelerator 96
8.4 Cyclotron and Betatron 97
8.5 Synchrotron and Collider 99

8.6 Spallation 102

8.7 Summary 103

8.8 Exercises 104

8.9 References 106

Chapter 9   Isotope Separators 109

9.1 Mass Spectrograph 109

9.2 Gaseous Diffusion Separator 110

9.3 Gas Centrifuge 116

9.4 Laser Isotope Separation 118

9.5 Separation of Deuterium 119

9.6 Summary 119

9.7 Exercises 120

9.8 References 122

Chapter 10   Radiation Detectors 125

10.1 Gas Counters 126

10.2 Neutron Detectors 127

10.3 Scintillation Counters 129

10.4 Solid State Detectors 130

10.5 Statistics of Counting 132

10.6 Pulse Height Analysis 133

10.7 Advanced Detectors 134

10.8 Detectors and Counterterrorism 135

10.9 Summary 136

10.10 Exercises 136

10.11 References 139

Chapter 11   Neutron Chain Reactions 141

11.1 Criticality and Multiplication 141

11.2 Multiplication Factors 142

11.3 Neutron Flux and Reactor Power 147

11.4 Reactor Types 148

11.5 Reactor Operation 152

11.6 The Natural Reactor 156

11.7 Summary 156

11.8 Exercises 156

11.9 References 159

Chapter 12   Nuclear Heat Energy 161

12.1 Methods of Heat Transmission 161

12.2 Heat Generation and Removal 161

12.3 Steam Generation and Electrical Power Production 167

12.4 Waste Heat Rejection. 169

12.5 Summary 173

12.6 Exercises 174

12.7 References 176

Chapter 13   Breeder Reactors. 179

13.1 The Concept of Breeding. 179

13.2 Isotope Production and Consumption. 181

13.3 The Fast Breeder Reactor 182

13.4 Integral Fast Reactor 186

13.5 Breeding and Uranium Resources 187

13.6 Summary 191

13.7 Exercises 191

13.8 References 192

Chapter 14   Fusion Reactors 195

14.1 Comparison of Fusion Reactions. 195

14.2 Requirements for Practical Fusion Reactors 197

14.3 Magnetic Confinement Machines 199

14.4 Inertial Confinement Machines. 203

14.5 Other Fusion Concepts 206

14.6 Prospects for Fusion 208

14.7 Summary 211

14.8 Exercises 211

14.9 References 212

Part III Nuclear Energy and Man

Chapter 15   The History of Nuclear Energy. 217

15.1 The Rise of Nuclear Physics. 217

15.2 The Discovery of Fission 218

15.3 The Development of Nuclear Weapons. 219

15.4 Reactor Research and Development 221

15.5 The Nuclear Controversy. 223

15.6 Summary 225

15.7 Exercises 226

15.8 References 226

Chapter 16   Biological Effects of Radiation. 229

16.1 Physiological Effects 230

16.2 Radiation Dose Units 231

16.3 Basis for Limits of Exposure 234

16.4 Sources of Radiation Dosage 238
16.5 Radiation and Terrorism 239
16.6 Summary 240
16.7 Exercises 240
16.8 References 241

Chapter 17 Information from Isotopes 245
17.1 Stable and Radioactive Isotopes 246
17.2 Tracer Techniques 246
17.3 Radiopharmaceuticals 248
17.4 Medical Imaging 249
17.5 Radioimmunoassay 250
17.6 Dating 252
17.7 Neutron Activation Analysis 253
17.8 Radiography 257
17.9 Radiation Gauges 259
17.10 Summary 261
17.11 Exercises 262
17.12 References 264

Chapter 18 Useful Radiation Effects 269
18.1 Medical Treatment 269
18.2 Radiation Preservation of Food 271
18.3 Sterilization of Medical Supplies 276
18.4 Pathogen Reduction 277
18.5 Crop Mutations 277
18.6 Insect Control 278
18.7 Applications in Chemistry 279
18.8 Transmutation Doping of Semiconductors 279
18.9 Neutrons in Fundamental Physics 280
18.10 Neutrons in Biological Studies 282
18.11 Research with Synchrotron X-rays 283
18.12 Summary 284
18.13 Exercises 284
18.14 References 285

Chapter 19 Reactor Safety and Security 289
19.1 Neutron Population Growth 289
19.2 Assurance of Safety 293
19.3 Emergency Core Cooling and Containment 300
19.4 Probabilistic Risk Assessment 303
19.5 The Three Mile Island Accident and Lessons Learned 306

19.6 The Chernobyl Accident 309
19.7 Philosophy of Safety 311
19.8 Nuclear Security 312
19.9 Summary 313
19.10 Exercises 314
19.11 References 318

Chapter 20 **Nuclear Propulsion** **323**
20.1 Reactors for Naval Propulsion 323
20.2 Space Reactors 325
20.3 Space Isotopic Power 327
20.4 Future Nuclear Space Applications 331
20.5 Summary 334
20.6 Exercises 334
20.7 References 335

Chapter 21 **Radiation Protection** **339**
21.1 Protective Measures 339
21.2 Calculation of Dose 341
21.3 Effects of Distance and Shielding 342
21.4 Internal Exposure 348
21.5 The Radon Problem 349
21.6 Environmental Radiological Assessment 350
21.7 Newer Radiation Standards 352
21.8 Summary 355
21.9 Exercises 355
21.10 References 358

Chapter 22 **Radioactive Waste Disposal** **361**
22.1 The Nuclear Fuel Cycle 361
22.2 Waste Classification 363
22.3 Spent Fuel Storage 365
22.4 Transportation 366
22.5 Reprocessing 369
22.6 High-Level Waste Disposal 372
22.7 Low-Level Waste Generation, Treatment, and Disposal 378
22.8 Environmental Restoration of Defense Sites 385
22.9 Nuclear Power Plant Decommissioning 386
22.10 Summary 387
22.11 Exercises 388
22.12 References 391

Chapter 23   Laws, Regulations, and Organizations **395**
   23.1 The Atomic Energy Acts 395
   23.2 The Environmental Protection Agency 396
   23.3 The Nuclear Regulatory Commission 397
   23.4 The Department of Energy. 399
   23.5 International Atomic Energy Agency. 400
   23.6 Institute of Nuclear Power Operations 401
   23.7 Department of Homeland Security 404
   23.8 Other Organizations 404
   23.9 Energy Policy Acts. 408
  23.10 Summary 409
  23.11 References 410

Chapter 24   Energy Economics **415**
   24.1 Components of Electrical Power Cost. 415
   24.2 Forecasts and Reality. 418
   24.3 Technical and Institutional Improvements. 423
   24.4 Effect of Deregulation and Restructuring. 425
   24.5 Advanced Reactors 427
   24.6 Nuclear Power Renaissance 429
   24.7 Summary 433
   24.8 Exercises 433
   24.9 References 434

Chapter 25   International Nuclear Power **437**
   25.1 Reactor Distribution 438
   25.2 Western Europe 441
   25.3 Eastern Europe and the CIS 443
   25.4 The Far East 445
   25.5 Other Countries 446
   25.6 Summary 448
   25.7 References 449

Chapter 26   Nuclear Explosions **453**
   26.1 Nuclear Power Versus Nuclear Weapons. 453
   26.2 Nuclear Explosives 454
   26.3 The Prevention of Nuclear War 460
   26.4 Nonproliferation and Safeguards. 463
   26.5 IAEA Inspections. 465
   26.6 Production of Tritium 466
   26.7 Management of Weapons Uranium and Plutonium. 467

26.8	Summary	469
26.9	Exercises	469
26.10	References	470

Chapter 27   The Future **473**

27.1	Dimensions	474
27.2	World Energy Use	475
27.3	Nuclear Energy and Sustainable Development	478
27.4	Greenhouse Effect and Global Climate Change	480
27.5	Perspectives	483
27.6	Desalination	487
27.7	Recycling and Breeding	488
27.8	The Hydrogen Economy	489
27.9	Next Generation Nuclear Plant	491
27.10	Summary	493
27.11	Exercises	493
27.12	References	494

Appendix **499**

Conversion Factors	499
Atomic and Nuclear Data	500
Answers to Exercises	504
Comments	509
Original Computer Programs	510

Index **513**

Preface

The prospects for continued and new nuclear power plants in the United States have improved greatly since the fifth edition was written. In what is called a nuclear renaissance, many utilities have made application to the Nuclear Regulatory Commission for license extension and approval for new reactor construction. Nuclear reactors are planned that combat global warming, conserve nuclear fuel, support desalination, and produce hydrogen for transportation.

Since the terrorist attacks of September 11, 2001, applications of nuclear processes have become more important. Detectors of nuclear materials entering the United States are being installed, and physical protection of nuclear facilities has been greatly enhanced.

The classical method of collecting information by consultation of books, reports, and articles in a library has been essentially replaced by use of Internet search engines, especially Google. Often, it is more convenient to try search words or phrases instead of typing in the URL of a Web site. The down side of Internet use is the omission of dates of posting, the lack of peer review of Web contents, and the tendency of sites to vanish. These problems are partly balanced by the wealth of information made available conveniently and instantly.

I am convinced that student learning is enhanced by performing calculations on nuclear quantities. The new edition provides Exercises, solvable by hand-held calculator, with Answers to Exercises given in the Appendix.

Computer programs in Qbasic, Excel, and MATLAB for the solution of computer exercises in the text can be found at http://elsevierdirect.com/companions/9780123705471.

For faculty who use the text in an academic course, instructor support materials, including solutions to exercises and PowerPoint slides, are available by registering at http://textbooks.elsevier.com.

As stated in the preface to the first edition (1975), the book "is designed for use by anyone who wishes to know about the role of nuclear energy in our society or to learn nuclear concepts for use in professional work." I hope that the book will benefit both future nuclear leaders and interested members of the public.

Each of the editions has dealt with events and trends. Included were the need for new and different sources of energy, United States government activities and reorganizations, the Arab oil embargo, the stagnation of nuclear power, the TMI-2 and Chernobyl accidents, the end of the Cold War, growth in applications of radioisotopes and radiation, the persistent nuclear waste problem, continued safe plant operation, and predictions of a brighter nuclear future.

Communication by e-mail (murray@eos.ncsu.edu) with teachers, students, and other users of the book will be most welcome.

Many persons have provided valuable ideas and information. They are recognized at appropriate places in the book. The author has appreciated the interaction with Dr. Randy J. Jost of Utah State University and welcomes his continued upgrading of computer programs and the creation of slides and other material for use by instructors. Thanks go to Dr. Keith E. Holbert of Arizona State University for his thorough review of the complete manuscript of this book. Special thanks are due Nancy Reid Baker for vital computer support, for preparation of artwork, and for formatting copy to ensure completeness and correctness.

Thanks are due members of the Elsevier team, who provided advice and prepared the text for publication—Joseph Hayton, Publisher; Maria Alonso, Assistant Editor; Eric DeCicco, Senior Designer; Anne McGee, Production Manager; and Suja Narayana, Project Manager.

I appreciated encouragement by my wife, Elizabeth Reid Murray.

RAYMOND L. MURRAY
Raleigh, North Carolina, 2008

About the Author

Raymond L. Murray (Ph.D., University of Tennessee) is Professor Emeritus in the Department of Nuclear Engineering of North Carolina State University. His technical interests include reactor analysis, nuclear criticality safety, radioactive waste management, and applications of computers.

Dr. Murray studied under J. Robert Oppenheimer at the University of California at Berkeley. In the Manhattan Project of World War II, he contributed to the uranium isotope separation process at Berkeley and Oak Ridge.

In the early 1950s, he helped found the first university nuclear engineering program and the first university nuclear reactor. During his 30 years of teaching and research in reactor analysis at North Carolina State, he taught many of our leaders in universities and industry throughout the world. He is the author of textbooks in physics and nuclear technology and the recipient of a number of awards, including the Eugene P. Wigner Reactor Physicist Award of the American Nuclear Society in 1994. He is a Fellow of the American Physical Society, a Fellow of the American Nuclear Society, and a member of several honorary, scientific, and engineering societies.

Since retirement from the university, Dr. Murray has been a consultant on criticality for the Three Mile Island Recovery Program, served as chairman of the North Carolina Radiation Protection Commission, and served as chairman of the North Carolina Low-Level Radioactive Waste Management Authority. He provides an annual lecture at MIT for the Institute of Nuclear Power Operations.

Basic Concepts

Energy

OUR MATERIAL world is composed of many substances distinguished by their chemical, mechanical, and electrical properties. They are found in nature in various physical states—the familiar solid, liquid, and gas, along with the ionic "plasma." However, the apparent diversity of kinds and forms of material is reduced by the knowledge that there are only a little more than 100 distinct chemical elements and that the chemical and physical features of substances depend merely on the strength of force bonds between atoms.

In turn, the distinctions between the elements of nature arise from the number and arrangement of basic particles—electrons, protons, and neutrons. At both the atomic and nuclear levels, the structure of elements is determined by internal forces and energy.

1.1 FORCES AND ENERGY

A limited number of basic forces exist—gravitational, electrostatic, electromagnetic, and nuclear. Associated with each of these is the ability to do work. Thus energy in different forms may be stored, released, transformed, transferred, and "used" in both natural processes and man-made devices. It is often convenient to view nature in terms of only two basic entities—particles and energy. Even this distinction can be removed, because we know that matter can be converted into energy and vice versa.

Let us review some principles of physics needed for the study of the release of nuclear energy and its conversion into thermal and electrical form. We recall that if a constant force F is applied to an object to move it a distance s, the amount of work done is the product Fs. As a simple example, we pick up a book from the floor and place it on a table. Our muscles provide the means to lift against the force of gravity on the book. We have done work on the object, which now possesses stored energy (potential energy), because it could do work if allowed to fall back to the original level. Now a force F acting on a mass m provides an acceleration a, given by Newton's law $F = ma$. Starting from rest, the object gains a speed v,

3

and at any instant has energy of motion (kinetic energy) in amount $E_k = \frac{1}{2}m \, v^2$. For objects falling under the force of gravity, we find that the potential energy is reduced as the kinetic energy increases, but the sum of the two types remains constant. This is an example of the principle of conservation of energy. Let us apply this principle to a practical situation and perform some illustrative calculations.

As we know, falling water provides one primary source for generating electrical energy. In a hydroelectric plant, river water is collected by a dam and allowed to fall through a considerable distance. The potential energy of water is thus converted into kinetic energy. The water is directed to strike the blades of a turbine, which turns an electric generator.

The potential energy of a mass m located at the top of the dam is $E_p = Fh$, being the work done to place it there. The force is the weight $F = mg$, where g is the acceleration of gravity. Thus $E_p = mgh$. For example, for 1 kg and 50 m height of dam, using $g = 9.8$ m/s^{2†}, E_p is $(1)(9.8)(50) = 490$ joules (J). Ignoring friction, this amount of energy in kinetic form would appear at the bottom. The water speed would be $v = \sqrt{2E_k/m} = 31.3$ m/s.

Energy takes on various forms, classified according to the type of force that is acting. The water in the hydroelectric plant experiences the force of gravity, and thus gravitational energy is involved. It is transformed into mechanical energy of rotation in the turbine, which is then converted to electrical energy by the generator. At the terminals of the generator, there is an electrical potential difference, which provides the force to move charged particles (electrons) through the network of the electrical supply system. The electrical energy may then be converted into mechanical energy as in motors, into light energy as in light bulbs, into thermal energy as in electrically heated homes, or into chemical energy as in a storage battery.

The automobile also provides familiar examples of energy transformations. The burning of gasoline releases the chemical energy of the fuel in the form of heat, part of which is converted to energy of motion of mechanical parts, while the rest is transferred to the atmosphere and highway. Electricity is provided by the automobile's generator for control and lighting. In each of these examples, energy is changed from one form to another but is not destroyed. The conversion of heat to other forms of energy is governed by two laws, the first and second laws of thermodynamics. The first states that energy is conserved; the second specifies inherent limits on the efficiency of the energy conversion.

Energy can be classified according to the primary source. We have already noted two sources of energy: falling water and the burning of the chemical fuel gasoline, which is derived from petroleum, one of the main fossil fuels. To these

---

[†]The standard acceleration of gravity is 9.80665 m/s^2. For discussion and simple illustrative purposes, numbers will be rounded off to two or three significant figures. Only when accuracy is important will more figures or decimals be used. The principal source of physical constants, conversion factors, and nuclear properties will be the *CRC Handbook of Chemistry and Physics* (see References), which is likely to be accessible to the faculty member, student, or reader.

we can add solar energy; the energy from winds, tides, or the sea motion; and heat from within the earth. Finally, we have energy from nuclear reactions (i.e., the "burning" of nuclear fuel).

1.2 THERMAL ENERGY

Of special importance to us is thermal energy, as the form most readily available from the sun, from burning of ordinary fuels, and from the fission process. First, we recall that a simple definition of the temperature of a substance is the number read from a measuring device such as a thermometer in intimate contact with the material. If energy is supplied, the temperature rises (e.g., energy from the sun warms the air during the day). Each material responds to the supply of energy according to its internal molecular or atomic structure, characterized on a macroscopic scale by the specific heat c. If an amount of thermal energy added to 1 gram of the material is Q, the temperature rise, ΔT, is Q/c. The value of the specific heat for water is $c = 4.18$ J/g-°C and thus it requires 4.18 J of energy to raise the temperature of 1 gram of water by 1 degree Celsius (1°C).

From our modern knowledge of the atomic nature of matter, we readily appreciate the idea that energy supplied to a material increases the motion of the individual particles of the substance. Temperature can thus be related to the average kinetic energy of the atoms. For example, in a gas such as air, the average energy of translational motion of the molecules $\bar{E}$ is directly proportional to the temperature T, through the relation $\bar{E} = \frac{3}{2} kT$, where k is Boltzmann's constant, 1.38×10^{-23} J/K. (Note that the Kelvin scale has the same spacing of degrees as does the Celsius scale, but its zero is at -273°C.)

To gain an appreciation of molecules in motion, let us find the typical speed of oxygen molecules at room temperature 20°C, or 293K. The molecular weight is 32, and because one unit of atomic weight corresponds to 1.66×10^{-27} kg, the mass of the oxygen (O_2) molecule is 5.30×10^{-26} kg. Now

$$\bar{E} = \frac{3}{2}\left(1.38 \times 10^{-23}\right)(293) = 6.07 \times 10^{-21} \text{J}$$

and thus the speed is

$$v = \sqrt{2\bar{E}/m} = \sqrt{2(6.07) \times 10^{-21})/(5.30 \times 10^{-26})} \cong 479 \; m/s.$$

Closely related to energy is the physical entity *power*, which is the rate at which work is done. To illustrate, let the flow of water in the plant of Section 1.1 be 2×10^6 kg/s. For each kg the energy is 490 J; the energy per second is (2×10^6) $(490) = 9.8 \times 10^8$ J/s. For convenience, the unit joule per second is called the watt (W). Our plant thus involves 9.8×10^8 W. We can conveniently express this in kilowatts (1 kW = 10^3 W) or megawatts (1 MW = 10^6 W). Such multiples of units are used because of the enormous range of magnitudes of quantities in nature—from the submicroscopic to the astronomical. The standard set of prefixes is given in Table 1.1.

Table 1.1 Prefixes for Numbers and Abbreviations

yotta	Y	10^{24}	deci	d	10^{-1}
zetta	Z	10^{21}	centi	c	10^{-2}
exa	E	10^{18}	milli	m	10^{-3}
peta	P	10^{15}	micro	m	10^{-6}
tera	T	10^{12}	nano	n	10^{-9}
giga	G	10^{9}	pico	p	10^{-12}
mega	M	10^{6}	femto	f	10^{-15}
kilo	k	10^{3}	atto	a	10^{-18}
hecto	h	10^{2}	zepto	z	10^{-21}
deca	da	10^{1}	yocto	y	10^{-24}

For many purposes we will use the metric system of units, more precisely designated as SI, Systeme Internationale. In this system (see References) the base units are the kilogram (kg) for mass, the meter (m) for length, the second (s) for time, the mole (mol) for amount of substance, the ampere (A) for electric current, the kelvin (K) for thermodynamic temperature, and the candela (cd) for luminous intensity. However, for understanding of the earlier literature, one requires a knowledge of other systems. The Appendix includes a table of useful conversions from British units to SI units.

The transition in the United States from British units to SI units has been much slower than expected. In the interest of ease of understanding by the typical reader, a dual display of numbers and their units appears frequently. Familiar and widely used units such as the centimeter, the barn, the curie, and the rem are retained.

In dealing with forces and energy at the level of molecules, atoms, and nuclei, it is conventional to use another energy unit, the *electron-volt* (eV). Its origin is electrical in character, being the amount of kinetic energy that would be imparted to an electron (charge 1.60×10^{-19} coulombs) if it were accelerated through a potential difference of 1 volt. Because the work done on 1 coulomb would be 1 J, we see that $1 \text{ eV} = 1.60 \times 10^{-19}$ J. The unit is of convenient size for describing atomic reactions. For instance, to remove the one electron from the hydrogen atom requires 13.5 eV of energy. However, when dealing with nuclear forces, which are very much larger than atomic forces, it is preferable to use the million-electron-volt unit (MeV). To separate the neutron from the proton in the nucleus of heavy hydrogen, for example, requires an energy of about 2.2 MeV (i.e., 2.2×10^{6} eV).

1.3 RADIANT ENERGY

Another form of energy is electromagnetic or radiant energy. We recall that this energy may be released by heating of solids, as in the wire of a light bulb; by electrical oscillations, as in radio or television transmitters; or by atomic interactions, as in the sun. The radiation can be viewed in either of two ways—as a wave or as a particle—depending on the process under study. In the wave view it is a combination of electric and magnetic vibrations moving through space. In the particle view it is a compact moving uncharged object, the photon, which is a bundle of pure energy, having mass only by virtue of its motion. Regardless of its origin, all radiation can be characterized by its frequency, which is related to speed and wavelength. Letting c be the speed of light, λ its wavelength, and v its frequency, we have $c = \lambda v$.[†] For example, if c in a vacuum is 3×10^8 m/s, yellow light of wavelength 5.89×10^{-7} m has a frequency of 5.1×10^{14} s^{-1}. X-rays and gamma rays are electromagnetic radiation arising from the interactions of atomic and nuclear particles, respectively. They have energies and frequencies much higher than those of visible light.

To appreciate the relationship of states of matter, atomic and nuclear interactions, and energy, let us visualize an experiment in which we supply energy to a sample of water from a source of energy that is as large and as sophisticated as we wish. Thus we increase the degree of internal motion and eventually dissociate the material into its most elementary components. Suppose in Figure 1.1 that the water is initially as ice at nearly absolute zero temperature, where water (H_2O) molecules are essentially at rest. As we add thermal energy to increase the temperature to 0°C or 32°F, molecular movement increases to the point at which the ice melts to become liquid water, which can flow rather freely. To cause a change from the solid state to the liquid state, a definite amount of energy (the heat of fusion) is required. In the case of water, this latent heat is 334 J/g. In the temperature range in which water is liquid, thermal agitation of the molecules permits some evaporation from the surface. At the boiling point, 100°C or 212°F at atmospheric pressure, the liquid turns into the gaseous form as steam. Again, energy is required to cause the change of state, with a heat of vaporization of 2258 J/g. Further heating, by use of special high temperature equipment, causes dissociation of water into atoms of hydrogen (H) and oxygen (O). By electrical means, electrons can be removed from hydrogen and oxygen atoms, leaving a mixture of charged ions and electrons. Through nuclear bombardment, the oxygen nucleus can be broken into smaller nuclei, and in the limit of temperatures in the billions of degrees, the material can be decomposed into an assembly of electrons, protons, and neutrons.

[†]We will need both Roman and Greek characters, identifying the latter by name the first time they are used, thus λ (lambda) and v (nu). The reader must be wary of symbols used for more than one quantity.

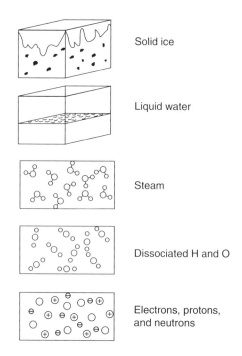

Solid ice

Liquid water

Steam

Dissociated H and O

Electrons, protons, and neutrons

FIGURE 1.1 Effect of added energy.

1.4 THE EQUIVALENCE OF MATTER AND ENERGY

The connection between energy and matter is provided by Einstein's theory of special relativity. It predicts that the mass of any object increases with its speed. Letting the mass when the object is at rest be m_0, the "rest mass," letting m be the mass when it is at speed v, and noting that the speed of light in a vacuum is $c = 3 \times 10^8$ m/s, then

$$m = \frac{m_0}{\sqrt{1 - (v/c)^2}}.$$

For motion at low speed (e.g., 500 m/s), the mass is almost identical to the rest mass, because v/c and its square are very small. Although the theory has the status of natural law, its rigor is not required except for particle motion at high speed (i.e., when v is at least several percent of c). The relation shows that a material object can have a speed no higher than c.

The kinetic energy imparted to a particle by the application of force according to Einstein is

$$E_k = (m - m_0)c^2.$$

(For low speeds, $v << c$, this is approximately $\frac{1}{2} m_0 v^2$, the classical relation.)

The implication of Einstein's formula is that any object has an energy $E_0 = m_0 c^2$ when at rest (its "rest energy"), and a total energy $E = mc^2$, the difference being E_k the kinetic energy (i.e., $E = E_0 + E_k$). Let us compute the rest energy for an electron of mass 9.1×10^{-31} kg

$$E_0 = m_0 c^2 = (9.1 \times 10^{-31})(3.0 \times 10^8)^2 = 8.2 \times 10^{-14} \text{ J}$$

$$E_0 = \frac{8.2 \times 10^{-14} \text{ J}}{1.60 \times 10^{-13} \text{ J/MeV}} = 0.51 \text{ MeV}.$$

For one unit of atomic mass, 1.66×10^{-27} kg, which is close to the mass of a hydrogen atom, the corresponding energy is 931 MeV.

Thus we see that matter and energy are equivalent, with the factor c^2 relating the amounts of each. This suggests that matter can be converted into energy and that energy can be converted into matter. Although Einstein's relationship is completely general, it is especially important in calculating the release of energy by nuclear means. We find that *the energy yield from a kilogram of nuclear fuel is more than a million times that from chemical fuel*. To prove this startling statement, we first find the result of the complete transformation of 1 kilogram of matter into energy, namely, $(1 \text{ kg})(3.0 \times 10^8 \text{ m/s})^2 = 9 \times 10^{16}$ J. The nuclear fission process, as one method of converting mass into energy, is relatively inefficient, because the "burning" of 1 kg of uranium involves the conversion of only 0.87 g of matter into energy. This corresponds to approximately 7.8×10^{13} J/kg of the uranium consumed. The enormous magnitude of this energy release can be appreciated only by comparison with the energy of combustion of a familiar fuel such as gasoline, 5×10^7 J/kg. The ratio of these numbers, 1.5×10^6, reveals the tremendous difference between nuclear and chemical energies.

Calculations involving Einstein's theory are made easy by use of a computer program ALBERT, described in Computer Exercise 1.A.

1.5 ENERGY AND THE WORLD

All of the activities of human beings depend on energy, as we realize when we consider the dimensions of the world's energy problem. The efficient production of food requires machines, fertilizer, and water, each making use of energy in a different way. Energy is vital to transportation, protection against the weather, and the manufacturing of all goods. An adequate long-term supply of energy is therefore essential for man's survival. The world energy problem has many dimensions: the increasing cost to acquire fuels as they become more scarce; the potential for global climate change resulting from burning fossil fuels; the effects on safety and health of the byproducts of energy consumption; the inequitable distribution of energy resources among regions and nations; and the discrepancies between current energy use and human expectations throughout the world.

1.6 SUMMARY

Associated with each basic type of force is an energy, which may be transformed to another form for practical use. The addition of thermal energy to a substance causes an increase in temperature, the measure of particle motion. Electromagnetic radiation arising from electrical devices, atoms, or nuclei may be considered to be composed of waves or of photons. Matter can be converted into energy and vice versa according to Einstein's formula $E = mc^2$. The energy of nuclear fission is millions of times as large as that from chemical reactions. Energy is fundamental to all human endeavors and, indeed, survival.

1.7 EXERCISES

1.1 Find the kinetic energy of a basketball player of mass 75 kg as he moves down the floor at a speed of 8 m/s.

1.2 Recalling the conversion formulas for temperature,

$$C = \frac{5}{9}(F - 32)$$

$$F = \frac{9}{5}C + 32$$

where C and F are degrees in respective systems, convert each of the following: $68°F$, $500°F$, $-273°C$, $1000°C$.

1.3 If the specific heat of iron is 0.45 J/g-°C, how much energy is required to bring 0.5 kg of iron from $0°C$ to $100°C$?

1.4 Find the speed corresponding to the average energy of nitrogen gas molecules (N_2, 28 units of atomic weight) at room temperature.

1.5 Find the power in kilowatts of an auto rated at 200 horsepower. In a drive for 4 h at average speed 45 mph, how many kWh of energy are required?

1.6 Find the frequency of a γ ray photon of wavelength 1.5×10^{-12} m.

1.7 (a) For very small velocities compared with the velocity of light, show that the relativistic formula for kinetic energy is $(1/2)m_0v^2$. Hint: use the series expansion

$$(1 + x)^n = 1 + nx + \ldots.$$

(b) Find the approximate relativistic mass increase of a car with rest mass 1000 kg moving at 20 m/s.

1.8 Noting that the electron-volt is 1.60×10^{-19} J, how many joules are released in the fission of one uranium nucleus, which yields 190 MeV?

1.9 Applying Einstein's formula for the equivalence of mass and energy, $E = mc^2$, where $c = 3 \times 10^8$ m/s, the speed of light, how many kilograms of matter are converted into energy in Exercise 1.8?

1.10 If the atom of uranium-235 has mass of $(235)(1.66 \times 10^{-27})$ kg, what amount of equivalent energy does it have?

1.11 Using the results of Exercises 1.8, 1.9, and 1.10, what fraction of the mass of a U-235 nucleus is converted into energy when fission takes place?

1.12 Show that to obtain a power of 1 W from fission of uranium, it is necessary to cause 3.3×10^{10} fission events per second. Assume that each fission releases 190 MeV of useful energy.

1.13 (a) If the fractional mass increase caused by relativity is $\Delta E/E_0$, show that

$$v/c = \sqrt{1 - (1 + \Delta E/E_0)^{-2}}.$$

(b) At what fraction of the speed of light does a particle have a mass that is 1% higher than the rest mass? 10%? 100%?

1.14 The heat of combustion of hydrogen by the reaction $2H + O = H_2O$ is quoted to be 34.18 kilogram calories per gram of hydrogen. (a) Find how many Btu per pound this is with the conversions 1 Btu = 0.252 kcal, 1 lb = 454 grams. (b) Find how many joules per gram this is noting 1 cal = 4.18 J. (c) Calculate the heat of combustion in eV per H_2 molecule. Note: Recall the number of particles per gram of molecular weight, Avogadro's number, $N_A = 6.02 \times 10^{23}$.

COMPUTER EXERCISES

1.A Properties of particles moving at high velocities are related in a complicated way according to Einstein's theory of special relativity. To obtain answers easily, the computer program ALBERT (after Dr. Einstein) can be used to treat the following quantities:

Velocity
Momentum
Total mass-energy
Kinetic energy
Ratio of mass to rest mass
Given one of the above, for a selected particle, ALBERT calculates the others. Test the program with various inputs, for example $v/c = 0.9999$ and $T = 1$ billion electron volts.

1.8 GENERAL REFERENCES

Encyclopedia Britannica online, http://www.britannica.com
Brief articles are free; full articles require paid membership.

Wikipedia, http://en.wikipedia.org
Millions of articles in free encyclopedia. Subject to edit by anyone and thus may contain
   misinformation.

McGraw-Hill Concise Encyclopedia of Physics, McGraw-Hill, New York, 2004.

Isaac Asimov, *Asimov's Biographical Encyclopedia of Science and Technology*, 2nd revised
   edition, Doubleday & Co., Garden City, NY, 1982. Subtitle: The Lives and Achievements of
   1510 Great Scientists from Ancient Times to the Present Chronologically Arranged.

Frank J. Rahn, Achilles G. Adamantiades, John E. Kenton, and Chaim Braun, *A Guide to Nuclear
   Power Technology: A Resource for Decision Making*, Krieger Publishing Co., Melbourne, FL,
   1991 (reprint of 1984 edition). A book for persons with some technical background. Almost
   a thousand pages of fine print. A host of tables, diagrams, photographs, and references.

Radiation Information Network
http://www.physics.isu.edu/radinf
Numerous links to sources. By Bruce Busby, Idaho State University.

American Nuclear Society publications
http://www.ans.org
Nuclear News, *Radwaste Solutions*, *Nuclear Technology*, *Nuclear Science and Engineering*,
   Fusion Science and Technology, and *Transactions of the American Nuclear Society*. Online
   Contents pages and selected articles or abstracts of technical papers.

American Nuclear Society public information
http://www.ans.org/pi
Essays on selected topics (e.g., radioisotope).

Glossary of Terms in Nuclear Science and Technology, American Nuclear Society, La Grange
   Park, IL, 1986.

Ronald Allen Knief, *Nuclear Engineering: Theory and Technology of Commercial Nuclear
   Power*, Taylor & Francis, Bristol, PA, 1992. Comprehensive textbook that may be found in
   technical libraries.

Robert M. Mayo, *Introduction to Nuclear Concepts for Engineers*, American Nuclear Society, La
   Grange Park, IL, 1998. College textbook emphasizing nuclear processes.

William D. Ehmann and Diane E. Vance, *Radiochemistry and Nuclear Methods of Analysis*,
   John Wiley & Sons, New York, 1991. Covers many of the topics of this book in greater
   length.

David R. Lide, Editor, *CRC Handbook of Chemistry and Physics*, 88th Edition, 2007–2008, CRC
   Press, Boca Raton, FL, 2008. A standard source of data on many subjects.

WWW Virtual Library
http://www.vlib.org
Links to Virtual Libraries in Engineering, Science, and other subjects.

WWW Virtual Library Nuclear Engineering
http://www.nuc.berkeley.edu/main/vir_library.html

Alsos Digital Library for Nuclear Issues
http://alsos.wlu.edu
Large collection of references.

How Things Work
http://howthingswork.virginia.edu
Information on many subjects by Professor Louis Bloomfield.

How Stuff Works
http://www.howstuffworks.com
Brief explanations by Marshall Brain of familiar devices and concepts, including many of the
    topics of this book.

Internet Detective
http://www.vts.intute.ac.uk/detective
A tutorial on browsing for quality Internet information.

Energy Quest
http://www.energyquest.co.gov/story/index.html
The Energy Story. From California Energy Commission.

1.9 REFERENCES

David Halliday, Jearl Walker, and Robert E. Resnick, *Fundamentals of Physics Extended* 7th Ed.,
    John Wiley & Sons, New York, 2007. Classic popular textbook for college science and
    engineering students.

Paul A. Tipler and Gene Mosca, *Physics for Scientists and Engineers*, 6th Ed., W. H. Freeman,
    San Francisco, 2007.
Calculus-based college textbook.

Raymond L. Murray and Grover C. Cobb, *Physics: Concepts and Consequences*. Prentice-Hall,
    Englewood Cliffs, NJ, 1970 (available from American Nuclear Society, La Grange Park, IL).
    Non-calculus text for liberal arts students.

The NIST Reference on Constants, Units, and Uncertainty
http://physics.nist.gov/cuu/
Information on SI units and fundamental physical constants.

American Physical Society
http://www.aps.org
Select Students & Educators/Physics Central.

The Particle Adventure: Fundamentals of Matter and Force
http://www.particleadventure.org
By Lawrence Berkeley National Laboratory.

PhysLink
http://www.physlink.com
Select Reference for links to sources of many physics constants, conversion factors, and
    other data.

Physical Science Resource Center
http://www.psrc-online.org
Links provided by American Association of Physics Teachers. Select Browse Resources.

Antimatter: Mirror of the Universe
http://livefromcern.web.cern.ch/livefromcern/antimatter

Albert Einstein
http://www.westegg.com/einstein
http://www.pbs.org/wgbh/nova/einstein
http://www.aip.org/history/einstein

Atoms and Nuclei

2

A COMPLETE understanding of the microscopic structure of matter and the exact nature of the forces acting is yet to be realized. However, excellent models have been developed to predict behavior to an adequate degree of accuracy for most practical purposes. These models are descriptive or mathematical, often based on analogy with large-scale processes, on experimental data, or on advanced theory.

2.1 ATOMIC THEORY

The most elementary concept is that matter is composed of individual particles— atoms—that retain their identity as elements in ordinary physical and chemical interactions. Thus a collection of helium atoms that forms a gas has a total weight that is the sum of the weights of the individual atoms. Also, when two elements combine to form a compound (e.g., if carbon atoms combine with oxygen atoms to form carbon monoxide molecules), the total weight of the new substance is the sum of the weights of the original elements.

There are more than 100 known elements. Most are found in nature; some are artificially produced. Each is given a specific number in the periodic table of the elements—examples are hydrogen (H) 1, helium (He) 2, oxygen (O) 8, and uranium (U) 92. The symbol Z is given to that *atomic number*, which is also the number of electrons in the atom and determines its chemical properties.

Computer Exercise 2.A describes the program ELEMENTS, which helps find atomic numbers, symbols, and names of elements in the periodic table.

Generally, the higher an element is in the periodic table, the heavier are its atoms. The *atomic weight M* is the weight in grams of a definite number of atoms, 6.02×10^{23}, which is Avogadro's number, N_a. For the preceding example elements, the values of M are approximately H, 1.008; He, 4.003; O, 16.00; and U, 238.0. Accurate values of atomic weights of all the elements are in the Appendix. We can easily find the number of atoms per cubic centimeter in a substance if its density ρ (rho) in grams per cubic centimeter is known. For example, if we had a

15

container of helium gas with density of 0.00018 g/cm^3, each cubic centimeter would contain a fraction 0.00018/4.003 of Avogadro's number of helium atoms (i.e., 2.7×10^{19}). This procedure can be expressed as a convenient formula for finding N, the number per cubic centimeter for any material

$$N = \frac{\rho}{M} N_a.$$

Thus in natural uranium with its density of 19 g/cm^3, we find $N = (19/238)$ $(6.02 \times 10^{23}) = 0.048 \times 10^{24}$ cm^{-3}. The relationship holds for compounds as well, if M is taken as the molecular weight. In water, H_2O, with $\rho = 1.0$ g/cm^3 and $M = 2$ $(1.008) + 16.00 \cong 18.0$, we have $N = (1/18)(6.02 \times 10^{23}) = 0.033 \times 10^{24}$ molecules/cm^{-3}. (The use of numbers times 10^{24} will turn out to be convenient later.)

2.2 GASES

Substances in the gaseous state are described approximately by the perfect gas law, relating pressure (p), volume (V), and absolute temperature (T),

$$pV = nkT,$$

where n is the number of particles and k is Boltzmann's constant. An increase in the temperature of the gas as a result of heating causes greater molecular motion, which results in an increase of particle bombardment of a container wall and thus of pressure on the wall. The particles of gas, each of mass m, have a variety of speeds v in accord with Maxwell's gas theory, as shown in Figure 2.1. The most probable speed, at the peak of this Maxwellian distribution, depends on temperature according to the relation

$$v_p = \sqrt{2kT/m}.$$

The kinetic theory of gases provides a basis for calculating properties such as the specific heat. By use of the fact from Chapter 1 that the average energy of gas molecules is proportional to the temperature, $\bar{E} = \frac{3}{2}kT$, we can deduce, as in Exercise 2.4, that the specific heat of a gas consisting only of atoms is $c = \frac{3}{2}k/m$, where m is the mass of one atom. Thus we see an intimate relationship between mechanical and thermal properties of materials.

2.3 THE ATOM AND LIGHT

Until the 20th century the internal structure of atoms was unknown, but it was believed that electric charge and mass were uniform. Rutherford performed some crucial experiments in which gold atoms were bombarded by charged particles. He deduced in 1911 that most of the mass and positive charge of an atom were

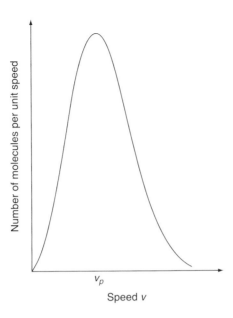

FIGURE 2.1 Distribution of molecular speeds.

concentrated in a *nucleus* of radius only approximately 10^{-5} times that of the atom, and thus occupying a volume of approximately 10^{-15} times that of the atom. (See Exercises 2.2 and 2.11.) The new view of atoms paved the way for Bohr to find an explanation for the production of light.

It is well known that the color of a heated solid or gas changes as the temperature is increased, tending to go from the red end of the visible region toward the blue end (i.e., from long wavelengths to short wavelengths). The measured distribution of light among the different wavelengths at a certain temperature can be explained by the assumption that light is in the form of photons. These are absorbed and emitted with definite amounts of energy E that are proportional to the frequency v, according to

$$E = hv,$$

where h is Planck's constant, 6.63×10^{-34} J-s. For example, the energy corresponding to a frequency of 5.1×10^{14} is $(6.63 \times 10^{-34})\ (5.1 \times 10^{14}) = 3.4 \times 10^{-19}$ J, which is seen to be a very minute amount of energy.

The emission and absorption of light from incandescent hydrogen gas was first explained by Bohr with a novel model of the hydrogen atom. He assumed that the atom consists of a single electron moving at constant speed in a circular orbit about a nucleus—the proton—as sketched in Figure 2.2. Each particle has an electric charge of 1.6×10^{-19} coulombs, but the proton has a mass that is 1836 times that of the electron. The radius of the orbit is set by the equality of

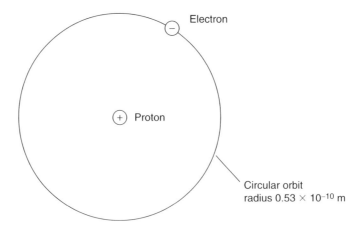

FIGURE 2.2 Hydrogen atom.

electrostatic force, attracting the two charges toward each other, to centripetal force, required to keep the electron on a circular path. If sufficient energy is supplied to the hydrogen atom from the outside, the electron is caused to jump to a larger orbit of definite radius. At some later time, the electron falls back spontaneously to the original orbit, and energy is released in the form of a photon of light. The energy of the photon $h\nu$ is equal to the difference between energies in the two orbits. The smallest orbit has a radius $R_1 = 0.53 \times 10^{-10}$ m, whereas the others have radii increasing as the square of integers (called quantum numbers). Thus if n is 1, 2, 3,..., the radius of the nth orbit is $R_n = n^2 R_1$. Figure 2.3 shows the allowed electron orbits in hydrogen. The energy of the atom system when the

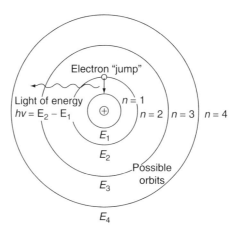

FIGURE 2.3 Electron orbits in hydrongen (Bohr theory).

electron is in the first orbit is $E_1 = -13.5$ eV, where the negative sign means that energy must be supplied to remove the electron to a great distance and leave the hydrogen as a positive ion. The energy when the electron is in the nth orbit is E_1 / n^2. The various discrete levels are sketched in Figure 2.4.

The electronic structure of the other elements is described by the shell model, in which a limited number of electrons can occupy a given orbit or shell. The atomic number Z is unique for each chemical element and represents both the number of positive charges on the central massive nucleus of the atom and the number of electrons in orbits around the nucleus. The maximum allowed numbers of electrons in orbits as Z increases for the first few shells are 2, 8, and 18. The number of electrons in the outermost, or valence, shell determines the chemical behavior of elements. For example, oxygen with $Z = 8$ has two electrons in the inner shell, six in the outer. Thus oxygen has an affinity for elements with two electrons in the valence shell. The formation of molecules from atoms by electron sharing is illustrated by Figure 2.5, which shows the water molecule.

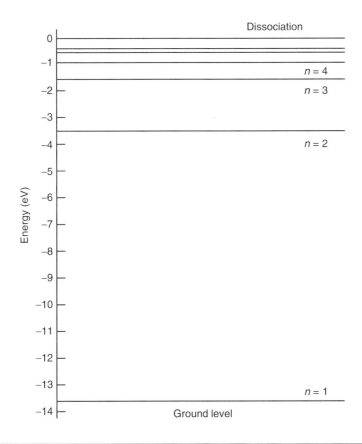

FIGURE 2.4 Energy levels in hydrogen atom.

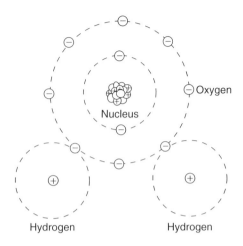

FIGURE 2.5 Water molecule.

The Bohr model of atoms is useful for visualization, but quantum mechanics provides a more rigorous view. There, the location of the electron in the H atom is described by a probability expression. A key feature of quantum mechanics is Heisenberg's uncertainty principle. It states that the precise values of both a particle's position and momentum cannot be known.

2.4 LASER BEAMS

Ordinary light as in the visible range is a mixture of many frequencies, directions, and phases. In contrast, light from a *laser* (*l*ight *a*mplified by *s*timulated *e*mission of *r*adiation) consists of a direct beam of one color and with the waves in step. The device consists of a tube of material to which energy is supplied, exciting the atoms to higher energy states. A photon of a certain frequency is introduced. It strikes an excited atom, causing it to fall back to the ground state and in so doing emit another photon of the same frequency. The two photons strike other atoms, producing four identical photons, and so on. The ends of the laser are partially reflecting, which causes the light to be trapped and to build up inside by a combination of reflection and stimulation. An avalanche of photons is produced that makes a very intense beam. Light moving in directions other than the long axis of the laser is lost through the sides, so that the beam that escapes from the end proceeds in only one direction. The reflection between the two end mirrors assures a coherent beam (i.e., the waves are in phase).

Lasers can be constructed from several materials. The original one (1960) was the crystalline gem ruby. Others use gases such as a helium-neon mixture, liquids with dye in them, or semiconductors. The external supply of energy can be chemical reactions, a discharge produced by accelerated electrons, energetic particles from nuclear reactions, or another laser. Some lasers operate continuously, whereas others produce pulses of energy as short as a fraction of a nanosecond (10^{-9} sec) with a power of a terawatt (10^{12} watts). Because of the high intensity, laser light, if viewed directly, can be hazardous to the eyes.

Lasers are widely used where an intense well-directed beam is required, as in metal cutting and welding, eye surgery and other medical applications, and accurate surveying and range finding. Newer applications are noise-free phonographs, holograms (3D images), and communication between airplane and submarine.

Later, we will describe some nuclear applications—isotope separation (Section 9.4) and thermonuclear fusion (Section 14.4).

2.5 NUCLEAR STRUCTURE

Most elements are composed of particles of different weight, called isotopes. For instance, hydrogen has three isotopes of weights in proportion 1, 2, and 3—ordinary hydrogen, heavy hydrogen (deuterium), and tritium. Each has atomic number $Z = 1$ and the same chemical properties, but they differ in the composition of the central nucleus, where most of the weight resides. The nucleus of ordinary hydrogen is the positively charged proton; the deuteron consists of a proton plus a neutron, a neutral particle of weight very close to that of the proton; the triton contains a proton plus two neutrons. To distinguish isotopes, we identify the mass number A, as the total number of nucleons, the heavy particles in the nucleus. A complete shorthand description is given by the chemical symbol with superscript A value and subscript Z value (e.g., $^{1}_{1}H, ^{2}_{1}H$ and $^{3}_{1}H$). Figure 2.6 shows the nuclear and atomic structure of the three hydrogen isotopes. Each has one electron in the outer shell, in accord with the Bohr theory described earlier.

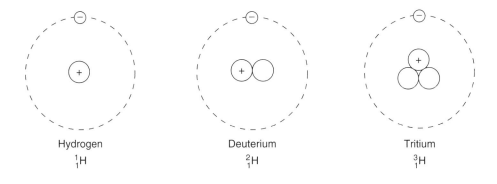

| Hydrogen | Deuterium | Tritium |
| $^{1}_{1}H$ | $^{2}_{1}H$ | $^{3}_{1}H$ |

FIGURE 2.6 Isotopes of hydrogen.

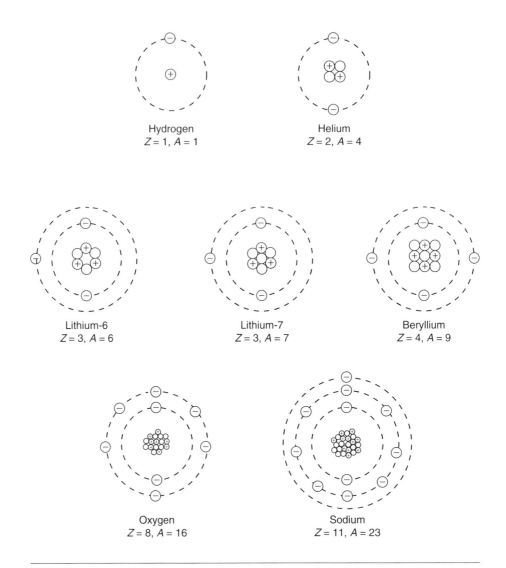

FIGURE 2.7 Atomic and nuclear structure.

The structure of some of the lighter elements and isotopes is sketched in Figure 2.7. In each case, the atom is neutral, because the negative charge of the Z electrons in the outer shell balances the positive charge of the Z protons in the nucleus. The symbols for the isotopes shown in Figure 2.7 are:

$$^{1}_{1}\text{H}, ^{4}_{2}\text{He}, ^{6}_{3}\text{Li}, ^{7}_{3}\text{Li}, ^{9}_{4}\text{Be}, ^{16}_{8}\text{O}, ^{23}_{11}\text{Na}.$$

In addition to the atomic number Z and the mass number A, we often need to write the neutron number N, which is, of course, $A-Z$. For the set of isotopes listed, N is 0, 2, 3, 4, 5, 8, and 12, respectively.

When we study nuclear reactions, it is convenient to let the neutron be represented by the symbol $^1_0$n, implying a mass comparable to that of hydrogen, $^1_1$H, but with no electronic charge, $Z = 0$. Similarly, the electron is represented by $^{\ }_{-1}$e, suggesting nearly zero mass in comparison with that of hydrogen, but with negative charge. An identification of isotopes frequently used in qualitative discussion consists of the element name and its A value, thus sodium-23 and uranium-235, or even more simply Na-23 and U-235.

2.6 SIZES AND MASSES OF NUCLEI

The dimensions of nuclei are found to be very much smaller than those of atoms. Whereas the hydrogen atom has a radius of approximately 5×10^{-9} cm, its nucleus has a radius of only approximately 10^{-13} cm. Because the proton weight is much larger than the electron weight, the nucleus is extremely dense. The nuclei of other isotopes may be viewed as closely packed particles of matter—neutrons and protons—forming a sphere whose volume, $\frac{4}{3}\pi R^3$, depends on A, the number of nucleons. A useful rule of thumb to calculate radii of nuclei is

$$R(\text{cm}) = 1.4 \times 10^{-13} A^{1/3}.$$

Because A ranges from 1 to approximately 250, we see that all nuclei are smaller than 10^{-12} cm.

The masses of atoms, labeled M, are compared on a scale in which an isotope of carbon $^{12}_6$C has a mass of exactly 12. For $^1_1$H, the atomic mass is $M = 1.007825$, for $^2_1$H, M = 2.014102, and so on. The atomic mass of the proton is 1.007276, of the neutron 1.008665, the difference being only about 0.1%. The mass of the electron on this scale is 0.000549. A list of atomic masses appears in the Appendix.

The atomic mass unit (amu), as $\frac{1}{12}$ the mass of $^{12}_6$C, corresponds to an actual mass of 1.660539×10^{-24} g ($\cong 1.66 \times 10^{-24}$ g). To verify this approximation merely divide 1 g by Avogadro's number 6.02×10^{23}. By use of Einstein's $E = mc^2$ with constants in the Appendix, we find 1 amu = 931.494 MeV ($\cong 931$ MeV). We can calculate the actual masses of atoms and nuclei by multiplying the mass in atomic mass units by the mass of 1 amu. Thus the mass of the neutron is $(1.008665)(1.660539 \times 10^{-24}) = 1.674928 \times 10^{-24}$ ($\cong 1.67 \times 10^{-24}$ g).

2.7 BINDING ENERGY

The force of electrostatic repulsion between like charges, which varies inversely as the square of their separation, would be expected to be so large that nuclei could not be formed. The fact that they do exist is evidence that there is an even larger force of attraction. The nuclear force is of very short range, as we can deduce from the preceding rule of thumb. As shown in Exercise 2.9, the radius of a nucleon is approximately 1.4×10^{-13} cm; the distance of separation of centers is about twice that. The nuclear force acts only when the nucleons are very close to each other and binds them into a compact structure. Associated with the net force is a potential energy of binding. To disrupt a nucleus and separate it into its component nucleons, energy must be supplied from the outside. Recalling Einstein's relation between mass and energy, this is the same as saying that a given nucleus is lighter than the sum of its separate nucleons, the difference being the binding mass-energy. Let the mass of an atom including nucleus and external electrons be M, and let m_n and m_H be the masses of the neutron and the proton plus matching electron. Then the binding energy is

$$B = \text{total mass of separate particles} - \text{mass of the atom}$$

or

$$B = Nm_n + Zm_H - M.$$

(Neglected in this relation is a small energy of atomic or chemical binding.) Let us calculate B for tritium, the heaviest hydrogen atom, ^3_1H. Figure 2.8 shows the dissociation that would take place if a sufficient energy were provided. Now $Z = 1$, $N = 2$, $m_n = 1.008665$, $m_H = 1.007825$, and $M = 3.016049$. Then

$$B = 2(1.008665) + 1(1.007825) - 3.016049$$

$$B = 0.009106 \text{ amu.}$$

Converting by use of the relation 1 amu $= 931$ MeV, the binding energy is $B = 8.48$ MeV. Calculations such as these are required for several purposes—to compare the stability of one nucleus with that of another, to find the energy release in a nuclear reaction, and to predict the possibility of fission of a nucleus.

We can speak of the binding energy associated with one particle such as a neutron. Suppose that M_1 is the mass of an atom and M_2 is its mass after

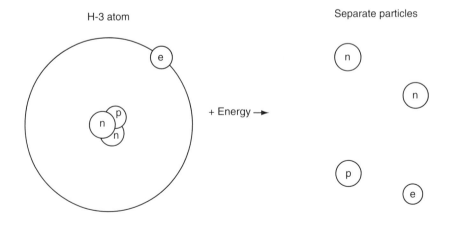

FIGURE 2.8 Dissociation of tritium.

absorbing a neutron. The binding energy of the additional neutron of mass m_n is then

$$B_n = M_1 + m_n - M_2.$$

Explanations of binding energy effects by means of physical logic and measured atomic masses have led to what are called "semi empirical formulas" for binding energy. The value of B for any nuclide is calculated approximately by an expression that accounts for (1) attraction of nucleons for each other, (2) electrostatic repulsion, (3) surface tension effects, and (4) the imbalance of neutrons and protons in the nucleus. Computer Exercise 2.B makes use of a program BINDING that estimates binding energy B, the internal energy per nucleon $E = - B/A$, and the atomic mass M for given mass number A and atomic number Z.

2.8 SUMMARY

All material is composed of elements whose chemical interaction depends on the number of electrons (Z). Light is absorbed and emitted in the form of photons when atomic electrons jump between orbits. Isotopes of elements differ according to the number of nucleons (A). Nuclei are much smaller than atoms and contain most of the mass of the atom. The nucleons are bound together by a net force in which the nuclear attraction forces exceed the electrostatic repulsion forces. Energy must be supplied to dissociate a nucleus into its components.

2.9 EXERCISES

2.1 Find the number of carbon ($_6^{12}$C) atoms in 1 cm^3 of graphite, density 1.65 g/cm^3.

2.2 Estimate the radius and volume of the gold atom, using the metal density of 19.3 g/cm^3 and atomic weight close to 197. Assume that atoms are located at corners of cubes and that the atomic radius is that of a sphere with volume equal to that of a cube.

2.3 Calculate the most probable speed of a "neutron gas" at temperature 20°C (293°K), noting that the mass of a neutron is 1.67×10^{-27} kg.

2.4 Prove that the specific heat of an atomic gas is given by $c_p = (3/2)(k/m)$, by use of the formula for average energy of a molecule.

2.5 Calculate the energy in electron volts of a photon of yellow light (see Section 2.3). Recall from Section 1.2 that 1 eV $= 1.60 \times 10^{-19}$ J.

2.6   What frequency of light is emitted when an electron jumps into the smallest orbit of hydrogen, coming from a very large radius (assume infinity)?

2.7   Calculate the energy in electron-volts of the electron orbit in hydrogen for which $n = 3$, and find the radius in centimeters. How much energy would be needed to cause an electron to go from the innermost orbit to this one? If the electron jumped back, what frequency of light would be observed?

2.8   Sketch the atomic and nuclear structure of carbon-14, noting Z and A values and the numbers of electrons, protons, and neutrons.

2.9   If A nucleons are visualized as spheres of radius r that can be deformed and packed tightly in a nucleus of radius R, show that $r = 1.4 \times 10^{-13}$ cm.

2.10   What is the radius of the nucleus of uranium-238 viewed as a sphere? What is the area of the nucleus, seen from a distance as a circle?

2.11   Find the fraction of the volume that is occupied by the nucleus in the gold-197 atom, by use of the relationship of radius R to mass number A. Recall from Exercise 2.2 that the radius of the atom is 1.59×10^{-8} cm.

2.12   Find the binding energy in MeV of ordinary helium, ^4_2He, for which $M = 4.002603$.

2.13   How much energy (in MeV) would be required to completely dissociate the uranium-235 nucleus (atomic mass 235.043923) into its component protons and neutrons?

2.14   Find the mass density of the nucleus, the electrons, and the atom of U-235, assuming spherical shapes and the following data:

Atomic radius	1.7×10^{-10} m
Nuclear radius	8.6×10^{-15} m
Electron radius	2.8×10^{-15} m
Mass of 1 amu	1.66×10^{-27} kg
Mass of electron	9.11×10^{-31} kg

Discuss the results.

2.15   Maxwell's formula for the number of molecules per unit speed is

$$n(v) = n_0 \, Av^2 \, \exp(-mv^2/2kT)$$

where n_0 is the total number of molecules and

$$A = 4\pi \left(\frac{m}{2\pi kT} \right)^{3/2}.$$

(a)   Verify by differentiation that the peak of the curve is at

$$v_p = \sqrt{2kT/m}.$$

(b)   Verify by integration that the average speed

$$\bar{v} = \int_0^\infty v\, n(v)dv/n_0$$

is given by

$$\bar{v} = \sqrt{\frac{8kT}{\pi m}}.$$

Hint: let $mv^2/2kT = x$.

2.16   The temperature of the surface of the sun is approximately 5,800 K. To what light frequency and wavelength does that correspond?

COMPUTER EXERCISES

2.A   The program ELEMENTS is a miniature "expert system" that gives information on elements in the periodic table. Three related quantities are listed—atomic number, symbol for the chemical element, and its name. If one of these is input, the other two are displayed. Run the program's options, being sure to try values of Z well above 100.

2.B   The program BINDING calculates the approximate binding energy B and atomic mass M for any nuclide, by use of a semi empirical formula containing six terms that depend on atomic number Z and mass number A. Run the program on these isotopes:

$$^{16}_{8}O,\ ^{17}_{8}O,\ ^{92}_{37}Rb,\ ^{140}_{55}Cs,\ ^{235}_{92}U,\ ^{238}_{92}U.$$

How do the results compare with the values listed in the Appendix?

2.10 REFERENCES

N. N. Greenwood and A. Earnshaw, *Chemistry of the Elements*, 2nd Ed., Butterworth-Heinemann, Oxford, 1997. Structures, properties, and reactions.

Robley D. Evans, *The Atomic Nucleus*, Krieger Publishing Co., Melbourne, FL, 1982. A reprint of the McGraw-Hill 1955 classic advanced textbook containing a wealth of information on nuclei, radioactivity, radiation, and nuclear processes.

Robert M. Mayo, *Introduction to Nuclear Concepts for Engineers*, American Nuclear Society, La Grange Park, IL, 1998. Thorough discussion of the atomic nucleus.

Educational Site from Berkeley Lab
http://csee.lbl.gov/WebBased.html
Select ABC's of Nuclear Science; Glenn Seaborg: His Life & Contributions.

Kinetic Theory of Gases: A Brief Review
http://www.phys.virginia.edu/classes/252/kinetic_theory.html
Derivations of pressure, the gas law, Maxwell's equation, etc. by Michael Fowler

How Things Work
http://howthingswork.virginia.edu
Select topic from menu.

How Stuff Works
http://www.howstuffworks.com
Search on "relativity."

Special relativity
http://www4.ncsu.edu/unity/lockers/users/f/felder/public/kenny/papers/relativity.html
An instructive and amusing explanation.

Nuclear Data
http://ie.lbl.gov/toi.html
Comprehensive source of information by Lawrence Berkeley Laboratory and Lunds Universitet
    (Sweden).

WebElements Periodic Table of the Elements
http://www.webelements.com
Mark Winter provides information about each element.

General Chemistry Online!
http://antoine.frostburg.edu/chem/senese/101
One of several good interactive chemistry courses.

Radioactivity

MANY NATURALLY occurring and man-made isotopes have the property of radioactivity, which is the spontaneous disintegration (decay) of the nucleus with the emission of a particle. The process takes place in minerals of the ground, in fibers of plants, in tissues of animals, and in the air and water, all of which contain traces of radioactive elements.

3.1 RADIOACTIVE DECAY

Many heavy elements are radioactive. An example is the decay of the main isotope of uranium, in the reaction

$$^{238}_{92}\mathrm{U} \rightarrow \, ^{234}_{90}\mathrm{Th} + \, ^{4}_{2}\mathrm{He}.$$

The particle released is the α (alpha) particle, which is merely the helium nucleus. The new isotope of thorium is also radioactive, according to

$$^{234}_{90}\mathrm{Th} \rightarrow \, ^{234}_{91}\mathrm{Pa} + \, ^{0}_{-1}\mathrm{e} + \nu.$$

The first product is the element protactinium. The second is an electron, which is called the β (beta) particle when it arises in a nuclear process. The nucleus does not contain electrons; they are produced in the reaction, as discussed in Section 3.2. The third is the neutrino, symbolized by ν (nu). It is a neutral particle that shares with the beta particle the reaction's energy release. On average, the neutrino carries $\frac{2}{3}$ of the energy, the electron, $\frac{1}{3}$. The neutrino has zero or possibly a very small mass and readily penetrates enormous thicknesses of matter. We note that the A value decreases by 4 and the Z value by 2 on emission of an α particle, whereas the A remains unchanged, but Z increases by 1 on emission of a β particle. These two events are the start of a long sequence or "chain" of disintegrations that produce isotopes of the elements radium, polonium, and bismuth, eventually yielding the stable lead isotope $^{206}_{82}\mathrm{Pb}$. Other chains found in nature start with $^{235}_{92}\mathrm{U}$ and $^{232}_{90}\mathrm{Th}$. Hundreds of "artificial" radioisotopes have been produced by bombardment of nuclei by charged particles or neutrons and by separation of the products of the fission process.

3.2 **THE DECAY LAW**

The rate at which a radioactive substance disintegrates (and thus the rate of release of particles) depends on the isotopic species, but there is a definite "decay law" that governs the process. In a given time period, say 1 second, each nucleus of a given isotopic species has the same chance of decay. If we were able to watch one nucleus, it might decay in the next instant, or a few days later, or even hundreds of years later. Such statistical behavior is described by a constant property of the atom called half-life. This time interval, symbolized by t_H, is the time required for half of the nuclei to decay, leaving half of them intact. We should like to know how many nuclei of a radioactive species remain at any time. If we start at time zero with N_0 nuclei, after a length of time t_H, there will be $N_0/2$; by the time $2t_H$ has elapsed, there will be $N_0/4$; etc. A graph of the number of nuclei as a function of time is shown in Figure 3.1. For any time t on the curve, the ratio of the number of nuclei present to the initial number is given by

$$\frac{N}{N_0} = \left(\frac{1}{2}\right)^{t/t_H}.$$

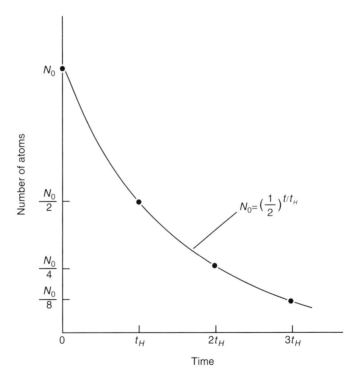

FIGURE 3.1 Radioactive decay.

Half-lives range from very small fractions of a second to billions of years, with each radioactive isotope having a definite half-life. Table 3.1 gives several examples of radioactive materials with their emissions, product isotopes, and half-lives. The β particle energies are maximum values; on average the emitted betas have only one-third as much energy. Included in the table are both natural and

Table 3.1 Selected Radioactive Isotopes[†]

Isotope	Half-life	Principal Radiations (Type, MeV)
Neutron	614 s	β, 0.782
Tritium (H-3)	12.33 y	β, 0.0186
Carbon-14	5715 y	β, 0.156
Nitrogen-16	7.13 s	β, 4.27; γ, 6.129
Sodium-24	14.96 h	β, 1.389; γ, 1.369, 2.754
Phosphorus-32	14.28 d	β, 1.710
Potassium-40	1.25×10^9 y	β, 1.312
Argon-41	1.82 h	β, 1.198; γ, 1.294
Cobalt-60	5.271 y	β, 0.315; γ, 1.173, 1.332
Krypton-85	10.73 y	β, 0.15; γ, 0.514
Strontium-90	29.1 y	β, 0.546
Technetium-99m	6.01 h	β, 0.142
Iodine-129	1.7×10^7 y	β, 0.15
Iodine-131	8.021 d	β, 0.606; γ, 0.284, 0.364
Xenon-135	9.10 h	β, 0.91; γ, 0.250
Cesium-137	30.2 y	β, 0.514; γ, 0.662
Radon-222	3.823 d	α, 5.490; γ, 0.510
Radium-226	1599 y	α, 4.870
Uranium-235	7.04×10^8 y	α, 4.152
Uranium-238	4.47×10^9 y	α, 4.040
Plutonium-239	2.410×10^4 y	α, 5.055

[†]David R. Lide, Editor, *CRC Handbook of Chemistry and Physics*, 88th Edition, 2007–2008, CRC Press, Boca Raton, FL, 2007. Values have been rounded.

man-made radioactive isotopes (also called radioisotopes). We note the special case of neutron decay according to

$$\text{Neutron} \rightarrow \text{Proton} + \text{Electron}.$$

A free neutron has a half-life of 10.3 min. The conversion of a neutron into a proton can be regarded as the origin of beta emission in radioactive nuclei. Most of the radioisotopes in nature are heavy elements. One exception is potassium-40, half-life 1.25×10^9 y, with abundance 0.0117% in natural potassium. Others are carbon-14 and hydrogen-3 (tritium), which are produced continuously in small amounts by natural nuclear reactions. All three radioisotopes are found in plants and animals.

In addition to the radioisotopes that decay by beta or alpha emission, there is a large group of artificial isotopes that decay by the emission of a positron, which has the same mass as the electron and an equal but positive charge. An example is sodium-22, which decays with 2.6 y half-life into a neon isotope as

$$^{22}_{11}\text{Na} \rightarrow ^{22}_{10}\text{Ne} + ^{0}_{+1}\text{e}.$$

Whereas the electron (sometimes called negatron) is a normal part of any atom, the positron is not. It is an example of what is called an antiparticle, because its properties are opposite to those of the normal particle. Just as particles form matter, antiparticles form antimatter.

The preceding Na-22 reaction can be regarded as involving the conversion of a proton into a neutron with the release of a positron, by use of excess energy in the parent nucleus. This is an example of the conversion of energy into mass. Usually, the mass appears in the form of pairs of particles of opposite charge. The positron-electron pair is one example. As discussed in Section 5.3(c), an electron and a positron will combine, and both be annihilated to form two γ (gamma) rays.

A nucleus can get rid of excess internal energy by the emission of a gamma ray, but in an alternate process called internal conversion, the energy is imparted directly to one of the atomic electrons, ejecting it from the atom. In an inverse process called K-capture, the nucleus spontaneously absorbs one of its own orbital electrons. Each of these processes is followed by the production of X-rays as the inner shell vacancy is filled.

The formula for N/N_0 is not very convenient for calculations except when t is some integer multiple of t_H. Defining the decay constant λ (lambda) as the chance of decay of a given nucleus each second, an equivalent *exponential formula*[†] for decay is

---

[†] If λ is the chance one nucleus will decay in a second, then the chance in a time interval dt is λdt. For N nuclei, the change in number of nuclei is $dN = -\lambda N dt$. Integrating, and letting the number of nuclei at time zero be N_0 yields the formula quoted. Note that if

$$e^{-\lambda t} = (1/2)^{t/t_H}$$

$$\lambda t = t/t_H \log_e 2 \text{ or } \lambda = (\log_e 2)/t_H.$$

$$\frac{N}{N_0} = e^{-\lambda t}.$$

We find that $\lambda = 0.693/t_H$. To illustrate, let us calculate the ratio N/N_0 at the end of 2 y for cobalt-60, half-life 5.27 y. This artificially produced radioisotope has many medical and industrial applications. The reaction is

$$^{60}_{27}\mathrm{Co} \rightarrow {}^{60}_{28}\mathrm{Ni} + {}^{0}_{-1}\mathrm{e} + 2\gamma,$$

where the gamma ray energies are 1.17 and 1.33 MeV and the maximum beta energy is 0.315 MeV. By use of the conversion 1 y $= 3.16 \times 10^7$ s, $t_H = 1.67 \times 10^8$ s. Then $\lambda = 0.693/(1.67 \times 10^8) = 4.15 \times 10^{-9}\,\mathrm{s}^{-1}$, and because t is 6.32×10^7 s, λt is 0.262 and $N/N_0 = e^{-0.262} = 0.77$.

The number of disintegrations per second (dps) of a radioisotope is called the activity, A. Because the decay constant λ is the chance of decay each second of one nucleus, for N nuclei the activity is the product

$$A = \lambda N.$$

For a sample of cobalt-60 weighing 1 µg, which is also 10^{16} atoms,

$$A = \left(4.15 \times 10^{-9}\right)\left(10^{16}\right) = 4.15 \times 10^7\,\mathrm{dps}.$$

The unit dps is called the becquerel (Bq), honoring the scientist who discovered radioactivity.

Another older and commonly used unit of activity is the curie (Ci) named after the French scientists Pierre and Marie Curie who studied radium. The curie is 3.7×10^{10} dps, which is an early measured value of the activity per gram of radium. Our cobalt sample has a "strength" of $(4.15 \times 10^7$ Bq$)/(3.7 \times 10^{10}$ Bq/Ci$) = 0.0011$ Ci or 1.1 mCi.

The half-life tells us how long it takes for half of the nuclei to decay, whereas a related quantity, the mean life, τ (tau), is the average time elapsed for decay of an individual nucleus. It turns out that τ is $1/\lambda$ and thus equal to $t_H/0.693$. For Co-60, τ is 7.6 y.

Computer Exercise 3.A calculates activities and 3.B displays formulas, calculations, and a graph of decay.

3.3 RADIOACTIVE CHAINS

Radionuclides arise in several processes. They may be produced by the bombardment of stable nuclei by charged particles as in an accelerator or by neutrons as in a nuclear reactor. Or, they may come from other radionuclides, in which the "parent" nuclide decays and produces a "daughter" isotope. Still more generally, there may be a sequence of decays between a series of radionuclides, called a "chain," leading eventually to a stable nucleus.

Let us examine the method of calculating yields of some of these processes. The easiest case is the generation rate that is constant in time. For example, suppose that neutrons absorbed in cobalt-59 create cobalt-60 at a rate g. The net rate of change with time of the number of cobalt-60 atoms is

$$\text{Rate of change} = \text{Generation rate} - \text{Decay rate}$$

which may be written in the form of a differential equation,

$$dN/dt = g - \lambda\, N.$$

If the initial number is zero, the solution is

$$N = (g/\lambda)(1 - e^{-\lambda t}).$$

The function rises linearly at the start, then flattens out. At long times, the exponential term goes toward zero, leaving $N \cong g/\lambda$. Numerical values of numbers of atoms and activity can be calculated with the program GROWTH, described in Computer Exercise 3.C.

In the decay of a parent radionuclide to form a daughter radionuclide, the generation rate g is an exponential function of time. Computer Exercise 3.D displays the solution of the differential equation and suggests tests of the computer program RADIOGEN for the decay of plutonium-241 into americium-241.

Natural radioactive isotopes such as uranium-238 (4.47×10^9 y) and thorium-232 (1.4×10^{10} y) were produced billions of years ago but still persist because of their long half-lives. Their products form a long chain of radionuclides, with the emission of α particles and β particles. Those forming the uranium series are:

$$^{238}_{92}\text{U}, \quad ^{234}_{90}\text{Th}, \quad ^{234}_{91}\text{Pa}, \quad ^{234}_{92}\text{U}, \quad ^{230}_{90}\text{Th},$$

$$^{226}_{88}\text{Ra}, \quad ^{222}_{86}\text{Rn}, \quad ^{218}_{84}\text{Po}, \quad ^{214}_{82}\text{Pb}, \quad ^{214}_{83}\text{Bi},$$

$$^{214}_{84}\text{Po}, \quad ^{210}_{82}\text{Pb}, \quad ^{210}_{83}\text{Bi}, \quad ^{210}_{84}\text{Po}, \quad ^{206}_{82}\text{Pb}.$$

Note that radium-226 (1599 y) is fairly far down the chain. The final product is stable lead-206. Because of the very long half-life of uranium-238, the generation rate of its daughters and their descendants are practically constant. Let us write $g \cong N_{238}\, \lambda_{238}$ and apply the expression for the number of atoms at long times to the radium-226, $N_{226} \cong g/\lambda_{226}$. Rearranging, the activities are approximately equal,

$$A_{238} \cong A_{226},$$

a condition called secular equilibrium.

In the preceding list is the alpha particle emitter polonium-210, half-life 138 days. It was the "poison" that caused the death in 2006 of the former Russian KGB agent Litvinenko.

3.4 MEASUREMENT OF HALF-LIFE

Finding the half-life of an isotope provides part of its identification needed for beneficial use or for protection against radiation hazard. Let us look at a method for measuring the half-life of a radioactive substance. As in Figure 3.2, a detector that counts the number of particles striking it is placed near the source of radiation. From the number of counts observed in a known short time interval, the counting rate is computed. It is proportional to the rates of emission of particles or rays from the sample and thus to the activity A of the source. The process is repeated after an elapsed time for decay. The resulting values of activity are plotted on semilog graph paper as in Figure 3.3, and a straight line drawn through the observed points. From any pairs of points on the line λ and $t_H = 0.693/\lambda$ can be calculated (see Exercise 3.8). The technique may be applied to mixtures of two radioisotopes. After a long time has elapsed, only the isotope of longer half-life will contribute counts. By extending its graph linearly back in time, one can find the counts to be subtracted from the total to yield the counts from the isotope of shorter half-life.

Activity plots cannot be used for a substance with long half-life (e.g., strontium-90, 29.1 y). The change in activity is almost zero over the span of time one is willing to devote to a measurement. However, if one knows the number of atoms present in the sample and measures the activity, the decay constant can be calculated from $\lambda = A/N$, from which t_H can be found.

The measurement of the activity of a radioactive substance is complicated by the presence of background radiation, which is due to cosmic rays from outside

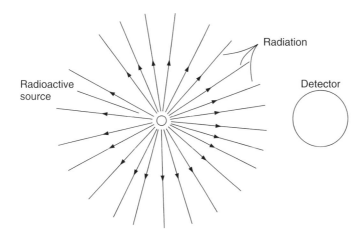

FIGURE 3.2 Measurement of radiation from radioactive source.

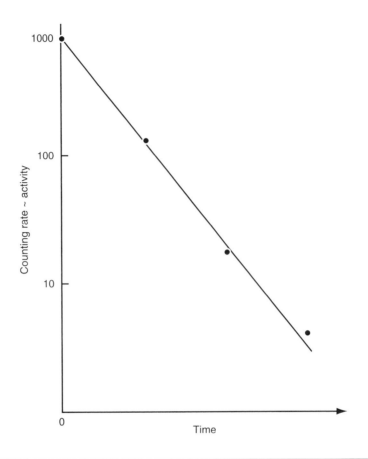

FIGURE 3.3 Activity plot.

the earth or from the decay of minerals in materials of construction or in the earth. It is always necessary to measure the background counts and subtract them from those observed in the experiment.

3.5 SUMMARY

Many elements that are found in nature or are man-made are radioactive, emitting α particles, β particles, and γ-rays. The process is governed by an exponential relation, such that half of a sample decays in a time called the half-life t_H. Values of t_H range from fractions of a second to billions of years among the hundreds of radioisotopes known. Measurement of the activity, as the disintegration rate of a sample, yields half-life values, of importance in radiation use and protection.

3.6 EXERCISES

3.1 Find the decay constant of cesium-137, half-life 30.2 y; then calculate the activity in becquerels and curies for a sample containing 3×10^{19} atoms.

3.2 Calculate the activity A for 1 g of radium-226 with the modern value of the half-life, and compare it with the definition of a curie.

3.3 The radioisotope sodium-24 ($^{24}_{11}$Na), half-life 15 h, is used to measure the flow rate of salt water. By irradiation of stable $^{23}_{11}$Na with neutrons, suppose that we produce 5 micrograms of the isotope. How much do we have at the end of 24 h?

3.4 For a 1-mg sample of Na-24, what is the initial activity and that after 24 h, in Bq and curies? See Exercise 3.3.

3.5 The isotope uranium-238 ($^{238}_{92}$U) decays successively to form $^{234}_{90}$Th, $^{234}_{91}$Pa, $^{234}_{92}$U, and $^{230}_{90}$Th, finally becoming radium-226 ($^{226}_{88}$Ra). What particles are emitted in each of these five steps? Draw a graph of this chain, using A and Z values on the horizontal and vertical axes, respectively.

3.6 A capsule of cesium-137, half-life 30.2 y, is used to check the accuracy of detectors of radioactivity in air and water. Draw a graph on semilog paper of the activity over a 10 y period of time, assuming the initial strength is 1 mCi. Explain the results.

3.7 There are approximately 140 grams of potassium in a typical person's body. From this weight, the abundance of potassium-40, and Avogadro's number, find the number of atoms. Find the decay constant in s^{-1}. How many disintegrations per second are there in the body? How many becquerels and how many microcuries is this?

3.8 (a) Noting that the activity of a radioactive substance is $A = \lambda N_0 e^{-\lambda t}$, verify that the graph of counting rate vs. time on semilog paper is a straight line and show that

$$\lambda = \frac{\log_e(C_1/C_2)}{t_2 - t_1}$$

where points 1 and 2 are any pair on the curve.
(b) With the following data, deduce the half-life of an "unknown," and suggest what isotope it is.

Time (s)	Counting Rate (/s)
0	200
1000	182
2000	162
3000	144
4000	131

3.9 By chemical means, we deposit 10^{-8} moles of a radioisotope on a surface and measure the activity to be 82,000 Bq. What is the half-life of the substance and what element is it (see Table 3.1)?

Computer Exercises

3.A Program DECAY1 is convenient for calculating the amount of decay of a radioactive sample in a given time. Program DECAY1 has input of the original number of curies and the half-life; it calculates the final number of curies. Load the program, examine its form, and look at the results for the decay in 100 y of cesium-137, half-life 30.2 y. Then change input (e.g., x = 302 y [10 half-lives]), or enter figures for another radionuclide such as cobalt-60.

3.B The details of a calculation of radioactive decay are presented in the program DECAY. By means of a set of menus, the equations and solution can be inspected, useful numbers noted, calculations carried out, and a graph of the results viewed. Run the program, using the menus, making choices as desired. Then modify the program to handle another radionuclide.

3.C GROWTH is a program that calculates the number of radioactive cobalt-60 atoms and their activity, assuming a constant generation rate.
  (a) Load and run the program, exploring its menus.
  (b) Modify the program to calculate the growth of sodium-24 (15 h) resulting from neutron capture in sodium-23.

3.D The number of atoms of a parent radioisotope initially is N_{p0}. At any time, the number as the result of decay is

$$N_p = N_{p0} \, E_p$$

where

$$E_p = \exp(-\lambda_p t).$$

Let k be the fraction of parents that decay into a particular daughter. Then the generation rate for the latter is

$$g = k \, N_p \, \lambda_p.$$

The solution of the differential equation

$$dN_d/dt = g - \lambda_d \, N_d$$

is

$$N_d = k \, \lambda_p \, N_{p0} (E_p - E_d)/(\lambda_d - \lambda_p)$$

where

$$E_{\rm d} = \exp(-\lambda_{\rm d}t).$$

Computer program RADIOGEN uses these formulas to calculate the number of atoms $N_{\rm p}$ and $N_{\rm d}$ as a function of time and their activities. Test the program for the decay by beta emission of 10 Ci of reactor-produced plutonium-241, (14.4 y) into americium-241 (432 y), with $k = 1$.

3.7 REFERENCES

Richard B. Firestone and Virginia S. Shirley, *Table of Isotopes*, 8th Ed., John Wiley & Sons, New York, 1998. Two volumes with nuclear structure and decay data for more than 3100 isotopes.

WebElements Periodic Table
http://www.webelements.com
Probably the best of its kind.

W. B. Mann, R. L. Ayres, and S. B. Garfinkel, *Radioactivity and its Measurement*, 2nd Ed., Pergamon Press, Oxford, 1980. History, fundamentals, interactions, and detectors.

Alfred Romer, Editor, *The Discovery of Radioactivity and Transmutation*, Dover Publications, New York, 1964. A collection of essays and articles of historical interest. Researchers represented are Becquerel, Rutherford, Crookes, Soddy, the Curies, and others.

Alfred Romer, Editor, *Radioactivity and the Discovery of Isotopes*, Dover Publications, New York, 1970. Selected original papers, with a thorough historical essay by the editor entitled, "The Science of Radioactivity 1896–1913; Rays, Particles, Transmutations, Nuclei, and Isotopes."

Robley D. Evans, *The Atomic Nucleus*, Krieger Publishing Co. Melbourne, FL, 1982. Classic advanced textbook.

Merril Eisenbud and Thomas F. Gesell, *Environmental Radioactivity*, 4th Ed., Academic Press, New York, 1987. Subtitle: From Natural, Industrial, and Military Sources. Includes information on the Chernobyl reactor accident.

Michael F. L'Annunziata, Editor, *Handbook of Radioactivity Analysis*, Academic Press, New York, 1998. An advanced reference of 771 pages. Many of the contributors are from commercial instrument companies.

Michael F. L'Annunciata, *Radioactivity: Introduction and History*, Elsevier, Oxford, 2007. "Answers questions on the origins, properties, detection, measurement, and applications".

ABC's of Nuclear Science
http://www.lbl.gov/abc
Radioactivity, radiations, and much more.

Radioactivity in Nature
http://www.physics.isu.edu/radinf/natural.htm
Facts and data on natural and manmade radioactivity and on radiation. From Idaho State University Physics Department.

Radiation and Health Physics
http://www.umich.edu/~radinfo
Select Introduction. From University of Michigan.

Radioactive Decay Series
http://www.ead.anl.gov/pub/doc/natural-decay-series.pdf
Diagrams of Uranium-238, Uranium-235, and Thorium-232 series

History of Radioactivity
http://www.accessexcellence.com/AE/AEC/CC
Biographies of scientists responsible for discoveries.

Environmental Radiation and Uranium: A Radioactive Clock
http://scifun.chem.wisc.edu/chemweek/Radiation/Radiation.html
Briefings on radiation and its effect; uranium decay series.

Nuclear Processes

NUCLEAR REACTIONS—those in which atomic nuclei participate—may take place spontaneously, as in radioactivity or may be induced by bombardment with a particle or ray. Nuclear reactions are much more energetic than chemical reactions, but they obey the same physical laws—conservation of momentum, energy, number of particles, and charge.

The number of possible nuclear reactions is extremely large, because there are approximately 2000 known isotopes and many particles that can either be projectiles or products—photons, electrons, protons, neutrons, alpha particles, deuterons, and heavy charged particles. In this chapter we will emphasize induced reactions, especially those involving neutrons.

4.1 TRANSMUTATION OF ELEMENTS

The conversion of one element into another, a process called transmutation, was first achieved in 1919 by Rutherford in England. He bombarded nitrogen atoms with α particles from a radioactive source to produce an oxygen isotope and a proton, according to the equation

$$\ce{^4_2He} + \ce{^{14}_7N} \rightarrow \ce{^{17}_8O} + \ce{^1_1H}.$$

We note that on both sides of the equation the A values add to 18 and the Z values add to 9. Figure 4.1 shows Rutherford's experiment. It is difficult for the positively charged α particle to enter the nitrogen nucleus because of the force of electrical repulsion between charged particles. The α particle thus must have several MeV of energy.

Nuclear transmutations can also be achieved by charged particles that are electrically accelerated to high speeds. The first such example discovered was the reaction

$$\ce{^1_1H} + \ce{^7_3Li} \rightarrow 2\ce{^4_2He}.$$

Another reaction,

$$\ce{^1_1H} + \ce{^{12}_6C} \rightarrow \ce{^{13}_7N} + \gamma,$$

Rutherford's experiment

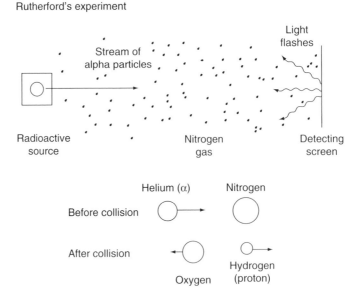

FIGURE 4.1 Transmutation by nuclear reaction.

yields a gamma ray and an isotope of nitrogen. The latter decays with a half-life of 10.3 min, releasing a positron, the positive counterpart of the electron.

Because the neutron is a neutral particle, it does not experience electrostatic repulsion and can readily penetrate a target nucleus. Neutrons are thus especially useful as projectiles to induce reactions. Several examples are chosen on the basis of interest or usefulness. The conversion of mercury into gold, the alchemist's dream, is described by

$$\,_{0}^{1}n + \,_{80}^{198}Hg \rightarrow \,_{79}^{198}Au + \,_{1}^{1}H.$$

The production of cobalt-60 is governed by

$$\,_{0}^{1}n + \,_{27}^{59}Co \rightarrow \,_{27}^{60}Co + \gamma,$$

where a capture gamma ray is produced. Neutron capture in cadmium, often used in nuclear reactor control rods, is given by

$$\,_{0}^{1}n + \,_{48}^{113}Cd \rightarrow \,_{48}^{114}Cd + \gamma.$$

A reaction that produces tritium, which may be a fuel for controlled fusion reactors of the future, is

$$\,_{0}^{1}n + \,_{3}^{6}Li \rightarrow \,_{1}^{3}H + \,_{2}^{4}He.$$

A shorthand notation is used to represent nuclear reactions. Let an incoming particle a strike a target nucleus X to produce a residual nucleus Y and an outgoing particle b, with equation $a + X = Y + b$. The reaction may be abbreviated $X(a, b)$ Y, where a and b stand for the neutron (n), alpha particle (α), gamma ray (γ), proton (p), deuteron (d), and so on. For example, Rutherford's experiment can be written $^{14}\text{N}(\alpha, \text{p})^{17}\text{O}$ and the reaction in control rods $^{113}\text{Cd}(\text{n}, \gamma)^{114}\text{Cd}$. The Z value can be omitted, because it is unique to the chemical element.

The interpretation of nuclear reactions often involves the concept of compound nucleus. This intermediate stage is formed by the combination of a projectile and target nucleus. It has extra energy of excitation and breaks up into the outgoing particle or ray and the residual nucleus.

Later, in Section 6.1, we will discuss the absorption of neutrons in uranium isotopes to cause fission.

The reaction equations can be used to calculate balances in properties such as mass-energy, visualizing conditions before and after. In place of the symbols, the atomic masses are inserted. Strictly the masses of the nuclei should be used, but in most reactions, the same number of electrons appears on both sides of the equation and cancels out. In the case of reactions that produce a positron, however, either nuclear masses or atomic masses with the subtraction of the mass-energy required to create an electron-positron pair should be used, 0.0011 amu or 1.02 MeV.

4.2 ENERGY AND MOMENTUM CONSERVATION

The conservation of mass-energy is a firm requirement for any nuclear reaction. Recall from Chapter 1 that the total mass is the sum of the rest mass m_0 and the kinetic energy E_k (in mass units). Let us calculate the energy released when a slow neutron is captured in hydrogen, according to

$$\tfrac{1}{0}\text{n} + \tfrac{1}{1}\text{H} \rightarrow \tfrac{2}{1}\text{H} + \gamma.$$

This process occurs in reactors that use ordinary water. Conservation of mass-energy says

Mass of neutron + Mass of hydrogen atom

= Mass of deuterium atom + Kinetic energy of products.

We use accurately known masses, as given in the Appendix, along with a conversion factor 1 amu = 931.49 MeV,

$$1.008665 + 1.007825 \rightarrow 2.014102 + E_k,$$

from which $E_k = 0.002388$ amu with an energy release per capture of 2.22 MeV. This energy is shared by the deuterium atom and the gamma ray, which has no rest mass.

A similar calculation can be made for the proton-lithium reaction of the previous section. Suppose that the target nucleus is at rest and that the incoming proton has a kinetic energy of 2 MeV, which corresponds to 2/931.49 = 0.002147 amu. The energy balance statement is

Kinetic energy of hydrogen + Mass of hydrogen + Mass of lithium

= Mass of helium + Kinetic energy of helium.

0.002147 + 1.007825 + 7.016004

= 2(4.002603) + E_k.

Then E_k = 0.02077 amu = 19.3 MeV. This energy is shared by the two α particles.

The calculations just completed tell us the total kinetic energy of the product particles but do not reveal how much each has or what the speeds are. To find this information we must apply the principle of conservation of momentum. Recall that the linear momentum p of a material particle of mass m and speed v is $p = mv$. This relation is correct in both the classical and relativistic senses. The total momentum of the interacting particles before and after the collision is the same.

For our problem of a very slow neutron striking a hydrogen atom at rest, we can assume the initial momentum is zero. If it is to be zero finally, the $^2_1$H and γ-ray must fly apart with equal magnitudes of momentum $p_d = p_γ$. The momentum of a γ-ray having the speed of light c may be written $p_γ = mc$ if we regard the mass as an effective value, related to the γ energy $E_γ$ by Einstein's formula $E = mc^2$. Thus

$$p_γ = \frac{E_γ}{c}.$$

Most of the 2.22-MeV energy release of the neutron capture reaction goes to the γ-ray, as shown in Exercise 4.5. Assuming that to be correct, we can estimate the effective mass of this γ-ray. It is close to 0.00238 amu, which is very small compared with 2.014 amu for the deuterium. Then from the momentum balance, we see that the speed of recoil of the deuterium is much smaller than the speed of light.

The calculation of the energies of the two α particles is a little complicated even for the case in which they separate along the same line that the proton entered. The particle speeds of interest are low enough that relativistic mass variation with speed is small, and thus the classical formula for kinetic energy can be used, $E_k = (1/2)m_0 v^2$. If we let m be the α particle mass and v_1 and v_2 be their speeds, with p_H the proton momentum, we must solve the two equations

$$mv_1 - mv_2 = p_H.$$

$$\frac{1}{2}mv_1^2 + \frac{1}{2}mv_2^2 = E_k.$$

4.3 REACTION RATES

When any two particles approach each other, their mutual influence depends on the nature of the force between them. Two electrically charged particles obey Coulomb's relation $F \sim q_1 q_2/r^2$, where the q's are the amounts of charge and

r is the distance of separation of centers. There will be some influence no matter how far they are apart. However, two atoms, each of which is neutral electrically, will not interact until they get close to one another ($\cong 10^{-10}$ m). The special force between nuclei is limited still further ($\cong 10^{-15}$ m).

Although we cannot see nuclei, we imagine them to be spheres with a certain radius. To estimate that radius, we need to probe with another particle—a photon, an electron, or a γ-ray. But the answer will depend on the projectile used and its speed, and thus it is necessary to specify the apparent radius and cross sectional area for the particular reaction. This leads to the concept of cross section as a measure of the chance of collision.

We can perform a set of imaginary experiments that will clarify the idea of cross section. Picture, as in Figure 4.2(A), a tube of end area 1 cm^2 containing only one target particle. A single projectile is injected parallel to the tube axis, but its exact location is not specified. It is clear that the chance of collision, labeled

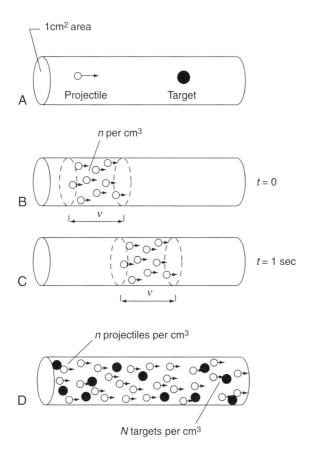

FIGURE 4.2 Particle collisions.

σ (sigma) and called the microscopic cross section, is the ratio of the target area to the area of the tube, which is 1 cm². In Computer Exercise 4.A, the programs MOVENEUT and CURRENT show graphically the flow of neutrons in a column.

Now let us inject a continuous stream of particles of speed v into the empty tube (see Figure 4.2[B]). In a time of 1 second, each of the particles has moved along a distance v cm (see Figure 4.2(C)). All of them in a column of volume (1 cm²) (v cm) = v cm³ will sweep past a point at which we watch each second. If there are n particles per cubic centimeter, then the number per unit time that crosses any unit area perpendicular to the stream direction is nv, called the current density.

Finally, in Figure 4.2(D) we fill each unit volume of the tube with N targets, each of area σ as seen by incoming projectiles (we presume that the targets do not "shadow" each other). If we focus attention on a unit volume, there is a total target area of $N\sigma$. Again, we inject the stream of projectiles. In a time of 1 second, the number of them that pass through the target volume is nv; and because the chance of collision of each with one target atom is σ, the number of collisions is $nv\,N\sigma$. We can thus define the reaction rate per unit volume,

$$R = nv\,N\sigma.$$

We let the current density nv be abbreviated by j and let the product $N\sigma$ be labeled Σ (capital sigma), the macroscopic cross section, referring to the large-scale properties of the medium. Then the reaction rate per cubic centimeter is simply $R = j\Sigma$. We can easily check that the units of j are cm^{-2} s^{-1} and those of Σ are cm^{-1}, so that the unit of R is cm^{-3} s^{-1}.

In a different experiment, we release particles in a medium and allow them to make many collisions with those in the material. In a short time, the directions of motion are random, as sketched in Figure 4.3. We will look only at particles of the same speed v, of which there are n per unit volume. The product nv in this situation is no longer called current density but is given a different name, the flux,

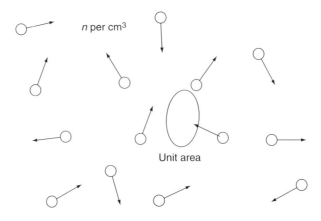

FIGURE 4.3 Particles in random motion.

symbolized by ϕ (phi). If we place a unit area anywhere in the region, there will be flows of particles across it each second from both directions, but it is clear that the current densities will now be less than nv. It turns out that they are each $nv/4$, and the total current density is $nv/2$. The rate of reaction of particles with those in the medium can be found by adding up the effects of individual projectiles. Each behaves the same way in interacting with the targets, regardless of direction of motion. The reaction rate is again $nv\ N\sigma$ or, for this random motion, $R = \phi\ \Sigma$.

The random motion of particles can be simulated mathematically by the use of random numbers, which form a collection of decimal fractions that are independent and are uniformly distributed over the range 0 to 1. They are useful for the study of neutron and gamma ray processes, both of which are governed by statistics. Computer Exercise 4.B describes their generation by three small programs RANDY, RANDY1, and RANDY2.

When a particle such as a neutron collides with a target nucleus, there is a certain chance of each of several reactions. The simplest is elastic scattering, in which the neutron is visualized as bouncing off the nucleus and moving in a new direction with a change in energy. Such a collision, governed by classical physics, is predominant in light elements. In the inelastic scattering collision, an important process for fast neutrons in heavy elements, the neutron becomes a part of the nucleus; its energy provides excitation; and a neutron is released. The cross section σ_s is the chance of a collision that results in neutron scattering. The neutron may instead be absorbed by the nucleus, with cross section σ_a. Because σ_a and σ_s are chances of reaction, their sum is the chance for collision or total cross section $\sigma = \sigma_a + \sigma_s$. Computer Exercise 4.A also introduces a program called CAPTURE related to neutron capture and another called HEADON describing a scattering collision of a neutron with a nucleus in which the neutron direction is exactly reversed.

Let us illustrate these ideas by some calculations. In a typical nuclear reactor used for training and research in universities, a large number of neutrons will be present with energies near 0.0253 eV. This energy corresponds to a most probable speed of 2200 m/s for the neutrons viewed as a gas at room temperature, 293 K. Suppose that the flux of such neutrons is 2×10^{12} cm^{-2}-s^{-1}. The number density is then

$$n = \frac{\phi}{v} = \frac{2 \times 10^{12} \mathrm{cm}^{-2} - \mathrm{s}^{-1}}{2.2 \times 10^5 \mathrm{cm/s}} = 9 \times 10^6 \mathrm{cm}^{-3}.$$

Although this is a very large number by ordinary standards, it is exceedingly small compared with the number of water molecules per cubic centimeter (3.3×10^{22}) or even the number of air molecules per cubic centimeter (2.7×10^{19}). The "neutron gas" in a reactor is almost a perfect vacuum.

Now let the neutrons interact with uranium-235 fuel in the reactor. The cross section for absorption σ_a is 681×10^{-24} cm^2. If the number density of fuel atoms is $N = 0.048 \times 10^{24}$ cm^{-3}, as in uranium metal, then the macroscopic cross section is

$$\Sigma_a = N\sigma_a = \left(0.048 \times 10^{24}\mathrm{cm}^{-3}\right)\left(681 \times 10^{-24}\mathrm{cm}^{-2}\right) = 32.7\mathrm{cm}^{-1}.$$

The unit of area 10^{-24} cm^2 is conventionally called the barn.* If we express the number of targets per cubic centimeter in units of 10^{24} and the microscopic cross section in barns, then $\Sigma_a = (0.048)(681) = 32.7$ cm^{-1} as previously shown. With a neutron flux $\phi = 3 \times 10^{13}$ cm^{-2}·s^{-1}, the reaction rate for absorption is

$$R = \phi\Sigma_a = (3 \times 10^{13} \text{cm}^{-2} - s^{-1})(32.7 \text{cm}^{-1}) = 9.81 \times 10^{14} \text{cm}^{-3} - s^{-1}.$$

This is also the rate at which uranium-235 nuclei are consumed.

The average energy of neutrons in a nuclear reactor used for electrical power generation is approximately 0.1 eV, almost four times the value used in our example. The effects of the high temperature of the medium (approximately 600°F) and of neutron absorption give rise to this higher value.

4.4 PARTICLE ATTENUATION

Visualize an experiment in which a stream of particles of common speed and direction is allowed to strike the plane surface of a substance as in Figure 4.4. Collisions with the target atoms in the material will continually remove projectiles from the stream, which will thus diminish in strength with distance, a process we label *attenuation*. If the current density incident on the substance at position $z = 0$ is labeled j_0, the current of those not having made any collision on penetrating to a depth z is given by[†]

$$j = j_0 e^{-\Sigma z},$$

where Σ is the macroscopic cross section. The similarity in form to the exponential for radioactive decay is noted, and one can deduce by analogy that the half-thickness, the distance required to reduce j to half its initial value, is $z_H = 0.693/\Sigma$. Another more frequently used quantity is the mean free path λ, the average distance a particle goes before making a collision. By analogy with the mean life for radioactivity, we can write[‡]

$$\lambda = 1/\Sigma.$$

This relation is applicable as well to particles moving randomly in a medium. Consider a particle that has just made a collision and moves off in some direction. On the

---

*As the story goes, an early experimenter observed that the cross section for U-235 was "as big as a barn."

[†]The derivation proceeds as follows. In a slab of material of unit area and infinitesimal thickness dz, the target area will be $N\sigma dz$. If the current at z is j, the number of collisions per second in the slab is $jN\sigma dz$, and thus the change in j on crossing the layer is $dj = -j\Sigma dz$, where the reduction is indicated by the negative sign. By analogy with the solution of the radioactive decay law, we can write the formula cited.

[‡]This relation can be derived directly by use of the definition of an average as the sum of the distances the particles travel divided by the total number of particles. When integrals are used, this is $\bar{z} = \int z \, dj / \int dj$.

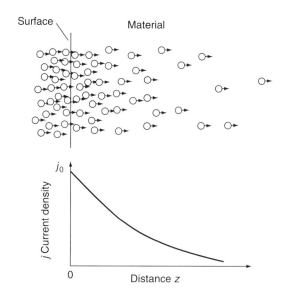

FIGURE 4.4 Neutron penetration and attenuation.

average, it will go a distance λ through the array of targets before colliding again. For example, we can find the mean free path of 1 eV neutrons in water, assuming that scattering by hydrogen with cross section 20 barns is the dominant process. Now the number of hydrogen atoms is $N_H = 0.0668 \times 10^{24}$ cm^{-3}, σ_s is 20×10^{-24} cm^2, and $\Sigma_s = 1.34$ cm^{-1}. Thus the mean free path for scattering λ_s is approximately 0.75 cm.

The cross sections for atoms interacting with their own kind at the energies corresponding to room temperature conditions are of the order of 10^{-15} cm^2. If we equate this area to $\pi\, r^2$, the calculated radii are of the order of 10^{-8} cm. This is in rough agreement with the theoretical radius of electron motion in the hydrogen atom 0.53×10^{-8} cm. On the other hand, the cross sections for neutrons interacting with nuclei by scattering collisions, those in which the neutron is deflected in direction and loses energy, are usually very much smaller than those for atoms. For the case of 1 eV neutrons in hydrogen with a scattering cross section of 20 barns (i.e., 20×10^{-24} cm^2), one deduces a radius of approximately 2.5×10^{-12} cm. These results correspond to our earlier observation that the nucleus is thousands of times smaller than the atom.

4.5 **NEUTRON CROSS SECTIONS**

The cross section for neutron absorption in materials depends greatly on the isotope bombarded and on the neutron energy. For consistent comparison and use, the cross section is often cited at 0.0253 eV, corresponding to neutron speed

Table 4.1 Selected Thermal Neutron Absorption Cross Sections (in order of increasing size)[†]

Isotope or Element	σ_a (Barns)
^4_2He	$\cong 0$
$^{16}_8\text{O}$	0.00019
^2_1H	0.00051
$^{12}_6\text{C}$	0.0035
Zr	0.19
^1_1H	0.332
$^{238}_{92}\text{U}$	2.7
Mn	13.3
In	197
$^{235}_{92}\text{U}$	681
$^{239}_{94}\text{Pu}$	1022
$^{10}_5\text{B}$	3840
$^{135}_{54}\text{Xe}$	2,650,000

[†]*CRC Handbook of Chemistry and Physics.*

2200 m/s. Values for absorption cross sections for a number of isotopes at that energy are listed in order of increasing size in Table 4.1. The dependence of absorption cross section on energy is of two types, one called $1/v$, in which σ_a varies inversely with neutron speed, the other called resonance, in which there is a very strong absorption at certain neutron energies. Many materials exhibit both variations. Figures 4.5 and 4.6 show the cross sections for boron and natural uranium. The use of the logarithmic plot enables one to display the large range of cross section over the large range of energy of interest. Neutron scattering cross sections are more nearly the same for all elements and have less variation with neutron energy. Figure 4.7 shows the trend of σ_s for hydrogen as in water. Over a large range of neutron energy the scattering cross section is nearly constant, dropping off in the million-electron-volt region. This high-energy range is of special interest, because neutrons produced by the fission process have such energy values.

The competition between scattering and capture for neutrons in a medium is statistical in nature. The number of scattering collisions that occur before an absorption removes the neutron may be none, one, a few, or many. Computer Exercise 4.C discusses the program ABSCAT, which simulates the statistical competition.

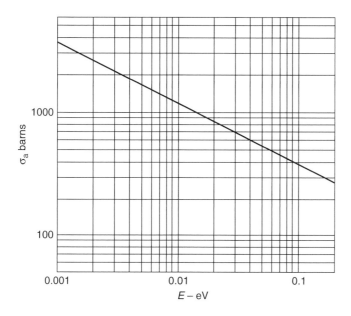

FIGURE 4.5 Absorption cross section for elemental boron.

4.6 NEUTRON MIGRATION

When fast neutrons, those of energy of the order of 2 MeV, are introduced into a medium, they make inelastic or elastic collisions with nuclei. On each elastic collision neutrons are deflected in direction, they lose energy, and they tend to migrate away from their origin. Each neutron has a unique history, and it is impractical to keep track of all of them. Instead, we seek to deduce average behavior. First, we note that the elastic scattering of a neutron with an initially stationary nucleus of mass number A causes a reduction in neutron energy from E_0 to E and a change of direction through an angle θ (theta), as sketched in Figure 4.8. The length of arrows indicates the speeds of the particles. The example shown is but one of a great variety of possible results of scattering collisions. For each final energy, there is a unique angle of scattering, and vice versa, but the occurrence of a particular E and θ pair depends on chance. The neutron may bounce directly backward, $\theta = 180°$, dropping down to a minimum energy αE_0, where $\alpha = (A-1)^2/(A+1)^2$, or it may be undeflected, $\theta = 0°$ and retain its initial energy E_0, or it may be scattered through any other angle, with corresponding energy loss. For the special case of a hydrogen nucleus as scattering target, $A = 1$ and $\alpha = 0$, so that the neutron loses all of its energy in a head-on collision. As we will see later, this makes water a useful material in a nuclear reactor.

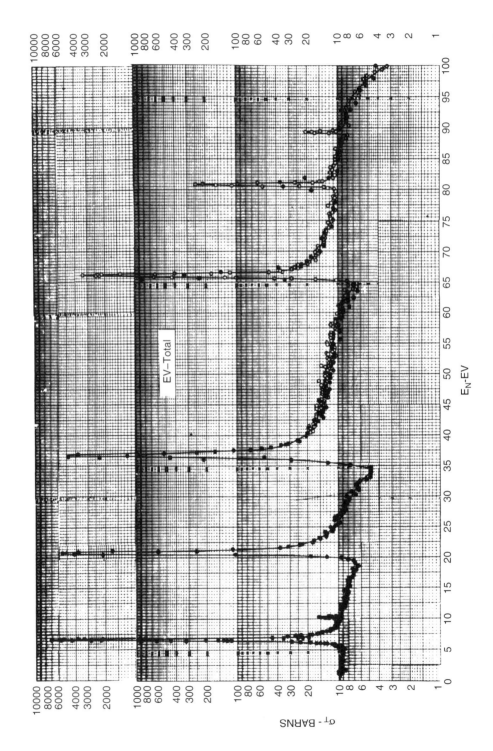

FIGURE 4.6 Cross section for natural uranium.

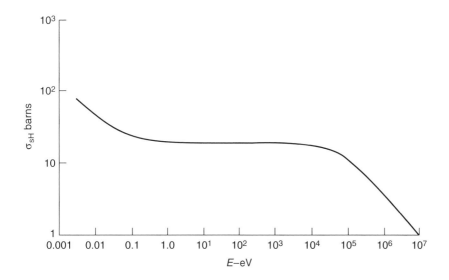

FIGURE 4.7 Scattering cross section for hydrogen.

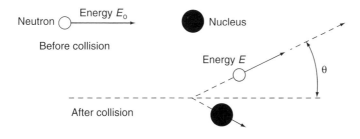

FIGURE 4.8 Neutron scattering and energy loss.

The process of neutron scattering with energy loss is graphically displayed by application of the program SCATTER, see Computer Exercise 4.D.

The average elastic scattering collision is described by two quantities that depend only on the nucleus, not on the neutron energy. The first is $\overline{\cos\theta}$, the average of the cosines of the angles of scattering, given by

$$\overline{\cos\theta} = \frac{2}{3A}.$$

For hydrogen it is 2/3, meaning that the neutron tends to be scattered in the forward direction; for a very heavy nucleus such as uranium it is near zero, meaning that the scattering is almost equally likely in each direction. Forward scattering results in an enhanced migration of neutrons from their point of appearance in a medium. Their free paths are effectively longer, and it is conventional to use the transport mean free path $\lambda_t = \lambda_s/(1 - \overline{\cos\theta})$ instead of λ_s to account for the effect. We note that λ_t is always the larger. Consider slow neutrons in carbon, for which $\sigma_s = 4.8$ barns and $N = 0.083$ (in units of 10^{24} as in Section 4.3), so that $\Sigma_s = 0.4$ cm^{-1} and $\lambda_s = 2.5$ cm. Now $\overline{\cos\theta} = 2/(3)(12) = 0.056$, $1 - \overline{\cos\theta} = 0.944$, and $\lambda_t = 2.5/0.994 = 2.7$ cm.

The second quantity that describes the average collision is ξ (xi), the average change in the natural logarithm of the energy, given by

$$\xi = 1 + \frac{\alpha \ln \alpha}{1 - \alpha}.$$

For hydrogen, it is exactly 1, the largest possible value, meaning that hydrogen is a good "moderator" for neutrons, its nuclei permitting the greatest neutron energy loss; for a heavy element it is $\xi \cong 2/(A + 2/3)$, which is much smaller than 1 (e.g., for carbon, $A = 12$, it is 0.16).

To find how many collisions C are required on the average to slow neutrons from one energy to another, we merely divide the total change in $\ln E$ by ξ, the average per collision. In going from the fission energy 2×10^6 eV to the thermal energy 0.025 eV, the total change is $\ln (2 \times 10^6) - \ln(0.025) = \ln(8 \times 10^7) = 18.2$. Then $C = 18.2/\xi$. For example in hydrogen, $\xi = 1$, C is 18, whereas in carbon $\xi = 0.16$, C is 114. Again, we see the virtue of hydrogen as a moderator. The fact that hydrogen has a scattering cross section of 20 barns over a wide range, whereas carbon has a σ_s of only 4.8 barns, implies that collisions are more frequent and the slowing takes place in a smaller region. The only disadvantage is that hydrogen has a larger thermal neutron absorption cross section, 0.332 barns vs. 0.0035 barns for carbon.

The statistical nature of the neutron slowing process is demonstrated in Computer Exercise 4.E, which uses the program ENERGY to calculate the number of collisions to go from fission energy to thermal energy in carbon.

The movement of individual neutrons through a moderator during slowing consists of free flights, interrupted frequently by collisions that cause energy loss. Picture, as in Figure 4.9, a fast neutron starting at a point and migrating outward. At some distance r away, it arrives at the thermal energy. Other neutrons become thermal at different distances, depending on their particular histories. If we were to measure all of their r values and form the average of r^2, the result would be $\overline{r^2} = 6\tau$, where τ (tau) is called the "age" of the neutron. Approximate values of the age for various moderators, as obtained from experiment, are listed below:

Moderator	τ, Age to Thermal (cm^2)
H_2O	26
D_2O	125
C	364

We thus note that water is a much better agent for neutron slowing than is graphite because of the larger scattering and energy loss.

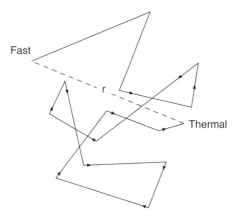

FIGURE 4.9 Neutron migration during slowing.

As neutrons slow into the energy region that is comparable to thermal agitation of the moderator atoms, they may either lose or gain energy on collision. Members of a group of neutrons have various speeds at any instant and thus the group behaves as a gas in Maxwellian distribution, as was shown in Figure 2.1 and discussed in Exercise 2.15. The neutron group has a temperature T that is close to that of the medium in which they are found. Thus if the moderator is at room temperature 20°C, or 293°K, the most likely neutron speed is approximately 2200 m/s, corresponding to a kinetic energy of 0.0253 eV. The neutrons are said to be thermal, in contrast to fast or intermediate.

Another parameter that characterizes neutron migration while at thermal energy is the diffusion length, symbolized by L. By analogy to the slowing process, the average square distance between origin and absorption is given by $\overline{r^2} = 6L^2$. Approximate values of L for three moderators are listed below.

Moderator	L, Diffusion Length (cm)
H_2O	2.85
D_2O	116
C	54

According to theory, $L = \sqrt{D/\Sigma_a}$, where $D = \lambda_t/3$. This shows that the addition of an absorber to pure moderator reduces the distance neutrons travel, as expected.

The process of diffusion of gas molecules is familiar to us. If a bottle of perfume is opened, the scent is quickly observed, because the molecules of the substance migrate away from the source. Because neutrons in large numbers behave as a gas, the descriptions of gas diffusion may be applied. The flow of neutrons through a medium at a location is proportional to the way the concentration of neutrons varies, in particular to the negative of the slope of the neutron number density. We can guess that the larger the neutron speed v and the larger the transport mean free path λ_t, the more neutron flow will take place. Theory and measurement show that if n varies in the z-direction, the net flow of neutrons across a unit area each second, the net current density, is

$$ j = \frac{-\lambda_t v}{3} \frac{dn}{dz}. $$

This is called Fick's law of diffusion, derived long ago for the description of gases. It applies if absorption is small compared with scattering. In terms of the flux $\phi = nv$ and the diffusion coefficient $D = \lambda_t /3$, this may be written compactly $j = - D\, \phi'$ where ϕ' is the slope of the neutron flux.

4.7 SUMMARY

Chemical and nuclear equations have similarities in the form of equations and in the requirements on conservation of particles and charge. The bombardment of nuclei by charged particles or neutrons produces new nuclei and particles. Final energies are found from mass differences and final speeds from conservation of momentum. The cross section for interaction of neutrons with nuclei is a measure of the chance of collision. Reaction rates depend mutually on neutron flows and macroscopic cross section. A stream of uncollided particles is reduced exponentially as it passes through a medium. Neutron absorption cross sections vary greatly with target isotope and with neutron energy, whereas scattering cross sections are relatively constant. Neutrons are slowed readily by collisions with light nuclei and migrate from their point of origin. On reaching thermal energy, they continue to disperse, with the net flow dependent on the spatial variation of flux.

4.8 EXERCISES

4.1   The energy of formation of water from its constituent gases is quoted to be 54500 cal/mole. Verify that this corresponds to 2.4 eV per molecule of H_2O.

4.2   Complete the following nuclear reaction equations:

$$ {}^{1}_{0}n + {}^{14}_{7}N \rightarrow {}^{(\ )}_{(\ )}(\ ) + {}^{1}_{1}H, $$

$$ {}^{2}_{1}H + {}^{9}_{4}Be \rightarrow {}^{(\ )}_{(\ )}(\ ) + {}^{1}_{0}n. $$

4.3   Using the accurate atomic masses listed below, find the minimum amount of energy an α particle must have to cause the transmutation of nitrogen to oxygen. (${}^{14}_{7}N$ 14.003074, ${}^{4}_{2}He$ 4.002603, ${}^{17}_{8}O$ 16.999132, ${}^{1}_{1}H$ 1.007825.)

4.4 Find the energy release in the reaction $^6_3\text{Li}(n, \alpha)^3_1\text{H}$, noting the masses ^1_0n 1.008665, ^3_1H 3.016049, ^4_2He 4.002603, and ^6_3Li 6.015123.

4.5 A slow neutron of mass 1.008665 amu is caught by the nucleus of a hydrogen atom of mass 1.007825 and the final products are a deuterium atom of mass 2.014102 and a γ-ray. The energy released is 2.22 MeV. If the γ-ray is assumed to have almost all of this energy, what is its effective mass in kg? What is the speed of the ^2_1H particle in m/s, using equality of momenta on separation? What is the recoil energy of ^2_1H in MeV? How does this compare with the total energy released? Was the assumption about the γ-ray reasonable?

4.6 Calculate the speeds and energies of the individual α particles in the reaction $^1_1\text{H} + ^7_3\text{Li} \rightarrow 2^4_2\text{He}$, assuming that they separate along the line of proton motion. Note that the mass of the lithium-7 atom is 7.016005.

4.7 Calculate the energy release in the reaction

$$^{13}_7\text{N} \rightarrow ^{13}_6\text{C} + ^0_{+1}\text{e}.$$

The atomic masses are $^{13}_7\text{N}$ 13.005739, $^{13}_6\text{C}$ 13.003355, and the masses of the positron and electron are 0.000549. Calculate (a) using nuclear masses, subtracting the proper number of electron masses from the atomic masses, and (b) using atomic masses with account for the energy of pair production.

4.8 Calculate the macroscopic cross section for scattering of 1 eV neutrons in water, using N for water as 0.0334×10^{24} cm^{-3} and cross sections 20 barns for hydrogen and 3.8 barns for oxygen. Find the mean free path λ_s.

4.9 Find the speed v and the number density of neutrons of energy 1.5 MeV in a flux 7×10^{13} cm^{-2} – s^{-1}.

4.10 Compute the flux, macroscopic cross section and reaction rate for the following data: $n = 2 \times 10^5$ cm^{-3}, $v = 3 \times 10^8$ cm/sec, $N = 0.04 \times 10^{24}$ cm^{-3}, $\sigma = 0.5 \times 10^{-24}$ cm^2.

4.11 What are the values of the average logarithmic energy change ξ and the average cosine of the scattering angle $\overline{\cos\theta}$ for neutrons in beryllium, $A = 9$? How many collisions are needed to slow neutrons from 2 MeV to 0.025 eV in Be-9? What is the value of the diffusion coefficient D for 0.025 eV neutrons if Σ_s is 0.90 cm^{-1}?

4.12 (a) Verify that neutrons of speed 2200 m/s have an energy of 0.0253 eV. (b) If the neutron absorption cross section of boron at 0.0253 eV is 760 barns, what would it be at 0.1 eV? Does this result agree with that shown in Figure 4.5?

4.13 Calculate the rate of consumption of U-235 and U-238 in a flux of 2.5×10^{13} cm^{-2} – s^{-1} if the uranium atom number density is

0.0223×10^{24} cm^{-3}, the atom number fractions of the two isotopes are 0.0072 and 0.9928, and cross sections are 681 barns and 2.7 barns, respectively. Comment on the results.

4.14 How many atoms of boron-10 per atom of carbon-12 would result in an increase of 50% in the macroscopic absorption cross section of graphite? How many ^{10}B atoms would there then be per million ^{12}C atoms?

4.15 Calculate the absorption cross section of the element zirconium using the isotopic data in the following table:

Mass number	Abundance (Atom %)	Cross Section (Barns)
90	51.45	0.014
91	11.22	1.2
92	17.15	0.2
94	17.38	0.049
96	2.80	0.020

Compare the result with the figure given in Table 4.1.

4.16 The total cross section for uranium dioxide of density 10 g/cm^3 is to be measured by a transmission method. To avoid multiple neutron scattering, which would introduce error into the results, the sample thickness is chosen to be much smaller than the mean free path of neutrons in the material. Using approximate cross sections for UO$_2$ of $\sigma_s = 15$ barns and σ_a of 7.6 barns, find the macroscopic cross section $\Sigma = \Sigma_a + \Sigma_s$. Then find the thickness of target t such that $t/\lambda = 0.05$. How much attenuation in neutron beam would that thickness give?

4.17 The manganese content of a certain stainless steel is to be verified by an activation measurement. The activity induced in a sample of volume V by neutron capture during a time t is given by

$$A = \phi \Sigma_a V[(1 - \exp(-\lambda t)].$$

A foil of area 1 cm^2 and thickness 2 mm is irradiated in a thermal neutron flux of 3×10^{12}/cm^2 – s for 2 h. Counts taken immediately yield an activity of 150 mCi for the induced Mn-56, half-life 2.58 h. Assuming that the atom number density of the alloy is 0.087 in units of 10^{24} and that the cross section for capture in Mn-55 is 13.3 barns, find the percent of Mn in the sample.

4.18 For fast neutrons in uranium-235 metal, find the density ρ, the number of atoms per cubic centimeter N, the macroscopic cross section Σ_a and Σ_t, the transport mean free path λ_t, the diffusion coefficient D, and the diffusion length L. NOTE: the density of natural U (99.3% U-238) is approximately

19.05 g/cm^3; for U-235, $\sigma_c = 0.25$ barns, $\sigma_f = 1.4$ barns, and $\sigma_t = 6.8$ barns (Report ANL-5800, p. 581).

4.19 When a projectile of mass m_1 and vector velocity $\mathbf{u_1}$ collides elastically with a target of mass m_2 and vector velocity $\mathbf{u_2}$, the final velocities are:

$$\mathbf{v_1} = [2\ m_2\mathbf{u_2} + (m_1 - m_2)\mathbf{u_1}]/(m_1 + m_2)$$

$$\mathbf{v_2} = [2\ m_1\mathbf{u} + (m_2 - m_1)\mathbf{u_2}]/(m_1 + m_2).$$

Find the velocities if $\mathbf{u_2} = 0$ and $m_2 \gg m_1$. Discuss the results.

4.20 A neutron of energy E_0 collides head-on with a heavy nucleus of mass number A. Using the velocity equations of Exercise 4.19, verify that the minimum neutron energy after collision is $E_1 = \alpha E_0$, where $\alpha = [(A-1)/(A+1)]^2$. Evaluate α and ξ for U-238.

4.21 Show for the case of $\mathbf{u_2} = 0$ (Exercise 4.19) that kinetic energy is conserved.

Computer Exercises

4.A Several computer programs provide visual images of neutron processes. MOVENEUT merely shows a moving particle; CURRENT gives a flow of many particles; CAPTURE allows a moving neutron to be captured by a stationary target nucleus. Run the programs to help visualize the processes. The program HEADON demonstrates an elastic collision in which neutron direction is reversed. Run the program with various choices of mass number A: 12 (carbon), 2 (deuterium), 238 (uranium), and 1 (hydrogen). Note and report on differences.

4.B Random variables are numbers between 0 and 1 that are statistically independent. They are at the heart of the method known as Monte Carlo (after the gambling casino in Monaco). Such numbers are produced by the command RND(X).

(a) Program RANDY generates and prints out a sequence of random numbers. Run the program two or three times to see results. Then delete the command RANDOMIZE TIMER and repeat. Comment on the effect.

(b) Program RANDY1 is the same as RANDY except that the average value is calculated. Run the program with increasing values of input NT, the total number of random numbers, to see what happens. What would you expect?

(c) Program RANDY2 is the same as RANDY1 except that additional statistical features are calculated. Run the program; note and comment on the results.

4.C On average, scattering and absorption of neutrons is determined by the macroscopic cross sections Σ_s, and Σ_a. For a given neutron, however, by chance the number of scatterings before being absorbed varies widely. The program ABSCAT uses random numbers to describe the process. Run the program several times to note the variation. Explain how the expected number of scatterings per absorption is calculated.

4.D The computer program SCATTER shows the general elastic collision of a neutron with a stationary nucleus, in which the neutron loses energy and moves off at an angle from the original direction, while the struck nucleus recoils in another direction. (a) Run the program several times to see the variety of final motions. (b) Change the mass ratio A (line 330) to 1, or 12, or 238, and observe differences.

4.E The energy loss of a neutron in an elastic collision with a nucleus can range from zero to $E_0 (1 - \alpha)$. Thus there is considerable statistical variation in the number of collisions C required to go between two energies. By use of random numbers, the computer program ENERGY shows a set of values of C for neutron slowing in carbon between fission and thermal energies. (a) Run the program to note the variation about the average of 114 collisions. (b) Change the A value to 238 as for U-238 and run again. (c) Repeat for $A = 1$ as for hydrogen. (d) Make a large change in the number of histories (e.g., decrease or increase by a factor of 10) and note the effect.

4.F Apply computer program ALBERT (See Chapter 1) to find a more accurate pair of numbers (e.g., 7 significant figures) than 2200 m/s and 0.0253 eV to describe room temperature 20°C neutrons at absolute temperature $T = 293.15$ K. NOTE: 1 ev $= 1.60217646 \times 10^{-19}$ J and Boltzmann's constant in $E = kT$ is $1.3806503 \times 10^{-23}$ J/K. What limits the accuracy of the result?

4.9 **REFERENCES**

Hans A. Bethe, Robert F. Bacher, and M. Stanley Livingston, *Basic Bethe: Seminal Articles on Nuclear Physics, 1936-1937*, American Institute of Physics, 1986. Reprints of classic literature on nuclear processes.

Neutron Cross Sections, Vol. 1, *Neutron Resonance Parameters and Thermal Cross Sections*, S. F. Mughabghab, Part B, $Z = 61 - 100$, 1981; S. F. Mughabghab, M. Divadeenam, and N. E. Holden, Part A, $Z = 1 - 60$, 1984; Vol. 2, *Neutron Cross Section Curves*, Victoria McLane, Charles L. Dunford, and Philip F. Rose, 1988; Academic Press, New York.

Tables of Thermal Cross Sections
http://ie.lbl.gov/ngdata/sig.txt.
Some of the same information as in the preceding book, at a Web site maintained by Lawrence Berkeley Laboratory.

Said F. Mughabghab, *Atlas of Neutron Resonances, 5th Ed.: Resonance Parameters and Thermal Cross Sections, Z = 1 − 100*, Elsevier Science, Amsterdam, 2006.
Explanations, theory, and cross section data. An updated version of *Neutron Cross Sections*.

National Nuclear Data Center (NNDC)
http://www.nndc.bnl.gov/masses
2003 atomic mass evaluation. Latest and perhaps last.

Atomic Mass Data Center
http://amdc.in2p3.fr
Select AME. Mirror site to BNL's.

The Isotope Project (Lawrence Berkeley National Laboratory)
http://isotopes.lbl.gov
Data centers and links to Web sites.

C. A. Bertulani and P. Danielewicz, *Introduction to Nuclear Reactions*, Taylor and Francis, 2004.

John R. Lamarsh and Anthony J. Baratta, *Introduction to Nuclear Engineering*, 3rd Ed., Prentice-Hall, Upper Saddle River, NJ, 2001. Update of classic textbook.

Raymond L. Murray, *Nuclear Reactor Physics*, Prentice-Hall, Englewood Cliffs, NJ, 1957. Elementary theory, analysis, and calculations.

John R. Lamarsh, *Introduction to Nuclear Reactor Theory*, Addison-Wesley, Reading, MA, 1972. Widely used textbook. Paperback reprint by American Nuclear Society, La Grange Park, IL, 2002.

Allen F. Henry, *Nuclear Reactor Analysis*, MIT Press, Cambridge, 1975. Advanced textbook.

James J. Duderstadt and Louis J. Hamilton, *Nuclear Reactor Analysis*, John Wiley & Sons, New York, 1976. Thorough treatment.

Weston M. Stacey, *Nuclear Reactor Physics*, John Wiley & Sons, New York, 2001. Advanced textbook.

Richard E. Faw and J. Kenneth Shultis, *Fundamentals of Nuclear Science and Engineering*, Marcel Dekker, New York, 2002.

David Bodansky, *Nuclear Energy: Principles, Practices, and Prospects*, 2nd Ed., Springer/AIP Press, New York, 2003.

Radiation and Materials

THE WORD "radiation" will be taken to embrace all particles, whether they are of material or electromagnetic origin. We include those particles produced by both atomic and nuclear processes and those resulting from electrical acceleration, noting that there is no essential difference between X-rays from atomic collisions and gamma rays from nuclear decay; protons can come from a particle accelerator, from cosmic rays, or from a nuclear reaction in a reactor. The word "materials" will refer to bulk matter, whether of mineral or biological origin, as well as to the particles of which the matter is composed, including molecules, atoms, electrons, and nuclei.

When we put radiation and materials together, a great variety of possible situations must be considered. Bombarding particles may have low or high energy; they may be charged, uncharged, or photons; and they may be heavy or light in the scale of masses. The targets may be similarly distinguished, but they may also exhibit degrees of binding that range from (a) none, as for "free" particles, to (b) weak, as for atoms in molecules and electrons in atoms, to (c) strong, as for nucleons in nuclei. In most interactions, the higher the projectile energy in comparison with the energy of binding of the structure, the greater is the effect.

Out of the broad subject we will select for review some of the reactions that are important in the nuclear energy field. Looking ahead, we will need to understand the effects produced by the particles and rays from radioactivity and other nuclear reactions. Materials affected may be in or around a nuclear reactor, as part of its construction or inserted to be irradiated. Materials may be of biological form, including the human body, or they may be inert substances used for protective shielding against radiation. We will not attempt to explain the processes rigorously but be content with qualitative descriptions based on analogy with collisions viewed on an elementary physics level.

5.1 EXCITATION AND IONIZATION BY ELECTRONS

These processes occur in the familiar fluorescent light bulb, in an X-ray machine, or in matter exposed to beta particles. If an electron that enters a material has a very low energy, it will merely migrate without affecting the molecules significantly.

If its energy is large, it may impart energy to atomic electrons as described by the Bohr theory (Chapter 2), causing excitation of electrons to higher energy states or producing ionization, with subsequent emission of light. When electrons of inner orbits in heavy elements are displaced, the resultant high-energy radiation is classed as X-rays. These rays, which are so useful for internal examination of the human body, are produced by accelerating electrons in a vacuum chamber to energies in the kilovolt range and allowing them to strike a heavy element target. In addition to the X-rays as a result of transitions in the electron orbits, a similar radiation called *bremsstrahlung* (German: braking radiation) is produced. It arises from the deflection and resulting acceleration of electrons as they encounter nuclei.

Beta particles as electrons from nuclear reactions have energies in the range 0.01 to 1 MeV and thus are capable of producing large amounts of ionization as they penetrate a substance. As a rough rule of thumb, approximately 32 eV of energy is required to produce one ion pair. The beta particles lose energy with each event and eventually are stopped. For electrons of 1 MeV energy, the range, as the typical distance of penetration, is no more than a few millimeters in liquids and solids or a few meters in air.

5.2 HEAVY CHARGED PARTICLE STOPPING BY MATTER

Charged particles such as protons, alpha particles, and fission fragment ions are classified as heavy, being much more massive that the electron. For a given energy, their speed is lower than that of an electron, but their momentum is greater, and they are less readily deflected on collision. The mechanism by which they slow down in matter is mainly electrostatic interaction with the atomic electrons and with nuclei. In each case the Coulomb force, varying as $1/r^2$ with distance of separation r, determines the result of a collision. Figure 5.1 illustrates the effect of the passage of an ion by an atom. An electron is displaced and gains energy in a large amount compared with its binding in the atom, creating an ion. Application of the collision formulas of Exercise 4.19 leads to the energy change when a heavy particle of mass m_H and energy E_0 collides head-on with an electron of mass m_e, as approximately $4(m_e/m_H) E_0$. For example, for an alpha particle of 5 MeV, the loss by the projectile and the gain by the target are $4(0.000549/4.00)\ 5 = 0.00274$ MeV or 2.74 keV. The electron is energetic enough to produce secondary ionization, whereas hundreds of collisions are needed to reduce the alpha particle's energy by as little as 1 MeV. As the result of primary and secondary processes, a great deal of ionization is produced by heavy ions as they move through matter.

In contrast, when a heavy charged particle comes close to a nucleus, the electrostatic force causes it to move in a hyperbolic path as in Figure 5.2. The projectile is scattered through an angle that depends on the detailed nature of the collision (i.e., the initial energy and direction of motion of the incoming ion relative to the target nucleus) and the magnitudes of electric charges of the

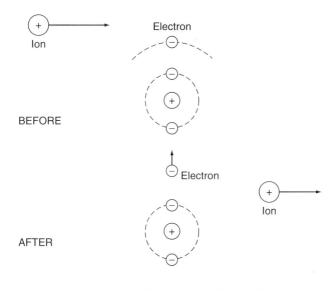

FIGURE 5.1 Interaction of heavy ion with electron.

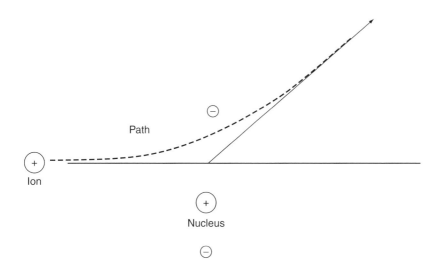

FIGURE 5.2 Interaction of heavy ion with nucleus.

interacting particles. The charged particle loses a significant amount of energy in the process, in contrast with the slight energy loss on collision with an electron. Unless the energy of the bombarding particle is very high and it comes within the short range of the nuclear force, there is a small chance that it can enter the nucleus and cause a nuclear reaction.

A measure of the rate of ion energy loss with distance traveled is the *stopping power*, symbolized by $-dE/dx$. It is also known as the linear energy transfer (LET). Two separate components, atomic and nuclear, add to give the total, as tabulated in the NIST Web site (see References). Theoretical formulas giving the dependence on electric charges, masses, and energy are given by Mayo (see References). A related quantity is the *range*, which is the maximum distance of travel of a projectile as it makes multiple collisions in matter. Integration of the reciprocal of the stopping power yields values of the range, also given by NIST. For example, in its ASTAR database, the range of 4 MeV alpha particles is listed as 3.147E-03 cm^2/g, and with an air density of 0.001293 g/cm^3, a distance of 2.43 cm. An alpha particle has a very small range in solid materials: a sheet of paper is sufficient to stop it and the outer layer of human skin provides protection for sensitive tissue.

5.3 GAMMA RAY INTERACTIONS WITH MATTER

We now turn to a group of three related processes involving gamma ray photons produced by nuclear reactions. These have energies as high as a few MeV. The interactions include simple scattering of the photon, ionization by it, and a special nuclear reaction known as pair production.

(a) Photon-Electron Scattering

One of the easiest processes to visualize is the interaction of a photon of energy $E = h\nu$ and an electron of rest mass m_0. Although the electrons in a target atom can be regarded as moving and bound to their nucleus, the energies involved are very small (eV) compared with those of typical gamma rays (keV or MeV). Thus the electrons may be viewed as free stationary particles. The collision may be treated by the physical principles of energy and momentum conservation. As sketched in Figure 5.3, the photon is deflected in its direction and loses energy, becoming a photon of new energy $E' = h\nu'$. The electron gains energy and moves away with high-speed υ and total mass-energy mc^2, leaving the atom ionized. In this Compton effect, named after its discoverer, one finds that the greatest photon energy loss occurs when it is scattered backward (180°) from the original direction. Then, if E is much larger than the rest energy of the electron $E_0 = m_0c^2 = 0.51$ MeV, it is found that the final photon energy E' is equal to $E_0/2$. On the other hand, if E is much smaller than E_0, the fractional energy loss of the photon is $2E/E_0$ (see also Exercise 5.3). The derivation of the photon energy loss in general is complicated by the fact that the special theory of relativity must be applied. The resulting formulas are displayed in the computer program COMPTON, which is used in several Computer Exercises to find photon energy losses.

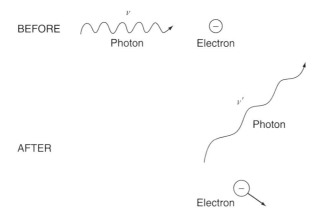

FIGURE 5.3 Photon electron scattering (Compton effect).

The probability of Compton scattering is expressed by a cross section, which is smaller for larger gamma energies as shown in Figure 5.4 for the element lead, a common material for shielding against X-rays or gamma rays. We can deduce that the chance of collision increases with each successive loss of energy by the photon, and eventually the photon disappears.

(b) *Photoelectric Effect*

This process is in competition with scattering. An incident photon of high enough energy dislodges an electron from the atom, leaving a positively charged ion. In so doing, the photon is absorbed and thus lost (see Figure 5.5). The cross section for the photoelectric effect decreases with increasing photon energy, as sketched in Figure 5.4 for the element lead.

The preceding two processes are usually treated separately even though both result in ionization. In the Compton effect, a photon of lower energy survives, but in the photoelectric effect, the photon is eliminated. In each case, the electron released may have enough energy to excite or ionize other atoms by the mechanism described earlier. Also, the ejection of the electron is followed by light emission or X-ray production, depending on whether an outer shell or inner shell is involved.

(c) *Electron-Positron Pair Production*

The third process to be considered is one in which the photon is converted into matter. This is entirely in accord with Einstein's theory of the equivalence of mass and energy. In the presence of a nucleus, as sketched in Figure 5.6, a gamma ray photon disappears and two particles appear—an electron and a positron. Because

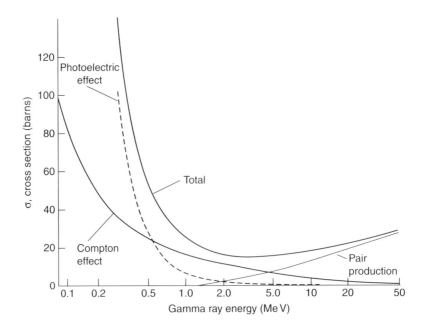

FIGURE 5.4 Gamma ray cross sections in lead, Pb. Plotted from data in National Bureau of Standards report NSRDS-NSB-29.

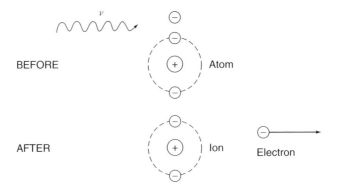

FIGURE 5.5 Photoelectric effect.

these are of equal charge but of opposite sign, there is no net charge after the reaction, just as before, the gamma ray having zero charge. The law of conservation of charge is thus met. The total new mass produced is twice the mass-energy of the electron, $2(0.51) = 1.02$ MeV, which means that the reaction can occur only if the gamma ray has at least this amount of energy. The cross section for

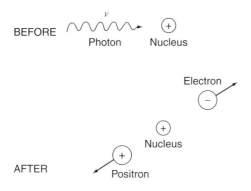

FIGURE 5.6 Pair production.

the process of pair production rises from zero as shown in Figure 5.4 for lead. The reverse process also takes place. As sketched in Figure 5.7, when an electron and a positron combine, they are annihilated as material particles, and two gamma rays of energy totaling at least 1.02 MeV are released. That there must be two photons is a consequence of the principle of momentum conservation.

The reverse process, in which two high-energy photons collide to form an electron-positron pair, is believed to have been common in early times after the Big Bang.

Figure 5.4 shows that the total gamma ray cross section curve for lead (Pb), as the sum of the components for Compton effect, photoelectric effect, and pair production, exhibits a minimum at approximately 3 MeV energy.

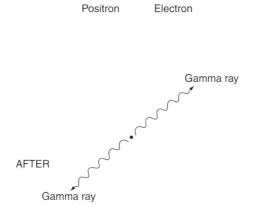

FIGURE 5.7 Pair annihilation.

This implies that gamma rays in this vicinity are more penetrating than those of higher or lower energy. In contrast with β particles and α particles, which have a definite range, a certain fraction of incident gamma rays can pass through any thickness of material. The exponential expression $e^{-\Sigma z}$ as used to describe neutron behavior can be carried over to the attenuation of gamma rays in matter. One can use the mean free path $\lambda = 1/\Sigma$ or, better, the half-thickness $0.693/\Sigma$, the distance in which the intensity of a gamma ray beam is reduced by a factor of two.

Cross section data for the interaction of photons with many elements are found in the NIST Web site (see References).

5.4 NEUTRON REACTIONS

For completeness, we review again the interaction of neutrons with matter. Neutrons may be scattered by nuclei elastically or inelastically, may be captured with resulting gamma ray emission, or may cause fission. If their energy is high enough, neutrons may induce (n, p) and (n, α) reactions as well.

We are now in a position to understand the connection between neutron reactions and atomic processes. When a high-speed neutron strikes the hydrogen atom in a water molecule, a proton is ejected, resulting in chemical dissociation of the H_2O. A similar effect takes place in molecules of cells in any biological tissue. The proton compared with the electron is a heavy charged particle. It passes through matter, slowing and creating ionization along its path. Thus two types of neutron radiation damage take place—primary and secondary.

After many collisions, the neutron arrives at a low enough energy that it can be readily absorbed. If it is captured by the proton in a molecule of water or some other hydrocarbon, a gamma ray is released, as discussed in Chapter 4. The resulting deuteron recoils with energy that is much smaller than that of the gamma ray but still is far greater than the energy of binding of atoms in the water molecule. Again dissociation of the compound takes place, which can be regarded as a form of radiation damage.

5.5 SUMMARY

Radiation of special interest includes electrons, heavy charged particles, photons, and neutrons. Each of the particles tends to lose energy by interaction with the electrons and nuclei of matter, and each creates ionization in different degrees. The ranges of beta particles and alpha particles are short, but gamma rays penetrate in accord with an exponential law. Gamma rays can also produce electron-positron pairs. Neutrons of both high and low energy can create radiation damage in molecular materials.

5.6 EXERCISES

5.1 The charged particles in a highly ionized electrical discharge in hydrogen gas—protons and electrons, mass ratio $m_p/m_e = 1836$—have the same energies. What is the ratio of the speeds v_p/v_e? Of the momenta p_p/p_e?

5.2 A gamma ray from neutron capture has an energy of 6 MeV. What is its frequency? Its wavelength?

5.3 For 180° scattering of gamma or X-rays by electrons, the final energy of the photon is

$$E' = \frac{1}{\dfrac{1}{E} + \dfrac{2}{E_0}}.$$

  (a) What is the final photon energy for the 6 MeV gamma ray of Exercise 5.2?
  (b) Verify that if $E \gg E_0$, then $E' \cong E_0/2$ and if $E \ll E_0$, $(E - E')/E \cong 2\,E/E_0$.
  (c) Which approximation should be used for a 6 MeV gamma ray? Verify numerically.

5.4 An electron-positron pair is produced by a gamma ray of 2.26 MeV. What is the kinetic energy imparted to each of the charged particles?

5.5 Estimate the thickness of paper required to stop 2 MeV alpha particles, assuming the paper to be of density 1.29 g/cm^3, about the same electronic composition as air, density 1.29×10^{-3} g/cm^3.

5.6 The element lead, $M = 206$, has a density of 11.3 g/cm^3. Find the number of atoms per cubic centimeter. If the total gamma ray cross section at 3 MeV is 14 barns, what is the macroscopic cross section Σ and the half-thickness $0.693/\Sigma$?

5.7 The range of beta particles of energy greater than 0.8 MeV is given roughly by the relation

$$R(\text{cm}) = \frac{0.55E(\text{MeV}) - 0.16}{\rho\,(\text{g/cm}^3)}.$$

Find what thickness of aluminum sheet (density 2.7 g/cm^3) is enough to stop the betas from phosphorus-32 (see Table 3.1).

5.8 A radiation worker's hands are exposed for 5 seconds to a 3×10^8 cm^{-2} s^{-1} beam of 1 MeV beta particles. Find the range in tissue of density 1.0 g/cm^3 and calculate the amounts of charge in coulombs (C) and energy deposition in C/cm^3 and J/g. Note that the charge on the electron is 1.60×10^{-19} C. For tissue, use the equation in Exercise 5.7.

5.9 Calculate the energy gain by an electron struck head-on by an alpha particle of energy 4 MeV. How many such collisions would it take to reduce the alpha particle energy to 1 MeV?

5.10 At a certain time after the Big Bang, high-speed photons collided to form electrons and positrons. Assuming energies of 0.51 MeV each, what temperature is implied?

Computer Exercises

5.A The scattering at any angle of a photon colliding with a free electron is analyzed by the program COMPTON, after Arthur Holly Compton's theory. (a) Run the program and use the menus. (b) Find the maximum and minimum photon energies of 50 keV X-rays passing through a thin aluminum foil and making no more than one collision.

5.B With the program COMPTON, compare the percent energy change of 10 keV and 10 MeV photons scattered at 90°. What conclusion do these results suggest?

5.C (a) Find the fractional energy loss for a 20 keV X-ray scattered from an electron at angle 180° and compare with $2E/E_0$. (b) Find the final energy for a 10 MeV gamma ray scattered from an electron at 180° and compare with $E_0/2$.

5.7 REFERENCES

Emilio Segrè, *Nuclei and Particles*, 2nd Ed., Benjamin-Cummings, Reading, MA, 1977. A classic book on nuclear theory and experiments for undergraduate physics students written by a Nobel Prize winner.

Robert M. Mayo, *Introduction to Nuclear Concepts for Engineers*, American Nuclear Society, La Grange Park, IL, 1998. Chapter 6 is devoted to the interaction of radiation with matter.

Hans A. Bethe, Robert F. Bacher, and M. Stanley Livingston, *Basic Bethe, Seminal Articles on Nuclear Physics, 1936-1937*, American Institute of Physics and Springer Verlag, New York, 1986. Reprints of classic literature on nuclear processes. Discussion of stopping power, p. 347 ff.

National Institute of Science and Technology (NIST)
http://www.nist.gov
For stopping powers and ranges of electrons, protons and alpha particles, use Search with keyword "stopping power."
For photon cross sections for many elements, use Search with keyword "XCOM." Data by Berger, Hubbell, et al.

Gary. S. Was, *Fundamentals of Radiation Materials Science*, Springer, New York, 2007.

J. Kenneth Shultis and Richard E. Faw, *Radiation Shielding*, Prentice-Hall, Upper Saddle River, NJ, 1996. Basics and modern analysis techniques.

Richard E. Faw and J. Kenneth Shultis, *Radiological Assessment: Sources and Doses*, American Nuclear Society, La Grange Park, IL, 1999. Includes fundamentals of radiation interactions.

Fission

OUT OF the many nuclear reactions known, that resulting in fission has at present the greatest practical significance. In this chapter we will describe the mechanism of the process, identify the byproducts, introduce the concept of the chain reaction, and look at the energy yield from the consumption of nuclear fuels.

6.1 THE FISSION PROCESS

The absorption of a neutron by most isotopes involves radiative capture, with the excitation energy appearing as a gamma ray. In certain heavy elements, notably uranium and plutonium, an alternate consequence is observed—the splitting of the nucleus into two massive fragments, a process called fission. Computer Exercise 6.A provides a graphic display of the process. Figure 6.1 shows the sequence of events by use of the reaction with U-235. In Stage A, the neutron approaches the U-235 nucleus. In Stage B, the U-236 nucleus has been formed in an excited state. The excess energy in some cases may be released as a gamma ray, but more frequently, the energy causes distortions of the nucleus into a dumbbell shape, as in Stage C. The parts of the nucleus oscillate in a manner analogous to the motion of a drop of liquid. Because of the dominance of electrostatic repulsion over nuclear attraction, the two parts can separate, as in Stage D. They are then called fission fragments, bearing most of the energy released. They fly apart at high speeds, carrying some 166 MeV of kinetic energy out of the total of approximately 200 MeV released in the whole process. As the fragments separate, they lose atomic electrons, and the resulting high-speed ions lose energy by interaction with the atoms and molecules of the surrounding medium. The resultant thermal energy is recoverable if the fission takes place in a nuclear reactor. Also shown in the diagram are the prompt gamma rays and fast neutrons that are released at the time of splitting.

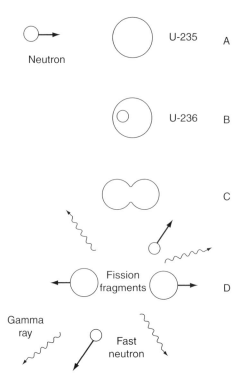

FIGURE 6.1 The fission process.

6.2 ENERGY CONSIDERATIONS

The absorption of a neutron by a nucleus such as U-235 gives rise to extra internal energy of the product, because the sum of masses of the two interacting particles is greater than that of a normal U-236 nucleus. We write the first step in the reaction

$$^{235}_{92}\text{U} + ^{1}_{0}\text{n} \rightarrow \left(^{236}_{92}\text{U} \right)^{*},$$

where the asterisk signifies the excited state. The mass in atomic mass units of (U-236)* is the sum $235.043923 + 1.008665 = 236.052588$. However, U-236 in its ground state has a mass of only 236.045562, lower by 0.007026 amu or 6.54 MeV. This amount of excess energy is sufficient to cause fission. Figure 6.2 shows these energy relationships. To achieve accurate results, we use many more significant figures and decimals than for typical scoping calculations.

The preceding calculation did not include any kinetic energy brought to the reaction by the neutron, on the grounds that fission can be induced by absorption in U-235 of very slow neutrons. Only one natural isotope, $^{235}_{92}\text{U}$, undergoes

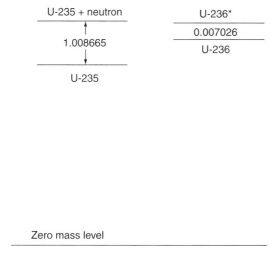

U-235 + neutron		U-236*
↑		0.007026
1.008665		U-236
↓		
U-235		

Zero mass level

FIGURE 6.2 Excitation energy caused by neutron absorption.

fission in this way, whereas $^{239}_{94}$Pu and $^{233}_{92}$U are the main artificial isotopes that do so. Most other heavy isotopes require significantly larger excitation energy to bring the compound nucleus to the required energy level for fission to occur, and the extra energy must be provided by the motion of the incoming neutron. For example, neutrons of at least 0.9 MeV are required to cause fission from U-238, and other isotopes require even higher energy. The precise terminology is as follows: fissile materials are those giving rise to fission with slow neutrons; many isotopes are fissionable, if enough energy is supplied. It is advantageous to use fast neutrons—of the order of 1 MeV energy—to cause fission. As will be discussed in Chapter 13, the fast reactor permits the "breeding" of nuclear fuel. In a few elements such as californium, spontaneous fission takes place. The isotope $^{252}_{98}$Cf, produced artificially by a sequence of neutron absorption, has a half-life of 2.645 y, decaying by alpha emission (96.9%) and spontaneous fission (3.1%).

A small but important amount of spontaneous fission occurs in plutonium-240 in competition with alpha decay.

It may be surprising that the introduction of only 6.5 MeV of excitation energy can produce a reaction yielding as much as 200 MeV. The explanation is that the excitation triggers the separation of the two fragments and the powerful electrostatic force provides them a large amount of kinetic energy. By conservation of mass-energy, the mass of the nuclear products is smaller than the mass of the compound nucleus from which they emerge.

6.3 BYPRODUCTS OF FISSION

Accompanying the fission process is the release of several neutrons, which are all-important for the practical application to a self-sustaining chain reaction. The numbers that appear v (nu) range from 1 to 7, with an average in the range 2 to 3, depending on the isotope and the bombarding neutron energy. For example, in U-235 with slow neutrons the average number v is 2.42. Most of these are released instantly, the so-called prompt neutrons, whereas a small percentage, 0.65% for U-235, appear later as the result of radioactive decay of certain fission fragments. These delayed neutrons provide considerable inherent safety and controllability in the operation of nuclear reactors, as we will see later.

The nuclear reaction equation for fission resulting from neutron absorption in U-235 may be written in general form, letting the chemical symbols for the two fragments be labeled F_1 and F_2 to indicate many possible ways of splitting. Thus

$$^{235}_{92}U + ^{1}_{0}n \rightarrow ^{A_1}_{Z_1}F_1 + ^{A_2}_{Z_2}F_2 + v^{1}_{0}n + \text{energy}.$$

The appropriate mass numbers and atomic numbers are attached. One example, in which the fission fragments are isotopes of krypton and barium, is

$$^{235}_{92}U + ^{1}_{0}n \rightarrow ^{90}_{36}Kr + ^{144}_{56}Ba + 2^{1}_{0}n + E.$$

Mass numbers ranging from 75 to 160 are observed, with the most probable at approximately 92 and 144 as sketched in Figure 6.3. The ordinate on this graph is the percentage yield of each mass number (e.g., approximately 6% for mass numbers 90 and 144). If the number of fissions is given, the number of atoms of those types is 0.06 as large. Computer Exercise 6.B describes the program YIELD, which calculates the fission yield for several mass numbers.

As a collection of isotopes, these byproducts are called fission products. The isotopes have an excess of neutrons or a deficiency of protons compared with naturally occurring elements. For example, the main isotope of barium is $^{137}_{56}Ba$, and a prominent element of mass 144 is $^{144}_{60}Nd$. Thus there are seven extra neutrons or four too few protons in the barium isotope from fission, and it is highly unstable. Radioactive decay, usually involving several emissions of beta particles and delayed gamma rays in a chain of events, brings the particles down to stable forms. An example is

$$^{90}_{36}Kr \underset{32.3\,s}{\rightarrow} ^{90}_{37}Rb \underset{2.6\,min}{\rightarrow} ^{90}_{38}Sr \underset{29.1y}{\rightarrow} ^{90}_{39}Y \underset{2.67\,d}{\rightarrow} ^{90}_{40}Zr.$$

The hazard associated with the radioactive emanations from fission products is evident when we consider the large yields and the short half-lives.

The total energy from fission, after all of the particles from decay have been released, is approximately 200 MeV. This is distributed among the various processes as shown in Table 6.1. The prompt gamma rays are emitted as a part of fission; the rest are fission product decay gammas. Neutrinos accompany the beta particle emission, but because they are such highly penetrating particles, their

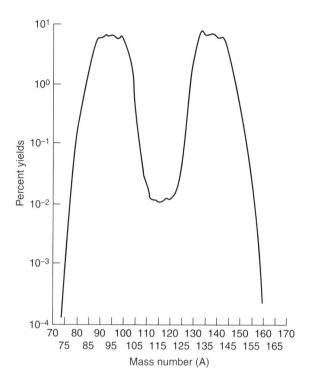

FIGURE 6.3 Yield of fission products according to mass number (Courtesy of T.R. England of Los Alamos National Laboratory).

Table 6.1 Energy from Fission, U-235	
	MeV
Fission fragment kinetic energy	166
Neutrons	5
Prompt gamma rays	7
Fission product gamma rays	7
Beta particles	5
Neutrinos	10
Total	**200**

energy cannot be counted as part of the useful thermal energy yield of the fission process. Thus only approximately 190 MeV of the fission energy is effectively available. However, several MeV of energy from gamma rays released from nuclei that capture neutrons can also be extracted as useful heat.

The average total neutron energy is noted to be 5 MeV. If there are approximately 2.5 neutrons per fission, the average neutron energy is 2 MeV. When one observes many fission events, the neutrons are found to range in energy from nearly 0 to more than 10 MeV, with a most likely value of 0.7 MeV. Computer Exercise 6.C discusses calculation of the fission neutron energy distribution according to a semi-empirical formula. We note that the neutrons produced by fission are fast, whereas the cross section for the fission reaction is high for slow neutrons. This fact serves as the basis for the use of a reactor moderator containing a light element that permits neutrons to slow down, by a succession of collisions, to an energy favorable for fission.

Although fission is the dominant process, a certain fraction of the absorptions of neutrons in uranium merely results in radiative capture, according to

$$^{235}_{92}\text{U} + {}^{1}_{0}\text{n} \rightarrow {}^{236}_{92}\text{U} + \gamma.$$

The U-236 is relatively stable, having a half-life of 2.34×10^7 y. Approximately 14% of the absorptions are of this type, with fission occurring in the remaining 86%. This means that η (eta), the number of neutrons produced per *absorption* in U-235 is lower than v, the number per *fission*. Thus by use of $v = 2.42$, η is $(0.86)(2.42) = 2.07$. The effectiveness of any nuclear fuel is sensitively dependent on the value of η. We find that η is larger for fission induced by fast neutrons than that by slow neutrons.

The possibility of a chain reaction was recognized as soon as it was known that neutrons were released in the fission process. If a neutron is absorbed by the nucleus of one atom of uranium and one neutron is produced, the latter can be absorbed in a second uranium atom, and so on. To sustain a chain reaction as in a nuclear reactor or in a nuclear weapon, the value of η must be somewhat greater than 1 because of processes that compete with absorption in uranium, such as capture in other materials and escape from the system. The size of η has two important consequences. First, there is a possibility of a growth of neutron population with time. After all extraneous absorption and losses have been accounted for, if one absorption in uranium ultimately gives rise to, say, 1.1 neutrons; these can be absorbed to give $(1.1)(1.1) = 1.21$, which produce 1.331, etc. The number available increases rapidly with time. Second, there is a possibility of the use of the extra neutron, over and above the one required to maintain the chain reaction, to produce new fissile materials. "Conversion" involves the production of some new nuclear fuel to replace that used up, whereas "breeding" is achieved if more fuel is produced than is used.

Out of the hundreds of isotopes found in nature, only one is fissile, $^{235}_{92}\text{U}$. Unfortunately, it is the less abundant of the isotopes of uranium, with weight percentage in natural uranium of only 0.711 in comparison with 99.3% of the heavier isotope $^{238}_{92}$U. The two other most important fissile materials, plutonium-239 and uranium-233, are "artificial" in the sense that they are man-made by use of neutron irradiation of two fertile materials, uranium-238 and thorium-232, respectively. The reactions by which $^{239}_{94}$Pu is produced are

$$^{238}_{92}U + {}^{1}_{0}n \rightarrow {}^{239}_{92}U,$$

$$^{239}_{92}U \xrightarrow[23.5\,\text{min}]{} {}^{239}_{93}Np + {}^{0}_{-1}e,$$

$$^{239}_{92}Np \xrightarrow[2.355\,\text{d}]{} {}^{239}_{94}Pu + {}^{0}_{-1}e,$$

whereas those yielding $^{233}_{92}$U are

$$^{232}_{90}Th + {}^{1}_{0}n \rightarrow {}^{233}_{90}Th,$$

$$^{233}_{90}Th \xrightarrow[22.3\,\text{min}]{} {}^{233}_{91}Pa + {}^{0}_{-1}e,$$

$$^{233}_{91}Pa \xrightarrow[27.0\,\text{d}]{} {}^{233}_{92}U + {}^{0}_{-1}e.$$

The half-lives for decay of the intermediate isotopes are short compared with times involved in the production of these fissile materials. For many purposes, these decay steps can be ignored. It is important to note that although uranium-238 is not fissile, it can be put to good use as a fertile material for the production of plutonium-239, as long as enough free neutrons are available.

6.4 ENERGY FROM NUCLEAR FUELS

The practical significance of the fission process is revealed by calculation of the amount of uranium that is consumed to obtain a given amount of energy. Each fission yields 190 MeV of useful energy, which is also (190 MeV) $(1.60 \times 10^{-13}$ J/MeV$) = 3.04 \times 10^{-11}$ J. Thus the number of fissions required to obtain 1 W-sec of energy is $1/(3.04 \times 10^{-11}) = 3.3 \times 10^{10}$. The number of U-235 atoms consumed in a thermal reactor is larger by the factor $1/0.86 = 1.16$ because of the formation of U-236 in part of the reactions.

In one day's operation of a reactor per megawatt of thermal power, the number of U-235 nuclei burned is

$$\frac{(10^6 W)(3.3 \times 10^{10}\text{fissions/W} - s)(86,400s/d)}{0.86\ \text{fissions/absorption}} = 3.32 \times 10^{21} \text{ absorptions/d.}$$

Then because 235 g corresponds to Avogadro's number of atoms 6.02×10^{23}, the U-235 weight consumed at 1 MW power is

$$\frac{(3.32 \times 10^{21}\text{d}^{-1})(235\text{g})}{6.02 \times 10^{23}} \cong 1.3\,\text{g/d.}$$

In other words, 1.3 g of fuel is used per megawatt-day (MW/d) of useful thermal energy released. In a typical reactor, which produces 3000 MW of thermal power, the U-235 fuel consumption is approximately 4 kg/day. To produce the same energy by the use of fossil fuels such as coal, oil, or gas, millions of times as much weight would be required.

6.5 SUMMARY

Neutron absorption by the nuclei of heavy elements gives rise to fission, in which heavy fragments, fast neutrons, and other radiations are released. Fissile materials are natural U-235 and the man-made isotopes Pu-239 and U-233. Many different radioactive isotopes are released in the fission process, and more neutrons are produced than are used, which makes possible a chain reaction and under certain conditions "conversion" and "breeding" of new fuels. Useful energy amounts to 190 MeV per fission, requiring only 1.3 g of U-235 to be consumed to obtain 1 MW/d of energy.

6.6 EXERCISES

6.1 Calculate the mass of the excited nucleus of plutonium-240 as the sum of the neutron mass 1.008665 and the Pu-239 mass 239.052157. How much larger is that sum than the mass of stable Pu-240, 240.053807? What energy in MeV is that?

6.2 If three neutrons and a xenon-133 atom ($^{133}_{54}$Xe) are produced when a U-235 atom is bombarded by a neutron, what is the second fission product isotope?

6.3 The total kinetic energy of the fission fragments is 166 MeV. (a) What are the energies of each if the mass ratio is 3/2? (b) What are the two mass numbers if three neutrons were released in fission? (c) What are the velocities of the fragments?

6.4 Calculate the energy yield from the reaction

$$^{235}_{92}U + ^{1}_{0}n \rightarrow ^{92}_{37}Rb + ^{140}_{55}Cs + 4^{1}_{0}n + E$$

with atomic masses 139.917277 for cesium and 91.919725 for rubidium.

6.5 The value of η for U-233 for thermal neutrons is approximately 2.30. By use of the cross sections for capture $\sigma_c = 47$ barns and fission $\sigma_f = 530$ barns, deduce the value of v, the number of neutrons per fission.

6.6 A mass of 8000 kg of slightly enriched uranium (2% U-235, 98% U-238) is exposed for 30 days in a reactor operating at heat power 2000 MW. Neglecting consumption of U-238, what is the final fuel composition?

6.7   The per capita consumption of electrical energy in the United States is approximately 50 kWh/d. If this were provided by fission with 2/3 of the heat wasted, how much U-235 would each person use per day?

6.8   Calculate the number of kilograms of coal, oil, and natural gas that must be burned each day to operate a 3000-MW thermal power plant, which consumes 4 kg/d of uranium-235. The heats of combustion of the three fuels (in kJ/g) are, respectively, 32, 44, and 50.

Computer Exercises

6.A   The fission process can be visualized by the computer program FISSION. It shows a neutron approaching a fissionable nucleus and the fragments emerging. Run the program several times, noting the variety of speeds and directions of the particles.

6.B   Program YIELD calculates the fission yield for several prominent long-lived radionuclides and their precursors by a summing process. Run the program selecting several mass numbers near the peaks of 92 and 144.

6.C   Program SPECTRUM gives a simple formula for the way fission neutrons are distributed in energy, shows a graph of the distribution, and calculates properties of the curve. Run the program with the menus.

6.7 REFERENCES

Basics of Nuclear Physics and Fission
http://www.ieer.org/reports/n-basics.html
Decay, binding energy, fission. By Institute for Energy and Environmental Research.

Ruth Lewin Sime, "Lise Meitner and the Discovery of Nuclear Fission," *Scientific American*, January 1998, p. 80. Contributions of the scientist who should have received the Nobel Prize.

Emilio Segrè, "The Discovery of Nuclear Fission," *Physics Today*, July 1998, p. 38.

Discovery of fission
http://aip.org/history/mod/fission/fission1/03.html
Includes quotations from the scientists. By American Institute of Physics.

Disintegration of Heavy Nuclei
1939 article in *Nature* by Niels Bohr
Google: liquid drop model fission

National Nuclear Data Center (NNDC)
http://www.nndc.bnl.gov/masses
2003 atomic mass evaluation. Latest and perhaps last.

The Isotope Project (Lawrence Berkeley National Laboratory)
http://isotopes.lbl.gov
Select Fission Home Page for data on fission product yields and spontaneous fission.

Physical Science Resource Center
http://www.psrc-online.org
Select Browse Resources/Modern Physics. By American Association of Physics Teachers.

John R. Lamarsh and Anthony J. Baratta, *Introduction to Nuclear Engineering*, 3rd Ed., Prentice-Hall, Upper Saddle River, NJ, 2001. Update of classic textbook.

David Bodansky, *Nuclear Energy: Principles, Practices, and Prospects*, 2nd Ed., Springer/AIP Press, New York, 2003.

Fusion

7

WHEN TWO light nuclear particles combine or "fuse" together, energy is released because the product nuclei have less mass than the original particles. Such fusion reactions can be caused by bombarding targets with charged particles, by use of an accelerator, or by raising the temperature of a gas to a high enough level for nuclear reactions to take place. In this chapter we will describe the interactions in the microscopic sense and discuss the phenomena that affect our ability to achieve a practical large-scale source of energy from fusion. Thanks are due to Dr. John G. Gilligan for his comments.

7.1 FUSION REACTIONS

The possibility of release of large amounts of nuclear energy can be seen by comparing the masses of nuclei of low atomic number. Suppose that one could combine two hydrogen nuclei and two neutrons to form the helium nucleus. In the reaction

$$2 \, {}_1^1\text{H} + 2 \, {}_0^1\text{n} \rightarrow {}_2^4\text{He},$$

the mass-energy difference (by use of atom masses) is

$$2(1.007825) + 2(1.008665) - 4.002603 = 0.030377 \text{ amu},$$

which corresponds to 28.3 MeV energy. A comparable amount of energy would be obtained by combining four hydrogen nuclei to form helium plus two positrons

$$4 \, {}_1^1\text{H} \rightarrow {}_2^4\text{He} + 2 \, {}_{+1}^0\text{e}.$$

This reaction in effect takes place in the sun and in other stars through the so-called carbon cycle, a complicated chain of events involving hydrogen and isotopes of the elements carbon, oxygen, and nitrogen. The cycle is extremely slow, however, and is not suitable for terrestrial application.

In the "hydrogen bomb," on the other hand, the high temperatures created by a fission reaction cause the fusion reaction to proceed in a rapid and uncontrolled manner. Between these extremes is the possibility of achieving a controlled fusion reaction that uses inexpensive and abundant fuels. As yet, a practical fusion device has not been developed, and considerable research and development will be required to reach that goal. Let us now examine the nuclear reactions that might be used. There seems to be no mechanism by which four separate nuclei can be made to fuse directly, and thus combinations of two particles must be sought.

The most promising reactions make use of the isotope deuterium, ^2_1H, abbreviated D. It is present in hydrogen as in water with abundance only 0.015% (i.e., there is one atom of ^2_1H for every 6700 atoms of ^1_1H), but because our planet has enormous amounts of water, the fuel available is almost inexhaustible. Four reactions are important:

$$^2_1\text{H} + {}^2_1\text{H} \rightarrow {}^3_1\text{H} + {}^1_1\text{H} + 4.03 \text{ MeV}$$

$$^2_1\text{H} + {}^2_1\text{H} \rightarrow {}^3_2\text{He} + {}^1_0\text{n} + 3.27 \text{ MeV}$$

$$^2_1\text{H} + {}^3_1\text{H} \rightarrow {}^4_2\text{He} + {}^1_0\text{n} + 17.6 \text{ MeV}$$

$$^2_1\text{H} + {}^3_2\text{He} \rightarrow {}^4_2\text{He} + {}^1_1\text{H} + 18.3 \text{ MeV}$$

The fusion of two deuterons—deuterium nuclei—in what is designated the D-D reaction results in two processes of nearly equal likelihood. The other reactions yield more energy but involve the artificial isotopes tritium, ^3_1H, abbreviated T, with the ion called the triton, and the rare isotope ^3_2He, helium-3. We note that the products of the first and second equations appear as reactants in the third and fourth equations. This suggests that a composite process might be feasible. Suppose that each of the reactions could be made to proceed at the same rate, along with twice the reaction of neutron capture in hydrogen

$$^1_1\text{H} + {}^1_0\text{n} \rightarrow {}^2_1\text{H} + 2.2 \text{ MeV}.$$

Adding twice this equation to the preceding four, we find that the net effect is to convert deuterium into helium according to

$$4{}^2_1\text{H} \rightarrow 2{}^4_2\text{He} + 47.7 \text{ MeV}.$$

The energy yield per atomic mass unit of deuterium fuel would thus be approximately 6 MeV, which is much more favorable that the yield per atomic mass unit of U-235 burned, which is only $190/235 = 0.81$ MeV.

Computer Exercise 7.A permits the exploration of possible nuclear reactions for fusion.

7.2 ELECTROSTATIC AND NUCLEAR FORCES

The reactions previously described do not take place merely by mixing the ingredients, because of the very strong force of electrostatic repulsion between the charged nuclei. Only by giving one or both of the particles a high speed can they be brought close enough to each other for the strong nuclear force to dominate the electrical force. This behavior is in sharp contrast to the ease with which neutrons interact with nuclei.

There are two consequences of the fact that the coulomb force between two charges of atomic numbers Z_1 and Z_2 varies with separation R according to $Z_1 Z_2/R^2$. First, we see that fusion is unlikely in elements other than those low in the periodic table. Second, the force and corresponding potential energy of repulsion is very large at the 10^{-15} m range of nuclear forces, and thus the chance of reaction is negligible unless particle energies are of the order of keV. Figure 7.1 shows the cross section for the D–D reaction. The strong dependence on energy is noted, with σ_{DD} rising by a factor of 1000 in the range 10 to 75 keV.

Energies in the kilo-electron-volt and million-electron-volt range can be achieved by a variety of charged particle accelerators. Bombardment of a solid or gaseous deuterium target by high-speed deuterons gives fusion reactions, but most of the particle energy goes into electrostatic interactions that merely heat up the bulk of the target. For a practical system, the recoverable fusion energy

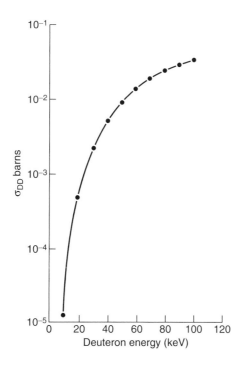

FIGURE 7.1 Cross section for D–D reaction.

must significantly exceed the energy required to operate the accelerator. Special equipment and processes are required to achieve that objective.

7.3 THERMONUCLEAR REACTIONS IN A PLASMA

A medium in which high particle energies are obtained is the *plasma*. It consists of a highly ionized gas as in an electrical discharge created by the acceleration of electrons. Equal numbers of electrons and positively charged ions are present, making the medium electrically neutral. The plasma is often called "the fourth state of matter." Through the injection of enough energy into the plasma, its temperature can be increased, and particles such as deuterons reach the speed for fusion to be favorable. The term thermonuclear is applied to reactions induced by high thermal energy, and the particles obey a speed distribution similar to that of a gas, as discussed in Chapter 2.

The temperatures to which the plasma must be raised are extremely high, as we can see by expressing an average particle energy in terms of temperature, by use of the kinetic relation

$$\bar{E} = \frac{3}{2}kT.$$

For example, even if $\bar{E}$ is as low as 10 keV, the temperature is

$$T = \frac{2}{3}\frac{\left(10^4 \text{eV}\right)\left(1.60 \times 10^{-19} \text{ J/eV}\right)}{1.38 \times 10^{-23} \text{ J/K}}$$

$$T = 77,000,000 \text{ K}.$$

Such a temperature greatly exceeds the temperature of the surface of the sun and is far beyond any temperature at which ordinary materials melt and vaporize. The plasma must be created and heated to the necessary temperature under some constraint provided by a physical force. In stars, gravity provides that force, but that is not sufficient on earth. Compression by reaction to ablation is designated as inertial confinement; restraint by electric and magnetic fields is called magnetic confinement. These methods will be discussed in Chapter 14. Such forces on the plasma are required to assure that thermal energy is not prematurely lost. Moreover, the plasma must remain intact long enough for many nuclear reactions to occur, which is difficult because of inherent instabilities of such highly charged media. Recalling from Section 2.2 the relationship $pV = nkT$, we note that even though the temperature T is very high, the particle density n/V is low, allowing the pressure p to be manageable.

The achievement of a practical energy source is further limited by the phenomenon of radiation losses. In Chapter 5 we discussed the *bremsstrahlung* radiation produced when electrons experience acceleration. Conditions are ideal for the generation of such electromagnetic radiation because the high-speed electrons in the plasma at elevated temperature experience continuous accelerations and decelerations as they interact with other charges. The radiation can readily escape from the region, because the number of target particles is very small. In typical plasma, the number density of electrons and deuterons is 10^{15}, which corresponds to a rarefied gas. The amount of radiation production (and loss) increases with temperature at a slower rate than does the energy released by fusion, as shown in Figure 7.2. At what is called the ignition temperature, the lines cross. Only for temperatures greater than that value, 400,000,000 K in the case of the D–D reaction, will there be a net energy yield, assuming that the radiation is lost. In a later chapter we will describe some of the devices that have been used to explore the possibility of achieving a fusion reactor.

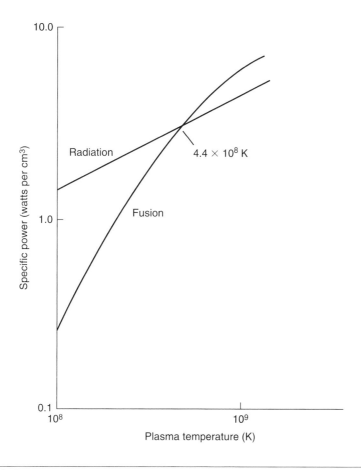

FIGURE 7.2 Fusion and radiation energies.

7.4 SUMMARY

Nuclear energy is released when nuclei of two light elements combine. The most favorable fusion reactions involve deuterium, which is a natural component of water and thus is a very abundant fuel. The reaction takes place only when the nuclei have a high enough speed to overcome the electrostatic repulsion of their charges. In a highly ionized electrical medium, the plasma, at temperatures of the order of 400,000,000 K, the fusion energy can exceed the energy loss because of radiation.

7.5 EXERCISES

7.1   Calculate the energy release in amu and MeV from the combination of four protons to form a helium nucleus and two positrons (each of mass 0.000549 amu).

7.2   Verify the energy yield for the reaction $^2_1\text{H} + ^3_2\text{He} \rightarrow ^4_2\text{He} + ^1_1\text{H} + 18.3$ MeV, noting atomic masses (in order) 2.014102, 3.016029, 4.002603, and 1.007825.

7.3   To obtain 3000 MW of power from a fusion reactor, in which the effective reaction is $2^2_1\text{H} \rightarrow ^4_2\text{He} + 23.8$ MeV, how many grams per day of deuterium would be needed? If all of the deuterium could be extracted from water, how many kilograms of water would have to be processed per day?

7.4   The reaction rate relation $nvN\sigma$ can be used to estimate the power density of a fusion plasma. (a) Find the speed v_D of 100 keV deuterons. (b) Assuming that deuterons serve as both target and projectile, such that the effective v is $v_D /2$, find what particle number density would be needed to achieve a power density of 1 kW/cm^3.

7.5   Estimate the temperature of the electrical discharge in a 120-volt fluorescent light bulb.

7.6   Calculate the potential energy in eV of a deuteron in the presence of another when their centers are separated by three nuclear radii (NOTE: $E_p = kQ_1Q_2/R$ where $k = 9 \times 10^9$, Q's are in coulombs, and R is in meters).

Computer Exercises

7.A   Program REACT1 displays the atomic masses for a number of light nuclides that are candidates as fusion projectiles and targets. Run the program and use Print Screen to obtain a paper copy of the table.

7.B   The reaction energy Q is the difference between masses of products and reactants. Program REACT2 calculates Q for an input of nuclei that might be involved and obtains the approximate distribution of energy between the product nuclei. (a) Test the program by use of the classic D-T reaction, with $A1 = 1$, $Z1 = 1$; $A2 = 2$, $Z2 = 1$; $A3 = 4$, $Z3 = 2$; $A4 = 1$, $Z4 = 0$. (b) Try the program with a few other reactions.

7.C   Program REACT3 surveys the array of light nuclei for potential fusion reactions. Run the program to find reactions with highest reaction energy, those that are neutron-free, and those that would require the lowest temperature on the basis of the product of Z1 and Z2.

7.6 REFERENCES

T. A. Heppenheimer, *The Man-Made Sun, The Quest for Fusion Power*, Little, Brown & Co., Boston, 1984. A narrative account of the fusion program of the United States, including personalities, politics, and progress to the date of publication. Good descriptions of equipment and processes.

Robin Herman, *Fusion: The Search for Endless Energy*, Cambridge University Press, New York, 1990. A well-written and interesting account.

Robert A. Gross, *Fusion Energy*, John Wiley & Sons, New York, 1985. A readable textbook. Main emphasis is on magnetic confinement fusion.

James J. Duderstadt and Gregory A. Moses, *Inertial Confinement Fusion*, John Wiley & Sons, New York, 1982. An excellent complement to the book by Gross.

Perspectives on Plasmas
http://www.plasmas.org
All aspects of plasma science and technology.

A. A. Harms, K. F. Schoepf, G. H. Miley, and D. R. Kingdon, *Principles of Fusion Energy*, World Scientific, Singapore, 2000. Subtitle: An Introduction to Fusion Energy for Students of Science and Engineering.

Kenro Miyamoto, *Plasma Physics and Controlled Nuclear Fusion*, Springer-Verlag, New York, 2005.

Jeffrey P. Freidberg, *Plasma Physics and Fusion Energy*, Cambridge University Press, New York, 2007.

Suzanne Pfalzner, *An Introduction to Inertial Confinement Fusion*, CRC Press, Boca Raton, FL, 2006.

Theoretical Principles of Plasma Physics
http://www.plasmaphysics.org
Highly technical but comprehensive. By Thomas Smid.

European Fusion Development Agreement (EFDA)
http://www.jet.efda.org
About the Joint European Torus (JET).

Educational Web Site Fusion Energy
http://fusioned.gat.com
Information on fusion from General Atomic.

Fusion Power Associates
http://fusionpower.org
Source of information on latest technical and political developments.

Educational Web Site Fusion Energy
http://fusioned.gat.com
Information on fusion from General Atomic.

Fusion Energy Educational Web Site
http://fusedweb.pppl.gov/cpep/chart.html
Select Fusion Basics. From Princeton Plasma Physics Laboratory.

Fusion Power Associates
http://fusionpower.org
A foundation that is a valuable source of information on current fusion research and political
status, with links to many other sites. Fusion Program Notes appear frequently as e-mail
messages.

PART

Nuclear Systems

2

Particle Accelerators

A DEVICE that provides forces on charged particles by some combination of electric and magnetic fields and brings the ions to high speed and kinetic energy is called an accelerator. Many types have been developed for the study of nuclear reactions and basic nuclear structure, with an ever-increasing demand for higher particle energy. In this chapter we will review the nature of the forces on charges and describe the arrangement and principle of operation of several important kinds of particle accelerators. In later chapters we describe some of the many applications.

8.1 ELECTRIC AND MAGNETIC FORCES

Let us recall how charged particles are influenced by electric and magnetic fields. First, visualize a pair of parallel metal plates separated by a distance d as in the sample capacitor shown in Figure 8.1. A potential difference V and electric field $\mathcal{E} = V/d$ are provided to the region of low gas pressure by a direct-current voltage supply such as a battery. If an electron of mass m and charge e is released at the negative plate, it will experience a force $\mathcal{E}e$, and its acceleration will be $\mathcal{E}e/m$. It will gain speed, and on reaching the positive plate it will have reached a kinetic energy $\frac{1}{2}mv^2 = Ve$. Thus its speed is $v = \sqrt{2Ve/m}$. For example, if V is 100 volts, the speed of an electron ($m = 9.1 \times 10^{-31}$ kg and $e = 1.60 \times 10^{-19}$ coulombs) is found to be 5.9×10^6 m/s.

Next, let us introduce a charged particle of mass m, charge e, and speed v into a region with uniform magnetic field B, as in Figure 8.2. If the charge enters in the direction of the field lines, it will not be affected, but if it enters perpendicularly to the field, it will move at constant speed on a circle. Its radius, called the radius of gyration, is $r = mv/eB$, such that the stronger the field or the lower the speed, the smaller will be the radius of motion. Let the angular speed be ω (omega) equal to v/r. By use of the formula for r, we find $\omega = eB/m$. If the charge enters at some other angle, it will move in a path called a helix, like a wire door spring.

Instead, let us release a charge in a region where the magnetic field B is changing with time. If the electron were inside the metal of a circular loop of wire of area A as in Figure 8.3, it would experience an electric force induced by the

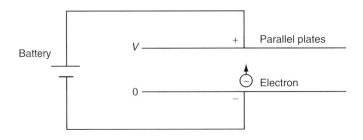

FIGURE 8.1 Capacitor as accelerator.

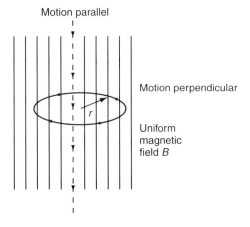

FIGURE 8.2 Electric charge motion in uniform magnetic field *B*.

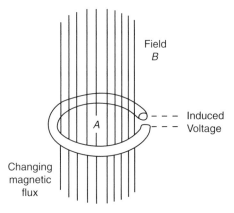

FIGURE 8.3 Magnetic induction.

change in magnetic flux BA. The same effect would take place without the presence of the wire, of course. Finally, if the magnetic field varies with position, there are additional forces on charged particles.

8.2 HIGH-VOLTAGE MACHINES

One way to accelerate ions to high speed is to provide a large potential difference between a source of charges and a target. In effect, the phenomenon of lightning, in which a discharge from charged clouds to the earth takes place, is produced in the laboratory. Two devices of this type are commonly used. The first is the voltage multiplier or Cockroft–Walton machine, Figure 8.4, which has a circuit that charges capacitors in parallel and discharges them in series. The second is the electrostatic generator or Van de Graaff accelerator, the principle of which is sketched in Figure 8.5. An insulated metal shell is raised to high potential by bringing it charge on a moving belt, permitting the acceleration of positive charges such as protons or deuterons. Particle energies of the order of 5 MeV are possible, with a very small spread in energy.

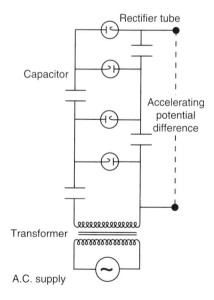

FIGURE 8.4 Cockroft–Walton circuit.

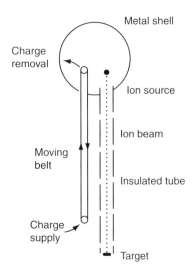

FIGURE 8.5 Van de Graaff accelerator.

8.3 LINEAR ACCELERATOR

Rather than giving a charge one large acceleration with a high voltage, it can be brought to high speed by a succession of accelerations through relatively small potential differences, as in the linear accelerator ("LINAC"), sketched in Figure 8.6. It consists of a series of accelerating electrodes in the form of tubes with alternating electric potentials applied as shown. An electron or ion gains energy in the gaps between tubes and "drifts" without change of energy while inside the tube, where the field is nearly zero. By the time the charge reaches the next gap, the voltage is again correct for acceleration. Because the ion is gaining speed along the path down the row of tubes, their lengths ℓ must be successively longer for the time of flight in each to be constant. The time to go a distance ℓ is ℓ/v, which is equal to the half-period of the voltage cycle $T/2$.

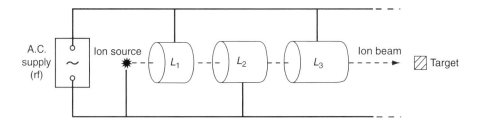

FIGURE 8.6 Simple linear accelerator.

The LINAC at the Stanford Linear Accelerator Center (SLAC) is 2 miles long. It produces electron and positron beams with energies up to 50 GeV (see References).

8.4 CYCLOTRON AND BETATRON

Successive electrical acceleration by electrodes and circular motion within a magnetic field are combined in the cyclotron, invented by Ernest O. Lawrence. As sketched in Figure 8.7, ions such as protons, deuterons, or alpha particles are provided by a source at the center of a vacuum chamber located between the poles of a large electromagnet. Two hollow metal boxes called "dees" (in the shape of the letter D) are supplied with alternating voltages in correct frequency and opposite polarity. In the gap between dees, an ion gains energy as in the linear accelerator, then moves on a circle while inside the electric-field–free region, guided by the magnetic field. Each crossing of the gap with potential difference V gives impetus to the ion with an energy gain Ve, and the radius of motion increases according to $r = v/\omega$, where $\omega = eB/m$ is the angular speed. The unique feature of the cyclotron is that the time required for one complete revolution, $T = 2\pi/\omega$, is independent of the radius of motion of the ion. Thus it is possible to use a synchronized alternating potential of constant frequency v, angular frequency $\omega = 2\pi v$, to provide acceleration at the right instant.

For example, in a magnetic field B of 0.5 Wb/m^2 (tesla) the angular speed for deuterons of mass 3.3×10^{-27} kg and charge 1.6×10^{-19} coulombs is

$$\omega = \frac{eB}{m} = \frac{(1.6 \times 10^{-19})(0.5)}{3.3 \times 10^{-27}} = 2.4 \times 10^7 /\text{s}$$

Equating this to the angular frequency for the power supply, $\omega = 2\pi v$, we find $v = (2.4 \times 10^7)/2\pi = 3.8 \times 10^6 \text{ s}^{-1}$, which is in the radiofrequency (RF) range.

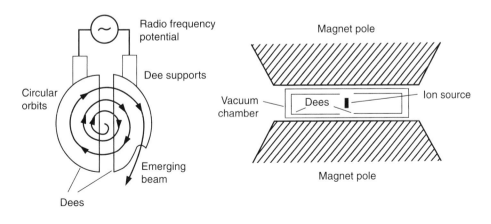

FIGURE 8.7 Cyclotron.

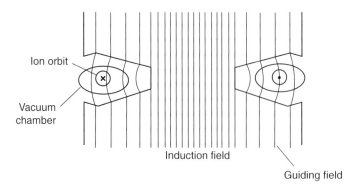

Ion orbit

Vacuum chamber

Induction field

Guiding field

FIGURE 8.8 Betatron.

The path of ions is approximately a spiral. When the outermost radius is reached and the ions have full energy, a beam is extracted from the dees by special electric and magnetic fields and allowed to strike a target, in which nuclear reactions take place.

The cyclotron is primitive compared with new machines but is still widely used in hospitals to produce radioisotopes.

In the betatron, the induction accelerator, electrons are brought to high speeds. A changing magnetic flux provides an electric field and a force on the charges while they are guided in a path of constant radius. Figure 8.8 shows the vacuum chamber in the form of a doughnut placed between specially shaped magnetic poles. The force on electrons of charge e is in the direction tangential to the orbit of radius r. The rate at which the average magnetic field within the loop changes is $\Delta B / \Delta t$, provided by varying the current in the coils of the electromagnet. The magnitude of the force is[†]

$$F = \frac{er}{2} \frac{\Delta B}{\Delta t}.$$

The charge continues to gain energy while remaining at the same radius if the magnetic field at that location is half the average field within the loop. The acceleration to energies in the million-electron-volt range takes place in the fraction of a second that it takes for the alternating magnetic current to go through a quarter-cycle.

The speeds reached in a betatron are high enough to require the use of relativistic formulas (Chapter 1). Let us find the mass m and speed v for an electron of kinetic energy $E_k = 1$ MeV. Rearranging the equation for kinetic energy, the ratio of m to the rest mass m_0 is

---

[†]To show this, note that the area within the circular path is $A = \pi r^2$ and the magnetic flux is $\Phi = BA$. According to Faraday's law of induction, if the flux changes by $\Delta \Phi$ in a time Δt, a potential difference around a circuit of $V = \Delta \Phi / \Delta t$ is produced. The corresponding electric field is $\mathcal{E} = V/2\pi r$, and the force is $e\mathcal{E}$. Combining, the relation quoted is obtained.

$$\frac{m}{m_0} = 1 + \frac{E_k}{m_0 c^2}.$$

Recalling that the rest energy $E_0 = m_0 c^2$ for an electron is 0.51 MeV, we find the ratio $m/m_0 = 1 + 1/0.51 = 2.96$. Solving Einstein's equation for the speed, $m/m_0 = 1/\sqrt{1 - (v/c)^2}$, we find that $v = c\sqrt{1 - (m_0/m)^2} = 0.94c$. Thus the 1 MeV electron's speed is close to that of light, $c = 3.0 \times 10^8$ m/s (i.e., $v = 2.8 \times 10^8$ m/s). If instead we impart a kinetic energy of 100 MeV to an electron, its mass increases by a factor 297 and its speed becomes 0.999995c.

Calculations of this type are readily made by use of the computer program ALBERT, introduced in Section 1.7. Some other applications to ion motion in modern accelerators are found in Computer Exercises 8.A and 8.B.

8.5 SYNCHROTRON AND COLLIDER

Over the past half-century, the science and engineering of accelerators has evolved dramatically, with ever-increasing beam currents and energy of the charged particles. A major step was the invention independently of the synchrotron by E. M. McMillan and V. I. Veksler. It consists of the periodic acceleration of the particles by radiofrequency electric fields, but with a time-varying magnetic field that keeps the charges on a circular path. Ions that are out of step are brought back into step (i.e., they are synchronized). Figure 8.9 shows schematically the Cosmotron, operated from 1953 to 1966 at Brookhaven National Laboratory. An ion source provided protons that were injected at 4 MeV into a vacuum chamber by a Van de Graaff accelerator. The inflector sent the charges into the magnet. There, the magnetic field rose to 1.4 tesla in one second to provide the constant radius condition $r = mv/eB$ as the protons gain energy. The field was shaped to assure proper focusing. The radiofrequency unit accelerated the particles with initial voltage 2000 V at frequency 2000 hertz. Ions at final energy 3 GeV struck an internal target to yield neutrons or mesons.

In a more modern version of synchrotron, the magnetic field that bends the particles in a circular orbit is provided by a series of separate magnets, like beads on a necklace. In between the magnets are quadrupole (2N and 2S) magnets that provide beam focusing, helping compensate for space charge spreading.

Most of the early accelerators involved charge bombardment of a fixed target. Recently, much larger energies are achieved by causing two oppositely circulating beams to collide in what is called a storage ring. The pairs of particles used in a "collider" are (a) electrons and positrons, (b) protons and antiprotons, or (c) protons and protons. The accelerating cavity of the electron-positron collider at the Thomas Jefferson Accelerator Laboratory is constructed of superconducting niobium to minimize energy losses. It provides a total energy of 4 GeV. The Large Electron Positron (LEP) collider at the European Laboratory for Particle Physics (CERN) gave particles of 209 GeV before being shut down in 2000.

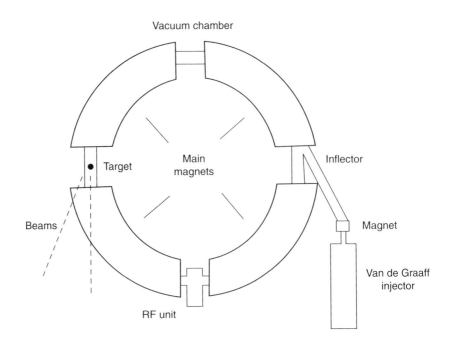

FIGURE 8.9 Cosmotron. Synchrotron at Brookhaven National Laboratory.

To reach high particle energies, a combination of accelerators of different types is used, as in the Tevatron at the Fermi National Accelerator Laboratory (Fermilab) near Chicago. The Tevatron involves a circular underground tunnel of diameter 3 m and length 6.3 km, containing the beam tube and a series of hundreds of magnets that provide ion bending. Negative hydrogen ions are first accelerated to 0.75 MeV by a Cockroft–Walton machine (Section 8.2) then raised to 200 MeV by a linear accelerator (Section 8.3). Electrons are stripped from the ions by a carbon foil, leaving protons. These are brought to 8 GeV by a small booster synchrotron. The ions are then injected into the Main Ring synchrotron and brought to 150 GeV. They are focused into short pulses and extracted to strike a copper target, creating large numbers of antiprotons. These are drawn off into a storage ring where they circulate and the beam is compressed, then transferred to an accumulator ring, and then put in the Tevatron ring. In the meanwhile a batch of protons from the Main Ring have also been put in the Tevatron ring. Along the path of that ring are 1000 superconducting magnets that use liquid nitrogen and helium for cooling. Finally, the two countercurrent beams, of diameter approximately 0.1 mm, are accelerated to their peak energy of nearly 1 TeV. Detection of the byproducts of collisions is by the Collider Detector Fermilab (CDF), a complex particle tracking device. Extensive additional information with photographs is found in the Fermilab Web site (see References).

Among the purposes of accelerators is the search for new particles in nature, which can be created only by transforming the energy of accelerated charges, in accord with Einstein's theory. Colliding high-energy beams of particles and anti-particles can create far more massive nuclear species than can simple ion bombardment of stationary targets. The reason is that a high-energy charge expends most of its energy in accelerating new particles to meet momentum conservation requirements. In contrast, when a particle collides with an antiparticle, the momentum is zero, allowing all of the energy to go into new mass.

One major accomplishment of high-energy machines was the discovery of the "top" quark (see References). Its existence is crucial to the correctness of the theory called the Standard Model. According to that picture, matter is composed of leptons (electrons, neutrinos, etc.) and quarks (types "up," "down," "charm," "strange," "top," and "bottom"), along with their antiparticle forms. The up quark has a charge 2/3, the down quark $-1/3$. Quarks are believed to have been free just after the Big Bang, forming what is viewed as a perfect liquid. They clustered together with the help of gluons to form protons and neutrons. The proton is made of two ups and one down, whereas the neutron is two downs and one up. In the collision of protons and antiprotons, it is actually the component quarks that collide. It is believed that the top quark existed in nature only in the first 10^{-16} second from the Big Bang that started the universe. Quarks can be freed with difficulty in the laboratory by collisions of very high-energy gold atoms. The Relativistic Heavy Ion Collider (RHIC) at Brookhaven National Laboratory has detected the products of quark combination. For a readily accessible discussion of quarks, go to Wikipedia.

Forces in nature are thought to be provided by the exchange of bosons, an example of which is the photon, for electromagnetic force. There are three other forces—weak (involved in radioactivity), strong (for binding in nuclei), and gravity. The electromagnetic and weak forces are viewed as different aspects of a more general "electroweak" force.

Studies of collisions of high-energy particles are intended to obtain information on the origin of mass, along with an answer why there is so much invisible mass ("dark matter") in the universe, along with the cause of accelerated expansion of the universe ("dark energy"). Questions to be addressed are the mass of neutrinos, the scarcity of antimatter, and extra dimensions of space. Also sought is a hypothetical heavy particle called the Higgs boson, which is thought to relate the vacuum of space to the existence of particles.

In the early 1990s, the United States had started to build in Texas a large super-conducting supercollider (SSC) to give a beam of 20 TeV, but the project was canceled by Congress because of excessive cost. With the demise of the SSC, a considerable part of high-energy particle research by United States physicists was shifted to CERN, the European Laboratory for Particle Physics (see References). The United States Department of Energy allocated funds to help construct the Large Hadron Collider (LHC) and the ATLAS detector (see References), which will analyze the products of proton–proton collisions. The LHC will make use of the existing 27 km circumference tunnel at the French–Swiss border. By use of superconducting magnets and advanced accelerator technology, it will be able to collide particles each of 7 TeV. Alternately, it will handle beams of heavy ions such as lead with total energy 1250 TeV.

Two extensions of particle accelerators have opened up new opportunities for research and industrial applications. The first is synchrotron radiation (SR), based on the fact that if an electric charge is given an acceleration, it radiates light. At each of the bending magnets of a synchrotron or storage ring, experimental beams of X-rays are available. The beams are very narrow, with an angle given by E_0/E_k, the ratio of rest energy and kinetic energy. An example of an SR facility is the National Synchrotron Light Source at Brookhaven National Laboratory (see References). The second is free electron laser (FEL), in which electrons are brought to high speed in a LINAC and injected into a tube with magnets along its length. These provide an alternating field that accelerates the electrons to radiate photons. The light is reflected back and forth by mirrors at the ends of the tube and interacts with the circulating electrons rather than with atoms as in a conventional laser. FELs can produce frequencies ranging from infrared to gamma rays. A Web site lists FELs around the world (see References).

8.6 SPALLATION

High-energy charged particles from an accelerator can disrupt nuclei of target materials. Experiments at California radiation laboratories showed that large neutron yields were achieved in targets bombarded by charged particles such as deuterons or protons of several hundred MeV energy. New dramatic nuclear reactions are involved. One is the stripping reaction, Figure 8.10(A), in which a deuteron is broken into a proton and a neutron by the impact on a target nucleus. Another is the process of spallation in which a nucleus is broken into pieces by an energetic projectile. Figure 8.10(B) shows how a cascade of nucleons is produced by spallation. A third is "evaporation" in which neutrons fly out of a nucleus with some 100 MeV of internal excitation energy, see Figure 8.10(C). The average energy of evaporation of neutrons is approximately 3 MeV. The excited nucleus may undergo fission, which releases neutrons, and further evaporation from the fission fragments can occur.

It has been predicted that as many as 50 neutrons can be produced by a single high-energy (500 MeV) deuteron. The large supply of neutrons can be used for a number of purposes: (a) physics and chemistry research; (b) production of new nuclear fuel, beneficial radioisotopes, or weapons tritium; and (c) burn unwanted plutonium or certain radioactive waste isotopes. Some of these applications will be discussed in later sections.

At Oak Ridge National Laboratory the Spallation Neutron Source (SNS) was put into operation in 2006. Design and construction of the Department of Energy facility was a cooperative effort of six laboratories (Argonne, Brookhaven, Lawrence Berkeley, Los Alamos, Oak Ridge, and Jefferson). A large linear accelerator produces

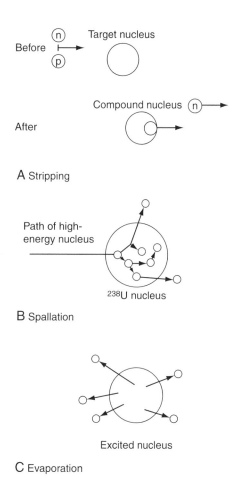

FIGURE 8.10 Nuclear reactions produced by very high energy charged particles.

high-speed protons to bombard a liquid mercury target. The particle energy is 1 GeV; the beam power is 1.4 MW. Neutrons are moderated by water and liquid hydrogen, and a time-of-flight device selects neutrons of desired energy. The SNS will serve many hundreds of researchers in neutron science from the United States and abroad, facilitating a great variety of programs, as discussed in Chapter 18.

8.7 SUMMARY

Charged particles such as electrons and ions of light elements are brought to high speed and energy by particle accelerators, which use electric and magnetic fields in various ways. In the high-voltage machines a beam of ions is accelerated directly through a large potential difference, produced by special voltage multi-plier circuits or by carrying charge to a positive electrode; in the linear acceler-ator, ions are given successive accelerations in gaps between tubes lined up in a row; in the cyclotron, the ions are similarly accelerated but move in circular orbits because of the applied magnetic field; in the betatron, a changing mag-netic field produces an electric field that accelerates electrons to relativistic speeds; in the synchrotron, both radiofrequency and time-varying magnetic field are used. High-energy nuclear physics research is carried out through the use of such accelerators. Through several spallation processes, high-energy charged particles can produce large numbers of neutrons which have a variety of applications.

8.8 EXERCISES

8.1 Calculate the potential difference required to accelerate an electron to speed 2×10^5 m/s.

8.2 What is the proper frequency for a voltage supply to a linear accelerator if the speed of protons in a tube of 0.6 m length is 3×10^6 m/s?

8.3 Find the time for one revolution of a deuteron in a uniform magnetic field of 1 Wb/m^2.

8.4 Develop a working formula for the final energy of cyclotron ions of mass m, charge q, exit radius R, in a magnetic field B. (Use nonrelativistic energy relations.)

8.5 What magnetic field strength (Wb/m^2) is required to accelerate deuterons in a cyclotron of radius 2.5 m to energy 5 MeV?

8.6 Performance data on the Main Ring proton synchrotron of Fermilab at Bata-via, Illinois (see References) were as follows:

> Diameter of ring 2 km
> Protons per pulse 6×10^{12}
> Number of magnets 954
> Initial proton energy 8 GeV
> Final proton energy 400 GeV
> Number of revolutions 200,000

(a) Find the proton energy gain per revolution.
(b) Find the speed of the protons at final energy by use of relativistic formulas of Sections 1.4 and 8.4 (or computer program ALBERT, see Chapter 1)
(c) Calculate the magnetic field at the final speed of the protons.

8.7 What is the factor by which the mass is increased and what fraction of the speed of light do protons of 200 billion-electron-volts have?

8.8 Calculate the steady deuteron beam current and the electric power required in a 500-GeV accelerator that produces 4 kg per day of plutonium-239. Assume a conservative 25 neutrons per deuteron.

8.9 By use of the relativistic formulas from Section 1.4, show that for very large particle energies the fractional difference in speed from that of light, $f = (c - v)/c$, is accurately approximated by $f = (1/2) (m_0/m)^2$. Find f for 50 GeV electrons of rest energy 0.511 MeV.

8.10 The velocities of protons and antiprotons in the 2 km diameter Tevatron ring are practically the same as the velocity of light, 299792458 m/s. Find the time for particles of final energy 1 TeV to traverse the circumference. How much error is there in this approximation?

8.11 The synchrotron radiation loss in joules of a charge e with rest mass m_0 moving in a circle of radius R is given by Cohen (see References) as

$$\Delta E = e^2 \gamma (\gamma^2 - 1)^{3/2} / (3\varepsilon_0 R)$$

where $\gamma = E/m_0 c^2$, with $E = mc^2$ and $\varepsilon_0 \cong 8.8542 \times 10^{-12}$ F/m. (a) Find an approximate formula for ΔE in keV for an electron as a function of energy in GeV and R in meters, when the speed is very close to the speed of light. (b) How much lower than the radiation from an electron is that from a proton of the same radius and energy? (c) Find a formula for the power radiated from an electron moving in a circle with speed much less than the speed of light, in terms of the acceleration.

Computer Exercises

8.A Verify with the computer program ALBERT (Chapter 1) that 1 TeV protons have a speed that seems to be the same as the velocity of light. Calculate the fractional difference between v and c with the formula derived in Exercise 8.9. Explain the discrepancy.

8.B The electron-positron collider at Hamburg, Germany, produces 23 TeV particles.

(a) What is the ratio of the electron's total energy to its rest energy (0.510998910 MeV). Check the result with the computer program ALBERT (Chapter 1) by supplying a kinetic energy of 2.3D7 (a double precision number).

(b) If 23 TeV electrons could be induced to travel around the earth (radius 6378 km), how far behind a light beam would they arrive? See Exercise 8.9 for a useful formula.

8.9 **REFERENCES**

Andrew Sessler and Edmund Wilson, *Engines of Discovery: A Century of Particle Accelerators*, World Scientific Publishing Co. Singapore, 2007. Accelerators for research, medicine, and industry.

Early Particle Accelerators-Ernest Lawrence and the Cyclotron
http://www.aip.org/history/lawrence/epa.htm
New systems, from the 1930s.

Stanley Humphries, Jr., *Principles of Charged Particle Acceleration*, John Wiley & Sons, New York, 1986.

Waldemar H. Scharf, *Biomedical Particle Accelerators*, AIP Press, New York, 1997.

Stephen M. Shafroth and James C. Austin (Eds.), *Accelerator-Based Atomic Physics Techniques and Applications*, AIP Press, Woodbury, NY, 1997.

Stanford Linear Accelerator Center
http://www2.slac.stanford.edu/vvc
Select Virtual Visitor Center

John R. Rees, "The Stanford Linear Collider," *Scientific American*, October 1990, p. 58.

Fermilab History and Archives Project
http://history.fnal.gov
Features Robert R. Wilson, first director.

Fermi National Accelerator Laboratory
http://www.fnal.gov
Select About Fermilab/Virtual Tour.

The Discovery of the Top Quark
http://www.hep.uiuc.edu/home/tml/SciAmTop.pdf
Scientific American feature article by Tony M. Liss and Paul L. Tipton.

Particle Accelerators Around the World
http://www-elsa.physik.uni-bonn.de/accelerator_list.html
Links to facilities sorted by location and by accelerator type.

European Laboratory for Particle Physics (CERN)
http://www.cern.ch
Select Search/LHC.

The ATLAS Experiment
http://atlas.ch
Select Virtual Tour

National Synchrotron Light Source
http://www.nsls.bnl.gov
Select About the NSLS/Visit the NSLS/Take an Online Tour.

E. Richard Cohen, *The Physics Quick Reference Guide*, American Institute of Physics, Woodbury, NY, 1995.

WWW Virtual Library: Free Electron Laser
http://sbfel3.ucsb.edu/www/vl_fel.html
Select Jefferson Lab or University of California at Santa Barbara.

Spallation Neutron Source
http://neutrons.ornl.gov
Extensive information about features of the system and prospective applications.

T. E. Mason, et al., "The Spallation Neutron Source: A powerful tool for materials research."
http://arxiv.org/abs/physics/0007068
A frequently cited article describing the equipment used in research. Select Download PDF.

Isotope Separators

ALL OF our technology is based on materials in various forms—elements, compounds, alloys, and mixtures. Ordinary chemical and mechanical processes can be used to separate many materials into components. In the nuclear field, however, individual isotopes such as uranium-235 and hydrogen-2 (deuterium) are required. Because isotopes of a given element have the same atomic number Z, they are essentially identical chemically, and thus a physical method must be found that distinguishes among particles on the basis of mass number A. In this chapter we will describe several methods by which isotopes of uranium and other elements are separated. Four methods that depend on differences in A are: (a) ion motion in a magnetic field, (b) diffusion of particles through a membrane, (c) motion with centrifugal force, and (d) atomic response to a laser beam. Calculations on the amounts of material that must be processed to obtain nuclear fuel will be presented, and estimates of costs will be given.

9.1 MASS SPECTROGRAPH

We recall from Chapter 8 that a particle of mass m, charge q, and speed v will move in a circular path of radius r if injected perpendicular to a magnetic field of strength B, according to the relation $r = mv/qB$. In the mass spectrograph (Figure 9.1), ions of the element whose isotopes are to be separated are produced in an electrical discharge and accelerated through a potential difference V to provide a kinetic energy $\frac{1}{2}\,mv^2 = qV$. The charges move freely in a chamber maintained at very low gas pressure, guided in semicircular paths by the magnetic field. The heavier ions have a larger radius of motion than the light ions, and the two may be collected separately. It is found (see Exercise 9.1) that the distance between the points at which ions are collected is proportional to the difference in the square roots of the masses. The spectrograph can be used to measure masses with some accuracy, to determine the relative abundance of isotopes in a sample, or to enrich an element in a certain desired isotope.

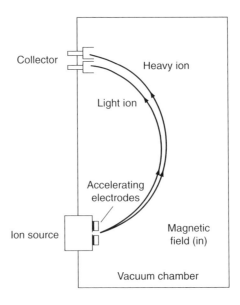

FIGURE 9.1 Mass spectrograph.

The electromagnetic process was used on uranium during World War II to obtain weapons material, using the "calutron" (named after the University of California at Berkeley, where it was developed). A total of 1152 units in the "Alpha" and "Beta" processes were operated at the Y-12 Plant at Oak Ridge, producing the enriched uranium for one atomic bomb by 1945. Because the cost of electrical power for the process is large, alternative processes such as gaseous diffusion and centrifuge are used to produce reactor fuels. However, for more than 50 years a few calutrons were maintained at Oak Ridge. These separated light-stable isotopes in small quantities needed for research and for targets for accelerator-produced radioisotopes. The system was shut down permanently in 1999 (see References). It has been reported that Iraq was developing its own electromagnetic process before the Gulf War (see References).

9.2 GASEOUS DIFFUSION SEPARATOR

The principle of this process can be illustrated by a simple experiment (Figure 9.2). A container is divided into two parts by a porous membrane, and air is introduced on both sides. Recall that air is a mixture of 80% nitrogen, $A = 14$, and 20% oxygen, $A = 16$. If the pressure on one side is raised, the relative proportion of nitrogen on the other side increases. The separation effect can be explained on the basis of particle speeds. The average kinetic energies of the heavy (H) and light (L) molecules in the gas mixture are the same, $E_H = E_L$, but because the masses are different, the typical particle speeds bear a ratio.

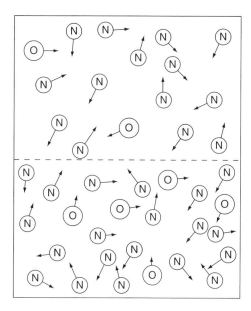

FIGURE 9.2 Gaseous diffusion separation of nitrogen and oxygen.

$$\frac{v_{\mathrm{L}}}{v_{\mathrm{H}}} = \sqrt{\frac{m_{\mathrm{H}}}{m_{\mathrm{L}}}}.$$

The number of molecules of a given type that hit the membrane each second is proportional to nv in analogy to neutron motion discussed in Chapter 4. Those with higher speed thus have a higher probability of passing through the holes in the porous membrane, called the "barrier."

The physical arrangement of one processing unit of a gaseous diffusion plant for the separation of uranium isotopes U-235 and U-238 is shown in Figure 9.3. A thin nickel alloy serves as the barrier material. In this "stage," gas in the form

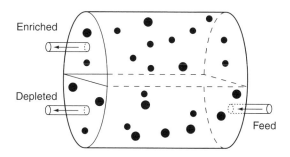

FIGURE 9.3 Gaseous diffusion stage.

of the compound uranium hexafluoride (UF_6) is pumped in as feed and removed as two streams. One is enriched and one depleted in the compound $^{235}UF_6$, with corresponding changes in $^{238}UF_6$. Because of the very small mass difference of particles of molecular weight 349 and 352, the amount of separation is small, and many stages in series are required in what is called a cascade.

Natural uranium has a small U-234 component, atom fraction 0.000055. For simplicity, we will ignore its effect except for Exercise 9.11.

Any isotope separation process causes a change in the relative numbers of molecules of the two species. Let n_H and n_L be the number of molecules in a sample of gas. Their abundance ratio is defined as

$$R = \frac{n_L}{n_H}.$$

For example, in ordinary air $R = 80/20 = 4$.

The effectiveness of an isotope separation process depends on a quantity called the separation factor r. If we supply gas at one abundance ratio R, the ratio R' on the low-pressure side of the barrier is given by

$$R' = rR.$$

If only a very small amount of gas is allowed to diffuse through the barrier, the separation factor is given by $r = \sqrt{m_H/m_L}$, which for UF_6 is 1.0043. However, for a more practical case, in which half the gas goes through, the separation factor is smaller, 1.0030 (see Exercise 9.2). Let us calculate the effect of one stage on natural uranium, 0.711% by weight, corresponding to a U-235 atom fraction of 0.00720, and an abundance ratio of 0.00725. Now

$$R' = rR = (1.0030)(0.00725) = 0.00727.$$

The amount of enrichment is very small. By processing the gas in a series of s stages, each one of which provides a factor r, the abundance ratio is increased by a factor r^s. If R_f and R_p refer to feed and product, respectively, $R_p = r^s R_f$. For $r = 1.0030$ we can easily show that 2375 enriching stages are needed to go from $R_f = 0.00725$ to highly enriched 90% U-235 (i.e., $R_p = 0.9/(1 - 0.9) = 9$). Figure 9.4 shows the arrangement of several stages in an elementary cascade, and indicates the value of R at various points. The feed is natural uranium, the product is enriched in U-235, and the waste is depleted in U-235.

Figure 9.5 shows a gaseous diffusion uranium isotope separation plant. Such a facility is very expensive, of the order of a billion dollars, because of the size and number of components such as separators, pumps, valves, and controls, but the process is basically simple. The plant runs continuously with a small number of operating personnel. The principal operating cost is for the electrical power to provide the pressure differences and to perform work on the gas. The Paducah, Ohio, plant is managed by the United States Enrichment Corporation (USEC), which is a business created by privatization of government-owned facilities (see References). That plant provides more than half of the United States market. USEC is participating in a program with Russia called "Megatons to Megawatts" involving dilution of highly enriched uranium to levels used in reactors.

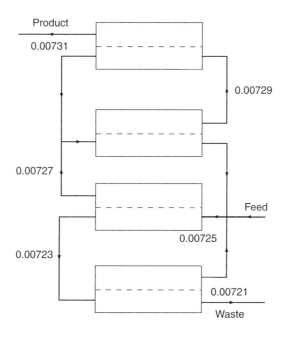

FIGURE 9.4 Gaseous diffusion cascade.

The flow of UF$_6$ and thus uranium through individual stages or the whole plant can be analyzed by the use of material balances. One could keep track of number of particles, moles, or kilograms, because the flow is continuous. It will be convenient to use kilograms per day as the unit of uranium flow for three streams: feed (F), product (P), and waste (W), also called "tails." Then,

$$F = P + W.$$

Letting x stand for the U-235 weight fractions in the flows, the balance for the light isotope is

$$x_f F = x_p P + x_w W.$$

(A similar equation could be written for U-238, but it would contain no additional information.) The two equations can be solved to obtain the ratio of feed and product mass rates. Eliminating W,

$$\frac{F}{P} = \frac{x_p - x_w}{x_f - x_w}.$$

For example, let us find the required feed of natural uranium to obtain 1 kg/day of product containing 3% U-235 by weight. The abbreviation w/o is typically used for weight percent. Assume that the waste is at 0.3 w/o. Now

$$\frac{F}{P} = \frac{0.03 - 0.003}{0.00711 - 0.003} = 6.57$$

FIGURE 9.5 Gaseous diffusion plant (Courtesy United States Enrichment Corporation).

and thus the feed is 6.57 kg/day. We note that W is 5.57 kg/day, which shows that large amounts of depleted uranium tails must be stored for each kilogram of U-235 produced. Depleted uranium is used for tank bullets. The U-235 content of the tails is too low for use in conventional reactors, but the breeder reactor can convert the U-238 into plutonium, as will be discussed in Chapter 13.

The cost of enrichment depends in part on the energy expended, which is measured in "separative work units" (SWU, pronounced "swoo") with units in kilograms. The method of calculating SWU is reserved for Computer Exercise 9.A. By use of a program called ENRICH3, Table 9.1 was developed. The feed w/o was taken as 0.711, corresponding to an atom percent of 0.720.

Let us use the table to find the amount of fuel needed and its cost to a utility. Assume that the fuel is to be enriched to 3 w/o. Thus each kg of fuel contains 30 grams of U-235 and 970 grams of U-238. The natural uranium feed required for the isotope separation process is 6.569 kg or 14.48 lb. It is easy to show (Exercise 9.8) that the U weight fraction in the U_3O_8 that would contain it is 0.848. Thus our feed becomes 6.569/0.848 = 7.75 kg of the oxide, or 17.1 lb.

Table 9.1 Nuclear Fuel Data

Weight Percent U-235	Ratio of Feed to Product	Separative Work Units (SWU)
0.711	1.000	0
0.8	1.217	0.070
1.0	1.703	0.269
2.0	4.136	1.697
3.0	6.569	3.425
5.0	11.436	7.198
10.0	23.601	17.284
20.0	47.932	38.315
90.0	218.248	192.938

The price of uranium oxide varies, but we assume $20/lb, giving $342 as the cost of U.

There is a cost for conversion of U_3O_8 into UF_6 for use in the enrichment process. Assuming $10/kg U, this amounts to $65.69. In column 3 of the table is found the SWU value of 3.425, and by use of a reasonable enrichment charge of $100/SWU, the cost is $342.50. A fuel element fabrication cost of say $275/kg adds for the 1 kg/day a sum of $275. Excluding transportation cost, the total of the preceding numbers is approximately $1025/kg. To fuel a nuclear reactor rated at 1000 MWe, an electric utility may need approximately 60,000 lb/y or 27,200 kg/y giving an annual fuel cost of $27.88 million. However, it can typically produce an average of 850 MW of electrical power over the 8760 hours per year, a total of 7.45×10^9 kWh. The basic fuel cost is thus 0.37 cents or 3.7 mills per kilowatt-hour.

The world picture on uranium enrichment has been changing in recent years, as more suppliers have appeared and United States utilities have diversified their sources. A large fraction of the natural uranium used in the United States comes from other countries such as Canada, Russia, and Australia. Approximately half of the enrichment services are provided by USEC, with the remainder from abroad (e.g., Eurodif, Urenco, and Tenex). A factor that renders the future situation uncertain is the amount and speed of reduction in weapons-grade uranium in the stockpiles of the United States and the Commonwealth of Independent States (CIS). Conversion of highly enriched uranium (HEU) into fuel suitable for reactor use as low-enrichment uranium (LEU) affects the supply situation significantly, including the mining and refining industries and the isotope separation process.

9.3 GAS CENTRIFUGE

This device for separating isotopes, also called the ultracentrifuge because of the very high speeds involved, has been known since the 1940s. It was tested and abandoned during World War II because materials that would withstand high rotation speeds were not available and existing bearings gave large power losses. Developments since have made centrifuges practical and economical. The centrifuge consists of a cylindrical chamber—the rotor—turning at very high speed in a vacuum as sketched in Figure 9.6(A).

The rotor is driven and supported magnetically. Gas is supplied and centrifugal force tends to compress it in the outer region, but thermal agitation tends to redistribute the gas molecules throughout the whole volume. Light molecules are favored in this effect, and their concentration is higher near the center axis. By various means a countercurrent flow of UF_6 gas is established that tends to carry the heavy and light isotopes to opposite ends of the rotor. Depleted and enriched streams of gas are withdrawn by scoop pipes, as shown schematically in Figure 9.6(B). More detailed diagrams are found in the References.

The theory of separation by centrifugal force starts with the formula for the gas density distribution in a gravitational field,

$$N = N_0 \exp(-mgb),$$

where the potential energy is mgb. Adapt the expression to a rotating gas, with kinetic energy at radius r being $\frac{1}{2} mv^2 = \frac{1}{2} m\omega^2 r^2$, where ω is the angular

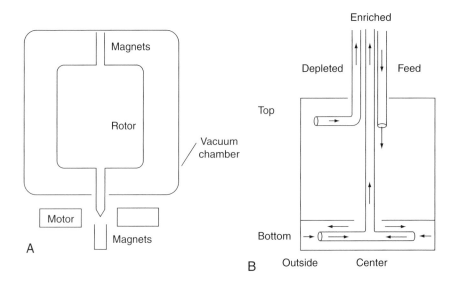

FIGURE 9.6 (A) Gas centrifuge. (B) Gas streams in centrifuge.

velocity, v/r. Apply to two gases of masses m_H and m_L to obtain the abundance ratio as a function of distance

$$R = R_0 \exp\left[(m_H - m_L)\omega^2 r^2/(2kT)\right].$$

Note that separation depends on the difference in masses rather than on their square roots as for gaseous diffusion.

Separation factors of 1.1 or better were obtained with centrifuges approximately 30 cm long, rotating at a rate such that the rotor surface speed is 350 m/s. The flow rate per stage of a centrifuge is much lower than that of gaseous diffusion, requiring large numbers of units in parallel.

The electrical power consumption for a given capacity is lower, however, by a factor of 6 to 10, giving a lower operating cost. In addition, the capital cost of a centrifuge plant is lower than that of a gaseous diffusion plant. European countries have taken advantage of the lower costs of centrifuge separation to challenge the former United States monopoly on enrichment services. In fact, several American utilities buy fuel from Europe. Examples of facilities are the French Eurodif operated by Cogema and the three plants of Urenco, Ltd. at Capenhurst in the United Kingdom, at Almelo in The Netherlands (see Figure 9.7), and at Gronau in West Germany.

FIGURE 9.7 One of the centrifuge enrichment halls of the plant at Almelo, The Netherlands. (Courtesy Urenco, Ltd.).

A centrifuge-based uranium enrichment plant is planned for Lea County, New Mexico, by Louisiana Energy Services (a project in Louisiana was abandoned). The partnership of four companies (see References) will install the latest and best equipment developed by Urenco's technology as used in Europe. Full capacity is expected to be 3 million SWU per year, available around 2010.

9.4 LASER ISOTOPE SEPARATION

A different technique for separating uranium isotopes uses laser light (see Section 2.4) to selectively ionize uranium-235 atoms, which can be drawn away from uranium-238 atoms. Research and development on the process, called atomic vapor laser isotope separation (AVLIS), was done in a cooperative program between Lawrence Livermore National Laboratory and Oak Ridge National Laboratory.

An element such as uranium has a well-defined set of electron orbits, similar to those described in Section 2.3, but much more complex because there are 92 electrons. The difference in masses of the nuclei of uranium-235 and uranium-238 results in subtle differences in the electronic orbit structure and corresponding energies required to excite or to ionize the two isotopes.

A laser can supply intense light of precise frequencies, and a fine-tuned laser beam can provide photons that ionize the U-235 and leave the U-238 unchanged. The ionization potential for U-235 is 6.1 volts. The method performs resonance stepwise excitation of an atom. In the AVLIS technique, three photons of approximately 2 eV achieve the ionization.

The virtue of the method is the almost perfect selection of the desired isotope. Of 100,000 atoms ionized by a laser beam, all but 1 are U-235. This permits enrichment from 0.7% to 3% in a single stage rather than thousands as with gaseous diffusion. One kilogram of enriched product comes from 6 kg of natural uranium. The metallic uranium is vaporized by a stream of electrons as in Figure 9.8. A yellow-green laser energizes ("pumps") a second orange-red laser. This irradiates the uranium vapor, with selective ionization of uranium-235 atoms. An electric field draws those ions off to condense on product collector plates. U-238 atoms pass through the laser beam and condense on the walls of the chamber to be removed as tails.

AVLIS was regarded as promising, but the USEC determined that is was less economic than gaseous diffusion or centrifuge separators. The R&D program was terminated in 1999. Some support was given to the Silex process (see References), but in 2003 that too was abandoned in favor of the American Centrifuge.

As a postscript, the French carried out tests with SILVA (AVLIS backward), obtaining 200 kg of low-enrichment uranium and a ton of depleted uranium. However, they concluded that the method would be useful only sometime in the future.

Thanks are due to James I. Davis of Lawrence Livermore National Laboratory and N. Haberman of the Department of Energy (DOE) for some of the information in this section.

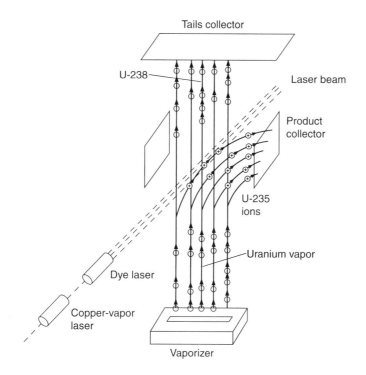

FIGURE 9.8 Atomic vapor laser isotopic separation.

9.5 SEPARATION OF DEUTERIUM

The heavy isotope of hydrogen ^2_1H, deuterium, has two principal nuclear applications: (a) as low-absorption moderator for reactors, especially those that use natural uranium, and (b) as a reactant in the fusion process. The differences between the chemical properties of light water and heavy water are slight, but sufficient to permit separation of ^1_1H and ^2_1H by several methods. Among these are electrolysis, in which the H_2O tends to be more readily dissociated; fractional distillation, which takes advantage of the fact that D_2O has a boiling point approximately $1°C$ higher than that of H_2O; and catalytic exchange, involving the passage of HD gas through H_2O to produce HDO and light hydrogen gas.

9.6 SUMMARY

The separation of isotopes requires a physical process that depends on mass. In the electromagnetic method, as used in a mass spectrograph, ions to be separated travel in circles of different radii. In the gaseous diffusion process, light molecules of a gas diffuse through a membrane more readily than do heavy molecules. The amount of enrichment in gaseous diffusion depends on the square root of the ratio of the masses and is small per stage, requiring a large number of stages. By the use of material balance equations, the amount of feed can be computed, and by the use of tables of separative work, costs of enriching uranium for reactor fuel can be found. An alternative separation device is the gas centrifuge, in which gases diffuse against the centrifugal forces produced by high speeds of rotation. Laser isotope separation involves the selective excitation of uranium atoms by lasers to produce chemical reactions. Several methods of separating deuterium from ordinary hydrogen are available.

9.7 EXERCISES

9.1 (a) Show that the radius of motion of an ion in a mass spectrograph is given by

$$r = \sqrt{\frac{2mV}{qB^2}}.$$

(b) If the masses of heavy (H) and light (L) ions are m_H and m_L, show that their separation at the plane of collection in a mass spectrograph is proportional to $\sqrt{m_H} - \sqrt{m_L}$.

9.2 The ideal separation factor for a gaseous diffusion stage is

$$r = 1 + 0.693\left(\sqrt{m_H/m_L} - 1\right).$$

Compute its value for $^{235}\text{UF}_6$ and $^{238}\text{UF}_6$, noting that $A = 19$ for fluorine.

9.3 (a) Verify that for particles of masses m_H and m_L the number fraction f_L of the light particle is related to the weight fractions w_H and w_L by

$$f_L = \frac{n_L}{n_L + n_H} = \frac{1}{1 + \frac{w_H m_L}{w_L m_H}}.$$

(b) Show that the abundance ratio of numbers of particles is either

$$R = \frac{n_L}{n_H} = \frac{f_L}{1 - f_L} \text{ or } \frac{w_L/m_L}{w_H/m_H}.$$

(c) Calculate the number fraction and abundance ratio for uranium metal that is 3% U-235 by weight.

9.4 The total fuel loading of a new research reactor is 2000 kg of uranium at 20 w/o U-235. From Table 9.1, find the amount of natural uranium feed and the SWUs required to fuel the reactor, assuming tails of 0.3 w/o.

9.5 A typical reactor that uses product uranium from an isotope separator at 3% enrichment burns 75% of the U-235 and 2.5% of the U-238. What percentage of the mined uranium is actually used for electrical power generation?

9.6 Find the amount of natural uranium feed (0.711% by weight) required to produce 1 kg/day of highly enriched uranium (90% by weight), if the waste concentration is 0.25% by weight. Assume that the uranium is in the form of UF_6.

9.7 How many enriching stages are required to produce uranium that is 3% by weight by use of natural UF_6 feed? Let the waste be 0.2%.

9.8 By use of the atomic weights of uranium and oxygen in the Appendix, verify that the weight fraction of U in U_3O_8 is 0.848.

9.9 The number density of molecules as the result of loss through a barrier can be expressed as $n = n_0 \exp(-c \, v \, t)$ where c is a constant, v is the particle speed, and n_0 and n are values before and after an elapsed time t. If half the heavy isotope is allowed to pass through, find the abundance ratio $R'/R = r$ in the enriched gas as a function of the ratio of molecular masses. Test the derived formula for the separation of uranium isotopes.

9.10 Depleted uranium (0.3% U-235) is processed by laser separation to yield natural uranium (0.711%). If the feed rate is 1 kg/day and all of the U-235 goes into the product, what amounts of product and waste are produced per day?

9.11 By use of natural uranium atom percents of 99.2745 for U-238, 0.7200 for U-235, and 0.0055 for U-234, and atomic masses given in the Appendix, calculate the atomic mass of natural U and the weight percents of each isotope. Suggestion: make a table of numbers.

9.12 A utility plans to increase the enrichment of its nuclear fuel from 3 w/o to 5 w/o, achieving an increase in capacity factor from 0.70 to 0.80 as the result of longer operating cycles. Estimate costs in the two cases and determine whether there is a net financial gain or loss, assuming that electricity is worth approximately 20 mills/kWh.

9.13 A certain country covertly builds production mass spectrographs to separate uranium isotopes. The objective is to obtain 50 kg of highly enriched uranium for a nuclear weapon, in 1 year of continuous operation. (a) Assuming optimistically that separation is perfect, what current of U^+ ions would be required? (b) Neglecting power needed for heating and magnets, what amount of electrical power at 50 kV is required? (c) Would the power source be difficult to conceal?

9.14 (a) Calculate the centrifugal acceleration $a = v^2/r$ in a centrifuge at radius $r = 0.1$ m with an angular speed of 5000 radians/second. By what factor is that larger than the acceleration of gravity 9.8 m/s^2? (b) Find the ratio R/R_0

for UF_6 of molecular weights 349 and 352 at 330 K, recalling $k = 1.38 \times 10^{-23}$ J/K and the mass of 1 amu $= 1.66 \times 10^{-27}$ kg.

Computer Exercises

9.A   The tails concentration of a gaseous diffusion separation process is typically 0.3 w/o. For a fixed product (e.g., 1 kg of 3 w/o fuel) study the variation of feed plus enrichment cost with the tails concentration, with (a) computer program ENRICH3 and some hand calculations, or (b) by adapting ENRICH3 to calculate costs.

9.B   Adapt computer program ENRICH3 to calculate costs as well as flows and SWU. Then, find the cost per gram of U-235 and cost per kilogram of U in product of 3 w/o, 20 w/o, and 90 w/o. Keep a constant tails assay of 0.3 w/o.

9.8 REFERENCES

Smyth Report on Separation of Uranium Isotopes (1945)
http://nuclearweaponarchive.org/Smyth/index.html
Reproduction of Chapters IX-XI.

Stelio Villani, *Isotope Separation*, American Nuclear Society, La Grange Park, IL, 1976.
A monograph that describes most of the techniques for separating isotopes, including theory, equipment, and data.

Willam E. Parkins, "The Uranium Bomb, the Calutron, and the Space Charge Problem," *Physics Today* 58, 45 (May 2005).

Uranium Enrichment
http://www.uic.com.au/nip33.htm
Facts and figures in a briefing paper by Uranium Information Centre (Australia).

Calutrons at Oak Ridge
http://www.ornl.gov/reporter/no1/calutron.htm
Final shutdown of machines.

Iraq's calutrons: 1991–2001
http://nuclearweaponarchive.org/Iraq/Calutron.html
Posting by Andre A. Gsponer, including report ISRI-95-03.

United States Enrichment Corporation
http://www.usec.com
Select Uranium Enrichment or The American Centrifuge.

Donald R. Olander, "The Gas Centrifuge," *Scientific American*, August 1978, p. 37.

Enrichment Technology
http://www.enritec.com
Urenco-Areva. Select Spotlight for gas centrifuge information.

Louisiana Energy Services (LES) Gas Centrifuge Facility
http://www.nrc.gov/materials/fuel-cycle-fac/lesfacility.html
Partners: Urenco, Exelon, Duke Power, and Energy.

Richard N. Zare, "Laser Separation of Isotopes," *Scientific American*, February 1977, p. 86.

Silex Isotope Separation
http://www.silex.com.au/s03_about_silex/s30_1_content.html
R&D with General Electric on laser method.

EIA–Nuclear Data
http://www.eia.doe.gov/fuelnuclear.html
Links to reports with data on nuclear fuel.

Uranium Enrichment Calculator
http://www.wise-uranium.org/nfcue.html
Reproduces calculations in Section 9.2.

Radiation Detectors[†]

10

MEASUREMENT OF radiation is required in all facets of nuclear energy—in scientific studies, in the operation of reactors for the production of electric power, and for protection from radiation hazard. Detectors are used to identify the radioactive products of nuclear reactions and to measure neutron flux. They determine the amount of radioisotopes in the air we breathe and the water we drink, or the uptake of a sample of radioactive material administered to the human body for diagnosis. In recent years detectors have become important in frustrating terrorism.

The kind of detector used depends on the particles to be observed—electrons, gamma rays, neutrons, ions such as fission fragments, or combinations. It depends on the energy of the particles. It also depends on the radiation environment in which the detector is to be used—at one end of the scale is a minute trace of a radioactive material and at the other a source of large radiation exposure. The type of measuring device, as in all applications, is chosen for the intended purpose and the accuracy desired.

The demands on the detector are related to what it is we wish to know: (a) whether there is a radiation field present; (b) the number of nuclear particles that strike a surface per second or some other specified period of time; (c) the type of particles present, and if there are several types, the relative number of each; (d) the energy of the individual particles; and (e) the instant a particle arrives at the detector. From the measurement of radiation we can deduce properties of the radiation such as ability to penetrate matter and to produce ionization. We can also determine properties of a radioactive source, including disintegration rate, half-life, and amount of material.

In this chapter we describe the important features of a few popular types of detectors. Most of them are based on the ionization produced by incoming radiation. The detector may operate in one of two modes: (a) current, in which an average electrical flow is measured, as with an ammeter; and (b) pulse, in which

---

[†]Suggestions by Glenn F. Knoll are recognized with appreciation.

the electrical signals produced by individual particles or rays are amplified and counted. A detector operating in this mode is known as a *counter*.

Because none of the five human senses will measure nuclear radiation, a detector serves us as a "sixth sense." A detector also makes it possible to reveal the existence of amounts of material much smaller than can be found by ordinary chemical tests.

10.1 GAS COUNTERS

Picture a gas-filled chamber with a central electrode (*anode*, electrically positive) and a conducting wall (*cathode*, negative). They are maintained at different potential, as shown in Figure 10.1. If a charged particle or gamma ray is allowed to enter the chamber, it may produce a certain amount of ionization in the gas. The resultant positive ions and electrons are attracted toward the negative and positive surfaces, respectively. A charge moves in the local field $\mathcal{E}$ with a drift velocity $v_D = \mu \mathcal{E}$, where the mobility μ depends on the time between collisions and the mean free path (see Sections 4.4 and 4.6). If a magnetic field is present, charges tend to execute circular paths interrupted by collisions. When the voltage across the tube is low, the charges merely migrate through the gas, they are collected, and a current of short duration (a pulse) passes through the resistor and the meter. More generally, amplifying circuits are required. The number of current pulses is a measure of the number of incident particles that enter the detector, which is designated as an *ionization chamber* when operated in this mode.

If the voltage is then increased sufficiently, electrons produced by the incident radiation through ionization are able to gain enough speed to cause further ionization in the gas. Most of this action occurs near the central electrode, where the electric field is highest. The current pulses are much larger than in the ionization chamber because of the amplification effect. The current is proportional to the original number of electrons produced by the incoming radiation, and the

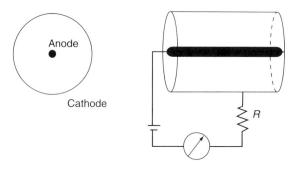

FIGURE 10.1 Basic detector.

detector is now called a *proportional counter.* One may distinguish between the passage of beta particles and alpha particles, which have a widely different ability to ionize. The time for collection is very short, of the order of microseconds.

If the voltage on the tube is raised still higher, a particle or ray of any energy will set off a discharge, in which the secondary charges are so great in number that they dominate the process. The discharge stops of its own accord because of the generation near the anode of positive ions, which reduce the electric field there to such an extent that electrons are not able to cause further ionization. The current pulses are then of the same size, regardless of the event that initiated them. In this mode of operation, the detector is called a *Geiger-Mueller* (GM) counter. Unlike the proportional counter, the magnitude of the pulses produced by a GM counter is independent of the original number of electrons released by the ionizing radiation. Therefore the counter provides no information about the type or energy of the radiation. There is a short period, the *dead time,* in which the detector will not count other incoming radiation. If the radiation level is very high, a correction of the observed counts to yield the "true" counts must be made to account for the dead time. In some gases, such as argon, there is a tendency for the electric discharge to be sustained, and it is necessary to include a small amount of foreign gas or vapor (e.g., alcohol) to "quench" the discharge. The added molecules affect the production of photons and resultant ionization by them.

A qualitative distinction between the preceding three types of counters is displayed graphically in Figure 10.2, which is a semilog plot of the charge collected as a function of voltage. We note that the current varies over several orders of magnitude.

10.2 NEUTRON DETECTORS

To detect neutrons, which do not create ionization directly, it is necessary to provide a means for generating the charges that can ionize a gas. Advantage is taken of the nuclear reaction involving neutron absorption in boron

$$\frac{1}{0}n + \frac{10}{5}B \rightarrow \frac{4}{2}He + \frac{7}{3}Li,$$

where the helium and lithium atoms are released as ions. One form of *boron counter* is filled with the gas boron trifluoride (BF_3) and operated as an ionization chamber or a proportional counter. It is especially useful for the detection of thermal neutrons because the cross section of boron-10 at 0.0253 eV is large, 3840 barns, as noted in Chapter 4. Most of the 2.8 MeV energy release goes to the kinetic energy of the product nuclei. The reaction rate of neutrons with the boron in BF_3 gas is independent of the neutron speed, as can be seen by forming the product $R = nvN\sigma_a$, where σ_a varies as $1/v$. The detector thus measures the number density n of an incident neutron beam rather than the flux. Alternately, the metal electrodes of a counter may be coated with a layer of boron that is thin

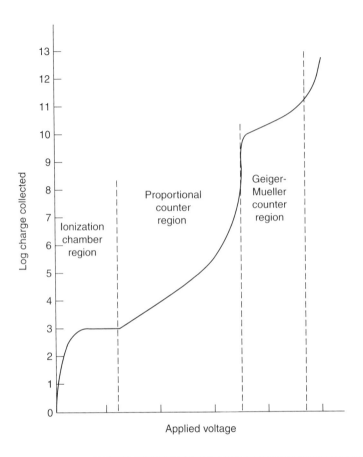

FIGURE 10.2 Collection of charge in counters.

enough to allow the alpha particles to escape into the gas. The counting rate in a boron-lined chamber depends on the surface area exposed to the neutron flux.

The *fission chamber* is often used for slow neutron detection. A thin layer of U-235, with high thermal neutron cross section, 681 barns, is deposited on the cathode of the chamber. Energetic fission fragments produced by neutron absorption traverse the detector and give the necessary ionization. Uranium-238 is avoided because it is not fissile with slow neutrons and because of its stopping effect on fragments from U-235 fission.

Neutrons in the thermal range can be detected by the radioactivity induced in a substance in the form of small foil or thin wire. Examples are manganese $^{55}_{25}\text{Mn}$, with a 13.3 barn cross section at 2200 m/s, which becomes $^{56}_{25}\text{Mn}$ with half-life 2.58 h; and dysprosium $^{164}_{66}\text{Dy}$, 1.7×10^3 barns, becoming $^{165}_{66}\text{Dy}$, half-life 2.33 h. For detection of neutrons slightly above thermal energy, materials with a high resonance cross section are used (e.g., indium) with a peak at 1.45 eV. To separate the effects of thermal neutron capture and resonance capture, comparisons are made between measurements made with thin foils of indium and those of indium covered with cadmium. The latter screens out low-energy neutrons (below 0.5 eV) and passes those of higher energy.

For the detection of fast neutrons, up in the MeV range, the *proton recoil* method is used. We recall from Chapter 4 that the scattering of a neutron by hydrogen results in an energy loss, which is an energy gain for the proton. Thus a hydrogenous material such as methane (CH_4) or H_2 itself may serve as the counter gas. The energetic protons play the same role as α particles and fission fragments in the counters discussed previously. Nuclear reactions such as $_2^3He(n, p)\,_1^3H$ can also be used to obtain detectable charged particles.

10.3 SCINTILLATION COUNTERS

The name of this detector comes from the fact that the interaction of a particle with some materials gives rise to a scintillation or flash of light. The basic phenomenon is familiar—many substances can be stimulated to glow visibly on exposure to ultraviolet light, and the images on a color television screen are the result of electron bombardment. Molecules of materials classed as phosphors are excited by radiation such as charged particles and subsequently emit pulses of light. The substances used in the scintillation detector are inorganic (e.g., sodium iodide or lithium iodide) or organic, in one of various forms—crystalline, plastic, liquid, or gas.

The amount of light released when a particle strikes a phosphor is often proportional to the energy deposited, and thus makes the detector especially useful for the determination of particle energies. Because charged particles have a short range, most of their energy appears in the substance. Gamma rays also give rise to an energy deposition through electron recoil in both the photoelectric effect and Compton scattering, and through the pair production-annihilation process. A schematic diagram of a detector system is shown in Figure 10.3. Some of the light released in the phosphor is collected in the photomultiplier tube, which consists of a set of electrodes with photosensitive surfaces. When a photon

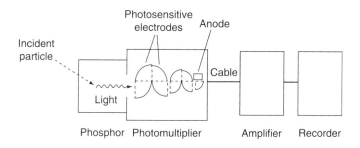

FIGURE 10.3 Scintillation detection system.

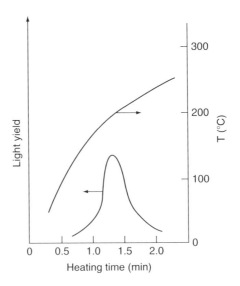

FIGURE 10.4 Glow curve of the phosphor CaF_2.

strikes the surface, an electron is emitted by the photoelectric effect, it is accelerated to the next surface where it dislodges more electrons, and so on, and a multiplication of current is achieved. An amplifier then increases the electrical signal
to a level convenient for counting or recording.

Radiation workers are required to wear personal detectors called dosimeters
to determine the amount of exposure to X- or gamma rays or neutrons. Among
the most reliable and accurate types is the thermoluminescent dosimeter (TLD),
which measures the energy of radiation absorbed. It contains crystalline materials
such as CaF_2 or LiF that store energy in excited states of the lattice called traps.
When the substance is heated, it releases light in a typical glow curve as shown
in Figure 10.4. The dosimeter reader consists of a small vacuum tube with a
coated cylinder that can be heated by a built-in filament when the tube is plugged
into a voltage supply. A photomultiplier reads the peak of the glow curve and
gives values of the accumulated energy absorbed (i.e., the dose). The device is
linear in its response over a very wide range of exposures; it can be used over
and over with little change in behavior.

10.4 SOLID STATE DETECTORS

The use of a solid rather than a gas in a detector has the advantage of compactness, because of the short range of charged particles. Also, when the solid is a
semiconductor, great accuracy in measurement of energy and arrival time is possible. The mechanism of ion motion in a solid detector is unique. Visualize a

crystal semiconductor, such as silicon or germanium, as a regular array of fixed atoms with their complement of electrons. An incident charged particle can dislodge an electron and cause it to leave the vicinity, which leaves a vacancy or "hole" that acts effectively as a positive charge. The electron-hole pair produced is analogous to negative and positive ions in a gas. Electrons can migrate through the material or be carried along by an electric field, while the holes "move" as electrons are successively exchanged with neighboring atoms. Thus, electrons and holes are negative and positive charge carriers, respectively.

The electrical conductivity of a semiconductor is very sensitive to the presence of certain impurities. Consider silicon, chemical valence 4 (with 4 electrons in the outer shell). Introduction of a small amount of valence 5 material such as phosphorus or arsenic gives an excess of negative carriers, and the new material is called *n*-type silicon. If instead a valence 3 material such as boron or gallium is added, there is an excess of positive carriers, and the substance is called *p*-type silicon. When two layers of *n*-type and *p*-type materials are put in contact and a voltage is applied, as in Figure 10.5, electrons are drawn one way and holes the other, leaving a neutral or "depleted" region. Most of the voltage drop occurs over that zone, which is very sensitive to radiation. An incident particle creates electron-hole pairs that are swept out by the internal electric field to register as a current pulse. High accuracy in measurement by an *n-p* junction comes from the fact that a low energy is needed to create an electron-hole pair (only 3 eV versus 32 eV for an ion pair in a gas). Thus a 100 keV photon creates a very large number of pairs, giving high statistical accuracy. The collection time is very short, approximately a billionth of a second, allowing precise time measurements.

One way to produce a semiconductor detector with a large active volume is to introduce lithium on one surface of a heated crystal and apply an electric field. This "drifts" the Li through the volume that compensates residual *p*-type impurities. This detector must be kept permanently at liquid nitrogen temperature ($-195.8°C$) to prevent redistribution of the lithium. A preferable detector for

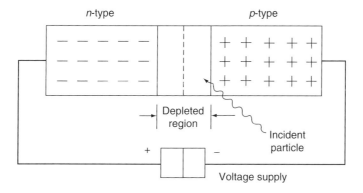

FIGURE 10.5 Solid-state *n-p* junction detector.

many applications is made of an ultrahigh-purity germanium, with impurity atoms reduced to 1 in approximately 10^{12}. A simple diode arrangement gives depletion depths of several centimeters. Such detectors still require liquid N_2 for operation, but they can be stored at room temperature.

10.5 STATISTICS OF COUNTING

The measurement of radiation has some degree of uncertainty because the basic processes, such as radioactive decay, are random in nature. From the radioactive decay law, Section 3.2, we can say that *on the average* in a time interval t a given atom in a large collection of atoms has a chance $\exp(-\lambda t)$ of not decaying, and thus it has a chance $1 - \exp(-\lambda t)$ of decaying. Because of the statistical nature of radioactivity, however, more or less than these numbers will actually be observed in a certain time interval. There is actually a small probability that either none or all would decay. In a series of identical measurements there will be a spread in the number of counts. Statistical methods may be applied to the data to estimate the degree of uncertainty or "error." The laws of probability may be applied. As discussed in texts on statistics and radiation detection (see References), the most rigorous expression is the binomial distribution (see Exercise 10.6), which must be used to interpret the decay of isotopes of very short half-life. A simple approximation to it is the Poisson distribution (see Exercise 10.7) required for the study of low-level environmental radioactivity. A further approximation is the widely used normal or Gaussian distribution, shown in Figure 10.6.

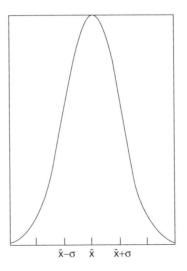

FIGURE 10.6 Gaussian distribution. The area between the limits $x \pm \sigma$ is 68% of the total.

Measured values of the number of counts x in repeated trials tend to fit the formula,

$$P(x) = \left(1/\sqrt{2\pi\bar{x}}\right)\exp\left(-(x-\bar{x})^2/(2\bar{x})\right)$$

where $P(x)$ is the probability of being in a unit range at x and $\bar{x}$ is the mean value of the counts. A measure of the width of the curve is the standard deviation, σ. For this function[†], $\sigma = \sqrt{\bar{x}}$. The area under the curve between $\bar{x} + \sigma$ and $\bar{x} - \sigma$ is 68% of the total, which indicates that the chance is 0.68 that a given measurement will lie in that range. The figure for 95% is $\pm 2\sigma$. It can be shown that the fractional error in count rate is inversely proportional to the square root of the total number of counts.

Because the calculation for plotting of the preceding statistical distribution is quite tedious, we have provided the computer program STAT, the use of which is described in Computer Exercises 10.A through 10.D. Also, program EXPOIS generates simulated counting data for study by use of the Poisson distribution.

10.6 PULSE HEIGHT ANALYSIS

The determination of the energy distribution of nuclear particles and rays is important for identifying radioactive species. If an incoming particle deposits all of its energy in the detector, the resulting voltage signal in the external electric circuit of Figure 10.7(A) can be used as a measure of particle energy. The particle ionizes the medium, a charge Q is produced, and a current flows, giving a time-varying voltage. If the time constant $\tau = RC$ of the circuit is short compared with the collection time, the voltage rises and drops to zero quickly, as in Figure 10.7(B). If τ is large, however, the voltage rises to a peak value $V_m = Q/C$, where C is the capacitance, and then because of the circuit characteristics declines slowly, as in Figure 10.7(C). The particle energy, proportional to charge, is thus obtained by a voltage measurement.

Suppose that there are two types of particles entering the detector, say alpha particles of 4 MeV and beta particles of 1 MeV. By application of a voltage bias, the pulses caused by beta particles can be eliminated, and the remaining counts represent the number of alpha particles. The circuit that performs that separation is called a *discriminator*.

[†]In general, for a series of trials, 1,2, 3,. . ., N, if count rates are x_i and the average is $\bar{x}$, the standard deviation is

$$\sigma = \sqrt{\sum_{i=1}^{N}(x_i - \bar{x})^2/(N-1)}.$$

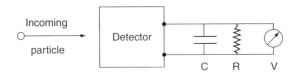

A Detector and electronic circuit

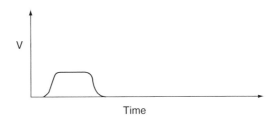

B Voltage variation with short time constant

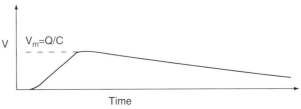

C Voltage variation with long time constant

FIGURE 10.7 Effect of circuit on pulse (after Knoll, see References).

The radiation from a given source will have some variation in particle energy and thus a series of pulses caused by successive particles will have a variety of heights. To find the energy distribution, a *single-channel analyzer* can be used. This consists of two adjustable discriminators and a circuit that passes pulses within a range of energy. The *multichannel analyzer* is a much more efficient and accurate device for evaluating an entire energy spectrum in a short time. Successive pulses are manipulated electronically and the signals are stored in computer memory according to energy. The data are displayed on an oscilloscope screen or are printed out.

10.7 ADVANCED DETECTORS

A number of specialized instruments have been developed in addition to basic detectors. They are used for precise measurements of the products of high-energy nuclear collisions. Examples are:

(a) Nuclear emulsion track detectors, originally used for cosmic ray studies. By application of the energy loss formula (see Section 5.2) information is obtained on particle energy, mass, and charge. Special etching techniques are used and the counting of tracks with a microscope is automated.

(b) Cerenkov counters, which measure the light produced when a particle has a speed higher than that of light in the medium. Cerenkov radiation gives the "blue glow" seen near a pool reactor core.

(c) Hadron calorimeters, which measure showers of hadrons (mesons and nucleons), protons, and neutrons produced by bombardment of materials of particles in the GeV range.

(d) Neutrino detectors, consisting of large volumes of liquid or metal in which the rare collisions resulting in a scintillation occur.

These specialized devices are discussed in the book by Kleinknecht (see References).

10.8 DETECTORS AND COUNTERTERRORISM

Radiation detectors play a vital role in protecting against terrorist action. Of greatest importance is the screening of shipments at their point of origin and on their arrival at an United States port. X-rays are helpful in finding well-shielded items in large shipping containers, prompting further inspection. Portable X-ray generators powered with batteries are available, giving 150 keV rays that can penetrate half an inch of steel.

To check for the presence of fissionable materials—enriched uranium or plutonium—several techniques are available. One involves an intense beam of 14 MeV neutrons from a D-T generator. Pulses of very short duration are introduced to cause fission and neutron release. Silicon carbide (SiC) semiconductor detectors measure the fission neutrons very quickly. Devices used in research are considered suitable for commercial production.

An alternative method to detect fissionable materials uses a pulsed photonuclear neutron detector. It consists of a portable accelerator that produces energetic photons. These cause fission, which releases neutrons that are detected externally. In tests, a sample vial shielded by wood, polyethylene, and lead was quickly and readily identified. By use of different photon energies, distinctions between enrichments of uranium or fissionable elements were achieved.

Amounts of weapons material that are significant are 25 kg for uranium and 4 kg for plutonium. The half-lives of U-235 and U-238 are too long for their inherent radiation to be detected. For the radiological dispersal device, the so-called dirty bomb, there is direct detection of their radiation. Isotopes that are likely to be used are cobalt-60, half-life 5.27 y, average 1.25 MeV gammas, and cesium-137, half-life 30.2 y, 0.662 MeV gammas. Other candidate radioisotopes are americium-241, californium-252, iridium-192, and strontium-90. Californium is a neutron emitter with half-life 2.65 y. In the case of Sr-90, the beta particles are readily shielded, but the *bremsstrahlung* emitted as the electrons slow down in matter can be detected. All of these isotopes have commercial applications, which makes them vulnerable for theft.

Nuclear forensics involves a signature method, the determination of the origin of radioactivity whether it is from a dirty bomb or from the explosion of a nuclear assembly. Search algorithms are developed that correlate the time dependence of isotope decay with reactor type and neutron irradiation history.

Research continues on the problem of finding the radiation dosage to individuals in emergency situations. Plans are being developed for the handling of a large number of people who are irradiated or contaminated.

Tracer studies in Manhattan are used to develop models for the dispersal of radioactivity in a city with streets between skyscrapers.

In methods that use or produce neutrons, consideration must be given to the "neutron ship effect." When muons in cosmic rays bombard iron as in bridges or ships there is a release of neutrons, which constitutes a competing background.

Much of the R&D on detectors is sponsored and supported by the Department of Homeland Security, formed in the wake of the September 11, 2001, attacks. The agency has a mammoth task in providing radiation detection equipment at the 1205 seaports and airports. The General Accountability Office (GAO) has reported slow progress (see References).

10.9 SUMMARY

The detection of radiation and the measurement of its properties are required in all aspects of the nuclear field. In gas counters, the ionization produced by incoming radiation is collected. Depending on the voltage between electrodes, counters detect all particles or distinguish between types of particles. Neutrons are detected indirectly by the products of nuclear reactions—for slow neutrons by absorption in boron or uranium-235, for fast neutrons by scattering in hydrogen. Scintillation counters release measurable light on bombardment by charged particles or gamma rays. Solid-state detectors generate a signal from the motion of electron-hole pairs created by ionizing radiation. Pulse-height analysis yields energy distributions of particles. Statistical methods are used to estimate the uncertainty in measured counting rates. Advanced specialized detectors are used in high-energy physics research. Detectors of nuclear radiation play a vital role in the United States counterterrorism programs.

10.10 EXERCISES

10.1 (a) Find the number density of molecules of BF_3 in a detector of 2.54 cm diameter to be sure that 90% of the thermal neutrons incident along a diameter are caught (σ_a for natural boron is 760 barns).

(b) How does this compare with the number density for the gas at atmospheric pressure, with density 3.0×10^{-3} g/cm^3?

(c) Suggest ways to achieve the high efficiency desired.

10.2 An incident particle ionizes helium to produce an electron and an He^{++}
ion halfway between two parallel plates with potential difference
between them. If the gas pressure is very low, estimate the ratio of the
times elapsed until the charges are collected. Discuss the effect of collisions on the collection time.

10.3 We collect a sample of gas suspected of containing a small amount of
radioiodine, half-life 8 days. If we observe in a period of 1 day a total
count of 50,000 in a counter that detects all radiation emitted, how many
atoms were initially present?

10.4 In a gas counter, the potential difference at any point r between a central
wire of radius r_1 and a concentric wall of radius r_2 is given by

$$V = V_0 \frac{\ln(r/r_1)}{\ln(r_2/r_1)},$$

where V_0 is the voltage across the tube. If $r_1 = 1$ mm and $r_2 = 1$ cm,
what fraction of the potential difference exists within a millimeter of
the wire?

10.5 How many electrodes would be required in a photomultiplier tube to
achieve a multiplication of 1 million if one electron releases four electrons?

10.6 The probability of x successful events in n trials, each of which has a
chance p, is given by the binomial distribution formula,

$$P(x) = n! \, p^x (1-p)^{n-x}/((n-x)!x!).$$

(a) Apply to flipping a coin 1, 2, and 3 times, finding the number of
times the result is heads, including zero. Check by simple logic.
(b) Apply to throwing a single die 1 or 2 times, finding the number of
sixes. Check.
(c) Repeat the preceding calculations with program STAT (see Computer Exercise 10.A).

10.7 For a situation in which the chance of success p is much smaller than 1,
the probability of x successes in n trials in the binomial formula of Exercise 10.6 is well approximated by the Poisson formula

$$P(x) \cong (\bar{x}^x/x!)\exp(-\bar{x}),$$

where $\bar{x} = pn$ is the mean value of x. What is the value of p in throwing
a single die? Find x for 1 or 2 throws of a die and calculate $P(x)$ for
each case.

10.8   Counts are taken for a minute from a microcurie source of cesium-137, half-life 30.2 years. (a) Assuming one count for each 50 disintegrations, find the expected counting rate and the number of counts for the interval. (b) Find the standard deviation in the counting rate. (c) Find the probability of decay of a given atom of cesium in the 1-minute interval.

10.9   A pair of dice is thrown $n = 10$ times. (a) Verify that the chance on one throw of getting a 7 is $p = 1/6$. (b) By use of the binomial distribution, find out the chance of getting a 7 exactly $x = 2$ times out of the 10. (c) Repeat with the Poisson distribution.

10.10 The cross section for absorption for low-energy neutrons of nuclides such as boron-10 varies as $1/v$, as discussed in Section 4.5. Formally, we may write

$$\sigma_a = \sigma_{a0} \, v_0/v$$

where σ_{a0} is the cross section at $v_0 = 2200$ m/s. A boron neutron detector is placed in a neutron speed distribution $n(v)$, with n_0 as the total number of neutrons per cm^3 and N as the number of boron nuclei per cm^3. Form the total reaction rate per cm^3 by integrating over the distribution, as a generalization of the equation in Section 4.3. Discuss the result in terms of what is being measured by the detector.

Computer Exercises

10.A   Program STAT calculates the probability distribution $P(x)$ with a choice of the Binomial, Poisson, or Gauss formulas. (a) What is the value of p for throwing a "six" with a single die? (b) Run the program with $n = 1, 2, 5,$ and 10 and note the probabilities of finding 0, 1, 2, . . . sixes. (c) Assuming that Binomial is exact, comment on the apparent accuracy of the other two methods.

10.B   An alpha particle detector for surface contamination is counted for 30 1-minute intervals, with a total of 225 counts. What is the value of p? With the Binomial and Poisson distributions in the computer program STAT, calculate $P(x)$ for $x = 0, 1, 2 . . ., 30$. How accurate is the Poisson formula?

10.C   (a) What is the chance that any given person's birthday is today? (b) If we select 1000 people at random, with the Poisson distribution in program STAT, what is the probability that none has a birthday today? (c) Calculate $P(x)$ for $x = 0$ to 10 and plot a bar graph of the results. What is the most likely number that has a birthday today and what is the mean value? (d) Run STAT in Binomial mode for 20 people at a party and show that the chance of two people having the same birthday is almost one-half.

10.D   Computer program EXPOIS calculates "experimental" particle counting data that can be analyzed by Poisson statistics. It uses random numbers with the command RND(N), where N is an assigned set of numbers. Run the program for a typical time from 10 to 30 minutes and compare the results graphically with Poisson data produced by the program STAT (Computer Exercise 10.A).

10.11 REFERENCES

Philip R. Bevington and D. Keith Robinson, *Data Reduction and Error Analysis for the Physical Sciences*, 2nd Ed., McGraw-Hill, New York, 1992. An update of a classic text on statistical methods, illustrated with examples from technology. Includes computer diskette.

Harry H. Ku, Editor, *Precision Measurement and Calibration—Statistical Concepts and Procedures*, National Bureau of Standards Publication 300, Volume 1, United States Government Printing Office, Washington, DC, 1969.

Geoffrey G. Eichholz and John W. Poston, *Principles of Nuclear Radiation Detection*, Ann Arbor Science, Ann Arbor, MI, 1979. A laboratory manual is also available.

Glenn F. Knoll, *Radiation Detection and Measurement*, 2nd Ed., John Wiley & Sons, New York, 1989. A very comprehensive, modern, and readable text that should be in every nuclear engineer's library.

Robley D. Evans, *The Atomic Nucleus*, Krieger, New York, 1982. Excellent treatment of statistical distributions. Reprint of 1955 classic book.

Nicholas Tsoulfanidis, *Measurement and Detection of Radiation*, 2nd Ed., Taylor & Francis, Washington, DC, 1995.

Konrad Kleinknecht, *Detectors for Particle Radiation*, 2nd Ed., Cambridge University Press, Cambridge, U.K., 1999.

Robert L. Fleischer, *Tracks to Innovation: Nuclear Tracks in Science and Technology*, Springer-Verlag, New York, 1998.

Dennis D. Wackerly, William Mendenhall III, and Richard L. Schaeffer, *Mathematical Statistics with Applications*, 6th Ed., Cengage Learning, Stanford, CT, 2001.

Michael F. L'Annunziata, Editor, *Handbook of Radioactivity Analysis*, 2nd Ed., Academic Press, San Diego, 2003.

The Smyth Report online
http://nuclearweaponarchive.org/Smyth
Appendix 1 on detectors.

Caltech Senior Physics Laboratory
http://www.pma.caltech.edu/~ph77
Nuclear measurements that prepare students for research. Select for example Gamma-and X-Ray Spectroscopy.

Commercial suppliers of nuclear radiation detectors
http://www.orau.gov/DDSC/instrument/instsuppliers.htm
Companies providing survey instrumentation.

Binomial and Poisson Statistics Functions
http://www.ciphersbyritter.com/JAVASCRP/BINOMPOI.HTM
Interactive computation. By Terry Ritter.

The Integrator
http://integrals.wolfram.com
Interactive integration of math functions. From Mathematica.

Combating Nuclear Smuggling (Report of GAO)
http://www.gao.gov/new.items/d061015.pdf
Challenge of providing protection.

Thomas E. Kiess, "Results and Characteristics of the Homeland Security Office of Research
and Development Radiological and Nuclear Countermeasures Program," *Transactions of
American Nuclear Society*, 95, 2006, p. 9.

IEEE 2006 Nuclear Science Symposium, October 2006.

Radiation threat
http://www.ready.gov/america/beinformed/radiation.html
Homeland Security advisory and information source

Neutron Chain Reactions

11

THE POSSIBILITY of a chain reaction involving neutrons in a mass of nuclear fuel such as uranium depends on (a) nuclear properties such as cross sections and neutrons produced per absorption (Section 6.3) and (b) the size, shape, and arrangement of the materials.

11.1 CRITICALITY AND MULTIPLICATION

To achieve a self-sustaining chain reaction, one needing no external neutron supply, a "critical mass" of uranium must be collected. To appreciate this requirement we visualize the simplest nuclear reactor, consisting of a metal sphere of uranium-235. Suppose that it consists of only one atom of U-235. If it absorbs a neutron and fissions, the resultant neutrons do nothing further, there being no more fuel. If instead we have a small chunk of uranium, say a few grams, the introduction of a neutron might set off a chain of several reactions, producing more neutrons, but most of them would escape through the surface of the body, a process called *leakage*. Such an amount of fuel is said to be "subcritical." Now if we bring together approximately 50 kg of U-235 metal, the neutron production balances the leakage losses, and the system is self-sustaining or "critical." The size is the critical volume and the amount of fuel is the critical mass. Neutrons had to be introduced to start the chain reaction, but the number is maintained without further additions. The term "critical mass" has become popular to describe any collection of entities large enough to operate independently.

Figure 11.1 shows the highly enriched metal assembly Lady Godiva, so named because it was "bare" (i.e., it had no surrounding materials). It was used for test purposes for a number of years at Los Alamos. If we add still more uranium to the 50 kg required for criticality, more neutrons are produced than are lost, the neutron population increases, and the reactor is "supercritical." Early nuclear weapons involved the use of such masses, in which the rapid growth of neutrons and resulting fission heat caused a violent explosion.

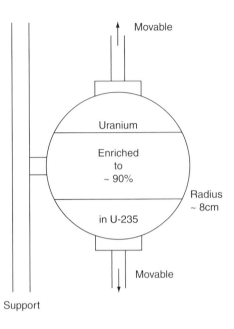

FIGURE 11.1 Fast metal assembly "Lady Godiva."

11.2 MULTIPLICATION FACTORS

The behavior of neutrons in a nuclear reactor can be understood through analogy with populations of living organisms; for example, of human beings. There are two ways to look at changes in numbers of people: as individuals and as a group. A person is born and throughout life has various chances of fatal illness or accident. On average the life expectancy at birth might be 75 years, according to statistical data. An individual may die without an heir, with one, or with many. If on average there is exactly 1, the population is constant. From the other viewpoint, if the rates of birth and death are the same in a group of people, the population is again steady. If there are more births than deaths by 1% per year, the population will grow accordingly. This approach emphasizes the competition of process rates.

The same ideas apply to neutrons in a multiplying assembly. We can focus attention on a typical neutron that was born in fission and has various chances of dropping out of the cycle because of leakage and absorption in other materials besides fuel. On the other hand we can compare the reaction rates for the processes of neutron absorption, fission, and leakage to see whether the number of neutrons is increasing, steady, or decreasing. Each of the methods has its merits for purposes of discussion, analysis, and calculation.

For any arrangement of nuclear fuel and other materials, a single number k tells the net number of neutrons per initial neutron, accounting for all losses and reproduction by fission. The state of the system can be summarized by:

$$k > 1 \text{ supercritical}$$
$$k = 1 \text{ critical}$$
$$k < 1 \text{ subcritical}$$

The design and operation of all reactors is focused on k or on related quantities such as $\delta k = k - 1$, called delta-k, or $\delta k/k$, called *reactivity*, symbolized by ρ. The choice of materials and size is made to assure that k has the desired value. For safe storage of fissionable material k should be well below 1. In the critical experiment, a process of bringing materials together with a neutron source present, observations on neutron flux are made to yield estimates of k. During operation, variations in k are made as needed by adjustments of neutron-absorbing rods or dispersed chemicals. Eventually, in the operation of the reactor, enough fuel is consumed to bring k below 1 regardless of adjustments in control materials, and the reactor must shut down for refueling.

We can develop a formula for k for our uranium metal assembly with the statistical approach. As in Figure 11.2(A), a neutron may escape on first flight from

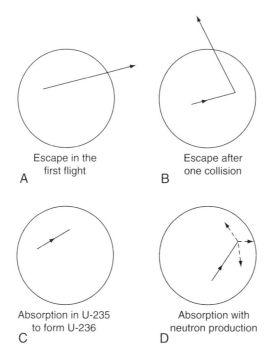

A   Escape in the
first flight

B   Escape after
one collision

C   Absorption in U-235
to form U-236

D   Absorption with
neutron production

FIGURE 11.2 Neutron histories.

the sphere, because mean free paths for fast neutrons are rather long. Another neutron (B) may make one scattering collision and then escape. Other neutrons may collide and be absorbed either (C) to form U-236 or (D) to cause fission, the latter yielding three new neutrons in this case. Still other neutrons may make several collisions before leakage or absorption takes place. The statistical nature of the process is revealed by the application of Computer Exercise 11.D, which involves the program SLOWINGS. Scattering, absorption, and escape are simulated with a Monte Carlo technique. A "flow diagram" as in Figure 11.3 is useful to describe the various fates. The boxes represent processes; the circles represent the numbers of neutrons at each stage.

The fractions of absorbed neutrons that form U-236 and that cause fission, respectively, are the ratios of the cross section for capture σ_c and fission σ_f to that for absorption σ_a. The average number of neutrons produced by fission is v. Now let η be the combination $v\sigma_f/\sigma_a$, and note that it is the number of neutrons per absorption in uranium. Thus letting $\mathcal{L}$ be the fraction *not* escaping by leakage,

$$k = \eta\mathcal{L}$$

The system is critical if $k = 1$, or $\eta\mathcal{L} = 1$. Measurements show that η is approximately 2.2 for fast neutrons, thus $\mathcal{L}$ must be $1/(2.2) = 0.45$, which says that as

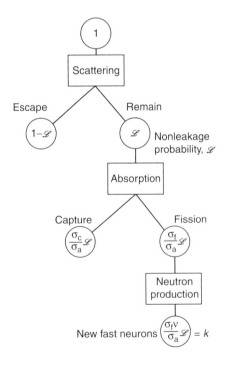

FIGURE 11.3 Neutron cycle in metal assembly.

many as 45% of the neutrons must remain in the sphere, whereas no more than 55% escape through its boundary.

Let us now examine more closely $\mathcal{L}$, the nonleakage factor, coming from the process of neutron loss through the surface of a reactor core without reflector. Leakage depends on scattering collisions and on the size and shape of the core. We would expect that the amount of neutron leakage depends on the ratio of surface to volume, because production occurs within the core and losses occur at the boundary. For a sphere, for example, the volume is $V = (4/3)\pi R^3$ and the surface area is $S = 4\pi R^2$, so the ratio is $S/V = 3/R$. As it turns out from the theory of neutron diffusion, the parameter that actually applies is $B = \pi/R$, the square of which, B^2, is called the "buckling". It is also logical that leakage should be larger, the greater the transport mean free path (recall Section 4.6), and the smaller the absorption cross section (Section 4.3). This is indeed the case, involving the use of the diffusion length, $L = \sqrt{D/\Sigma_a}$ as used in Section 4.6. The nonleakage factor for one neutron energy group in a bare reactor is thus

$$\mathcal{L} = 1/(1 + B^2 L^2).$$

Bucklings for three important shapes are as listed.

Sphere, radius R: $B^2 = (\pi/R)^2$
Parallelepiped, L, W, H: $B^2 = (\pi/L)^2 + (\pi/W)^2 + (\pi/H)^2$
Cylinder, H, R: $B^2 = (\pi/H)^2 + (j_0/R)^2$, where $j_0 = 2.40483$.

Critical conditions for more complex situations including mixtures of fuels can be analyzed by use of program CRITICAL, discussed in Computer Exercise 11.A.

The effect of flux variation with position is illustrated by Computer Exercise 11.B, dealing with the program MPDQ.

The presence of large amounts of neutron-moderating material such as water in a reactor greatly changes the neutron distribution in energy. Fast neutrons slow down by means of collisions with light nuclei, with the result that most of the fissions are produced by low-energy (thermal) neutrons. Such a system is called a "thermal" reactor in contrast with a system without moderator, a "fast" reactor, operating principally with fast neutrons. The cross sections for the two energy ranges are widely different, as noted in Exercise 11.4. Also, the neutrons are subject to being removed from the multiplication cycle during the slowing process by strong resonance absorption in elements such as U-238. Finally, there is competition for the neutrons between fuel, coolant, structural materials, fission products, and control absorbers.

The description of the multiplication cycle for a thermal reactor is somewhat more complicated than that for a fast metal assembly, as seen in Figure 11.4. The set of reactor parameters are (a) the fast fission factor ε (epsilon), representing the immediate multiplication because of fission at high neutron energy, mainly in U-238; (b) the fast nonleakage probability $\mathcal{L}_f$, being the fraction remaining in the core during neutron slowing; (c) the resonance escape probability p, the

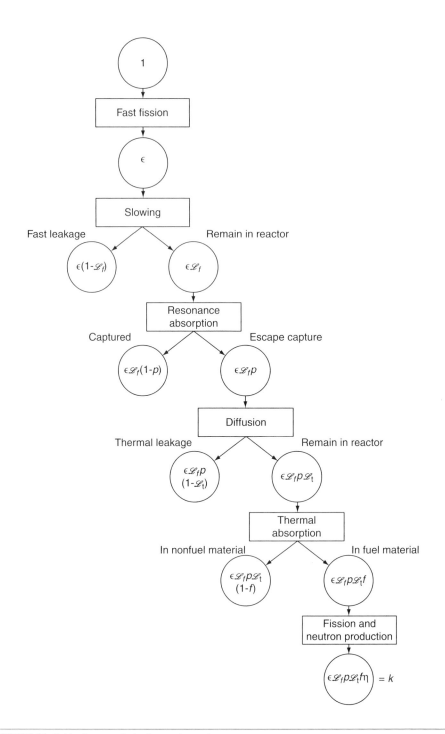

FIGURE 11.4 Neutron cycle in thermal reactor.

fraction of neutrons not captured during slowing; (d) the thermal nonleakage probability $\mathcal{L}_t$, the fraction of neutrons remaining in the core during diffusion at thermal energy; (e) the thermal utilization f, the fraction of thermal neutrons absorbed in fuel; and (f) the reproduction factor η, as the number of new fission neutrons per absorption in fuel. At the end of the cycle starting with one fission neutron, the number of fast neutrons produced is seen to be $\varepsilon\, pf\eta\, \mathcal{L}_f \mathcal{L}_t$, which may be also labeled k, the effective multiplication factor. It is convenient to group four of the factors to form $k_\infty = \varepsilon\, pf\eta$, the so-called infinite multiplication factor that would be identical to k if the medium were infinite in extent, without leakage. If we form a composite nonleakage probability $\mathcal{L} = \mathcal{L}_f \mathcal{L}_t$ then we may write

$$k = k_\infty \mathcal{L}.$$

For a reactor to be critical, k must equal 1, as before.

To provide some appreciation of the sizes of various factors, let us calculate the values of the composite quantities for a thermal reactor, for which $\varepsilon = 1.03$, $p = 0.71$, $\mathcal{L}_f = 0.97$, $\mathcal{L}_t = 0.99$, $f = 0.79$, and $\eta = 1.8$. Now $k_\infty = (1.03)\, (0.71)$ $(1.8)\, (0.79) = 1.04$, $\mathcal{L} = (0.97)\, (0.99) = 0.96$, and $k = (1.04)\, (0.96) = 1.00$. For this example, the various parameters yield a critical system. In Section 11.4 we will describe the physical construction of typical thermal reactors.

11.3 NEUTRON FLUX AND REACTOR POWER

The power developed by a reactor is a quantity of great interest for practical reasons. Power is related to the neutron population and to the mass of fissile material present. First, let us look at a typical cubic centimeter of the reactor, containing N fuel nuclei, each with cross section for fission σ_f at the typical neutron energy of the reactor, corresponding to neutron speed v. Suppose that there are n neutrons in the volume. The rate at which the fission reaction occurs is thus $R_f = nvN\sigma_f$ fissions per second. If each fission produces an energy w, then the power per unit volume is $p = w\, R_f$. For the whole reactor, of volume V, the rate of production of thermal energy is $P = pV$. If we have an average flux $\bar{\phi} = nv$ and a total number of fuel atoms $N_T = NV$, the total reactor power is seen to be

$$P = \bar{\phi} N_T \sigma_f\, w.$$

Thus we see that the power depends on the product of the number of neutrons and the number of fuel atoms. A high flux is required if the reactor contains a small amount of fuel, and conversely. All other things equal, a reactor with a high fission cross section can produce a required power with less fuel than one with small σ_f. We recall that σ_f decreases with increasing neutron energy. Thus for given power P, a fast reactor, operating with neutron energies principally in the vicinity of 1 MeV, requires either a much larger flux or a larger fissile fuel mass than does the thermal reactor, with neutrons of energy less than 0.1 eV.

The power developed by most familiar devices is closely related to fuel consumption. For example, a large car generally has a higher gasoline consumption rate than a small car, and more gasoline is used in operation at high speed than at low speed. In a reactor, it is necessary to add fuel very infrequently because of the very large energy yield per unit mass, and the fuel content remains essentially constant. From the formula relating power, flux, and fuel, we see that the power can be readily raised or lowered by changing the flux. By manipulation of control rods, the neutron population is allowed to increase or decrease to the proper level.

Power reactors used to generate electricity produce approximately 3000 megawatts of thermal power (MWt), and with an efficiency of approximately 1/3, give 1000 MW of electrical power (MWe).

11.4 **REACTOR TYPES**

Although the only requirement for a neutron chain reaction is a sufficient amount of a fissile element, many combinations of materials and arrangements can be used to construct an operable nuclear reactor. Several different types or concepts have been devised and tested over the period since 1942, when the first reactor started operation, just as various kinds of engines have been used—steam, internal combustion, reciprocating, rotary, jet, etc. Experience with individual reactor concepts has led to the selection of a few that are most suitable by use of criteria such as economy, reliability, and ability to meet performance demands.

In this Section we will identify these important reactor features, compare several concepts, and then focus attention on the components of one specific power reactor type. We will then examine the processes of fuel consumption and control in a power reactor.

A general classification scheme for reactors has evolved that is related to the distinguishing features of the reactor types. These features are listed in the following sections

(a) *Purpose*

Most reactors in operation or under construction have as purpose the generation of large blocks of commercial electric power. Others serve training or radiation research needs, and many provide propulsion power for submarines. At various stages of development of a new concept, such as the breeder reactor, there will be constructed both a prototype reactor, one that tests feasibility, and a demonstration reactor, one that evaluates commercial possibilities.

(b) *Neutron Energy*

A fast reactor is one in which most of the neutrons are in the energy range 0.1 to 1 MeV, below but near the energy of neutrons released in fission. The neutrons remain at high energy because there is relatively little material present to cause them to slow down. In contrast, the thermal reactor contains a good neutron-moderating material, and the bulk of the neutrons have energy less than 0.1 eV.

(c) *Moderator and Coolant*

In some reactors, one substance serves two functions—to assist in neutron slowing and to remove the fission heat. Others involve one material for moderator and another for coolant. The most frequently used materials are listed in the following:

Moderators	Coolants
Light water	Light water
Heavy water	Carbon dioxide
Graphite	Helium
Beryllium	Liquid sodium

The condition of the coolant serves as a further identification. The *pressurized water reactor* provides high-temperature water to a heat exchanger that generates steam, whereas the *boiling water reactor* supplies steam directly.

(d) *Fuel*

Uranium with U-235 content varying from natural uranium ($\cong 0.7\%$) to slightly enriched ($\cong 3\%$ to 5%) to highly enriched ($\cong 90\%$) is used in various reactors, with the enrichment depending on what other absorbing materials are present. The fissile isotope $^{239}_{94}Pu$ is produced and consumed in reactors containing significant amounts of U-238. Plutonium serves as fuel for fast breeder reactors and can be recycled as fuel for thermal reactors. A few reactors have been built with fertile Th-232, producing fissile U-233.

The fuel may have various physical forms—a metal, or an alloy with a metal such as aluminum, or a compound such as the oxide UO_2 or carbide UC.

(e) *Arrangement*

In most modern reactors, the fuel is isolated from the coolant in what is called a *heterogeneous* arrangement. The alternative is a homogeneous mixture of fuel and moderator or fuel and moderator-coolant.

(f) *Structural Materials*

The functions of support, retention of fission products, and heat conduction are provided by various metals. The main examples are aluminum, stainless steel, and zircaloy, an alloy of zirconium.

By placing emphasis on one or more of the preceding features of reactors, reactor concepts are identified. Some of the more widely used or promising power reactor types are the following:

Pressurized water reactor (PWR), a thermal reactor with light water at high pressure (2200 psi) and temperature (600°F) serving as moderator-coolant, and a heterogeneous arrangement of slightly enriched uranium fuel.

Boiling water reactor (BWR), similar to the PWR except that the pressure and temperature are lower (1000 psi and 550°F).

High temperature gas-cooled reactor (HTGR) that uses graphite moderator, highly enriched uranium, and helium coolant (1430°F and 600 psi).

Canadian deuterium uranium (CANDU) that uses heavy water moderator and natural uranium fuel that can be loaded and discharged during operation.

Liquid metal fast breeder reactor (LMFBR), with no moderator, liquid sodium coolant, and plutonium fuel, surrounded by natural or depleted uranium.

Table 11.1 amplifies on the principal features of the preceding five main power reactor concepts. A description of the RBMK, exemplified by the ill-fated Chernobyl-4 reactor, appears in Section 19.6. Other reactors include the Magnox and AGR of the United Kingdom and several concepts that were tested but abandoned (see encyclopedia article in References).

Table 11.1 Power Reactor Materials

	Pressurized Water (PWR)	Boiling Water (BWR)	Natural U Heavy Water (CANDU)	High Temp. Gas-cooled (HTGR)	Liquid Metal Fast Breeder (LMFBR)
Fuel form	UO_2	UO_2	UO_2	UC, ThC_2	PuO_2, UO_2
Enrichment	3% U-235	2.5 % U-235	0.7 % U-235	93 % U-235	15 wt. % Pu-239
Moderator	water	water	heavy water	graphite	none
Coolant	water	water	heavy water	helium gas	liquid sodium
Cladding	zircaloy	zircaloy	zircaloy	graphite	stainless steel
Control	B_4C or Ag-In-Cd rods	B_4C crosses	moderator level	B_4C rods	tantalum or B_4C rods
Vessel	steel	steel	steel	prestressed concrete	steel

The large-scale reactors used for the production of thermal energy that is con-verted to electrical energy are much more complex than the fast assembly described in Section 11.1. To illustrate, we can identify the components and their functions in a modern PWR. Figure 11.5 gives some indication of the sizes of the various parts. To gain some appreciation of the physical arrangement of fuel in power reactors, try out the graphics programs in Computer Exercises 11.E (ASSEMBLY) and 11.F (BWRASEM).

The fresh fuel installed in a typical PWR consists of cylindrical pellets of slightly enriched (3% U-235) uranium oxide (UO_2) of diameter approximately 3/8 in (~1 cm) and length approximately 0.6 in (~1.5 cm). A zircaloy tube of wall thickness 0.025 in (~0.6 mm) is filled with the pellets to an "active length" of

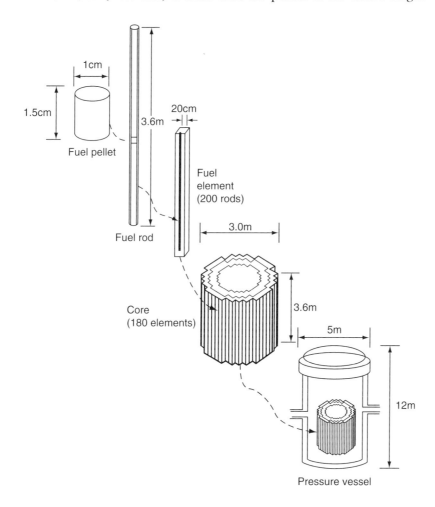

FIGURE 11.5 Reactor construction.

12 ft (365 cm) and sealed to form a *fuel rod* (or pin). The metal tube serves to provide support for the column of pellets, to provide cladding that retains radioactive fission products, and to protect the fuel from interaction with the coolant. Approximately 200 of the fuel pins are grouped in a bundle called a *fuel element* of approximately 8 in (~20 cm) on a side, and approximately 180 elements are assembled in an approximately cylindrical array to form the *reactor core*. This structure is mounted on supports in a steel *pressure vessel* of outside diameter approximately 16 ft (~5 m), height 40 ft (~12 m), and walls up to 12 in (~30 cm) thick. *Control rods*, consisting of boron carbide or an alloy of cadmium, silver, and indium, provide the ability to change the amount of neutron absorption. The rods are inserted in some vacant fuel pin spaces and magnetically connected to drive mechanisms. On interruption of magnet current, the rods enter the core through the force of gravity. The pressure vessel is filled with light water, which serves as neutron moderator, as coolant to remove fission heat, and as *reflector*, the layer of material surrounding the core that helps prevent neutron escape. The water also contains in solution the compound boric acid, H_3BO_3, which strongly absorbs neutrons in proportion to the number of boron atoms and thus inhibits neutron multiplication (i.e., "poisons" the reactor). The term "soluble poison" is often used to identify this material, the concentration of which can be adjusted during reactor operation. To keep the reactor critical as fuel is consumed, the boron content is gradually reduced. A shield of concrete surrounds the pressure vessel and other equipment to provide protection against neutrons and gamma rays from the nuclear reactions. The shield also serves as an additional barrier to the release of radioactive materials.

We have mentioned only the main components, which distinguish a nuclear reactor from other heat sources such as one burning coal. An actual system is much more complex than described earlier. It contains equipment such as spacers to hold the many fuel rods apart; core support structures; baffles to direct coolant flow effectively; guides, seals, and motors for the control rods; guide tubes and electrical leads for neutron-detecting instruments, brought through the bottom of the pressure vessel and up into certain fuel assemblies; and bolts to hold down the vessel head and maintain the high operating pressure.

The power reactor is designed to withstand the effects of high temperature, erosion by moving coolant, and nuclear radiation. The materials of construction are chosen for their favorable properties. Fabrication, testing, and operation are governed by strict procedures.

11.5 REACTOR OPERATION

The generation of energy from nuclear fuels is unique in that a rather large amount of fuel must be present at all times for the chain reaction to continue. In contrast, an automobile will operate even though its gasoline tank is practically empty.

There is a subtle relationship between reactor fuel and other quantities such as consumption, power, neutron flux, criticality, and control.

The first and most important consideration is the energy production, which is directly related to fuel consumption. Let us simplify the situation by assuming that the only fuel consumed is U-235 and that the reactor operates continuously and steadily at a definite power level. Because each atom "burned" (i.e., converted into either U-236 or fission products by neutron absorption) has an accompanying energy release, we can find the amount of fuel that must be consumed in a given period.

Let us examine the fuel use in a simplified PWR that uses 20 w/o fuel and operates at 100 MWe or 300 MWt, as in a test reactor or a propulsion reactor. The initial fuel loading into a single zone is 1000 kg U. We apply the rule of thumb that 1.3 grams of U-235 is consumed for each megawatt-day of thermal energy, assuming that all fissions are due to U-235. In 1 year, the amount of U-235 consumed is (300 MWt) (1.3 g/MWt-day) (365 days) $= 1.42 \times 10^5$ g or 142 kg. We see that a great deal of the original 200 kg of U-235 has been consumed, with a final enrichment of 6.8 w/o. Let us assume that a completely new core is installed at the end of a year's operation. If we carry out the calculations as in Section 9.2, the fuel cost excluding fabrication and transport is $4.59 million. Most of that is for enrichment. The electricity produced is

$$(10^5 \text{ kW})(8760 \text{ h/y}) = 8.76 \times 10^8 \text{ kWh,}$$

making the unit cost of fuel (4.59×10^6)/(8.76×10^8) = $0.0052 or approximately half a cent per kWh. In Chapter 19 we will analyze fuel cycles in a large power reactor that has several zones with different enrichments and shuts down periodically to remove, rearrange, and install fuel.

We continue studying the operating features of our small PWR. Because no fuel is added during the operating cycle of the order of a year, the amount to be burned must be installed at the beginning. First, the amount of uranium needed to achieve criticality is loaded into the reactor. If then the "excess" is added, it is clear that the reactor would be supercritical unless some compensating action were taken. In the PWR, the excess fuel reactivity is reduced by the inclusion of control rods and boron solution.

The reactor is brought to full power, operating temperature, and pressure by means of rod position adjustments. Then, as the reactor operates and fuel begins to burn out, the concentration of boron is reduced. By the end of the cycle, the extra fuel is gone, all of the available control absorption has been removed, and the reactor is shut down for refueling. The trends in fuel and boron are shown in Figure 11.6, neglecting the effects of certain fission product absorption and plutonium production. The graph represents a case in which the power is kept constant. The fuel content thus linearly decreases with time. Such operation characterizes a reactor that provides a "base load" in an electrical generating system that also includes fossil fuel plants and hydroelectric stations.

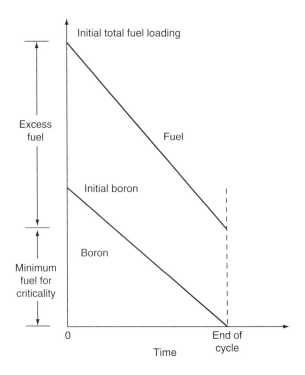

FIGURE 11.6 Reactor control during fuel consumption in power reactor.

The power level in a reactor was shown in Section 11.3 to be proportional to neutron flux. However, in a reactor that experiences fuel consumption the flux must increase in time, because the power is proportional also to the fuel content.

The amount of control absorber required at the beginning of the cycle is proportional to the amount of excess fuel added to permit burn up for power production. For example, if the fuel is expected to go from 3% to 1.5% U-235, an initial boron atom number density in the moderator is approximately 1.0×10^{-4} (in units of 10^{24}). For comparison, the number of water molecules per cubic centimeter is 0.0334. The boron content is usually expressed in parts per million (i.e., micrograms of an additive per gram of diluent). For our example, with 10.8 and 18.0 as the molecular weights of boron and water, there are $10^{6}(10^{-4})$ $(10.8)/(0.0334)(18.0) = 1800$ ppm.

The description of the reactor process just completed is somewhat idealized. Several other phenomena must be accounted for in design and operation.

At the periodic shutdowns for fuel replacement and rearrangement, a great deal of maintenance work is done. There is an economic premium for careful and thorough outage management to minimize the time the reactor is not producing power. Down times as low as 3 weeks have been achieved. Replacement of a steam generator requires several months. The capacity factor (CF) is an important parameter in the performance of a power reactor and the nuclear industry as a whole. It is the percent of output compared with the maximum. Included in its calculation are the outage for refueling and any other shutdown time. The median CF for 3-year periods in the United States has risen from 59% in 1974–1976 to more than 90% in 2004–2006. For data on individual reactors and analysis of trends, see References.

If a reactor is fueled with natural uranium or slightly enriched uranium, the generation of plutonium tends to extend the cycle time. The fissile Pu-239 helps maintain criticality and provides part of the power. Small amounts of higher plutonium isotopes are also formed: Pu-240, fissile Pu-241 (14.4 year half-life), and Pu-242. These isotopes and those of elements farther up the periodic table are called transuranic materials or actinides. They are important as fuels, poisons, or nuclear wastes. (See Exercise 11.14).

Neutron absorption in the fission products has an effect on control requirements. The most important of these is a radioactive isotope of xenon, Xe-135, which has a cross section at 0.0253 eV of 2.65 *million* barns. Its yield in fission is high, $y=0.06$, meaning that for each fission, one obtains 6% as many atoms of Xe-135. In steady operation at high neutron flux, its rate of production is equal to its consumption by neutron absorption. Hence

$$N_X \sigma_{aX} = N_F \sigma_{fF} y.$$

With the ratio σ_f / σ_a for U-235 of 0.86, we see that the absorption rate of Xe-135 is $(0.86)(0.06) = 0.05$ times that of the fuel itself. This factor is approximately 0.04 if the radioactive decay ($t_H = 9.10$ h) of xenon-135 is included (see Exercise 11.8). The time-dependent variation of neutron absorption in xenon-135 is the subject of Computer Exercise 11.C, which describes the program XETR.

It might appear from Figure 11.6 that the reactor cycle could be increased to as long a time as desired merely by adding more U-235 at the beginning. There are limits to such additions, however. First, the more the excess fuel that is added, the greater must be the control by rods or soluble poison. Second, radiation and thermal effects on fuel and cladding materials increase with life. The amount of allowable total energy extracted from the uranium, including all fissionable isotopes, is expressed as the number of megawatt-days per metric ton (MWd/ton).[†] We can calculate its value for 1 year's operation of a 3000 MWt power reactor with initial U-235 fuel loading of 2800 kg; with an enrichment of 0.03, the *uranium* content was 2800/0.03 = 93,000 kg or 93 tons. With the energy yield of (3000 MW)(365 days) ≅1,100,000 MWd, we find 12,000 MWd/ton. Taking account of plutonium and the management of fuel in the core, a typical average exposure is actually

---

[†]The metric ton (ton) is 1000 kg.

30,000 MWd/ton. It is desirable to seek larger values of this quantity to prolong the cycle and thus minimize the costs of fuel, reprocessing, and fabrication.

11.6 THE NATURAL REACTOR

Until the 1970s, it had been assumed that the first nuclear reactor was put into operation by Enrico Fermi and his associates in 1942. It seems, however, that a natural chain reaction involving neutrons and uranium took place in the African state of Gabon, near Oklo, some 2 billion years ago (see References). At that time, the isotope concentration of U-235 in natural uranium was higher than it is now because of the differences in half-lives: U-235, 7.04×10^8 years; U-238, 4.47×10^9 years. The water content in a rich vein of ore was sufficient to moderate neutrons to thermal energy. It is believed that this "reactor" operated off and on for thousands of years at power levels of the order of kilowatts. The discovery of the Oklo phenomenon resulted from the observations of an unusually low U-235 abundance in the mined uranium. The effect was confirmed by the presence of fission products.

11.7 SUMMARY

A self-sustaining chain reaction involving neutrons and fission is possible if a critical mass of fuel is accumulated. The value of the multiplication factor k indicates whether a reactor is subcritical (<1), critical ($=1$), or supercritical (>1). The reactor power, which is proportional to the product of flux and the number of fuel atoms, is readily adjustable. A thermal reactor contains a moderator and operates on slowed neutrons. Reactors are classified according to purpose, neutron energy, moderator and coolant, fuel, arrangement, and structural material. Principal types are the PWR, the BWR, the HTGR, and LMFBR. Excess fuel is added to a reactor initially to take care of burning during the operating cycle, with adjustable control absorbers maintaining criticality. Account must be taken of fission product absorbers such as Xe-135 and of limitations related to thermal and radiation effects. Approximately 2 billion years ago, deposits of uranium in Africa had a high enough concentration of U-235 to become natural chain reactors.

11.8 EXERCISES

11.1 Calculate the reproduction factor η for fast neutrons, with $\sigma_f = 1.40$, $\sigma_a = 1.65$, and $v = 2.60$ (ANL-5800, p.581).

11.2 If the power developed by the Godiva-type reactor of mass 50 kg is 100 watts, what is the average flux? Note that the energy of fission is $w = 3.04 \times 10^{-11}$ W–s.

11.3   Find the multiplication factors k_∞ and k for a thermal reactor with $\varepsilon = 1.05$ $p = 0.75$, $\mathcal{L}_f = 0.90$, $\mathcal{L}_f = 0.98$, $f = 0.85$, and $\eta = 1.75$. Evaluate the reactivity ρ.

11.4   The value of the reproduction factor η in uranium containing both U-235 (1) and U-238 (2), is given by

$$\eta = \frac{N_1 \sigma_{f1} v_1 + N_2 \sigma_{f2} v_2}{N_1 \sigma_{a1} + N_2 \sigma_{a2}}.$$

Calculate η for three reactors (a) thermal, with 3% U-235, $N_1/N_2 = 0.0315$; (b) fast, with the same fuel; (c) fast, with pure U-235. Comment on the results. Note values of constants:

	Thermal	Fast
σ_{f1}	586	1.40
σ_{a1}	681	1.65
σ_{f2}	0	0.095
σ_{a2}	2.70	0.255
v_1	2.42	2.60
v_2	0	2.60

11.5   By means of the formula and thermal neutron numbers from Exercise 11.4, find η, the number of neutrons per absorption in fuel, for uranium oxide in which the U-235 atom fraction is 0.2, regarded as a practical lower limit for nuclear weapons material. Would the fuel be suitable for a research reactor?

11.6   How many individual fuel pellets are there in the PWR reactor described in the text? Assuming a density of uranium oxide of 10 g/cm^3, estimate the total mass of uranium and U-235 in the core in kilograms. What is the initial fuel cost?

11.7   The core of a PWR contains 180 fuel assemblies of length 4 m, width 0.2 m. (a) Find the core volume and radius of equivalent cylinder. (b) If there are 200 fuel rods per assembly with pellets of diameter 0.9 cm, what is the approximate UO$_2$ volume fraction of the core?

11.8   (a) Taking account of Xe-135 production, absorption, and decay, show that the balance equation is

$$N_X(\phi \sigma_{aX} + \lambda_X) = \phi N_F \sigma_{fF} y.$$

(b)   Calculate λ_X and the ratio of absorption rates in Xe-135 and fuel if ϕ is 2×10^{13} cm^{-2}–s^{-1}.

11.9   The initial concentration of boron in a 10,000-ft^3 reactor coolant system is 1500 ppm, (the number of micrograms of additive per gram of diluent). What volume of solution of concentration 8000 ppm should be added to achieve a new value of 1600 ppm?

11.10 An adjustment of boron content from 1500 to 1400 ppm is made in the reactor described in Exercise 11.9. Pure water is pumped in and then mixed coolant and poison are pumped out in two separate steps. For how long should the 500 ft^3/min pump operate in each of the operations?

11.11 Find the ratio of weight percentages of U-235 and U-238 at a time 1.9 billion years ago, assuming the present 0.711/99.3.

11.12 Constants for a spherical fast uranium-235 metal assembly are: diffusion coefficient $D = 1.02$ cm; macroscopic absorption cross section $\Sigma_a = 0.0795$ cm^{-1}; effective radius $R = 10$ cm. Calculate the diffusion length L, the buckling B^2, and the nonleakage factor $\mathcal{L}$.

11.13 The neutron flux in a reactor varies with position. In a simple core such as a bare uranium metal sphere of radius R, it varies as $\phi = \phi_c$ (sin x)/x, where $x = \pi r/R$. At the center of the sphere the flux is ϕ_c. Calculate and plot the flux distribution for a core with radius 10 cm and central flux 5×10^{11}/cm^2–s.

11.14 A reactor is loaded with 90,000 kg of U at 3 w/o U-235. It operates for a year at 75% of its rated 3000 MWt capacity. (a) Apply the rule of thumb 1.3 g/MWt-day to find the consumption of U-235. What is the final enrichment of the fuel? (b) If instead one third of the energy came from plutonium, what would the final U-235 enrichment be? Note thermal cross sections, all in barns: U-235 $\sigma_f = 586$, $\sigma_c = 95$; Pu-239 $\sigma_f = 752$, $\sigma = 270$.

Computer Exercises

11.A The evaluation of critical conditions for a variety of spherical metal assemblies can be made with the program CRITICAL. It uses a one-neutron group model with cross sections deduced from early critical experiments related to weapons. CRITICAL can handle any combination of uranium and plutonium. Run the program, choosing U enrichment and Pu content. Suggested configurations:
(a) Pure U-235.
(b) Godiva (93.9% U-235, experimental U-235 mass 48.8 kg).
(c) Jezebel (pure plutonium, experimental mass 16.28 kg).
(d) Natural U (0.0072 atom fraction U-235, should not be possible to be made critical).
(e) Depleted U (0.003 atom fraction U-235).
(f) Elementary breeder reactor (Pu-239 volume 10%, depleted U).

11.B A miniature version of a classic computer code PDQ is called MPDQ. It finds the amount of critical control absorber in a core of the form of an unreflected slab, by solution of difference equations.
(a) Load the program and study the listing.
(b) Run the program and study the displays, then compare the results of choosing a linear or sine trial fast flux function.
(c) With the constants given in Exercise 11.1, modify the program to calculate the critical control for a metal assembly as a slab of width 5 cm.

11.C The amount of xenon in a reactor varies with time, especially when large changes in neutron flux occur, as at startup or shutdown. Program XETR (Xenon Transient) solves differential equations for the content of iodine-135 and xenon-135.
(a) Load the program and examine the input constants and conditions. Study the trend in the output as the reactivity ρ (see Section 11.2) vs. time.
(b) Use the concentrations of I-135 and Xe-135 calculated for long times after start up as initial concentrations, and set the flux equal to zero to simulate a sudden shutdown of the reactor. Note and discuss the trend in xenon with time.

11.D Competition among three neutron processes—scattering, absorption, and leakage—is illustrated by the program SLOWINGS. It simulates the release of a series of neutrons at the center of a carbon sphere, and by use of slowing theory and random numbers, finds the number of neutrons absorbed and escaping.
(a) Run the program several times to note statistical variations.
(b) Increase the absorption cross section by a factor of 200 as if considerable boron were added to the sphere, and note the effect.

11.E The program ASSEMBLY displays a pressurized water reactor fuel assembly, an array of 14×14 fuel rods. Run the program.

11.F The program BWRASEM displays four boiling water reactor fuel assemblies with a cross-shaped control rod between them. Run the program to inspect.

11.9 REFERENCES

Anthony V. Nero, Jr., *A Guidebook to Nuclear Reactors*, University of California Press, Berkeley, 1979.

Ronald Allen Knief, *Nuclear Engineering: Theory and Technology of Commercial Nuclear Power*, Taylor & Francis, Bristol, PA, 1992.

Samuel Glasstone and Alexander Sesonske, *Nuclear Reactor Engineering*, 4th Ed., Vol. 1, Reactor Design Basics, Vol. 2, Reactor Systems Engineering, Chapman and Hall, New York, 1994.

R. E. Peterson and G. A. Newby, "An Unreflected U-235 Critical Assembly," *Nuclear Science & Engineering* 1, 1956, p. 112. Description of Godiva.

Raymond L. Murray, "Nuclear Reactors," *Kirk-Othmer Encyclopedia of Chemical Technology*, John Wiley & Sons, New York, 2007.

WWW Virtual Library: Nuclear Engineering
http://www.nuc.berkeley.edu/main/vir_library.html
Links to nearly everything.

Uranium Information Centre (Australia)
http://www.uic.com.au
Briefing papers on a variety of nuclear topics, and many links.

International Nuclear Safety Center
http://www.insc.anl.gov
Database on Reactor Material Properties. By Argonne National Laboratory.

Nuclear Power Reactors
http://www.world-nuclear.org/
Select Information Papers/Nuclear Power Facts/Nuclear Power Reactors.

From World Nuclear Association.

Issue on Outage Management, Nuclear News, April 2007.

E. Michael Blake, "U.S. Capacity Factors: A Small Gain to an Already Large Number," *Nuclear News*, May 2007, p. 27.

Thorium and breeding of U-233
http://www.world-nuclear.org/
Search on Thorium.

The Virtual Nuclear Tourist
http://www.nucleartourist.com/
Comprehensive coverage of nuclear power. Explore links in Table of Contents. By Joseph Gonyeau.

Robert G. Cochran, Nicholas Tsoulfanidis, and W. F. Miller, *Nuclear Fuel Cycle: Analysis and Management*, 2nd Ed., American Nuclear Society La Grange Park, IL, 1993.

George A. Cowan, "A Natural Fission Reactor," *Scientific American*, July 1976, p. 36.

The Natural Reactor
http://www.physics.isu.edu/radinf/Files/Okloreactor.pdf
The ancient reactor in Oklo, Gabon. By Andrew Karam. Enter Google with Natural Reactor for DOE/OCRWM site.

Nuclear Heat Energy

12

MOST OF the energy released in fission appears as kinetic energy of a few high-speed particles. As these pass through matter, they slow down by multiple collisions and impart thermal energy to the medium. It is the purpose of this chapter to discuss the means by which this energy is transferred to a cooling agent and transported to devices that convert mechanical energy into electrical energy. Methods for dealing with the large amounts of waste heat generated will be considered.

12.1 METHODS OF HEAT TRANSMISSION

We learned in basic science that heat, as one form of energy, is transmitted by three methods—conduction, convection, and radiation. The physical processes for the methods are different. In *conduction*, molecular motion in a substance at a point at which the temperature is high causes motion of adjacent molecules, and a flow of energy toward a region of low temperature takes place. The rate of flow is proportional to the slope of the temperature (i.e., the temperature gradient). In *convection*, molecules of a cooling agent such as air or water strike a heated surface, gain energy, and return to raise the temperature of the coolant. The rate of heat removal is proportional to the difference between the surface temperature and that of the surrounding medium and also depends on the amount of circulation of the coolant in the vicinity of the surface. In *radiation*, molecules of a heated object emit and receive electromagnetic radiations, with a net transfer of energy that depends on the temperatures of the body and the adjacent regions, specifically on the difference between the temperatures raised to the fourth power. For reactors, this mode of heat transfer is generally of less importance than are the other two.

12.2 HEAT GENERATION AND REMOVAL

The transfer of heat by *conduction* in a flat plate (insulated on its edges) is reviewed. If the plate has a thickness x and cross-sectional area A, and the

161

temperature difference between its faces is $\triangle T$, the rate of heat flow through the plate, Q, is given by the relation

$$Q = kA \frac{\triangle T}{x},$$

where k is the thermal conductivity, with typical units J/s-°C-cm. For the plate, the slope of the temperature is the same everywhere. In a more general case, the slope may vary with position, and the rate of heat flow per unit area Q/A is proportional to the slope or gradient written as $\triangle T/\triangle x$.

The conductivity k varies somewhat with temperature, but we treat k as constant for the following analysis of conduction in a single fuel rod of a reactor (see Section 11.4). Let the rate of supply of thermal energy by fission be uniform throughout the rod. If the rod is long in comparison with its radius R, or if it is composed of a stack of pellets, most of the heat flow is in the radial direction. If the surface is maintained at a temperature T_s by the flow of coolant, the center of the rod must be at some higher temperature T_0. As expected, the temperature difference is large if the rate of heat generation per unit volume q or the rate of heat generation per unit length $q_1 = \pi R^2 q$ is large. We can show[†] that

$$T_0 - T_s = \frac{q_1}{4\pi k},$$

and that the temperature T is in the shape of a parabola within the rod. Figure 12.1 shows the temperature distribution.

Let us calculate the temperature difference $T_0 - T_s$ for a reactor fuel rod of radius 0.5 cm, at a point where the power density is $q = 200$ W/cm³. This corresponds to a linear heat rate $q_1 = \pi R^2 q = \pi(0.25)(200) = 157$ W/cm (or 4.8 kW/ft). Letting the thermal conductivity of UO_2 be $k = 0.062$ W/cm-°C, we find $T_0 - T_s = 200°C$ (or 360°F). If we wish to keep the temperature low along the centerline of the fuel, to avoid structural changes or melting, the conductivity k should be high, the rod size small, or the reactor power level low. In a typical reactor there is a small gap between the fuel pin and the inside surface of the cladding. During operation, this gap contains gases, which are poor heat conductors and thus there will be a rather large temperature drop across the gap. A smaller drop will occur across the cladding that is thin and has a high thermal conductivity.

We have so far assumed that the thermal conductivity is constant. It actually varies with temperature and thus with position in the fuel pin. A more general calculation of k is possible with the program CONDUCT discussed in Computer Exercise 12.A, and the temperature distribution may be found with a program TEMPLOT in Computer Exercise 12.B.

[†]The amount of energy supplied within a region of radius r must flow out across the boundary. For a unit length of rod with volume πr^2 and surface area $2\pi r$, the generation rate is $\pi r^2 q$, equal to the flow rate $- k(dT/dr)2\pi r$. Integrating from $r = 0$, where $T = T_0$, we have $T = T_0 - qr^2/4k$. At the surface $T_s = T_0 - qR^2/4k$.

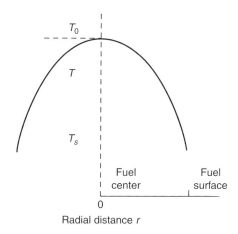

FIGURE 12.1 Temperature in fuel.

Convective cooling depends on many factors such as the fluid speed, the size and shape of the flow passage, and the thermal properties of the coolant, as well as the area exposed and the temperature difference between surface and coolant $T_s - T_0$. Experimental measurements yield the *heat transfer coefficient* h, appearing in a working formula for the rate of heat transfer Q across the surface S,

$$Q = hS(T_s - T_0).$$

The units of h are typically W/cm^2 – °C. To keep the surface temperature low to avoid melting of the metal cladding of the fuel or to avoid boiling if the coolant is a liquid, a large surface area is needed or the heat transfer coefficient must be large, a low-viscosity coolant of good thermal conductivity is required, and the flow speed must be high.

As coolant flows along the many channels surrounding fuel pins in a reactor, it absorbs thermal energy and rises in temperature. Because it is the reactor power that is being extracted, we may apply the principle of conservation of energy. If the coolant of specific heat c enters the reactor at temperature T_c (in) and leaves at T_c (out), with a mass flow rate M, then the reactor thermal power P is

$$P = cM[T_c(\text{out}) - T_c(\text{in})] = cM\Delta T.$$

For example, let us find the amount of circulating water flow to cool a reactor that produces 3000 MW of thermal power. Let the water enter at 300°C (572°F) and leave at 325°C (617°F). Assume that the water is at 2000 psi and 600°F. At these conditions the specific heat is 6.06×10^3 J/kg – °C and the specific gravity is 0.687. Thus the mass flow rate is

$$M = P/(c\Delta T) = (3000 \times 10^6)/[(6.06 \times 10^3)(25)] = 19,800 \,\text{kg/s}.$$

This corresponds to a volume flow rate of

$$V = (19,800\,\text{kg/s})/(687\,\text{kg/m}^3) = 28.8\,\text{m}^3/\text{s},$$

that is also 1,730,000 liters per minute. To appreciate the magnitude of this flow, we can compare it with that from a garden hose of 40 liters/min. The water for cooling a reactor is not wasted, of course, because it is circulated in a closed loop.

The temperature of coolant as it moves along any channel of the reactor can also be found by application of the preceding relation. In general, the power produced per unit length of fuel rod varies with position in the reactor because of the variation in neutron flux shape. For a special case of a *uniform* power along the z-axis with origin at the bottom as in Figure 12.2(A), the power per unit length is $P_1 = P/H$, where H is the length of fuel rod. The temperature rise of coolant at z with channel mass flow rate M is then

$$T_c(z) = T_c(\text{in}) + \frac{P_1 z}{cM},$$

that shows that the temperature increases with distance along the channel as shown in Figure 12.2(B). The temperature difference between coolant and fuel surface is the same at all points along the channel for this power distribution, and the temperature difference between the fuel center and fuel surface is also uniform. We can plot these as in Figure 12.2(C). The highest temperatures in this case are at the end of the reactor.

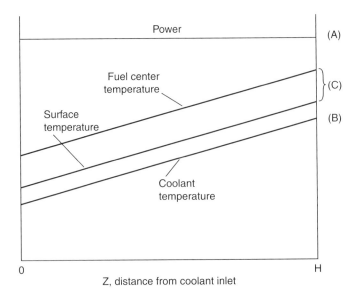

FIGURE 12.2 Temperature distributions along axis of reactor with uniform power.

If instead, the axial power were shaped as a sine function, see Figure 12.3(A) with $P \sim \sin(\pi z/H)$, the application of the relations for conduction and convection yields temperature curves as sketched in Figure 12.3(B). For this case, the highest temperatures of fuel surface and fuel center occur between the halfway point and the coolant exit. In the design of a reactor, a great deal of attention is given to the determination of which channels have the highest coolant temperature and at which points on the fuel pins "hot spots" occur. Ultimately, the power of the reactor is limited by conditions at these channels and points. The mechanism of heat transfer from metal surfaces to water is quite sensitive to the temperature difference. As the latter increases, ordinary convection gives way to *nucleate boiling*, in which bubbles form at points on the surface, and eventually *film boiling* can occur, in which a blanket of vapor reduces heat transfer and permits hazardous melting. The value of the heat flux Q at which film boiling occurs is called the critical heat flux. A parameter called "departure from nucleate boiling ratio" (DNBR) is used to indicate how close the heat flux is to the critical value. For example, a DNBR of 1.3 implies a safety margin of 30%. Figure 12.4 indicates maximum temperature values for a typical PWR reactor.

To achieve a water temperature of 600°F (approximately 315°C) requires that a very high pressure be applied to the water coolant-moderator. Figure 12.5 shows the behavior of water in the liquid and vapor phases. The curve that

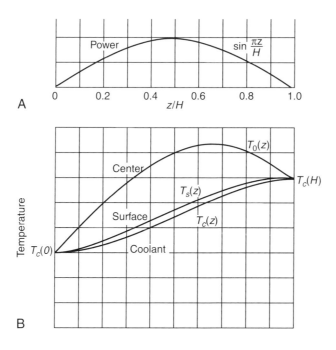

FIGURE 12.3 Temperature distributions along channel with sine function power.

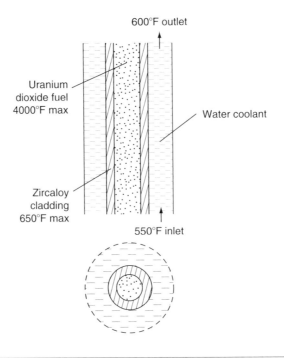

FIGURE 12.4 Reactor channel heat removal.

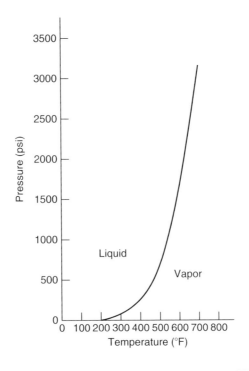

FIGURE 12.5 Relationship of pressure and temperature for water.

separates the two phase regions describes what are called saturated conditions. Suppose that the pressure vessel of the reactor contains water at 2000 psi and 600°F and the temperature is raised to 650°F. The result will be considerable steam formation (flashing) within the liquid. The two-phase condition could lead to inadequate cooling of the reactor fuel. If instead the pressure were allowed to drop, say to 1200 psi, the vapor region is again entered and flashing would occur. However, it should be noted that deliberate two-phase flow conditions are used in BWRs, providing efficient and safe cooling.

12.3 STEAM GENERATION AND ELECTRICAL POWER PRODUCTION

Thermal energy in the circulating reactor coolant is transferred to a working fluid such as steam by means of a heat exchanger or steam generator. In simplest construction, this device consists of a vessel partly filled with water, through which many tubes containing heated water from the reactor pass, as in Figure 12.6. At a number of nuclear plants the steam generator has failed prematurely because of corrosion that created holes in tubes, requiring plugging or repair. In some cases replacement of the steam generator was required, with corresponding outage, cost, and loss of revenue. Details on the problem appear in an Nuclear Regulatory Commission Technical Issue Paper (see References). Steam from the generator flows to a turbine, while the water returns to the reactor. The conversion of thermal energy of steam into mechanical energy of rotation of a turbine and then to electrical energy from a generator is achieved by

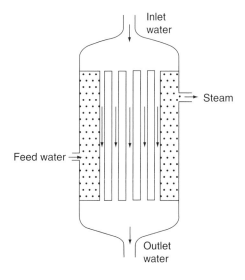

FIGURE 12.6 Heat exchanger or steam generator.

conventional means. Steam at high pressure and temperature strikes the blades of the turbine, which drives the generator. The exhaust steam is passed through a heat exchanger that serves as condenser, and the condensate is returned to the steam generator as feed water. Cooling water for the condenser is pumped from a nearby body of water or cooling tower as discussed in Section 12.4.

Figures 12.7 and 12.8 show the flow diagrams for the reactor systems of the PWR and BWR type. In the PWR, a pressurizer maintains the pressure in the system at the desired value. It uses a combination of immersion electric heaters and a water spray system to control the pressure. Figure 12.9 shows the Diablo Canyon nuclear power plant operated by Pacific Gas & Electric Company in San Luis Obispo County, California. The two Westinghouse PWR reactors were put into operation in 1985 and 1986.

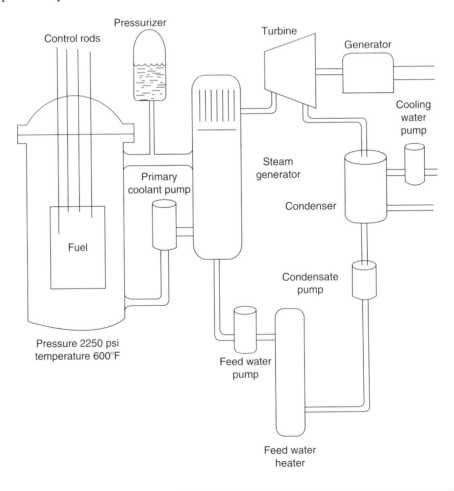

FIGURE 12.7 PWR system flow diagram.

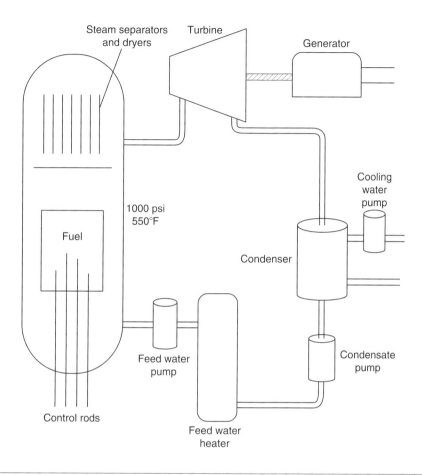

FIGURE 12.8 BWR system flow diagram.

12.4 **WASTE HEAT REJECTION**

The generation of electric power by consumption of any fuel is accompanied by the release of large amounts of waste heat. For any conversion process the thermal efficiency, e, the ratio of work done to thermal energy supplied, is limited by the temperatures at which the system operates. According to the second law of thermodynamics, an ideal cycle has the highest efficiency value,

$$e = 1 - T_1/T_2,$$

where T_1 and T_2 are the lowest and highest absolute temperatures (Kelvin, °C + 273; Rankine, °F + 460). For example, if the steam generator produces

FIGURE 12.9 A nuclear power plant, Diablo Canyon, on the California coast (Courtesy Pacific Gas & Electric Company).

steam at 300°C and cooling water for the condenser comes from a source at 20°C, we find the maximum efficiency of

$$e = 1 - 293/573 = 0.49.$$

The overall efficiency of the plant is lower than this because of heat loss in piping, pumps, and other equipment. The efficiency of a typical nuclear power plant is approximately 0.33. Thus twice as much energy is wasted as is converted into useful electrical energy. Fossil fuel plants can operate at higher steam temperatures, giving overall efficiencies of approximately 0.40.

A nuclear plant operating at electrical power 1000 MWe would have a thermal power of $1000/0.33 = 3030$ MWt and must reject a waste power of $P = 2030$ MWt. We can calculate the condenser cooling water mass flow rate M required to limit the temperature rise to a typical figure of $\Delta T = 12°C$ with a specific heat of $c = 4.18 \times 10^3$ J/kg-°C,

$$M = \frac{P}{c\Delta T} = \frac{2.03 \times 10^9}{(4.18 \times 10^3)(12)} = 4.05 \times 10^7 \text{kg/s}.$$

This corresponds to a flow of 925 million gallons per day. Smaller power plants in past years were able to use the "run of the river" (i.e., to take water from a stream, pass it through the condenser, and discharge heated water downstream). Stream flows of the order of a billion gallons a day are rare, and the larger power plants must dissipate heat by use of a large lake or cooling towers. Either method involves some environmental effects.

If a lake is used, the temperature of the water at the discharge point may be too high for certain organisms. It is common knowledge, however, that fishing is especially good where the heated water emerges. Means by which heat is removed from the surface of a lake are evaporation, radiation, and convection because of air currents. Regulations of the Environmental Protection Agency limit the rise in temperature in bodies of water. Clearly, the larger the lake and the wider the dispersal of heated water, the easier it is to meet requirements. When the thermal discharge goes into a lake, the ecological effects are frequently called "thermal pollution," especially when the high temperatures damage plants and animals. Other effects are the deaths of aquatic animals by striking screens, or passing through the system, or being poisoned by chemicals used to control the growth of undesirable algae. However, the environmental effects are mixed. Warm water is attractive to various fish and favors growth in their population.

Many nuclear plants have had to adopt the cooling tower for disposal of waste heat into the atmosphere. In fact, the hyperboloid shape (see Figure 12.10) is so common that many people mistake it for the reactor. A cooling tower is basically a large heat exchanger with airflow provided by natural convection or by blowers. In a "wet" type (see Figure 12.10(A)), the surface is kept saturated with moisture, and evaporation provides cooling. Water demands by this model may be excessive. In a "dry" type (see Figure 12.10(B)), analogous to an automobile radiator, the cooling is by convection and requires more surface area and airflow. It is therefore larger and more expensive. A hybrid wet/dry cooling tower is used to minimize effects of vapor plumes in cold weather and to conserve water in hot weather. The relationship of methods of reactor cooling is highlighted by the problem facing Dominion Energy as it considers an additional reactor at North Anna, Virginia. The water drawn from and returned to a lake—a once-through system—would be too hot for wildlife so one or more cooling towers will be required. Because water is lost by evaporation, reactors in areas experiencing drought may have to cut back on power.

Waste heat can be viewed as a valuable resource. If it can be used in any way, it reduces the need for oil and other fuels. Some of the actual or potential beneficial uses of waste heat are the following:

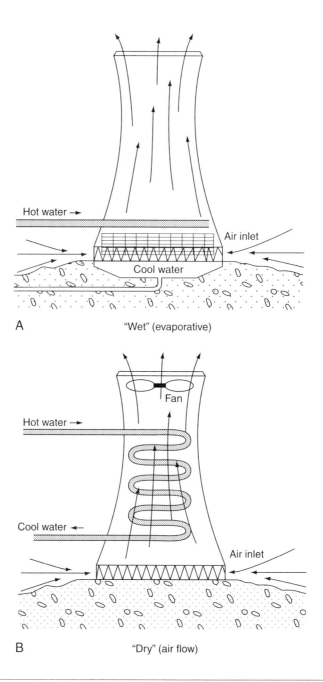

FIGURE 12.10 Cooling tower (From "Thermal Pollution and Aquatic Life" by John R. Clark, *Scientific American*, March 1969).

1. District heating. Homes, business offices, and factories of whole towns in Europe are heated in this way.
2. Production of fish. Warm water can be used to stimulate growth of the food fish need.
3. Extension of plant growth season. For colder climates, use of water to warm the soil in early spring would allow crops to be grown for a longer period.
4. Biological treatment. Higher temperatures may benefit water treatment and waste digestion.
5. Desalination. Removal of salt from seawater or brackish water. See Chapter 27 for details.
6. Production of hydrogen gas. Use of a dedicated reactor or combined heat and electricity source to isolate H_2. See Chapter 27 for details.

Each of these applications has merit, but there are two problems: (a) the need for heat is seasonal, so the systems must be capable of being bypassed in summer or, if buildings are involved, they must be designed to permit air conditioning; and (b) the amount of heat is far greater than any reasonable use that can be found. It has been said that the waste heat from electrical plants was sufficient to heat all of the homes in the United States. If all homes within practical distances of power plants were so heated, there still would be a large amount of unused waste heat.

A few reactors around the world have been designed or adapted to produce both electrical power and useful heat for space heating or process steam. The abbreviation CHP for combined heat and power is applied to these systems. It can be shown (see Exercise 12.11) that if half the turbine steam of a reactor with thermal efficiency 1/3 is diverted to useful purposes, the efficiency is doubled, neglecting any adjustment in operating conditions.

A practice called *cogeneration* is somewhat the reverse of waste heat use. A boiler used for producing steam can be connected to a turbine to generate electricity and to provide process heat. Typical steam users are refineries, chemical plants, and paper mills. In general, cogeneration is any simultaneous production of electrical or mechanical energy and useful thermal energy, but it is regarded as a way to save fuel. For example, an oil-fired system uses 1 barrel (bbl) of oil to produce 750 kWh of electricity, and a process-steam system uses 2 bbl of oil to give 8700 lb of steam, but cogeneration requires only 2.4 bbl to provide both products.

12.5 SUMMARY

The principal modes by which fission energy is transferred in a reactor are conduction and convection. The radial temperature distribution in a fuel pellet is approximately parabolic. The rate of heat transfer from fuel surface to coolant by convection is directly proportional to the temperature difference. The allowed power level of a reactor is governed by the temperatures at local "hot spots." Coolant flow along channels extracts thermal energy and delivers it to an external circuit consisting of a heat exchanger (PWR), a steam turbine that drives an electrical generator, a steam condenser, and various pumps. Large amounts of waste heat are discharged by electrical power plants because of inherent limits on efficiency. Typically, a billion gallons of water per day must flow through the steam condenser to limit the temperature rise of the environment. When rivers and lakes are not available or adequate, waste heat is dissipated by cooling towers. Potential beneficial uses of the waste thermal energy include space heating and stimulation of growth of fish and plants. Some nuclear facilities produce and distribute both steam and electricity.

12.6 EXERCISES

12.1 Show that the temperature varies with radial distance in a fuel pin of radius R according to

$$T(r) = T_s + (T_0 - T_s)\left[1 - (r/R^2)\right],$$

where the center and surface temperatures are T_0 and T_s, respectively. Verify that the formula gives the correct results at $r = 0$ and $r = R$.

12.2 Explain the advantage of a circulating fuel reactor, in which fuel is dissolved in the coolant. What disadvantages are there?

12.3 If the power density of a uranium oxide fuel pin, of radius 0.6 cm, is 500 W/cm^3, what is the rate of energy transfer per centimeter across the fuel pin surface? If the temperatures of pin surface and coolant are 300°C and 250°C, what must the heat transfer coefficient h be?

12.4 A reactor operates at thermal power of 2500 MW, with water coolant mass flow rate of 15,000 kg/s. If the coolant inlet temperature is 275°C, what is the outlet temperature?

12.5 A power reactor is operating with coolant temperature 500°F and pressure 1500 psi. A leak develops and the pressure falls to 500 psi. How much must the coolant temperature be reduced to avoid flashing?

12.6 The thermal efficiencies of a PWR converter reactor and a fast breeder reactor are 0.33 and 0.40, respectively. What are the amounts of waste heat for a 900 MWe reactor? What percentage improvement is achieved by going to the breeder?

12.7 As sketched, water is drawn from a cooling pond and returned at a temperature 14°C higher to extract 1500 MW of waste heat. The heat is dissipated by water evaporation from the pond with an absorption of 2.26×10^3 J/g. How many kilograms per second of makeup water must be supplied from an adjacent river? What percentage is this of the circulating flow to the condenser?

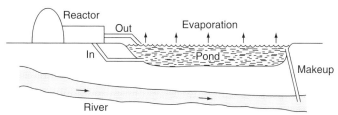

12.8 As a rough rule of thumb, it takes 1 to 2 acres of cooling lake per megawatt of installed electrical capacity. If one conservatively uses the latter figure, what is the area for a 1000-MWe plant? Assuming 35% efficiency, how much energy in joules is dissipated per square meter per hour from the water? NOTE: 1 acre = 4047 m^2.

12.9 How many gallons of water have to be evaporated each day to dissipate the waste thermal power of 2030 MWt from a reactor? Note that the heat of vaporization is 539.6 cal/g, the mechanical equivalent of heat is 4.18 J/cal, and 1 gal is 3785 cm^3.

12.10 Verify that approximately 1.6 kg of water must be evaporated to dissipate 1 kWh of energy.

12.11 A plant produces power both as useful steam, S, and electricity, E, from an input heat Q. Develop a formula for the overall efficiency e', expressed in terms of the ordinary efficiency $e = E/Q$ and x, the fraction of waste heat used for steam. Show that e' is 2/3 if $e = 1/3$ and $x = 1/2$. Find e' for $e = 0.4$ and $x = 0.6$.

Computer Exercises

12.A If the thermal conductivity of UO_2 used as reactor fuel pins varies with temperature, it can be shown that the linear heat rate q_1 (W/cm) is 4π times the integral of k (W/cm-°C) with respect to temperature T (°C). (a) With computer program CONDUCT, which calculates the integral from 0 to T, verify that the integral is approximately 93 W/cm when T is the melting point of UO_2, 2800°C. (b) Find the linear heat rate with the maximum temperature $TM = 2800$°C and surface temperature $TS = 315$°C.

12.B The temperature distribution within a reactor fuel pin for variable k can be calculated with the integrals of k over temperature (Exercise 12.A).

In the program TEMPLOT, by specifying maximum allowed center temperature and the expected surface temperature for a fuel pellet of radius RO, the linear heat rate is calculated and used to obtain values of radius R as a function of temperature T. Test the program with typical inputs such as $R = 0.5$ cm, $TM = 2300°$C, and $TS = 300°$C, plotting the resulting temperature distribution.

12.7 REFERENCES

Martin Becker, *Heat Transfer: A Modern Approach*, Plenum Press, New York, 1986. Uses spreadsheet techniques and analytic methods.

M. M. El-Wakil, *Nuclear Heat Transport*, American Nuclear Society, La Grange Park, IL, 1978.

William M. Kays and Michael E. Crawford, *Convective Heat and Mass Transfer*, 4th Ed., McGraw-Hill, New York, 2005.

Coolants for Nuclear Reactors
http://www.nuc.berkeley.edu/thyd/ne161/rahmed/coolants.html
Data on coolants and lists of advantages and disadvantages of various reactor types.

Frank P. Incropera and David P. DeWitt, *Introduction to Heat Transfer*, 4th Ed., John Wiley and Sons, New York, 2001.

W. J. Minkowycz, E. M. Sparrow, and J. Y. Murthy, *Handbook of Numerical Heat Transfer*, John Wiley & Sons, Hoboken, NJ, 2006.

Joseph H. Keenan, *Steam Tables: Thermodynamic Properties of Water Including Vapor, Liquid, and Solid Phases/With Charts*, Krieger Publishing Co., Malabar, FL, 1992. A reprint of a classic work.

Steam Generator Tube Issues
http://www.nrc.gov/reading-rm/doc-collections/fact-sheets/steam-gen.html
Technical Issue Paper by Nuclear Regulatory Commission.

Steam Generators
http://www.nucleartourist.com
Search on Steam Generators. Diagrams and descriptions. By Joseph Gonyeau.

Neil E. Todreas and Mujid S. Kazimi, *Nuclear Systems I: Thermal Hydraulic Fundamentals; Nuclear Systems II: Elements of Thermal Hydraulic Design.*, Hemisphere Publishing Corp., New York, 1990. Advanced but understandable texts on thermal analysis. Comprehensive coverage with many numerical examples.

L. S. Tong and Joel Weisman, *Thermal Analysis of Pressurized Water Reactors*, 3rd Ed., American Nuclear Society, La Grange Park, IL, 1996. A graduate level book with principles, engineering data, and new information.

R. T. Lahey and F. J. Moody, *The Thermal-Hydraulics of a Boiling Water Nuclear Reactor*, 2nd Ed., American Nuclear Society, La Grange Park, IL, 1993. Emphasizes understanding of physical phenomena.

Nicholas P. Cheremisinoff and Paul N. Cheremisinoff, *Cooling Towers: Selection, Design and Practice*, Ann Arbor Science, Ann Arbor, MI, 1981.

G. B. Hill, E. J. Pring, and Peter D. Osborn, *Cooling Towers: Principles and Practice*, 3rd Ed., Butterworth-Heinemann, London, 1990.

Bernd Becher and Hilla Becher, *Cooling Towers*, MIT Press, Cambridge, MA, 2006.

Cooling Towers
http://www.nucleartourist.com
Search on Cooling Towers. Diagrams and descriptions. By Joseph Gonyeau.

Breeder Reactors

THE MOST important feature of the fission process is, of course, the enormous energy release from each reaction. Another significant fact, however, is that for each neutron absorbed in a fuel such as U-235, more than two neutrons are released. To maintain the chain reaction, only one is needed. Any extra neutrons available can thus be used to produce other fissile materials such as Pu-239 and U-233 from the "fertile" materials, U-238 and Th-232, respectively. The nuclear reactions yielding the new isotopes were described in Section 6.3. If losses of neutrons can be reduced enough, the possibility exists for new fuel to be generated in quantities as large, or even larger than the amount consumed, a condition called "breeding." Several fuel cycles exist, which are distinguished by the amount of recycling. In the once-through cycle, all spent fuel is discarded as waste. Partial recycling makes use of separated plutonium, which can be combined with low-enrichment uranium to form mixed oxide. In the ultimate and ideal breeder cycle, all materials are recycled. As discussed later, there is a revival of interest in some level of recycling to help reduce radioactive waste and to use all fuel energy values.

In this chapter we shall (a) examine the relationship between the reproduction factor and breeding, (b) describe the physical features of the LMFBR, and (c) look into the compatibility of uranium fuel resources and requirements.

13.1 THE CONCEPT OF BREEDING

The ability to convert significant quantities of fertile materials into useful fissile materials depends crucially on the magnitude of the reproduction factor, η, which is the number of neutrons produced per neutron absorbed in fuel. If v neutrons are produced per fission, and the ratio of fission to absorption in fuel is σ_f/σ_a, then the number of neutrons per absorption is

$$\eta = \frac{\sigma_f}{\sigma_a} v.$$

Table 13.1 Values of Reproduction Factor η

	Neutron Energy	
Isotope	Thermal	Fast
U-235	2.07	2.3
Pu-239	2.11	2.7
U-233	2.30	2.45

The greater its excess above 2, the more likely is breeding. It is found that both v and the ratio σ_f/s_a increase with neutron energy and thus η is larger for fast reactors than for thermal reactors. Table 13.1 compares values of η for the main fissile isotopes in the two widely differing neutron energy ranges designated as thermal and fast. Inspection of the table shows that it is more difficult to achieve breeding with U-235 and Pu-239 in a thermal reactor, because the 0.07 or 0.11 neutrons are very likely to be lost by absorption in structural materials, moderator, and fission product poisons.

A thermal reactor that uses U-233 is a good prospect, but the fast reactor that uses Pu-239 is the most promising candidate for breeding. Absorption of neutrons in Pu-239 consists of both fission and capture, the latter resulting in the isotope Pu-240. If the latter captures a neutron, the fissile isotope Pu-241 is produced.

The ability to convert fertile isotopes into fissile isotopes can be measured by the conversion ratio (CR), which is defined as

$$CR = \frac{\text{fissile atoms produced}}{\text{fissile atoms consumed}}.$$

The fissile atoms are produced by absorption in fertile atoms; the consumption includes fission and capture.

We can compare values of CR for various systems. First is a "burner" fueled only with U-235. With no fertile material present, CR $= 0$. Second is a highly thermal reactor with negligible resonance capture, in which fuel as natural uranium, 99.28% U-238 and 0.72%U-235, is continuously supplied and consumed. Pu-239 is removed as fast as it is created. Here CR is the ratio of absorption in U-238 and U-235, and because they experience the same flux, CR is simply the ratio of macroscopic cross sections, $\Sigma_{a238}/\Sigma_{a235}$. Inserting the cross section ratio 2.7/681 and the atom ratio (ignoring U-234) 0.9928/0.0072, we obtain CR $= 0.547$. Third, we ask what CR is needed to completely consume both U isotopes in natural U as well as the Pu-239 produced? It is easy to show that CR is equal to the isotopic fraction of U-238 (viz., 0.9928). Fourth, we can derive a more general relationship from the neutron cycle of Figure 11.4. The result for initial operation of a critical reactor, before any Pu is produced, is

$$CR = \Sigma_{a238}/\Sigma_{a235} + \eta_{235}\varepsilon\mathcal{L}_f(1-p),$$

where η_{235} is the value for pure U-235 (i.e., 2.07). For a natural U reactor with $\mathcal{L}_f = 0.95$, $p = 0.9$. and $\varepsilon = 1.03$, we find

$$CR = 0.547 + 0.203 = 0.750.$$

It is clear that reducing fast neutron leakage and enhancing resonance capture are favorable to the conversion process. An alternative simple formula, obtained by considerable manipulation as in Exercise 13.6, is

$$CR = \eta_{235}\varepsilon - 1 - \ell$$

where ℓ is the total amount of neutron loss by leakage and by non-fuel absorption, per absorption in U-235.

If unlimited supplies of uranium were available at very small cost, there would be no particular advantage in seeking to improve CRs. One would merely burn out the U-235 in a thermal reactor and discard the remaining U-238. Because the cost of uranium goes up as the accessible reserves decline, it is desirable to use the U-238 as well as the U-235. Similarly, the exploitation of thorium reserves is worthwhile.

When the CR is larger than 1, as in a fast breeder reactor, it is instead called the breeding ratio (BR), and the breeding gain (BG) = BR − 1 represents the extra plutonium produced per atom burned. The doubling time (DT) is the length of time required to accumulate a mass of plutonium equal to that in a reactor system, and thus provide fuel for a new breeder. The smaller the inventory of plutonium in the cycle and the larger the BG, the quicker will doubling be accomplished. The technical term "specific inventory" is introduced, as the ratio of plutonium mass in the system to the electrical power output. Values of this quantity of 2.5 kg/MWe are sought. At the same time, a very long fuel exposure is desirable (e.g., 100,000 MWd/tonne) to reduce fuel fabrication costs. BG of 0.4 would be regarded as excellent, but a gain of only 0.2 would be very acceptable.

13.2 ISOTOPE PRODUCTION AND CONSUMPTION

The performance of a breeder reactor involves many isotopes of fertile and fissionable materials. In addition to the U-235 and U-238, there is short-lived neptunium-239 (2.355 d), Pu-239 (2.411×10^4 y), Pu-240 (6537 y), Pu-241 (14.4 y), and Pu-242 (3.76×10^5 y), as well as americium and curium isotopes resulting from multiple neutron capture. The idea of a chain of reactions is evident. To find the amount of any of these nuclides present at a given time, it is necessary to solve a set of connected equations, each of the general type

Rate of change = Generation rate − Removal rate,

which is similar to the statement in Section 3.3 except that "removal" is more general than "decay" in that absorption (consumption or burnup) is included.

We can illustrate the approach to solving the balance equations as differential equations. Consider a simplified three-component system of nuclides, using a shorthand for the full names of the isotopes: $1 =$ U-235, $2 =$ U-238, and $3 =$ Pu-239. Because all of their radioactive half-lives are long in comparison with the time of irradiation in a reactor, true decay can be ignored. However, it will be convenient to draw an analogy between decay and burnup. The equation for U-235 is

$$dN_1/dt = -\phi N_1 \sigma_{a1},$$

and if we let $\phi \, \sigma_{a1} = \lambda_{a1}$, the equation is the same as that for decay, the solution of which is

$$N_1(t) = N_{10} E_1, \text{with} \ \ E_1 = \exp(-\lambda_{a1} t).$$

A similar solution may be written for U-238,

$$N_2(t) = N_{20} E_2, \text{with} \ \ E_2 = \exp(-\lambda_{a2} t).$$

The growth equation for Pu-239 is

$$dN_3/dt = g - \phi N_3 \sigma_{a3}, \text{where} \ g = \phi N_2 \sigma_{c2},$$

where only the capture in U-238 gives rise to Pu-239, not the fission. Assuming that there is already some plutonium present when the fuel is loaded in the reactor, in amount N_{30}, the solution is

$$N_3(t) = N_{30} E_3 + N_{20} \lambda_{c2} (E_3 - E_2)/(\lambda_{c2} - \lambda_{a3}),$$

where $E_3 = \exp(-\lambda_{a3} t)$. The first term on the right describes the burnup of initial Pu-239; the second term represents the net of production and consumption. Note the similarity in form of the equations to those in Computer Exercise 3.D related to parent–daughter radioactivity processes.

It is straightforward to calculate the numbers of nuclei, but time-consuming and tedious if one wishes to vary parameters such as the reactor power and neutron flux level or the initial proportions of the different nuclides. To make such calculations easier, refer to Computer Exercise 13.A, in which the programs BREED and BREEDGE are applied.

A one-neutron group model is not adequate to analyze the processes in a fast breeder reactor, where cross sections vary rapidly with energy. The accurate calculation of multiplication requires the use of several neutron energy groups, with neutrons supplied to the groups by fission and removed by slowing and absorption. In Computer Exercise 13.B the analysis is displayed and a simple fast reactor is computed by the program FASTR.

13.3 THE FAST BREEDER REACTOR

LMFBRs have been operated successfully throughout the world. In the United States the Experimental Breeder Reactor I at Idaho Falls was the first power reactor to generate electricity in 1951. Its successor, EBR II, was used from 1963 to

1994 to test equipment and materials. An important feature was its closed fuel cycle, in which used fuel was removed, chemically processed, and refabricated. To accomplish these operations under conditions of high radioactivity, unique handling equipment was devised. In September 1969, the power reached its design value of 62.5 MWt (see References).

The Fermi I reactor built near Detroit was the first intended for commercial application. It was started in 1963 but was damaged by blockage of coolant flow passages and only operated briefly after being repaired.

The 400 MWt Fast Flux Test Facility (FFTF) at Richland, Washington, now shut down, did not generate electricity but provided valuable data on the performance of fuel, structural materials, and coolant (see References). After a number of years of design work and construction, the United States government canceled the demonstration fast power reactor called Clinch River Breeder Reactor Project (CRBRP). There was a great deal of debate in the United States before CRBRP was abandoned. One argument for stopping the project was that increased prices of fuel, being only approximately one fifth of the cost of producing electricity, would not cause converter reactors to shut down or warrant switching to the newer technology except on a long-term basis. This political decision shifted the leadership for breeder development from the United States to other countries.

France took the initiative in the development of the breeder for the production of commercial electric power in cooperation with other European countries. The reactor "Superphenix" was a full-scale pool-type breeder constructed with partial backing by Italy, West Germany, The Netherlands, and Belgium. Because of sodium leaks, and great public opposition, the reactor was shut down permanently. The lower power Phenix was shut down for 6 years but restarted in 2003 with excellent operation.

With the suspension of operation of Superphenix, the lead in breeder reactor development again shifted, this time to Japan, which placed its 280 MWe loop-type sodium-cooled MONJU into operation in 1993. It was part of Japan's long-range plan to construct of a number of breeders starting around 2020. In 1995 the reactor suffered a sodium leak (see References) and was shut down. Renewed interest in breeding prompted a restart of MONJU.

The largest remaining LMFBR in the world is the BN-600 at the Beloyarskiy plant in Russia. Supplying electricity since 1981, it has operated more successfully than any other reactor in that country. Some of its pertinent features are listed in Table 13.2.

India is preparing a 1200 MWt fast breeder reactor to go into operation around 2010. It will complement their proposed Advanced Heavy Water Reactor, a thermal breeder that uses fertile Th-232 to produce fissile U-233.

The use of liquid sodium as coolant ensures that there is little neutron moderation in the fast reactor. The element sodium melts at 208°F (98°C), boils at 1618°F (883°C), and has excellent heat transfer properties. With such a high melting point, pipes containing sodium must be heated electrically and thermally insulated to prevent freezing. The coolant becomes radioactive by neutron absorption

Table 13.2 BN-600 Liquid Metal Fast Breeder Reactor, Beloyarskiy Unit #3, Russia From *Nuclear Engineering International* Magazine

Electric power	560 MW
Sodium coolant temperatures	377°C, 550°C
Core fuel height	1.03 m
Core diameter	2.05 m
Vessel height, diameter	12.6 m, 12.86 m
Fuel (w/o U-235)	UO_2 (17, 21, 26)
Pin o.d.	6.9 mm
Cladding	stainless steel
Clad thickness	0.4 mm
Pin pitch (triangular)	9.82 mm
Pins per assembly	127
Number of assemblies	369
Number of B_4C rods	27
Average power density	413 kWt/1
Cycle length	5 months

in Na-23, producing the 15-h Na-24. Great care must be taken to prevent contact between sodium and water or air, which would result in a serious fire, accompanied by the spread of radioactivity. To avoid such an event, an intermediate heat exchanger is used, in which heat is transferred from radioactive sodium to nonradioactive sodium.

Two physical arrangements of the reactor core, pumps, and heat exchanger are possible, shown schematically in Figures 13.1 and 13.2. The "loop" type is similar to the thermal reactor system, whereas in the "pot" type all of the components are immersed in a pool of liquid Na. There are advantages and disadvantages to each concept, but both are practical.

To obtain maximum BRs in the production of new fertile material, more than one fuel zone is needed. The neutron-multiplying core of the breeder reactor is composed of mixed oxide (MOX) fuel as a mixture of U and Pu. Surrounding the core is a natural uranium oxide "blanket" or "breeding blanket." In early designs, the blanket acted as a reflector for a homogeneous core, but modern designs involve blanket rings both inside and outside the core, rendering the system heterogeneous. The new arrangement is predicted to have enhanced safety as well.

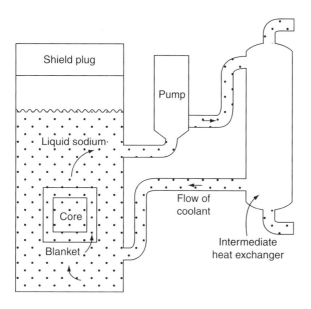

FIGURE 13.1 Loop system for LMFBR.

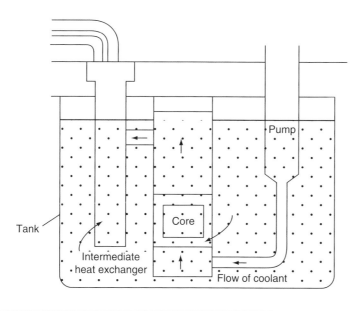

FIGURE 13.2 Pot system for LMFBR.

Deployment of breeder reactors demands recycling of the plutonium. This in turn requires reprocessing, which involves physical and chemical treatment of irradiated fuel to separate uranium, plutonium, and fission products. We reserve discussion of reprocessing until Section 22.5, in connection with waste disposal. The United States abandoned commercial reprocessing caused by concerns about the diversion of plutonium and is unlikely to resume the practice for the present generation of power reactors.

13.4 INTEGRAL FAST REACTOR

After CRBRP was canceled, the United States continued development of breeder reactors. The success of the Experimental Breeder Reactor-II prompted an extension called the Integral Fast Reactor (IFR). It was a fast breeder reactor coupled with a pyrometallurgical process that allowed all fuels and products to remain in the system. The fuel cycle included fuel fabrication, power generation, reprocessing, and waste treatment. Through the isolation of plutonium in a highly radioactive environment, the problem of weapons proliferation was eliminated. The reactor burned all actinides, elements 89 and above. No additional fuel was introduced and the predicted lifetime was 60 y.

Fuel in the IFR was metal, as an alloy of U, Pu, and Zr, with a much higher thermal conductivity than ceramic uranium oxide. Thus in operation the centers of fuel rods were cooler than in conventional reactors at the same power. Coolant in the reactor was liquid sodium that operates at atmospheric pressure and is not corrosive. With the reactor located within a pool, the problem of loss of coolant is eliminated, there being adequate natural convection cooling.

The reactor was found to be inherently safe, as was proved by experiments with EBR-II. In one test, the power to cooling systems was cut off when the reactor was at full power. The core temperature rose slightly as convective cooling in the pool-type vessel took over. The reactor went subcritical and shut itself down.

Use of the IFR with its closed cycle and consumption of plutonium would have eliminated the concern about the spent fuel repositories being "plutonium mines," with reactor grade plutonium available at some future date for some low-yield explosions. As such, IFR was highly proliferation resistant. The use of uranium was essentially complete, in contrast with a few percent in conventional power plants.

By the recycling of transuranic elements, the waste consisted only of fission products, which would need to be stored for a relatively short time. The need for extra waste repositories would vanish if IFR-type reactors were deployed. The reactor system was expected to be less expensive to build because of its simplicity. Most of the complex systems of water-cooled reactors are not needed, and the cost of basic fuel is essentially zero because of large inventory of uranium-238 in the depleted uranium from years of operation of separation plants.

The reactor serves as a source of experience and data and a starting point for the proposed Advanced Fast Reactor of the 21st century. Finally, it shares the virtue of all nuclear reactors in not releasing gases that could contribute to climate change.

Even though there were many potential benefits to the continuation of R&D on the IFR, Congress chose not to provide funds in 1994. In a Q&A session by George Stanford (see References), "Well-meaning but ill-informed people...convinced so many administrators and legislators that the IFR was a proliferation threat that the program was killed." There remains a possibility that the concept will be revived, in view of the extensive knowledge base that was developed and the promise that IFR has to solve many of nuclear's perceived problems.

A commercial outgrowth of the IFR was the Advanced Liquid Metal Reactor (ALMR) or PRISM, involving Argonne National Laboratory and General Electric (see References). That design project was also terminated.

Although the principal attention throughout the world has been given to the liquid metal–cooled fast breeder that uses U and Pu, other breeder reactor concepts might someday become commercially viable. The thermal breeder reactor, which uses thorium and uranium-233, has always been an attractive option. Favorable values of CRs are obtained with thorium as fertile material (see References). Neutron bombardment of Th-232 in a thermal reactor yields fissile U-233 as noted in Section 6.3. However, there is a special radiation hazard involved. Two neutron absorptions in U-233 yield 69 y half-life U-232 that decays into Tl-208, a 2.6 MeV gamma emitter. India, which has large reserves of thorium, is especially interested in a U-233 breeder cycle.

One extensive test of that type of reactor was the Molten Salt Reactor Experiment at Oak Ridge, an outgrowth of the aircraft nuclear program of the 1960s (see Section 20.6). The reactor demonstrated the feasibility of the circulating fuel concept with salts of lithium, beryllium, and zirconium as solvent for uranium and thorium fluorides. Other concepts are (a) uranium and thorium fuel particles suspended in heavy water, (b) a high-temperature gas-cooled graphite-moderated reactor containing beryllium, in which the (n,2n) reaction enhances neutron multiplication.

13.5 BREEDING AND URANIUM RESOURCES

From the standpoint of efficient use of uranium to produce power, it is clearly preferable to use a breeder reactor instead of a converter reactor. The breeder has the ability to use nearly all of the uranium rather than a few percent. Its impact can be viewed in two different ways. First, the demand for natural uranium would be reduced by a factor of approximately 30, cutting down on fuel costs while reducing the environmental effect of uranium mining. Second, the supply of fuel would last longer by the factor of 30. For example, instead of a mere 40 y for use of inexpensive fuel, we would have 1200 y. It is less clear, however, as to when a well-tested version of the breeder would actually be needed. A simplistic answer is, "when uranium gets very expensive." Such a situation is not imminent, because there has been an oversupply of uranium for a number of years, and all analyses show that breeders are more expensive to build and operate than converters. A reversal in trend is not expected until some time well into the 21st century. The urgency to develop a commercial breeder has lessened as the result of slower adoption of nuclear power than anticipated, with the smaller rate of depletion of resources. Another key factor is the availability in the United States and the former U.S.S.R of large quantities of surplus weapons plutonium, which can be used as fuel in the form of MOX.

Uranium resource data as of 2006 from the "Red Book" (see References) are reported by the World Information Source on Energy (WISE) (see References). Recoverable amounts of uranium are given for costs in $/kgU <40, <80, and <130. We show in Table 13.3 the sums by main country sources of two categories: reasonably assured resources (RAR) and inferred resources (IR). The world total is 2,643,340 tonnes. The table also gives the annual uranium requirements by principal country users. The world total per year is 66,877 tonnes. Simple arithmetic tells us that these resources would last 40 y, assuming constant fuel requirements.

The use of global figures obscures the problem of distribution. In Table 13.3 we list the top countries in the categories demand and resources. Some surprising disparities are seen. The leading potential uranium suppliers, Australia and Kazakhstan, are not on the list of users. On the other hand, the second highest user, Japan, has negligible U resources. Thus there must be a great deal of import/export trade to meet fuel needs. At some time in the future, in place of the Organization of Petroleum Exporting Countries (OPEC), there is the possibility of an "OUEC" cartel. Alternately, it means that for assurance of uninterrupted production of nuclear power, some countries are much more interested than others in seeing a breeder reactor developed.

Some data on United States uranium production are shown in Table 13.4. Not included are byproducts of phosphate and copper mining, or the large stockpile of depleted uranium as tails from the uranium isotope separation process. Such material is as valuable as natural uranium for use in a blanket to breed plutonium. The principal United States deposits in order of size are in Wyoming, New Mexico, Colorado, Texas (coastal plain), and near the Oregon–Nevada border. The greatest concentration of estimated additional resources are in Utah and Arizona. Most of the ores come from sandstone; approximately 30 uranium mills are available. Exploration by surface drilling has tapered off continually since the middle 1970s when nuclear power was expected to grow rapidly.

There is considerable sentiment in the nuclear community for storing spent fuel from converter reactors rather than burying it as waste, in anticipation of an energy shortage in the future as fossil fuels become depleted. If such a policy were adopted, the plutonium contained in the spent fuel could be recovered in a leisurely manner. The plutonium would provide the initial loading of a new

Table 13.3 Uranium Demand and Resources (in 1000s of tonnes) From OECD-IAEA Report (see References)

Country	Annual Demand (2006 est.)	Country	Reasonably Assured Resources to $80/kg
United States	22.875	Australia	714.00
Japan	8.670	Kazakhstan	378.29
France	7.185	Canada	345.20
Russia	4.465	Niger	180.47
Korea	3.400	South Africa	177.15
Germany	2.900	Brazil	157.70
Ukraine	2.350	Namibia	151.32
Canada	1.700	Russia	131.75
China	1.565	United States	102.00
United Kingdom	1.500	Uzbekistan	59.74
Belgium	1.455	Ukraine	58.50
Sweden	1.400	Mongolia	46.20
Spain	1.140	China	38.02
Bulgaria	0.840	Jordan	30.37
Taiwan	0.830	Algeria	19.50
Czech Republic	0.700	Malawi	8.77
Finland	0.557	Turkey	7.39
Slovakia	0.450	Portugal	6.00
Brazil	0.450	Bulgaria	5.87
India	0.380	Argentina	4.88
Hungary	0.370	Italy	4.80
Mexico	0.355	Spain	2.46
South Africa	0.280	Congo	1.35
Switzerland	0.270	Zimbabwe	1.35
Lithuania	0.190	Peru	1.22

(*Continued*)

Table 13.3 Uranium Demand and Resources (in 1000s of tonnes) From OECD-IAEA Report (see References)—*cont...*

Country	Annual Demand (2006 est.)	Country	Reasonably Assured Resources to $80/kg
Slovenia	0.160	Slovenia	1.21
Argentina	0.120	Greece	1.00
Romania	0.100	Czech Republic	0.51
Armenia	0.090	Indonesia	0.32
Pakistan	0.065		
Netherlands	0.065		
Total	**66.877**	**Total**	**2643.34**

Table 13.4 United States Uranium (2006) DOE/EIA (see References)

Estimated reserves at $30/lb 74 million tonnes, at $50/lb 424 million tonnes

Annual uranium mine production 4.7 million pounds U_3O_8

Total mines and sources 11

Employment (person-years) 755

Total expenditures $221 million

generation of fast breeder reactors, and the recovered uranium would serve as blanket material.

The energy content of uranium is so high that the cost of fuel for nuclear power plants is relatively small, of the order of 5% of the operating cost. However, if shortages of inventory occur because of inadequacy of supply, the price may rise significantly.

Around 1980 a peak price of $46/lb U_3O_8 was reached, but in subsequent decades dropped to a figure in 2000 of only $7/lb. The reason was the appearance of secondary sources such as depletion of inventories and released weapons uranium. Lesser amounts of secondary fuel can come from recycling, re-enrichment of separation tails, and surplus plutonium. These alternatives are expected to decline in the coming years. As the renaissance of nuclear power occurs in the

United States and countries such as France, India, South Korea, and China expand their nuclear capabilities, the price of uranium is likely to increase.

Coupled with the increased demand for uranium will be a growth in exploration and mining, but this will be hampered by the lack of qualified personnel. There is said to be ample resources, but investment in retrieval may be slow in coming until a stable price structure is established.

Finally, there is a large delay between discovery of new resources and the availability of nuclear fuel, in part because of a hodgepodge of regulatory oversight and the effect of public opposition. Thorough discussion of these issues appears in articles in *Nuclear News*, March 2006.

It is not possible to predict the rate of adoption of fast breeder reactors for several reasons. The capital costs and operating costs for full-scale commercial systems are not firmly established. The existence of the satisfactory LWR and the ability of a country to purchase slightly enriched uranium or MOX tends to delay the installation of breeders. It is conceivable, however, that the conventional converter reactors could be replaced by breeders in the present century because of fuel resource limitations. It is possible that the breeder could buy the time needed to fully develop alternative sources such as nuclear fusion, solar power, and geothermal energy. In the next chapter the prospects for fusion are considered.

13.6 SUMMARY

If the value of the neutron reproduction factor η is larger than 2 and losses of neutrons are minimized, breeding can be achieved, with more fuel produced than is consumed. The CR measures the ability of a reactor system to transform a fertile isotope (e.g., U-238) into a fissile isotope (e.g., Pu-239). Complete conversion requires a value of CR of nearly 1. Fast breeder reactors that use liquid sodium with BRs greater than 1 have been built and operated, but several development programs have been canceled. One large-scale breeder continues to operate in Russia. There is a great disparity between uranium resources and uranium use among the countries of the world. Application of the breeder could stretch the fission power option from a few decades to centuries.

13.7 EXERCISES

13.1 What are the largest conceivable values of the CR and the BG?

13.2 An "advanced converter" reactor is proposed that will use 50% of the natural uranium supplied to it. Assuming all the U-235 is used, what must the CR be?

13.3 Explain why the use of a natural uranium "blanket" is an important feature of a breeder reactor.

13.4 Compute η and BG for a fast Pu-239 reactor if $v = 2.98$, $\sigma_f = 1.85$, $\sigma_c = 0.26$, and $\ell = 0.41$. (Note that the fast fission factor ε need not be included.)

13.5 With a BR $= 1.20$, how many kilograms of fuel will have to be burned in a fast breeder reactor operating only on plutonium to accumulate an extra 1260 kg of fissile material? If the power of the reactor is 1250 MWt, how long will it take in days and years, noting that it requires approximately 1.3 g of plutonium per MWd?

13.6 (a) By use of the neutron cycle, Figure 11.4, find a formula for ℓ as defined in Section 13.1.
(b) Calculate the value of ℓ and verify that the alternative formula gives the same answer as in the text, CR $= 0.750$.

Computer Exercises

13.A A breeder reactor is successful if it produces more fissionable material than it consumes. To test that possibility apply computer programs BREED and BREEDGE. The first of these uses cross sections for U-235, U-238, and Pu-239 as deduced from early critical experiments on weapons material assemblies. The second uses more modern cross sections, appropriate to a power reactor design. (a) Run the programs, varying parameters, to explore trends. (b) Use the following common input on both programs: U-235 atom fraction 0.003 (depleted U), plutonium volume fraction 0.123, fast flux 4.46×10^{15} cm^{-2} s^{-1}. (c) Discuss observations of trends and seek to explain in terms of assumed cross section sets.

13.B Program FASTR solves the neutron balance equations for a fast reactor with classic 16-group Hansen-Roach cross sections prepared by Los Alamos. Those input numbers are found in the report *Reactor Physics Constants*, ANL-5800, 1963, page 568 ff. Run the program with the menus, observing input data and calculated results. Compare results for the case of pure U-235 with those obtained in Computer Exercise 11.A, with program CRITICAL.

13.8 REFERENCES

Karl Wirtz, *Lectures on Fast Reactors*, American Nuclear Society, La Grange Park, IL, 1976.

George A. Vendreyes, "Superphenix: A Full-Scale Breeder Reactor," *Scientific American*, W. H. Freeman Co., San Francisco, March 1977, p. 26.

A. M. Judd, *Fast Breeder Reactors: An Engineering Introduction*, Pergamon Press, Oxford, 1981.

Alan E. Waltar and Albert B. Reynolds, *Fast Breeder Reactors*, Pergamon Press, New York, 1981.

C. Pierre Zaleski, "Fast Breeder Reactor Economics," in Karl O. Ott and Bernard I. Spinrad, Editors, *Nuclear Energy: A Sensible Alternative*, Plenum Press, New York and London, 1985.

Charles E. Stevenson, *The EBR-II Fuel Cycle Story*, American Nuclear Society, La Grange Park, IL, 1987.

EBR-II: http://www.nuc.berkeley.edu/designs/ifr/ebr.html
Experimental Breeder Reactor at Argonne National Laboratory.

Thomas B. Cochran, *The Liquid Metal Fast Breeder Reactor*, Resources for the Future, Johns Hopkins University Press, Baltimore, MD, 1974.

Liquid Metal Fast Breeder
http://www.atomicinsights.com/oct95/LMFBR_oct95.html
History and status as viewed by Rod Adams in issue of Atomic Energy Insights.

"An Introduction to Argonne National Laboratory's Integral Fast Reactor (IFR) Program."
http://www.nuc.berkeley.edu/designs/ifr/anlw.html

George S. Stanford, "Integral Fast Reactors: Source of Safe, Abundant, Non-Polluting Power."
http://www.nationalcenter.org/NPA378.html

A double issue of *Progress in Nuclear Energy*, Vol. 31, Nos. 1–2 (1997) with title "The Technology of the Integral Fast Reactor and its Associated Fuel Cycle."

Introduction to the ALMR/PRISM
http://www.nuc.berkeley.edu/~gav/almr/01.intro.html
Description provided by University of California.

Status of liquid metal cooled fast reactor technology
http://www.iaea.org/inisnkm/nkm/aws/fnss/fulltext/30023917.pdf
1999 IAEA report (544 pages).

Fast Neutron Reactors
http://www.uic.com.au/nip98.htm
Comprehensive essay.

Uranium 2005 Resources, Production and Demand, OECD Nuclear Energy Agency and International Atomic Agency, Paris, 2006.
http://www.nea.fr/html/ndd/reports/2006/uranium2005-english.pdf
A biennial publication. Executive summary of "Red Book" (the cover is red).

Uranium Information Centre (Australia)
http://www.uic.com.au
Select Briefing Papers. More than 100 articles on all topics.

A. M. Perry and A. M. Weinberg, "Thermal Breeder Reactors," *Annual Review of Nuclear and Particle Science* 22, 1972, p. 317.

Jungmin Kang and Frank N. von Hippel, "U-232 and the Proliferation-Resistance of U-233 in Spent Fuel," *Science & Global Security* 9, 2001, p. 1.
Article is also available on the Web at http://www.princeton.edu/~globsec/publications/pdf/9_1kang.pdf

World Information Source on Energy Uranium Project
http://www.wise-uranium.org

The Web site for WISE includes a Nuclear Fuel Supply Calculator and listings of organizations handling fuel.

United States Uranium Reserves Estimates
http://www.eia.doe.gov/cneaf/nuclear/page/reserves/ures.html

Domestic Uranium Production Report
http://www.eia.doe.gov/cneaf/nuclear/dupr/dupr.html

Fusion Reactors

A DEVICE that permits the controlled release of fusion energy is designated as a fusion reactor in contrast with one yielding fission energy, the fission reactor. As discussed in Chapter 7, the potentially available energy from the fusion process is enormous. The possibility of achieving controlled thermonuclear power on a practical basis has not yet been demonstrated, but progress in recent years gives encouragement that fusion reactors can be in operation in the 21st century. In this chapter we will review the choices of nuclear reaction, study the requirements for feasibility and practicality, and describe the physical features of machines that have been tested. Suggestions on this chapter by John G. Gilligan are recognized with appreciation.

14.1 COMPARISON OF FUSION REACTIONS

The main nuclear reactions that combine light isotopes to release energy, as described in Section 7.1, are the D-D, D-T, and D-^{3}He. There are advantages and disadvantages of each. The reaction involving only deuterium uses an abundant natural fuel available from water by isotope separation. However, the energy yields from the two equally likely reactions are low (4.03 and 3.27 MeV). Also the reaction rate as a function of particle energy is lower for the D-D case than for the D-T case, as shown in Figure 14.1. The quantity $\overline{\sigma v}$, dependent on cross section and particle speed, is a more meaningful variable than the cross section alone.

The D-T reaction yields a helium ion and a neutron with energies as indicated:

$$^2_1\text{H} + ^3_1\text{H} \rightarrow ^4_2\text{He}(3.5\text{MeV}) + ^1_0\text{n}(14.1\text{MeV}).$$

The cross section is large and the energy yield is favorable. The ideal ignition temperature (Section 7.3) for the D-T reaction is only 4.4 keV in contrast with 48 keV for the D-D reaction, making the achievement of practical fusion with the former far easier. One drawback, however, is that the artificial isotope tritium

195

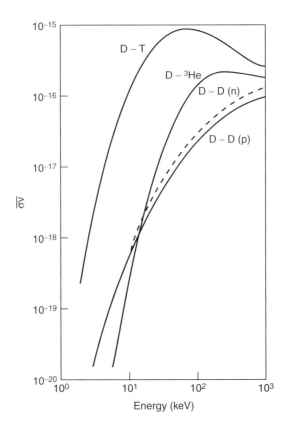

FIGURE 14.1 Reaction rates for fusion reactions. The quantity $\overline{\sigma v}$, the average over a Maxwellian distribution of cross section times speed, when multiplied by particle densities gives the fusion rate per unit volume.

is required. Tritium can be generated by neutron absorption in lithium, according to the two reactions

$$^{6}_{3}\text{Li} + {}^{1}_{0}\text{n} \rightarrow {}^{3}_{1}\text{H} + {}^{4}_{2}\text{He} + 4.8\text{MeV}$$

$$^{7}_{3}\text{Li} + {}^{1}_{0}\text{n} \rightarrow {}^{3}_{1}\text{H} + {}^{4}_{2}\text{H} + {}^{1}_{0}\text{n} - 2.5\text{MeV}.$$

The neutron can come from the D-T fusion process itself in a breeding cycle similar to that in fission reactors. Liquid lithium can thus be used as a coolant and a breeding blanket.

The fact that the D-T reaction gives a neutron as a byproduct is a partial disadvantage in a fusion machine. Wall materials are readily damaged by bombardment by 14.1 MeV neutrons, requiring frequent wall replacement. Also, materials of construction become radioactive as the result of neutron capture. These are engineering and operating difficulties, whereas the achievement of the high enough energy to use neutron-free reactions would be a major scientific challenge.

In the long run, use of the D-T reaction is limited by the availability of lithium, which is not as abundant as deuterium. All things considered, the D-T fusion reactor is the most likely to be operated first, and its success might lead to the development of a D-D reactor.

14.2 REQUIREMENTS FOR PRACTICAL FUSION REACTORS

The development of fusion as a new energy source involves several levels of accomplishment. The first is the performance of laboratory experiments to show that the process works on the scale of individual particles and to make measurements of cross sections and yields. The second is to test various devices and systems intended to achieve an energy output that is at least as large as the input and to understand the scientific basis of the processes. The third is to build and operate a machine that will produce net power of the order of megawatts. The fourth is to refine the design and construction to make the power source economically competitive. The first of these levels has been reached for some time, and the second is in progress with considerable promise of success. The third and fourth steps remain for achievement in the 21st century.

The hydrogen bomb was the first application of fusion energy, and it is conceivable that deep underground thermonuclear explosions could provide heat sources for the generation of electricity, but environmental concerns and international political aspects rule out that approach. Two methods involving machines have evolved. One consists of heating to ignition a plasma that is held together by electric and magnetic forces, the magnetic confinement fusion (MCF) method. The other consists of bombarding pellets of fuel with laser beams or charged-particle beams to compress and heat the material to ignition, the inertial confinement fusion (ICF) method. Certain conditions must be met for each of these approaches to be considered successful.

The first condition is achievement of the ideal ignition temperature of 4.4 keV for the D-T reaction. A second condition involves the fusion fuel particle number density n and a confinement time for the reaction, τ. It is called the *Lawson criterion* and is usually expressed as:

$$n\tau \geq 10^{14}\,\text{s/cm}^3.$$

A formula of this type can be derived for MCF by looking at energy and power in the plasma. Suppose that the numbers of particles per cm^3 are n_D deuterons, n_T tritons, and n_e electrons. Furthermore, let the total number of heavy particles be $n = n_D + n_T$ with equal numbers of the reacting nuclei, $n_D = n_T$, and $n_e = n$ for electrical neutrality. The reaction rate of the fusion fuel particles is written with Section 4.3 as $n_D\, n_T \sigma\, v$, and if E is the energy yield per reaction, the fusion power density is

$$p_f = n^2 \sigma v E/4,$$

proportional to the square of the ion number density.

Now the power loss rate can be expressed as the quotient of the energy content $(n_D + n_T)\, (3kT/2)$ and the confinement time τ, i.e.,

$$p_1 = 3nkT/\tau.$$

Equating the powers and solving,

$$n\tau = \frac{12kT}{\sigma v E}.$$

Insert the ideal ignition energy of $kT = 4.4$ keV, the fusion energy $E = 17.6$ MeV, and let $\overline{\sigma v}$ be equal to the value of $\overline{\sigma v}$ from Figure 14.1 of approximately 10^{-17}. The result is 3×10^{14}, of the correct order of magnitude. The Lawson criterion, however, is only a rough rule of thumb to indicate fusion progress through research and development. Detailed analysis and experimental testing are needed to evaluate any actual system.

Similar conditions must be met for ICF. An adequate ion temperature must be attained. The Lawson criterion takes on a little different form, relating the density ρ and the radius r of the compressed fuel pellet,

$$\rho r > 3 \text{ g/cm}^2.$$

The numerical value is set in part by the need for the radius to be larger than the range of α particles to take advantage of their heating effect. For example, suppose that 1 mm radius spheres of a mixture of D and T in liquid form, density 0.18 g/cm^3, are compressed by a factor of 2500. The radius is reduced by a factor of $(2500)^{1/3} = 13.6$, and the density is increased to $(2500)\, (0.18) = 450$ g/cm^3. Then $\rho r = 3.3$, which meets the objective.

It is interesting to note that the factors that go into the products $n\tau$ are very different for the two types of fusion. For MCF typically $n = 10^{14}$/cm^3 and $\tau = 1$ s, whereas for ICF $n = 10^{24}$/cm^3 and $\tau = 10^{-10}$ s.

The analysis of fusion reactors involves many other parameters of physics and engineering. A useful collection of formulas and methods of calculating are discussed in Computer Exercise 14.A.

Progress toward practical fusion can be measured by the parameter Q, which is the ratio of energy output to energy input. Four stages of plasma can be identified. In the first, more energy must be supplied than is produced, $Q < 1$. In the second, the breakeven case, fusion power equals input power, $Q = 1$. In the third, for an operating fusion power plant, Q is considerably larger than 1 (e.g., 10). In the fourth, the burning plasma, which results from ignition, heats itself without external input, and Q is infinity.

14.3 MAGNETIC CONFINEMENT MACHINES

A number of complex MCF machines have been devised to generate a plasma and to provide the necessary electric and magnetic fields to achieve confinement of the discharge. We will examine a few of these to illustrate the variety of possible approaches.

First, however, consider a simple discharge tube consisting of a gas-filled glass cylinder with two electrodes as in Figure 14.2(A). This is similar to the familiar fluorescent light bulb. Electrons accelerated by the potential difference cause excitation and ionization of atoms. The ion density and temperature of the plasma that is established are many orders of magnitude below that needed for fusion. To reduce the tendency for charges to diffuse to the walls and be lost, a current-carrying coil can be wrapped around the tube, as sketched in Figure 14.2(B). This produces a magnetic field directed along the axis of the tube, and charges move in paths described by a helix, the shape of a stretched coil spring. The motion is quite similar to that of ions in the cyclotron (Section 8.4) or the mass spectrograph (Section 9.1). The radii in typical magnetic fields and plasma temperatures are the order of 0.1 mm for electrons and near 1 cm for heavy ions (see Exercise 14.1). To further improve charge density and stability, the current along the tube is increased to take advantage of the *pinch effect*, a phenomenon related to the electromagnetic attraction of two wires that carry current in the same direction. Each of the charges that move along the length of the tube constitutes a tiny current, and the mutual attractions provide a constriction in the discharge.

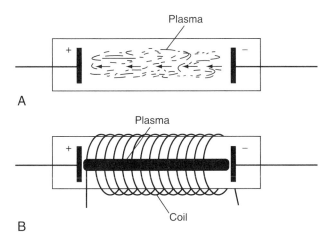

FIGURE 14.2 Electrical discharges.

Neither of the preceding magnetic effects prevent charges from moving freely along the discharge tube, and losses of both ions and electrons are experienced at the ends. Two solutions of this problem have been tried. One is to wrap extra current-carrying coils around the tube near the ends, increasing the magnetic field there. This causes charges to be forced back into the region of weak field (i.e., to be reflected). This "mirror machine" is not perfectly reflecting. Another approach is to create endless magnetic field lines by bending the vacuum chamber and the coils surrounding it into the shape of a figure eight. An early version of this arrangement, called a "stellarator," is still being considered as a favorable system because it does not depend on internal currents for plasma confinement. It could operate continuously rather than in pulses.

A completely different solution to the problem of charge losses is to produce the discharge in a doughnut-shaped tube, a torus, as shown in Figure 14.3. The first successful ring-shaped fusion machine was developed by scientists in the U.S.S.R. around 1960. They called it *tokamak*, an acronym in Russian for toroid–chamber–magnet–coil. Because the tube has no ends, the magnetic field lines produced by the coils are continuous. The free motion of charges along the circular lines does not result in losses. However, there is a variation in this toroidal magnetic field over the cross section of the tube that causes a small particle migration toward the wall. To prevent such migration, a current is passed through the plasma, generating a poloidal magnetic field. The field lines are circles around the current and tend to cancel electric fields that cause migration. Vertical magnetic fields are also used to stabilize the plasma.

Plasmas of MCF machines must be heated to reach the necessary high temperature. Various methods have been devised to supply the thermal energy. The first method, used by the tokamak, is resistance (ohmic) heating. A changing current in the coils surrounding the torus induces a current in the plasma. The power

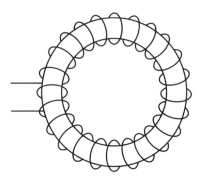

FIGURE 14.3 Plasma confinement in torus.

associated with a current through a resistance is I^2R. The resistivity of a "clean" hydrogen plasma, one with no impurity atoms, is comparable to that of copper. Impurities increase the resistivity by a factor of four or more. There is a limit set by stability on the amount of ohmic heating possible.

The second method of heating is neutral particle injection. The sequence of events is as follows: (a) a gas composed of hydrogen isotopes is ionized by an electron stream; (b) the ions of hydrogen and deuterium produced in the source are accelerated to high speed through a vacuum chamber by a voltage of approximately 100 kV; (c) the ions pass through deuterium gas and by charge exchange are converted into directed neutral atoms; (d) the residual slow ions are drawn off magnetically, whereas the neutral ions cross the magnetic field lines freely to deliver energy to the plasma.

The third method uses microwaves in a manner similar to their application to cooking. The energy supply is a radiofrequency (RF) generator. It is connected by a transmission line to an antenna next to the plasma chamber. The waves enter the chamber and die out there, delivering energy to the charges. If the frequency is right, resonant coupling to natural circular motions of electrons or ions can be achieved. The phrase electron (or ion) cyclotron radiofrequency, ECRF (or ICRF), comes from the angular frequency of a charge q with mass m in a magnetic field B, proportional to qB/m as discussed in Section 8.1.

Because the fusion reactions burn the deuterium-tritium fuel, new fuel must be introduced to the plasma as a puff of gas, as a stream of ions, or as particles of liquid or solid. The latter method seems best, despite the tendency for the hot plasma to destroy the pellet before it gets far into the discharge. It seems that particles that come off the pellet surface form a protective cloud. Compressed liquid hydrogen pellets of approximately 10^{20} atoms moving at 80 m/s are injected at a rate of 40 per second.

The mathematical theory of electromagnetism is used to deduce the magnetic field shape that gives a stable arrangement of electric charges. However, any disturbance can change the fields and in turn affect the charge motion, resulting in an instability that may disrupt the field configuration. The analysis of such behavior is more complicated than that of ordinary fluid flow because of the presence of charges. In a liquid or gas, the onset of turbulence occurs at a certain value of the Reynolds number. In a plasma with its electric and magnetic fields, many additional dimensionless numbers are needed, such as the ratio of plasma pressure to magnetic pressure (β) and ratios to the plasma size of the mean free path, the ion orbits, and the Debye length (a measure of electric field penetration into a cloud of charges). Several of the instabilities such as the "kink" and the "sausage" are well understood and can be corrected by assuring certain conditions.

Stability of the plasma is not sufficient to assure a practical fusion reactor because of various materials engineering problems. The lining of the vacuum chamber containing the plasma is subjected to radiation damage by the 14-MeV neutrons from the D-T reaction. Also, when the plasma is disrupted, the electric forces cause "runaway electrons" to bombard the chamber wall, generating large amounts of heat. Materials will be selected to minimize the effects on what are called plasma-facing components and reduce the frequency of need for replacement. An example is a graphite fiber composite similar to those used to protect the surface of the space shuttle on reentry. Other possible wall materials are silicon carbide, beryllium, tungsten, and zirconium, with the latter metals possibly enriched in an isotope that does not absorb neutrons. Some self-protection of the chamber lining is provided by vaporization of materials, with energy absorbed by a "vapor shield."

The eventual practical fusion reactor will require a system to generate tritium. As an alternative to the use of liquid lithium in a breeding blanket, consideration is given to a molten salt composed of fluorine, lithium, and beryllium (Li_2BeF_4 called "flibe"). The (n,2n) reaction in Be would enhance the breeding of tritium. Another possibility is the use of the ceramic lithium oxide (Li_2O).

A number of tokamaks have been built at research facilities around the world. Prominent examples are:

(a) The Tokamak Fusion Test Reactor (TFTR) at Princeton, now shut down, that achieved very high plasma temperatures.

(b) The Joint European Torus (JET) at Abingdon, England, a cooperative venture of several countries, which has used the D-T reaction. Figure 14.4 shows the interior of JET with a person inside to provide scale.

FIGURE 14.4 Interior of tokamak fusion reactor Joint European Torus at Culham, U.K. (Courtesy Joint European Torus).

(c) The Japanese Atomic Energy Research Institute Tokamak-60 (JT-60 Upgrade) used to study plasma physics. The National Institute for Fusion Sciences also operates the Large Helical Device, a modern stellarator.

(d) The DIII-D of General Atomic in San Diego is a modification of Doublet III. It involves science studies of turbulence, stability, and interactions, along with the role of the diverter, a magnetic method of removing debris from a fusion reaction.

(e) The Alcator-C-Mod of MIT, a compact machine with high general performance.

Concepts other than the tokamak have been studied. Princeton Plasma Physics Laboratory operates the National Spherical Torus Experiment, in which a hole passes through a spherical plasma (see References). The National Compact Stellarator Experiment, in which the chamber is in the shape of the figure eight (see References) was cancelled in 2008 by the Department of Energy.

14.4 INERTIAL CONFINEMENT MACHINES

Another approach to practical fusion is ICF, which uses very small pellets of a deuterium and tritium mixture as high-density gas or as ice. The pellets are heated by laser light or by high-speed particles. They act as miniature hydrogen bombs, exploding and delivering their energy to a wall and cooling medium. Figure 14.5 shows a quarter coin with some of the spheres. Their diameter is approximately 0.3 millimeters. To cause the thermonuclear reaction, a large number of beams of laser light or ions are trained on a pellet from different directions. A pulse of energy of the order of a nanosecond is delivered by what is called the "driver." The mechanism is believed to be as follows: the initial energy evaporates some material from the surface of the microsphere in a manner similar to the ablation of the surface of a spacecraft entering the earth's atmosphere. The particles that are driven off form a plasma around the sphere that can absorb further energy. Electrons are conducted through the sphere to heat it and cause more ablation. As particles leave the surface, they impart a reaction momentum to the material inside the sphere, just as a space rocket is propelled by escaping gases. A shock wave moves inward, compressing the D-T mixture to many thousands of times normal density and temperature. At the center, a spark of energy approximately 1 keV sets off the thermonuclear reaction. A burn front involving alpha particles moves outward, consuming the D-T fuel as it goes. Energy is shared by the neutrons, charged particles, and electromagnetic radiation, all of which will eventually be recovered as thermal energy. Consistent numbers are: 1 milligram of D-T per pellet, 5 million joules driver energy, an energy gain (fusion to driver) of approximately 60, and a frequency of 10 bursts per second.

FIGURE 14.5 Gold microshells containing high-pressure D-T gas for use in laser fusion (Courtesy Los Alamos National Laboratory, No. CN 76-6442).

In an alternate indirect method of heating, laser light or ions bombard the walls of a pellet cavity called a hohlraum, producing X-rays that drive the pellet target. One advantage besides high-energy efficiency is insensitivity to focus of the illuminating radiation.

The energy released in the series of microexplosions is expected to be deposited in a layer of liquid such as lithium that is continuously circulated over the surface of the container and out to a heat exchanger. This isolation of the reaction from metal walls is expected to reduce the amount of material damage. Other candidate wall protectors are liquid lead and flibe. It may not be necessary to replace the walls frequently or to install special resistant coatings. Figure 14.6 shows a schematic arrangement of a laser-fusion reactor.

Research on ICF is carried out at several locations in the United States:

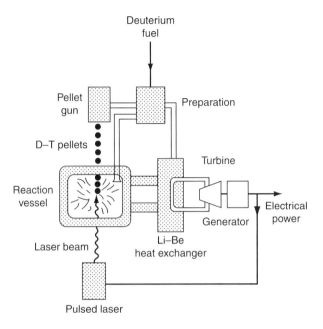

Deuterium
fuel

Pellet gun

Preparation

D–T pellets

Turbine

Reaction
vessel

Electrical
power

Generator

Laser beam

Li–Be
heat exchanger

Pulsed laser

FIGURE 14.6 Laser-fusion reactor.

(a) Lawrence Livermore National Laboratory (LLNL) operated Nova from 1985 to 1999. It used a neodymium-glass laser, with 10 separate beams. Nova could deliver 40 kJ of 351-nm light in a 1-ns pulse. It was the first ICF machine to exceed the Lawson criterion. Experiments are discussed in a comprehensive Web site (see References). LLNL is also the site of the National Ignition Facility (NIF), which has a dual purpose. The first is to provide information on target physics for the United States research program in ICF. The second is to simulate conditions in thermonuclear weapons as an alternative to underground testing actual devices (also see Chapter 26). NIF will have 192 beam lines focused on a target fuel capsule. The design permits either direct or indirect heating. One beam line called Beamlet was tested successfully, then transferred to Sandia National Laboratories. See References for additional information.

(b) The University of Rochester's Laboratory for Laser Energetics (LLE) operates the facility OMEGA, which has had impressive success, see Figure 14.7(B). Also, the effect on ICF of magnetic fields is being studied.

(c) Sandia National Laboratories first demonstrated with its Particle Beam Fusion Accelerator (PBFA) that targets could be heated with a proton beam. The equipment was converted into the Z-accelerator, which uses a pulse of current to create a powerful pinch effect (see Section 14.3).

The energy from the electrical discharge goes into accelerating electrons that create X-rays that heat the DT capsules. Power levels of near 300 trillion watts have been achieved (see References).

(d) Lawrence Berkeley National Laboratory tests methods of accelerating heavy ions such as potassium to serve as driver for ICF.

(e) General Atomics provides inertial fusion targets—spheres and hohlraums—for other laboratories.

(f) Los Alamos National Laboratory had an excimer (excited molecular) laser facility Aurora. It was followed by Mercury, which produced energy of approximately 50 J.

A number of conceptual inertial fusion reactor designs have been developed by national laboratories, universities, and companies to highlight the needs for research and development. These designs are intended to achieve power outputs comparable to those of fission reactors. They include both laser-driven and ion-driven devices. Examples are HIBALL-II (University of Wisconsin), HYLIFE-II and Cascade (Lawrence Livermore), Prometheus (McDonnell Douglas), and OSIRIS and SOMBRERO (W. J. Shafer). A considerable gap remains between performance required in these designs and that obtained in the laboratory to date.

14.5 OTHER FUSION CONCEPTS

Over the years since 1950 when research on fusion was begun in earnest, there have been many ideas for processes and systems. One was the "hybrid" reactor with a fusion core producing 14-MeV neutrons that would be absorbed in a uranium or thorium blanket, producing new fissile material. It was proposed as a stepping-stone to pure fusion, but seems unlikely to be considered.

Of the approximately 100 fusion reactions with light isotopes, some do not involve neutrons. If a "neutron-free" reaction could be harnessed, the problems of maintenance of activated equipment and disposal of radioactive waste could be eliminated. One example is proton bombardment of the abundant boron isotope, according to

$$ {}^1_1\text{H} + {}^{11}_5\text{B} \rightarrow 3\,{}^4_2\text{He} + 8.68\,\text{MeV}. $$

Because $Z = 5$ for boron, the electrostatic repulsion of the reactants is five times as great as the for D-T reaction, resulting in a much lower cross section. The temperature of the medium would have to be quite high. On the other hand, the elements are abundant and the boron-11 isotope is the dominant one in boron.

Another neutron-free reaction uses the rare isotope helium-3,

$$ {}^2_1\text{H} + {}^3_2\text{He} \rightarrow {}^4_2\text{He}(3.6\text{MeV}) + {}^1_1\text{H}(14.7\text{MeV}). $$

The D-^{3}He electrostatic force is twice as great as the D-T force, but because the products of the reaction are both charged, energy recovery would be more favorable. The process might be operated in such a way that neutrons from the D-D reaction could be minimized. This would reduce neutron bombardment to the vacuum chamber walls. A D-^{3}He fusion reactor thus could use a permanent first wall, avoiding the need for frequent replacement and at the same time reducing greatly the radioactive waste production by neutron activation.

The principal difficulty with use of the reaction is the scarcity of ^{3}He. One source is the atmosphere, but helium is present only to 5 ppm by volume of air and the helium-3 content is only 1.4 atoms per million of helium. Neutron bombardment of deuterium in a reactor is a preferable source. The decay of tritium in nuclear weapons could be a source of a few kilograms a year, but not enough to sustain an electrical power grid. Extraterrestrial sources are especially abundant but of course difficult to tap. Studies of moon rocks indicate that the lunar surface has a high ^{3}He content as the result of eons of bombardment by solar wind. Its ^{3}He concentration is 140 ppm in helium. It has been proposed that mining, refining, and isotope separation processes could be set up on the moon, with space shuttle transfer of equipment and product. The energy payback is estimated to be 250, the fuel cost for fusion would be 14 mills/kWh, and the total energy available is approximately 10^7 GWe-y. If space travel is further perfected, helium from the atmospheres of Jupiter and Saturn could be recovered in almost inexhaustible amounts.

A fusion process that is exotic physically but might be simple technically involves muons, negatively charged particles with mass 210 times that of the electron, and half-life 2.2 m/s. Muons can substitute for electrons in the atoms of hydrogen but with orbits that are 210 times smaller than the normal 0.53×10^{-10} m (see Exercise 14.5). They can be produced by an accelerator and directed to a target consisting of a deuterium-tritium compound such as lithium hydride. The beam of muons interacts with deuterons and tritons, forming DT molecules, with the muon playing the same role as an electron. However, the nuclei are now close enough together that some of them will fuse, releasing energy and allowing the muon to proceed to another molecule. Several hundred fusion events can take place before the muon decays. The system would appear not to need complicated electric and magnetic fields or large vacuum equipment. However, the concept has not been tested sufficiently to be able to draw conclusions about its feasibility or practicality.

Two researchers in 1989 reported the startling news that they had achieved fusion at room temperature, a process called "cold fusion." The experiments received a great deal of media attention because if the phenomenon were real, practical fusion would be imminent. Their equipment consisted of a heavy water electrolytic cell with cathode of metal palladium, which can absorb large amounts of hydrogen. They claimed that application of a voltage resulted in an enormous energy release. Attempts by others to confirm the experiments failed, and cold fusion is not believed to exist. Under certain conditions, there may be a release of large amounts of stored chemical energy, and research is continuing.

A scientific breakthrough whose effect is not yet determined is the discovery of materials that exhibit electrical superconductivity at relatively high temperatures, well above that of liquid helium. Fusion machines that use superconducting magnets will, at a minimum, be more energy efficient.

14.6 PROSPECTS FOR FUSION

Research on controlled thermonuclear processes has been underway for more than 50 years at several national laboratories, universities, and commercial organizations. The results of the studies include an improved understanding of the processes, the ability to calculate complex magnetic fields, the invention and testing of many devices and machines, and the collection of much experimental data. Over that period, there has been an approach to breakeven conditions, but progress has been painfully slow, involving decades rather than years. Various reasons have been suggested for this. First, and probably most important, is the fact that fusion is an extremely complex process from both the scientific and engineering standpoints. Second are policy decisions (e.g., emphasis on fundamental plasma physics rather than building large machines to reveal the true dimensions of the problem). In the case of ICF, the United States security classification related to weapons inhibited free international exchange of research information. Finally, there have been inconsistencies in funding allocations.

Figure 14.7(A) shows accomplishments of the MCF machines being tested. The plots give the Lawson criterion product of number density n and confinement time τ as a function of ion temperature T expressed as an energy in keV. Also noted on the diagram are the goals of breakeven and ignition. Although breakeven has been achieved, there still is a considerable way to go to approach ignition.

Figure 14.7(B) shows the progress by ICF machines. The plot relates the ion temperature to the product of density and radius as discussed in Section 14.2. OMEGA is expected to come near ignition and NIF to exceed it.

Predictions have repeatedly been made that practical fusion was only 20 y away. Two events provide some encouragement that the elusive 20-y figure might be met. The first is the discovery of a new tokamak current. As noted earlier, current flow in the plasma is induced by the changing external magnetic field. Because that field cannot increase indefinitely, it would be necessary to shut down and start over. In 1971 it had been predicted that there was an additional current in a plasma, but not until 1989 was that verified in several tokamaks. That "bootstrap" current amounts to up to 80% of the total, such that its contribution would allow essentially continuous operation.

The second event was a breakthrough in late 1997 in fusion energy release. Most fusion research had been conducted with the D-D reaction rather than the D-T reaction, to avoid the complication of contamination of equipment by

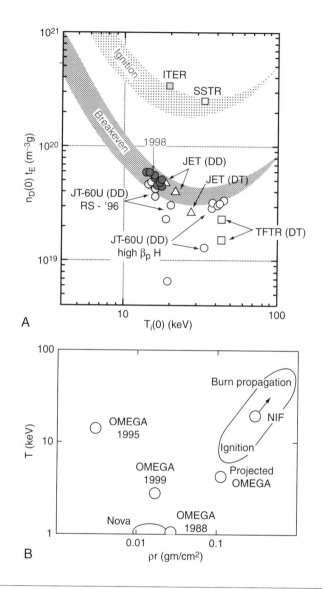

FIGURE 14.7 (A) Progress towards a practial MCF reactor. (Courtesy Japan Atomic Energy Research Institue. Thanks are due Robert Heeter). (B) Progress towards a practial ICF reactor. (Courtesy Lawrence Livermore National Laboratory. Thanks are due to John Soures and Alan Wootton.)

radioactive tritium. At the JET in England, tritium was injected as a neutral beam into a plasma. A series of records was set, ultimately giving 21 MJ of fusion energy, a peak power of 16 MW, and a ratio of fusion power to input power of 0.65. These results greatly exceeded those from D-D reactions.

In 1985, progress in tokamak performance over the years prompted planning for a large machine with the acronym ITER. Its objective is to demonstrate that fusion can be used to generate electrical power. Scientists will study conditions expected in a fusion power plant.

Participants in the project are the European Union, Japan, China, India, South Korea, Russia, and the United States (which withdrew in 1999 but reentered in 2001). ITER will be built at Cadarache in the South of France, with leadership provided by Japan. On November 21, 2006, an agreement was signed that established the ITER International Organization.

The tokamak is expected to produce more power than it consumes, with a Q value greater than 1. Technologies used include superconducting magnets, heat-resistant materials, remote handling systems for radioactive components, and breeding of tritium from lithium. The design was completed in 2001. Some of the features are a large plasma volume of roughly elliptical shape, a blanket to absorb neutron energy, and a diverter to extract the energy of charged particles and the helium ash. Selected parameters are:

Plasma major radius	6.2 m
Plasma minor radius	2.0 m
Magnetic field	5.3 T
Plasma current	15 MA
Fusion power	500 MW
Burn time	>400 s
Power amplification	>10
Temperature	$10^7 \,^\circ\mathrm{C}$

The schedule calls for the first plasma by 2016 and an operating period for approximately 20 y. The construction cost is estimated to be 5 billion Euros (1 EUR $\cong$ 1.45 USD), with another 5 billion over the 20 y. ITER maintains a comprehensive Web site (see References).

DEMO is a proposed fusion power plant that follows ITER's expected success. It is intended to be comparable in output to current fission power plants, being able to make the tritium it needs. There is a possibility that some magnetic confinement concept other than the tokamak will be used.

Most of the R&D on magnetic fusion has been focused on the tokamak mode. There is a possibility that some magnetic confinement concept other than the tokamak will be used. The United States participates in and supports the ITER program but continues to explore other concepts. The Department of Energy's Office of Fusion Energy Sciences provides research funds and receives recommendations from the Fusion Energy Sciences Advisory Committee (FESAC). In the Innovative Concepts Conferences (ICC), alternatives are discussed. Inspection of a typical meeting program reveals the diversity of fusion science and technology (see References).

It seems that practical fusion reactors still will not be available soon unless there is an unanticipated breakthrough or a completely new idea arises that changes the prospects dramatically. There is yet much to understand about plasma processes and a great deal of time is required to carry out research, development, and testing of a system that will provide competitive electric power.

From time to time, the wisdom of pursuing a vigorous and expensive research program in controlled fusion has been questioned in light of the uncertainty of success in achieving affordable fusion power. An excellent answer is the statement attributed by Robin Herman (see References) to the fusion pioneer Lyman Spitzer, "A fifty percent probability of getting a power source that would last a billion years is worth a great deal of enthusiasm."

14.7 SUMMARY

A fusion reactor, yet to be developed, would provide power that uses a controlled fusion reaction. Of the many possible nuclear reactions, the one that will probably be used first involves deuterium and tritium (produced by neutron absorption in lithium). A D-T reactor that yields net energy must exceed the ignition temperature of approximately 4.4 keV and have a product $n\tau$ above approximately 10^{14}, where n is the fuel particle number density and τ is the confinement time. Several experimental machines have been tested, involving an electrical discharge (plasma) that is constrained by electric and magnetic fields. One promising fusion machine, the tokamak, achieves magnetic confinement in a doughnut-shaped structure. ITER is a large fusion facility being built in France by an international consortium. Research is also underway on inertial confinement in which laser beams or charged particle beams cause the explosion of miniature D-T pellets. A neutron-free reaction involving deuterium and helium-3 would be practical if the moon could be mined for helium.

14.8 EXERCISES

14.1 Noting that the radius of motion R of a particle of charge q and mass m in a magnetic field B is $R = mv/qB$ and that the kinetic energy of rotation in the x-y plane is $(1/2)mv^2 = kT$, find the radii of motion of electrons and deuterons if B is 10 Wb/m^2 and kT is 100 keV.

14.2 Show that the effective nuclear reaction for a fusion reactor that uses deuterium, tritium, and lithium-6 is

$$^2_1H + ^6_3Li \rightarrow 2^4_2He + 22.4\text{MeV}.$$

14.3 Verify the statement that in the D-T reaction the ^4_2He particle will have 1/5 of the energy.

14.4 (a) Assuming that in the D-D fusion reaction the fuel consumption is 0.151 g/MWd (Exercise 7.3), find the energy release in J/kg. By how large a factor is the value larger or smaller than that for fission?

(b) If heavy water costs $100/kg, what is the cost of deuterium per kilogram?

(c) Noting 1 kWh $= 3.6 \times 10^6$ J, find from (a) and (b) the energy cost in mills/kWh.

14.5 (a) With the formula for the radius of the smallest electron orbit in hydrogen,

$$R = (10^7/m)(\hbar/ec)^2$$

where $\hbar = h/2\pi$ and the basic constants in the Appendix, verify that R is 0.529×10^{-10} m.

(b) Show that the rest energy of the muon, 105.66 MeV, is approximately 207 times the rest energy of the electron.

(c) What is the radius of the orbit of the muon about hydrogen in the muonium atom?

(d) The lengths of the chemical bonds in H_2 and in other compounds formed from hydrogen isotopes are all approximately 0.74×10^{-10} m. Estimate the bond in molecules where the muon replaces the electron.

(e) How does the distance in (d) compare with the radii of the nuclei of D and T (see Section 2.6)?

Computer Exercise

14.A Computer program FUSION describes a collection of small modules that calculate certain parameters and functions required in the analysis of a plasma and a fusion reactor. Among the properties considered are the theoretical fusion reaction cross sections, the Maxwellian distribution and characteristic velocities, the impact parameter for 90 ° ion scattering, the Debye length, cyclotron and plasma frequencies, magnetic field parameters, and electrical and thermal conductivities. The filenames of the modules are MAXWELL, VELOCITY, DEBYE, IMPACT, RADIUS, MEAN-PATH, TRANSIT, and CROSECT. Explore the modules with the menus provided and the sample input numbers.

14.9 REFERENCES

T. Kenneth Fowler, *The Fusion Quest*, Johns Hopkins University Press, Baltimore, 1997. Highly readable, nonmathematical treatment of both MCF and ICF.

T. J. Dolan, *Fusion Research*, Pergamon Press, New York, 1982.

http://www.fusionnow.org/dolan.html
Corrected electronic version INIS 2001.

Weston M. Stacey, *Fusion: An Introduction to the Physics and Technology of Magnetic Confinement Fusion*, John Wiley & Sons, New York, 1984.

John D. Lindl, *Inertial Confinement Fusion: The Quest for Ignition and Energy Gain Using Indirect Drive*, AIP Press, Springer-Verlag, New York, 1998.

J. Raeder, et al., *Controlled Nuclear Fusion: Fundamentals of its Utilization for Energy Supply*, John Wiley & Sons, Ltd., Chichester, England, 1986.

Keishiro Niu, *Nuclear Fusion*, Cambridge University Press, Cambridge, England, 1989. Treats both magnetic confinement by tokamaks and inertial confinement by lasers and ions.

S. Pfalzner, *An Introduction to Inertial Confinement Fusion*, Taylor and Francis and CRC Press, New York, 2006.

A. A. Harms, K. F. Schoepf, G. H. Miley, and D. R. Kingdon, *Principles of Fusion Energy*, World Scientific, Singapore, 2000.

General Atomics Fusion Energy Research
http://fusion.gat.com
Select various links in Research, e.g., DIII-D.

ITER (International Tokamak Reactor)
http://www.iter.org
Full description of the project and the device.

United States Fusion Energy Sciences Program
http://www.science.doe.gov/ofes
Includes links to fusion research activities around the world.

NSTX (National Spherical Torus Experiment)
http://nstx.pppl.gov
Select Overview/NSTX overview.

NCSX (National Compact Stellarator Experiment)
http://ncsx.pppl.gov
Select About NCSX.

JET (Joint European Torus)
http://www.jet.efda.org/index.html
Located in the U.K., it serves 20 European countries.

Sandia National Laboratories
http://www.sandia.gov/pulsedpower
Select Pulsed Power Facilities.

Nova Laser Experiments and Stockpile Stewardship
https://www.llnl.gov/str/Remington.html
Relationship of NIF to weapons tests.

National Ignition Facility Project (LLNL)
https://lasers.llnl.gov
Photos, videos, and explanations.

Laboratory for Laser Energetics
http://www.lle.rochester.edu
Information on OMEGA and research projects.

Alternative fusion methods
www.iccworkshops.org/icc2007/program/php
Innovative Confinement Concepts conference at University of Maryland.

Fusion Power Associates
http://fusionpower.org
Source of information on latest technical and political developments. Select Fusion Library.

Robin Herman, *Fusion: The Search for Endless Energy*, New York, Cambridge University Press, 1990.

Nuclear Energy and Man

3

The History of Nuclear Energy

15

THE DEVELOPMENT of nuclear energy exemplifies the consequences of scientific study, technological effort, and commercial application. We will review the history for its relation to our cultural background, which should include man's endeavors in the broadest sense. The author subscribes to the traditional conviction that history is relevant. Present understanding is grounded in recorded experience, and although we cannot undo errors, we can avoid them in the future. It is to be hoped that we can establish concepts and principles about human attitudes and capabilities that are independent of time to help guide future action. Finally, we can draw confidence and inspiration from the knowledge of what human beings have been able to accomplish.

15.1 THE RISE OF NUCLEAR PHYSICS

The science on which practical nuclear energy is based can be categorized as classical, evolving from studies in chemistry and physics for the last several centuries, and modern, which relates to investigations over the past hundred years into the structure of the atom and nucleus. The modern era begins in 1879 with Crookes' achievement of ionization of a gas by an electric discharge. Thomson in 1897 identified the electron as the charged particle responsible for electricity. Roentgen in 1895 discovered penetrating X-rays from a discharge tube, and Becquerel in 1896 found similar rays—now known as γ-rays—from an entirely different source, the element uranium, which exhibited the phenomenon of radioactivity. The Curies in 1898 isolated the radioactive element radium. As a part of his revolutionary theory of motion, Einstein in 1905 concluded that the mass of any object increased with its speed and stated his now-famous formula $E = mc^2$, which expresses the equivalence of mass and energy. At that time, no experimental verification was available, and Einstein could not have foreseen the implications of his equation.

217

In the first third of the 20th century, a host of experiments with the various particles coming from radioactive materials led to a rather clear understanding of the structure of the atom and its nucleus. It was learned from the work of Rutherford and Bohr that the electrically neutral atom is constructed from negative charge in the form of electrons surrounding a central positive nucleus, which contains most of the matter of the atom. Through further work by Rutherford in England around 1919, it was revealed that even though the nucleus is composed of particles bound together by forces of great strength, nuclear transmutations could be induced (e.g., the bombardment of nitrogen by helium yields oxygen and hydrogen).

In 1930, Bothe and Becker bombarded beryllium with α particles from polonium and found what they thought were γ-rays but which Chadwick in 1932 showed to be neutrons. A similar reaction is now used in nuclear reactors to provide a source of neutrons. Artificial radioactivity was first reported in 1934 by Curie and Joliot. Particles injected into nuclei of boron, magnesium, and aluminum gave new radioactive isotopes of several elements. The development of machines to accelerate charged particles to high speeds opened up new opportunities to study nuclear reactions. The cyclotron, developed in 1932 by Lawrence, was the first of a series of devices of ever-increasing capability.

15.2 THE DISCOVERY OF FISSION

During the 1930s, Enrico Fermi and his coworkers in Italy performed a number of experiments with the newly discovered neutron. He reasoned correctly that the lack of charge on the neutron would make it particularly effective in penetrating a nucleus. Among his discoveries was the great affinity of slow neutrons for many elements and the variety of radioisotopes that could be produced by neutron capture. Breit and Wigner provided the theoretical explanation of slow neutron processes in 1936. Fermi made measurements of the distribution of both fast and thermal neutrons and explained the behavior in terms of elastic scattering, chemical binding effects, and thermal motion in the target molecules. During this period, many cross sections for neutron reactions were measured, including that of uranium, but the fission process was not identified.

It was not until January 1939 that Hahn and Strassmann of Germany reported that they had found the element barium as a product of neutron bombardment of uranium. Frisch and Meitner made the guess that fission was responsible for the appearance of an element that is only half as heavy as uranium and that the fragments would be very energetic. Fermi then suggested that neutrons might be emitted during the process, and the idea was born that a chain reaction that releases great amounts of energy might be possible. The press picked up the idea, and many sensational articles were written. The information on fission, brought to the United States by Bohr on a visit from Denmark, prompted a flurry of activity at several universities, and by 1940 nearly a hundred papers had appeared in the technical literature. All of the qualitative characteristics of the chain reaction were soon learned—the moderation of neutrons by light elements, thermal and resonance capture, the existence of fission in U-235 by thermal neutrons, the large energy of fission fragments, the release of neutrons, and the possibility of producing transuranic elements, those beyond uranium in the periodic table.

15.3 THE DEVELOPMENT OF NUCLEAR WEAPONS

The discovery of fission, with the possibility of a chain reaction of explosive violence, was of especial importance at this particular time in history, since World War II had begun in 1939. Because of the military potential of the fission process, a voluntary censorship of publication on the subject was established by scientists in 1940. The studies that showed U-235 to be fissile suggested that the new element plutonium, discovered in 1941 by Seaborg, might also be fissile and thus also serve as a weapon material. As early as July 1939, four leading scientists—Szilard, Wigner, Sachs, and Einstein—had initiated a contact with President Roosevelt, explaining the possibility of an atomic bomb based on uranium. As a consequence, a small grant of $6000 was made by the military to procure materials for experimental testing of the chain reaction. Before the end of World War II, a total of $2 billion had been spent, an almost inconceivable sum in those times. After a series of studies, reports, and policy decisions, a major effort was mounted through the United States Army Corps of Engineers under General Leslie Groves. The code name "Manhattan District" (or "Project") was devised, with military security mandated on all information.

Although a great deal was known about the individual nuclear reactions, there was great uncertainty as to the practical behavior. Could a chain reaction be achieved at all? If so, could Pu-239 in adequate quantities be produced? Could a nuclear explosion be made to occur? Could U-235 be separated on a large scale? These questions were addressed at several institutions, and design of production plants began almost concurrently, with great impetus provided by the involvement of the United States in World War II after the attack on Pearl Harbor in December 1941 by the Japanese. The distinct possibility that Germany was actively engaged in the development of an atomic weapon served as a strong stimulus to the work of American scientists, most of whom were in universities. They and their students dropped their normal work to enlist in some phase of the project.

As it was revealed by the Alsos Mission (see References), a military investigation project, Germany had actually made little progress toward an atomic bomb. A controversy has developed as to the reasons for its failure (see References). One theory is that an overestimate was made of the critical mass of enriched uranium—as tons rather than kilograms—with the conclusion that such amounts were not achievable. The other theory is that scientist Werner Heisenberg, the leader of the German effort, had deliberately stalled the project to prevent Hitler from having a nuclear weapon to use against the Allies.

The Manhattan Project consisted of several parallel endeavors. The major effort was in the United States, with cooperation from the United Kingdom, Canada, and France.

An experiment at the University of Chicago was crucial to the success of the Manhattan Project and also set the stage for future nuclear developments. The team under Enrico Fermi assembled blocks of graphite and embedded spheres of uranium oxide and uranium metal into what was called a "pile." The main control rod was a wooden stick wrapped with cadmium foil. One safety rod would automatically drop on high neutron level; one was attached to a weight with a rope, ready to be cut with an axe if necessary. Containers of neutron-absorbing cadmium-salt solution were ready to be dumped on the assembly in case of emergency. On December 2, 1942, the system was ready. The team gathered for the key experiment as in Figure 15.1, an artist's recreation of the scene. Fermi calmly made calculations with his slide rule and called for the main control rod to be withdrawn in steps. The counters clicked faster and faster until it was necessary to switch to a recorder, whose pen kept climbing. Finally, Fermi closed his slide rule and said, "The reaction is self-sustaining. The curve is exponential." The sixtieth anniversary of the startup of the Chicago pile was celebrated in 2002 by *Nuclear News* with an issue containing an account of the event (see References). Comments by prominent nuclear leaders are also featured.

This first man-made chain reaction gave encouragement to the possibility of producing weapons material and was the basis for the construction of several nuclear reactors at Hanford, Washington. By 1944, these were producing plutonium in kilogram quantities.

FIGURE 15.1 The first man-made chain reaction, December 2, 1942. Painting *The Birth of the Atomic Age* by Gary Sheahan (Courtesy Chicago Historical Society).

At the University of California at Berkeley, under the leadership of Ernest O. Lawrence, the electromagnetic separation "calutron" process for isolating U-235 was perfected, and government production plants at Oak Ridge, Tennessee, were built in 1943. At Columbia University, the gaseous diffusion process for isotope separation was studied, forming the basis for the present production system, the first units of which were built at Oak Ridge. At Los Alamos, New Mexico, a research laboratory was established under the direction of J. Robert Oppenheimer. Theory and experiment led to the development of the nuclear weapons, first tested at Alamogordo, New Mexico, on July 16, 1945, and later used at Hiroshima and Nagasaki in Japan.

The brevity of this account fails to describe adequately the dedication of scientists, engineers, and other workers to the accomplishment of national objectives or the magnitude of the design and construction effort by American industry. Two questions are inevitably raised. Should the atom bombs have been developed? Should they have been used? Some of the scientists who worked on the Manhattan Project have expressed their feeling of guilt for having participated. Some insist that a lesser demonstration of the destructive power of the weapon should have been arranged, which would have been sufficient to end the conflict. Many others believed that the security of the United States was threatened and that the use of the weapon shortened World War II greatly and thus saved a large number of lives on both sides. Many surviving military personnel scheduled to invade Japan have expressed gratitude for the action taken.

In the ensuing years the buildup of nuclear weapons continued despite efforts to achieve disarmament. The dismantlement of excess weapons will require many years. It is of some comfort, albeit small, that the existence of nuclear weapons has served for several decades as a deterrent to a direct conflict between major powers.

The discovery of nuclear energy has a potential for the betterment of mankind through fission and fusion energy resources and through radioisotopes and their radiation for research and medical purposes. The benefits can outweigh the detriments if mankind is intelligent enough not to use nuclear weapons again.

15.4 REACTOR RESEARCH AND DEVELOPMENT

One of the first important events in the United States after World War II ended was the creation of the United States Atomic Energy Commission (AEC). This civilian federal agency was charged with the management of the nation's nuclear programs, including military protection and development of peaceful uses of the atom. Several national laboratories were established to continue nuclear research, sites such as Oak Ridge, Argonne (near Chicago), Los Alamos, and Brookhaven (on Long Island). A major objective was to achieve practical commercial nuclear power through research and development. Oak Ridge first studied a gas-cooled reactor and later planned a high-flux reactor fueled with highly enriched uranium alloyed with and clad with aluminum that used water as moderator and coolant. A reactor was eventually built in Idaho as the Materials Testing Reactor. The submarine reactor described in Section 20.1 was adapted by Westinghouse Electric Corporation for use as the first commercial power plant at Shippingport, Pennsylvania. It began operation in 1957 at an electric power output of 60 MW. Uranium dioxide pellets as fuel were first introduced in this pressurized water reactor (PWR) design.

In the decade of the 1950s, several reactor concepts were tested and dropped for various reasons (see References). One used an organic liquid diphenyl as a coolant on the basis of a high boiling point. Unfortunately, radiation caused deterioration of the compound. Another was the homogeneous aqueous reactor, with a uranium salt in water solution that was circulated through the core and heat exchanger. Deposits of uranium led to excess heating and corrosion of wall materials. The sodium-graphite reactor had liquid metal coolant and carbon moderator. Only one commercial reactor of this type was built. The high-temperature gas-cooled reactor, developed by General Atomics, has not been widely adopted but is a potential alternative to light water reactors by virtue of its graphite moderator, helium coolant, and uranium-thorium fuel cycle.

Two other reactor research and development programs were underway at Argonne over the same period. The first program was aimed at achieving power plus breeding of plutonium by use of the fast reactor concept with liquid sodium coolant. The first electric power from a nuclear source was produced in late 1951 in the Experimental Breeder Reactor, and the possibility of breeding was demonstrated. The second program consisted of an investigation of the possibility of allowing water in a reactor to boil and generate steam directly. The principal concern was with the fluctuations and instability associated with the boiling. Tests called BORAX were performed that showed that a boiling reactor could operate safely, and work proceeded that led to electrical generation in 1955. The General Electric Company then proceeded to develop the boiling water reactor (BWR) concept further, with the first commercial reactor of this type put into operation at Dresden, Illinois, in 1960.

On the basis of the initial success of the PWR and BWR, and with the application of commercial design and construction know-how, Westinghouse and General Electric were able, in the early 1960s, to advertise large-scale nuclear plants of power approximately 500 MWe that would be competitive with fossil fuel plants in the cost of electricity. Immediately thereafter, there was a rapid move on the part of the electric utilities to order nuclear plants, and the growth in the late 1960s was phenomenal. Orders for nuclear steam supply systems for the years 1965–1970 inclusive amounted to approximately 88,000 MWe, which was more than a third of all orders, including fossil-fueled plants. The corresponding nuclear electric capacity was approximately a quarter of the total United States capacity at the end of the period of rapid growth.

After 1970, the rate of installation of nuclear plants in the United States declined, for a variety of reasons: (a) the very long time required—greater than

10 years—to design, license, and construct nuclear facilities; (b) the energy conservation measures adopted as a result of the Arab oil embargo of 1973–1974, which produced a lower growth rate of demand for electricity; and (c) public opposition in some areas. The last order for nuclear plants was in 1978; a number of orders were canceled; and construction was stopped before completion on others. The total nuclear power capacity of the 104 United States reactors in operation by 2007 was 102,056 MW, representing more than 20% of the total electrical capacity of the country. In other parts of the world there were 339 reactors in operation with a 274,285 MW capacity.

This large new power source was put in place in a relatively brief period of 40 y after the end of World War II. The endeavor revealed a new concept—that large-scale national technological projects could be undertaken and successfully completed by the application of large amounts of money and the organization of the efforts of many sectors of society. The nuclear project in many ways served as a model for the United States space program of the 1960s. The important lesson that the history of nuclear energy development may have for us is that urgent national and world problems can be solved by wisdom, dedication, and cooperation.

For economic and political reasons, considerable uncertainty developed about the future of nuclear power in the United States and many other countries of the world. In the next section we will discuss the nuclear controversy and later describe the dimensions of the problem and its solution in coming decades.

15.5 THE NUCLEAR CONTROVERSY

The popularity of nuclear power decreased during the decades of the 1970s and 1980s, with adverse public opinion threatening to prevent the construction of new reactors. We can attempt to analyze this situation, explaining causes and assessing effects.

In the 1950s, nuclear power was heralded by the AEC and the press as inexpensive, inexhaustible, and safe. Congress was highly supportive of reactor development, and the general public seemed to feel that great progress toward a better life was being made. In the 1960s, however, a series of events and trends raised public concerns and began to reverse the favorable opinion.

First was the youth movement against authority and constraints. In that generation's search for a simpler and more primitive or "natural" lifestyle, the use of wood and solar energy was preferred to energy based on the high technology of the "establishment." Another target for opposition was the military–industrial complex, blamed for the generally unpopular Viet Nam War. A 1980s version of the antiestablishment philosophy advocated decentralization of government and industry, favoring small locally controlled power units based on renewable resources.

Second was the 1960s environmental movement, which revealed the extent to which industrial pollution in general was affecting wildlife and human beings, with its related issue of the possible contamination of air, water, and land by accidental releases of radioactivity from nuclear reactors. Continued revelations about the extent of improper management of hazardous chemical waste had a side effect of creating adverse opinion about radioactive wastes.

Third was a growing loss of respect for government, with public disillusionment becoming acute as an aftermath of the Watergate affair. Concerned observers cited actions taken by the AEC or the DOE without informing or consulting those affected. Changes in policy about radioactive waste management from one administration to another resulted in inaction, interpreted as evidence of ignorance or ineptness. A common opinion was that no one knew what to do with the nuclear wastes.

A fourth development was the confusion created by the sharp differences in opinion among scientists about the wisdom of developing nuclear power. Nobel prize winners were arrayed on both sides of the argument; the public understandably could hardly fail to be confused and worried about where the truth lay.

The fifth was the fear of the unknown hazard represented by reactors, radioactivity, and radiation. It may be agreed that an individual has a much greater chance of dying in an automobile accident than from exposure to fallout from a reactor accident. But because the hazard of the roads is familiar, and believed to be within the individual's control, it does not evoke nearly as great concern as does a nuclear event.

The sixth was the association between nuclear power and nuclear weapons. This is in part inevitable, because both involve plutonium, use the physical process of fission with neutrons, and have radioactive byproducts. On the other hand, the connection has been cultivated by opponents of nuclear power, who stress the similarities rather than the differences.

As with any subject, there is a spectrum of opinions. At one end are the dedicated advocates, who believe nuclear power to be safe, badly needed, and capable of success only if opposition can be reduced. A large percentage of physical scientists and engineers fall in this category, believing that technical solutions for most problems are possible.

Next are those who are technically knowledgeable but are concerned about the ability of man to avoid reactor accidents or to design and build safe waste facilities. Depending on the strength of their concerns, they may believe that consequences outweigh benefits.

Next are average citizens who are suspicious of government and who believe in "Murphy's law," being aware of failures such as Love Canal, Three Mile Island, the 1986 space shuttle, and Chernobyl. They have been influenced as well by strong antinuclear claims and tend to be opposed to further nuclear power development, although they recognize the need for continuous electric power generation.

At the other end of the spectrum are ardent opponents of nuclear power who actively speak, write polemics, intervene in licensing hearings, lead demonstrations, or take physical action to try to prevent power plants from coming into being.

There are a variety of attitudes among representatives of the news and entertainment media—newspapers, magazines, radio, television, and movies—but there is an apparent tendency toward skepticism. Nuclear advocates are convinced that any incident involving reactors or radiation is given undue emphasis by the media. They believe that if people were adequately informed they would find nuclear power acceptable. This view is only partially accurate, for two reasons: (a) some technically knowledgeable people are strongly antinuclear; and (b) irrational fears cannot be removed by additional facts. Many people have sought to analyze the phenomenon of nuclear fear, but the study by Weart (see References) is one of the best.

Nevertheless, in recent years there has been a growing public acceptance of nuclear power in the United States for several reasons: (a) the industry has maintained an excellent nuclear safety record, through actions by utilities, the Nuclear Regulatory Commission, and the Institute of Nuclear Power Operations; (b) increased awareness of energy needs, related to the continued demand for expensive and uncertain foreign oil; and (c) realization that the generation of electricity by fission does not release greenhouse gases that contribute to global warming. Polls indicate that two thirds of the public favor the construction of new nuclear plants and some communities welcome them.

15.6 SUMMARY

A series of investigations in atomic and nuclear physics in the period 1879–1939 led to the discovery of fission. New knowledge was developed about particles and rays, radioactivity, and the structures of the atom and the nucleus. The existence of fission suggested that a chain reaction involving neutrons was possible and that the process had military significance. A major national program was initiated in the United States during World War II. The development of uranium isotope separation methods, of nuclear reactors for plutonium production, and of weapons technology culminated in the use of atomic bombs to end the war.

In the post-war period, emphasis was placed on maintenance of nuclear protection and on peaceful applications of nuclear processes under the AEC. Four reactor concepts—the pressurized water, boiling water, fast breeder, and gas-cooled—evolved through work by national laboratories and industry. The first two concepts were brought to commercial status in the 1960s.

Public support for nuclear power waned for a variety of reasons in the late 20th century but has increased markedly in recent years.

15.7 EXERCISES

15.1   Enter into an Internet search engine the phrase "Nuclear Age Timeline" and consult several sources to develop a list of what seem to be the most important single events in each decade.

15.2   Enter into Google the phrase Einstein letter Roosevelt and read the letter of August 2, 1939.

15.8 REFERENCES

Historical Figures in Nuclear Science
http://www.accessexcellence.org/AE/AEC/CC/historical_background.html
Biographies of Roentgen, Becquerel, Mme. Curie, and Rutherford.

Henry De Wolfe Smyth, *Atomic Energy for Military Purposes*, Princeton University Press, Princeton, NJ. 1945. Reprinted by AMS Press, New York, 1978 and reprinted with additional commentary by Stanford University Press, Palo Alto, CA 1989. The first unclassified account of the nuclear effort of World War II.

The Smyth Report online
http://nuclearweaponarchive.org/Smyth
Chapters I, II, IX–XIII, and Appendices

Cynthia C. Kelly, Editor, *The Manhattan Project: The Birth of the Atomic Bomb in the Words of its Creators, Eyewitnesses, and Historians*, Blackdog & Leventhal, New York, 2007.

The Manhattan Project: An Interactive History
http://www.cfo.doe.gov/me70/manhattan/about.htm
Comprehensive account of WWII project.

The History of Nuclear Energy
http://www.ne.doe.gov/pdfFiles/History.pdf
Includes timeline.

Atomic Heritage Foundation
http://childrenofthemanhattanproject.org
Information on the nuclear effort of WWII.

Robert Jungk, *Brighter Than a Thousand Suns*, Harcourt Brace & Co., New York, 1958. A very readable history of nuclear developments from 1918 to 1955, with emphasis on the atomic bomb. Based on conversations with many participants.

Lt. General Leslie R. Groves, *Now it Can be Told*, Harper & Row, New York, 1962. The account of the Manhattan Project by the person in charge.

Richard G. Hewlett and Oscar E. Anderson, Jr., *The History of the United States Atomic Energy Commission*. The New World, Vol. I, 1939/1946; Richard G. Hewlett and Francis Duncan, Atomic Shield, Vol. II, 1947/1952, United States Atomic Energy Commission, Washington DC, 1972. The first volume starts with the discovery of fission and covers the Manhattan Project in great detail.

The United States Department of Energy, 1977–1994
http://energy.gov/media/Summary_History.pdf
Includes material from 1939 to present. By Terrence R. Fehner and Jack M. Holl.

Nuclear Age Timeline
http://www.em.doe.gov/Publications/timeline.aspx
Events from the pre-40s to 1993, with sources of additional information.

The Manhattan Project: Making the Atomic Bomb
http://www.osti.gov/accomplishments/pdf/DE99001330/DE99001330.pdf
January 1999 edition by F. G. Gosling.

Nuclear Technology Milestones: 1942–1998
http://www.nei.org
Search on Timeline. By Nuclear Energy Institute.

Corbin Allardice and Edward R. Trapnell, "The First Pile," *Nuclear News*, November 2002, p.34.

Raymond L. Murray, "The Etymology of 'Scram'," *Nuclear News*, La Grange Park, IL August 1988. pp. 105–107. An article on the origin of the word based on correspondence with Norman Hilberry, the Safety Control Rod Axe Man in the Chicago experiment.

Boris T. Pash, *The Alsos Mission*, Universal Publishing & Distribution Corp., New York, 1970. This book, long out of print, describes the adventures of the military unit that entered Germany in World War II to find out about the atom bomb effort.

Samuel A. Goudsmit, *Alsos*, Henry Schuman, Inc., New York, 1947. Reprinted in 1983 by Tomash Publishers and American Institute of Physics, with a new introduction by R. V. Jones and supplemental photographs. The technical leader of the Alsos mission describes his experiences and assessment of the German atom bomb program.

Mark Walker, *German National Socialism and the Quest for Nuclear Power, 1939–1949*, Cambridge University Press, Cambridge, UK, 1989.

Mark Walker, "Heisenberg, Goudsmit, and the German Atomic Bomb," *Physics Today*, January 1990, p. 52. This article prompted a series of letters to the editor in the issues of May 1991 and February 1992.

Sir Frank Charles, *Operation Epsilon: The Farm Hall Transcripts*, University of California Press, Berkeley, 1993. Recorded conversations among German scientists captured at the end of World War II.

Paul Lawrence Rose, *Heisenberg and the Nazi Atomic Bomb Project: A Study in German Culture*, University of California Press, Berkeley, 1998. Thorough investigation with voluminous bibliography. Believes that the critical mass was overestimated by error.

Stephane Groueff, *Manhattan Project*, Little, Brown & Co., Boston, 1967. Subtitle: The Untold Story of the Making of the Atomic Bomb. The author benefited from material published right after WWII and interviews with participants. The book was praised by both General Leslie Groves and AEC Commissioner Glenn Seaborg.

Kai Bird and Martin J. Sherwin, *American Prometheus: The Triumph and Tragedy of J. Robert Oppenheimer*, Knopf, New York, 2005. The best book on the subject.

Nuell Pharr Davis, *Lawrence and Oppenheimer*, Simon & Schuster, New York, 1968. (Also available in paperback from Fawcett Publications). The roles of the two atomic leaders, Ernest O. Lawrence and J. Robert Oppenheimer, and their conflict are described. The book presents an accurate portrayal of the two men.

Bertrand Goldschmidt, *The Atomic Complex*, American Nuclear Society, La Grange Park, IL, 1982. A technical and political history of nuclear weapons and nuclear power. The author participated in developments in the United States and France.

Richard Rhodes, *The Making of the Atomic Bomb*, Simon & Schuster, New York, 1986. A detailed fascinating account of the Manhattan Project.

Richard Rhodes, *Dark Sun: The Making of the Hydrogen Bomb*, Simon & Schuster, New York, 1995. Research and development by both the United States and U.S.S.R. Describes espionage.

John W. Simpson, *Nuclear Power from Underseas to Outer Space*, American Nuclear Society, La Grange Park, IL, 1995. Personalized technical history of nuclear submarine, early nuclear power, and nuclear rocket by a Westinghouse executive.

Frank G. Dawson, *Nuclear Power: Development and Management of A Technology*, University of Washington Press, Seattle, 1976. Covers nuclear power development and regulations from 1946 to around 1975.

Raymond L. Murray, "Nuclear Reactors," in *Kirk-Othmer Encyclopedia of Chemical Technology*, John Wiley & Sons, New York, 2007.

The Virtual Nuclear Tourist
http://www.nucleartourist.com
A wealth of information with links to history sites. By Joseph Gonyeau.

Spencer R. Weart, *Nuclear Fear: A History of Images*, Harvard University Press, Cambridge, MA, 1988. An analysis of attitudes toward nuclear power.

Stephen Hilgartner, Richard C. Bell, and Rory O'Connor, *Nukespeak*, Sierra Club Books, San Francisco, 1982. Subtitle: Nuclear Language, Visions, and Mindset.

Bernard L. Cohen, *Before It's Too Late*, Plenum Press, New York, 1983. Subtitle: A Scientist's Case FOR Nuclear Energy. Discusses risks on nuclear power and public perception of them.

History of the Department of Energy's National Laboratories
http://www.osti.gov/accomplishments/nuggets/historynatlabs.html
Histories of 13 individual national nuclear laboratories.

Biological Effects of Radiation

16

ALL LIVING species are exposed to a certain amount of natural radiation in the form of particles and rays. In addition to the sunlight, without which life would be impossible to sustain, all beings experience cosmic radiation from space outside the earth and natural background radiation from materials on the earth. Rather large variations occur in the radiation from one place to another, depending on mineral content of the ground and on the elevation above sea level. Man and other species have survived and evolved within such an environment despite the fact that radiation has a damaging effect on biological tissue. The situation has changed somewhat by the discovery of the means to generate high-energy radiation that uses various devices such as X-ray machines, particle accelerators, and nuclear reactors. In the assessment of the potential hazard of the new artificially generated radiation, comparison is often made with levels in naturally occurring background radiation.

We will now describe the biological effect of radiation on cells, tissues, organs, and individuals; identify the units of measurement of radiation and its effect; and review the philosophy and practice of setting limits on exposure. Special attention will be given to regulations related to nuclear power plants.

A brief summary of modern biological information will be useful in understanding radiation effects. As we know, living beings represent a great variety of species of plants and animals; they are all composed of cells, which carry on the processes necessary to survival. The simplest organisms such as algae and protozoa consist of only one cell, whereas complex beings such as man are composed of specialized organs and tissues that contain large numbers of cells, examples of which are nerve, muscle, epithelial, blood, skeletal, and connective. The principal components of a cell are the *nucleus* as control center, the *cytoplasm* containing vital substances, and the surrounding *membrane* as a porous cell wall. Within the nucleus are the *chromosomes*, which are long threads containing hereditary material. The growth process involves a form of cell multiplication called *mitosis* in which the chromosomes separate to form two new cells identical to the original one. The reproduction process involves a cell division process called *meiosis* in which germ cells are produced with only half the necessary complement of chromosomes, such that the union of sperm and egg creates a complete new entity. The laws of heredity are based on this process. The genes are the distinct regions on the chromosomes that are responsible for inheritance of certain body characteristics. They are constructed of a universal molecule called DNA, a very long spiral staircase structure, with the stair steps consisting of paired molecules of four types (see References). Duplication of cells in complete detail involves the splitting of the DNA molecule along its length, followed by the accumulation of the necessary materials from the cell to form two new ones. In the case of man, 46 chromosomes are present, containing approximately 4 billion of the DNA molecule steps in an order that describes each unique person.

16.1 PHYSIOLOGICAL EFFECTS

The various ways that moving particles and rays interact with matter discussed in earlier chapters can be reexamined in terms of biological effect. Our emphasis previously was on what happened to the radiation. Now, we are interested in the effects on the medium, which are viewed as "damage" in the sense that disruption of the original structure takes place, usually by *ionization*. We saw that energetic electrons and photons are capable of removing electrons from an atom to create ions; that heavy charged particles slow down in matter by successive ionizing events; that fast neutrons in slowing impart energy to target nuclei, which in turn serve as ionizing agents; and that capture of a slow neutron results in a gamma ray and a new nucleus. In Section 5.2 we defined linear energy transfer (LET). A distinction is made between low LET (electrons and gamma rays) and high LET (alpha particles and neutrons).

As a good rule of thumb, 32 eV of energy is required on average to create an ion pair. This figure is rather independent of the type of ionizing radiation, its energy, and the medium through which it passes. For instance, a single 4-MeV alpha particle would release approximately 10^5 ion pairs before stopping. Part of the energy goes into molecular excitation and the formation of new chemicals. Water in cells can be converted into free radicals such as H, OH, H_2O_2, and HO_2. Because the human body is largely water, much of the effect of radiation can be attributed to the chemical reactions of such products. In addition, direct damage can occur, in which the radiation strikes certain molecules of the cells, especially the DNA that controls all growth and reproduction. Turner (see References) displays computer-generated diagrams of ionization effects.

The most important point from the biological standpoint is that the bombarding particles have energy, which can be transferred to atoms and molecules of living cells, with a disruptive effect on their normal function. Because an organism is composed of very many cells, tissues, and organs, a disturbance of one atom is likely to be imperceptible, but exposure to many particles or rays can alter the function of a group of cells and thus affect the whole system. It is usually assumed that damage is cumulative, even though some accommodation and repair take place.

The physiological effects of radiation may be classified as *somatic*, which refers to the body and its state of health, and *genetic*, involving the genes that transmit hereditary characteristics. The somatic effects range from temporary skin reddening when the body surface is irradiated, to a life shortening of an exposed individual because of general impairment of the body functions, to the initiation of cancer in the form of tumors in certain organs or as the blood disease, leukemia. The term "radiation sickness" is loosely applied to the immediate effects of exposure to very large amounts of radiation. The genetic effect consists of mutations, in which progeny are significantly different in some respect from their parents, usually in ways that tend to reduce the chance of survival. The effect may extend over many generations.

Although the amount of ionization produced by radiation of a certain energy is rather constant, the biological effect varies greatly with the type of tissue involved. For radiation of low penetrating power such as α particles, the outside skin can receive some exposure without serious hazard, but for radiation that penetrates tissue readily such as X-rays, gamma rays, and neutrons, the critical parts of the body are bone marrow as blood-forming tissue, the reproductive organs, and the lenses of the eyes. The thyroid gland is important because of its affinity for the fission product iodine, whereas the gastrointestinal tract and lungs are sensitive to radiation from radioactive substances that enter the body through eating or breathing.

If a radioactive substance enters the body, radiation exposure to organs and tissues will occur. However, the foreign substance will not deliver all of its energy to the body because of partial elimination. If there are N atoms present, the physical decay rate is λN and the biological elimination rate is $\lambda_b N$. The total rate is $\lambda_e N$, where the effective decay constant is

$$\lambda_e = \lambda + \lambda_b.$$

The corresponding relation between half-lives is

$$1/t_e = 1/t_H + 1/t_b.$$

For example, iodine-131 has an 8-day physical half-life and a 4-day biological half-life for the thyroid gland. Thus its effective half-life is $2\frac{2}{3}$ days.

16.2 RADIATION DOSE UNITS

A number of specialized terms need to be defined for discussion of biological effects of radiation. First is the absorbed *dose* (*D*). This is the amount of energy in joules imparted to each kilogram of exposed biological tissue, and it appears as excitation or ionization of the molecules or atoms of the tissue. The SI unit of dose is the *gray* (Gy), which is 1 J/kg. To illustrate, suppose that an adult's gastrointestinal tract weighing 2 kg receives energy of amount 6×10^{-5} J as the result of ingesting some radioactive material. The dose would be

$$D = (6 \times 10^{-5} \text{ J})/(2 \text{ kg}) = 3 \times 10^{-5} \text{ J/kg} = 3 \times 10^{-5} \text{ Gy}.$$

An older unit of energy absorption is the *rad*, which is 0.01 J/kg (i.e., 1 Gy = 100 rads). The preceding dose to the GI tract would be 0.003 rad or 3 millirads.

The biological effect of energy deposition may be large or small depending on the type of radiation. For instance a rad dose caused by fast neutrons or alpha particles is much more damaging than a rad dose by X-rays or gamma rays. In general, heavy particles create a more serious effect than do photons because of the greater energy loss with distance and resulting higher concentration of ionization. The *dose equivalent* (*H*) as the biologically important quantity takes account of those differences by scaling the energy absorption up by a *quality factor* (*QF*), with values as in Table 16.1. Thus

$$H = (D)(QF).$$

If the D is expressed in Gy, then H is in *sieverts* (Sv); if the D is in rad, then H is in *rems*. Suppose that the gastrointestinal tract dose were due to plutonium, an alpha particle emitter. The equivalent dose would then be (20) $(3 \times 10^{-5}) = 6 \times 10^{-4}$ Sv or 0.6 mSv. Alternately, the H would be (20)(0.003) = 0.06 rem or 60 millirems. In scientific research and the analysis of biological effects of radiation, the SI units gray and sievert are used; in nuclear plant operation, rads and rems are more commonly used. Summarizing, conversion factors commonly needed:

1 gray (Gy) = 100 rads
0.01 Gy = 1 rad
1 sievert (Sv) = 100 rems
10 mSv = 1 rem
1 mSv = 100 mrems
10 μSv = 1 mrem.

Table 16.1 Quality Factors (NRC 10CFR20, see References)	
X-rays, gamma rays, beta particles	1
Thermal neutrons (0.025 eV)	2
Neutrons of unknown energy	10
High-energy protons	10
Heavy ions, including alpha particles	20

The great variety of radioactivity and radiation units is confusing and a source of much time and effort to convert between systems. Although it would be desirable to switch completely to the newer units, it is unrealistic to expect it to happen. The United States at least will long be burdened with a dual system of units. We will frequently include the newer units in parentheses. As a memory device, let sieverts be $ and rems be ¢.

Computer Exercise 16.B makes use of the program RADOSE to conveniently translate numbers from a technical article.

The long-term effect of radiation on an organism also depends on the rate at which energy is deposited. Thus the *dose rate*, expressed in convenient units such as rads per hour or millirems per year, is used. Note that if dose is an energy, the dose rate is a power.

We will describe the methods of calculating dosage in Chapter 21. For perspective, however, we can cite some typical figures. A single sudden exposure that gives the whole body of a person 20 rems (0.2 Sv) will give no perceptible clinical effect, but a dose of 400 rems (4 Sv) will probably be fatal; the typical annual natural radiation exposure, including radon, of the average citizen is 295 millirems; medical and dental applications give another 54, with all other sources 11, giving a total of 360 millirems (3.6 Sv). Figure 16.1 shows the distribution by percentages. Earlier literature on radiation protection cited typical annual dose figures of 100 mrems (0.1 rem), but in recent times the effect of radon amounting to approximately 200 mrems/y has been included. Computer Exercise 16.A addresses the buildup of radon in an enclosed space without ventilation.

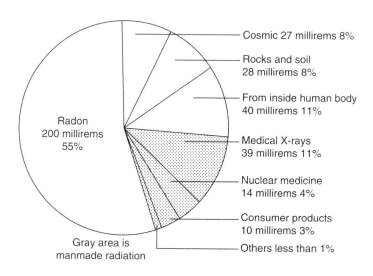

FIGURE 16.1 Annual average radiation exposure to an individual in the United States. The total is 360 millirems (NCRP 93, 1988).

A wide variation of annual dose exists. The radiation level in many parts of the world is larger than the average annual United States figure of 360 mrems (3.6 mSv) and greatly exceeds the Nuclear Regulatory Commission (NRC) regulatory limit of 0.1 mrem/y for members of the public and 5 rems/y for nuclear workers. According to Eisenbud (see References, Chapter 3), in countries such as India the presence of thorium gives exposures of approximately 600 mrems/y. Many waters at health spas give rates that are orders of magnitude higher. Other examples are Brazil with an annual dose of 17,500 mrems (175 mSv) and the city of Ramsar, Iran, where hot springs bring up radium-226, with an annual figure of 26,000 mrems (260 mSv). The frequency of cancer and life span of people in that area is not noticeably different from other populations.

The amounts of energy that result in biological damage are remarkably small. A gamma dose of 400 rems, which is very large in terms of biological hazard, corresponds to 4 J/kg, which would be insufficient to raise the temperature of a kilogram of water as much as 0.001 °C. This fact shows that radiation affects the function of the cells by action on certain molecules, not by a general heating process.

16.3 BASIS FOR LIMITS OF EXPOSURE

A typical bottle of aspirin will specify that no more than two tablets every 4 hours should be administered, implying that a larger or more frequent "dose" would be harmful. Such a limit is based on experience accumulated over the years with many patients. Although radiation has a medical benefit only in certain treatment, the idea of the need for a limit is similar.

As we seek to clean up the environment by controlling emissions of waste products from industrial plants, cities, and farms, it is necessary to specify water or air concentrations of materials such as sulfur or carbon monoxide that are below the level of danger to living beings. Ideally, there would be zero contamination, but it is generally assumed that some releases are inevitable in an industrialized world. Again, limits on the basis of the knowledge of effects on living beings must be set.

For the establishment of limits on radiation exposure, agencies have been in existence for many years. Examples are the International Commission on Radiological Protection (ICRP) and the National Council on Radiation Protection and Measurements (NCRP). Their general procedure is to study data on the effects of radiation and to arrive at practical limits that take account of both risk and benefit of the use of nuclear equipment and processes.

Extensive studies of the survival of colonies of cells exposed to radiation have led to the conclusion that double-strand breaks in DNA are responsible for cell damage. Hall (see References) shows diagrams of various types of breaks. Much of the research was prompted by the need to know the best way to administer radiation for the treatment of cancer. A formula for the number of breaks N as a function of dose D is

$$N = aD + bD^2$$

where the first term refers to the effect of a single particle, the second to that of two successive particles. This is the so-called linear-quadratic model. The fraction S of the cells surviving a dose D is deduced to be

$$S = \exp(-pN)$$

where p is the probability that a break causes cell death. The formula is somewhat analogous to that for radioactive decay or the burn up of an isotope. Cell survival data are fitted to graphs where near zero dose, the curve is linear.

There have been many studies of the effect of radiation on animals other than human beings, starting with early observations of genetic effects on fruit flies. Small mammals such as mice provide a great deal of data rapidly. Because controlled experiments on humans are unacceptable, most of the available information on somatic effects comes from improper practices or accidents. Data are available, for example, on the incidence of sickness and death from exposure of workers who painted radium on luminous-dial watches or of doctors who used X-rays without proper precautions. The number of serious radiation exposures in the nuclear industry is too small to be of use on a statistical basis. The principal source of information is the comprehensive study of the victims of the atomic bomb explosions in Japan in 1945. Continued studies of effects are being made (see RERF in References). The incidence of fatalities as a function of dose is plotted on a graph similar to Figure 16.2 where the available data are seen to lie only in the high dosage range. In the range below 10 rads, there is no statistical indication of any increase in incidence of fatalities over the number in unexposed populations.

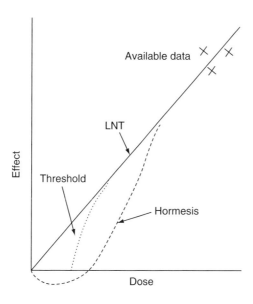

FIGURE 16.2 Radiation hazard analysis.

The nature of the graph of effect versus dose in the low-dose range is unknown. One can draw various curves starting with one based on the linear-quadratic model. Another would involve a threshold, below which there is no effect. To be conservative (i.e., to overestimate effects in the interest of providing protection) organizations such as NCRP and NRC support a linear extrapolation to zero as sketched in Figure 16.2. This assumption is given the acronym LNT (linear no-threshold). Other organizations such as the American Nuclear Society and BELLE (see References) believe that there is insufficient evidence for the LNT curve. Critics such as Radiation, Science, Health (RSH) (see References) believe that the insistence on conservatism and the adoption by the NRC of the LNT recommendation causes an unwarranted expense for radiation protection. One writer (see References) calls the ethics of the use of the LNT into question.

There is considerable support for the existence of a process called *hormesis*, in which a small amount of a substance such as aspirin is beneficial, whereas a large amount can cause bleeding and even death. When applied to radiation, the curve of effect versus dose labeled "hormesis" in Figure 16.2 would dip below the horizontal axis near zero. The implication is that a small amount of radiation can be beneficial to health. A definitive physiological explanation for the phenomenon is not available, but it is believed that small doses stimulate the immune system of organisms, a cellular effect. There also may be molecular responses involving DNA repair or free radical detoxification. A book by Luckey (see References) is devoted to the subject of hormesis.

The National Academies sponsored a study designed to settle the matter. In report BEIR VII (see References), the linear no-threshold theory is preserved. The report states that it is unlikely that there is a threshold in the dose-effect relation. The authors of BEIR VII dismiss reports supporting hormesis by saying they are "... based on ecological studies ...." However, research on the subject is recommended. No explanation is given in the report for the low cancer incidence in regions of the United States with much higher radiation dosage than the average.

There are significant economic implications of the distinction among models of radiation effect. The LNT version leads to regulations on the required degree of cleanup of radioactivity-contaminated sites and the limits on doses to workers in nuclear facilities.

There is evidence that the biological effect of a given dose administered almost instantly is greater than if it were given over a long period of time. In other words, the hazard is less for low dose rates, presumably because the organism has the ability to recover or adjust to the radiation effects. If, for example (see Exercise 16.2), the effect actually varied as the square of the dose, the linear curve would overestimate the effect by a factor of 100 in the vicinity of 1 rem. Although the hazard for low dose rates is small, and there is no clinical evidence of permanent injury, it is *not* assumed that there is a threshold dose (i.e., one below which no biological damage occurs). Instead, it is assumed that there is always some risk. The linear hypothesis is retained despite the likelihood that it is overly conservative. The basic question then faced by standards-setting bodies is "what is the maximum acceptable upper limit for exposure?" One answer is zero, on the grounds that any radiation is deleterious. The view is taken that it is unwarranted to demand zero, as both maximum and minimum, because of the benefit from the use of radiation or from devices that have potential radiation as a byproduct.

The limits adopted by the NRC for use starting January 1, 1994, are 5 rems/y (0.05 Sv/y) for total body dose of adult occupational exposure. Alternate limits for worker dose are 50 rems/y (0.5 Sv/y) to any individual organ or tissue other than the eye, 15 rems/y (0.15 Sv/y) for the eye, and 50 rems/y (0.5 Sv/y) to the skin or any extremity. In contrast, the limits for individual members of the public are set at 0.1 rem/y (1 mSv/y) (i.e., 2% of the worker dose). These figures take account of all radiation sources and all affected organs.

For the special case of the site boundary of a low-level radioactive waste disposal facility, NRC specifies a lower figure for the general public, 25 mrems/y (0.25 mSv/y), and for a nuclear power plant, a still lower 3 mrems/y (0.03 mSv/y).

The occupational dose limits are considerably higher than the average United States citizen's background dose of 0.36 rem/y, whereas those for the public are only a fraction of that dose. The National Academy of Sciences Committee on the Biological Effects of Ionizing Radiation analyzes new data and prepares occasional reports, such as BEIR V (see References). In the judgment of that group, the lifetime increase in risk of a radiation-induced cancer fatality for workers when the official dose limits are used is 8×10^{-4} per rem, and the NRC and other organizations assume half of that figure, 4×10^{-4} per rem. However, because the practice of maintaining doses as low as reasonably achievable (ALARA) in nuclear facilities keeps doses well below the limit, the increase in chance of cancer is only a few percent. Measured dose figures have decreased considerably over the years, as reported by the Institute of Nuclear Power Operations. Table 16.2 shows the trend.

For the general public, the radiation exposure from nuclear power plants is negligible compared with other hazards of existence.

It has been said that knowledge about the origins and effects of radiation is greater than that for any chemical contaminant. The research over decades has led to changes in acceptable limits. In the very early days, soon after radioactivity and X-rays were discovered, no precautions were taken, and indeed radiation was thought to be healthful, hence the popularity of radioactive caves and springs that one might frequent for health purposes. Later, reddening of the skin was a crude indicator of exposure. Limits have decreased a great deal in recent decades, making the older literature outdated. A further complication is the development cycle: research and analysis of effects; discussion, agreement, and publication of conclusions as by ICRP and NCRP; and proposal, review, and adoption of rules by an agency such as the NRC. This cycle requires considerable time.

Table 16.2 Median Radiation Dose in Nuclear Power Plants	
Year	**Dose (Person-rems Per Unit)**
1984	591
1986	344
1988	320
1990	331
1992	251
1994	217
1996	162
1998	119
2000	105
2002	95
2004	88
2006	91

For example, recommendations made in 1977 were not put into effect until 1994, leaving some later suggested modifications in limbo. The time lag can sometimes be different for various applications, leading to apparent inconsistencies.

16.4 SOURCES OF RADIATION DOSAGE

The term "radiation" has come to imply something mysterious and harmful. We will try to provide here a more realistic perspective. The key points are that (a) people are more familiar with radiation than they believe; (b) there are sources of natural radiation that parallel the man-made sources; and (c) radiation can be both beneficial and harmful.

First, solar radiation is the source of heat and light that supports plant and animal life on earth. We use its visible rays for sight; the ultraviolet rays provide vitamin D, cause tanning, and produce sunburn; the infrared rays give us warmth; and finally, solar radiation is the ultimate source of all weather. Man-made devices produce electromagnetic radiation that is identical physically to solar and has the same biological effect. Familiar equipment includes microwave ovens, radio and TV transmitters, infrared heat lamps, ordinary light bulbs and fluorescent lamps, ultraviolet tanning sources, and X-ray machines. The gamma rays from nuclear processes have higher frequencies and thus greater penetrating power than X-rays but are no different in kind from other electromagnetic waves.

In recent years, concern has been expressed about a potential cancer hazard because of electromagnetic fields (EMF) from 60 Hz sources such as power lines or even household circuits or appliances. Biological effects of EMF on lower organisms have been demonstrated, but research on physiological effects on humans is inconclusive and is continuing. More recently, concerns have arisen about the possibility of brain tumors caused by cell phone use.

Human beings are continually exposed to gamma rays, beta particles, and α particles from radon and its daughters. Radon gas is present in homes and other buildings as a decay product of natural uranium, a mineral occurring in many types of soil. Neutrons as a part of cosmic radiation bombard all living things.

If is often said that all nuclear radiation is harmful to biological organisms. There is evidence, however, that the statement is not quite true. First, there seems to be no increase in cancer incidence in the geographic areas where natural radiation background is high. Second, in the application of radiation for the treatment of disease such as cancer, advantage is taken of differences in response of normal and abnormal tissue. The net effect in many cases is of benefit to the patient. Third, it is possible that the phenomenon of hormesis occurs with small doses of radiation, as discussed in Section 16.3.

In Chapter 21 we will discuss radiation protective measures and the application of regulatory limits on exposure.

16.5 RADIATION AND TERRORISM

One of the weapons that terrorists could use is the "dirty bomb" or technically "radiological dispersal device (RDD)." It would consist of a radioactive substance that is a gamma emitter such as cobalt-60 (5.26 y) or cesium-137 (30.2 y), combined with an explosive such as dynamite. An explosion would kill or maim those nearby but the dispersed radioactivity would harm relatively few people.

The resulting damage would be very small compared with that of a nuclear bomb. The main effects would be to create fear and panic, to cause serious property damage, and to require extensive decontamination over an area such as a city block. It can be called a "weapon of mass disruption." The effect would be psychological because of public fear of radiation. Also, as noted by Mark M. Hart (see References) "... effectiveness [of an RDD] can be unintentionally enhanced by professionals and public officials." The news media could also contribute to terror. With the public in panic, the intent of terrorists would be achieved.

To reduce the chance of terrorist action, tight control must be maintained over radiation sources used in research, processing, and medical treatment. The NRC addresses the main features of dirty bombs (see References), and the Centers for Disease Control and Prevention (CDC) has a Web site containing answers to questions (see References).

16.6 SUMMARY

When radiation interacts with biological tissue, energy is deposited and ionization takes place that causes damage to cells. The effect on organisms is somatic, related to body health, and genetic, related to inherited characteristics. Radiation dose equivalent as a biologically effective energy deposition per gram is usually expressed in rems, with natural background giving approximately 0.36 rem/y in the United States. Exposure limits are set by use of data on radiation effects at high dosages with a conservative linear hypothesis applied to predict effects at low dose rates. Such assumptions have been questioned. The terrorist use of a radiological dispersal device (dirty bomb) is of concern.

16.7 EXERCISES

16.1 A beam of 2-MeV alpha particles with current density 10^6 cm^{-2}–s^{-1} is stopped in a distance of 1 cm in air, number density 2.7×10^{19} cm^{-3}. How many ion pairs per cm^3 are formed? What fraction of the targets experience ionization?

16.2 If the chance of fatality from radiation dose is taken as 0.5 for 400 rems, by what factor would the chance at 2 rems be overestimated if the effect varied as the square of the dose rather than linearly?

16.3 A worker in a nuclear laboratory receives a whole-body exposure for 5 minutes by a thermal neutron beam at a rate 20 millirads per hour. What dose (in mrads) and dose equivalent (in mrems) does he receive? What fraction of the yearly dose limit of 5000 mrems/y for an individual is this?

16.4 A person receives the following exposures in millirems in a year: 1 medical X-ray, 100; drinking water, 50; cosmic rays, 30; radon in house, 150; K-40 and other isotopes, 25; airplane flights, 10. Find the percentage increase in exposure that would be experienced if he also lived at a reactor site boundary, assuming that the maximum NRC radiation level existed there.

16.5 A plant worker accidentally breathes some stored gaseous tritium, a beta emitter with maximum particle energy 0.0186 MeV. The energy absorbed by the lungs, of total weight 1 kg, is 4×10^{-3} J. How many millirems dose equivalent was received? How many millisieverts? (NOTE: The average beta energy is one third of the maximum).

16.6 If a radioisotope has a physical half-life t_H and a biological half-life t_b, what fraction of the substance decays within the body? Calculate that fraction for 8-days I-131, biological half-life 4 days.

Computer Exercises

16.A A room with concrete walls is constructed with sand with a small uranium content, such that the concentration of radium-226 (1599 y) is 10^6 atoms per cm^3. Normally, the room is well ventilated so the gaseous radon-222 (3.82 d) is continually removed, but during a holiday the room is closed up. With the parent-daughter computer program RADIOGEN (Chapter 3), calculate the trend in air activity caused by Rn-222 over a week's period, assuming that half of the radon enters the room. Data on the room: 10 ft × 10 ft × 10 ft, walls 3 in. thick.

16.B A mixture of radiation and radioactivity units are used in an article on high natural doses (*IAEA Bulletin*, Vol. 33, No. 2, 1991, p. 36), as follows:

(a) Average radiation exposure in the world, 2.4 mSv/y.
(b) Average radiation exposure in S.W. India, 10 mGy/y.
(c) High outdoor dose in Iran, 9 mrems/h.
(d) Radon concentration at high altitudes in Iran, 37 kBq/m^3.
(e) Radon concentration in Czech houses, 10 kBq/m^3.
(f) High outdoor dose in Poland, 190 nGy/h.

With the computer program RADOSE, which converts numbers between units, find what the numbers mean in the familiar United States units mrems/y or pCi/1.

16.8 REFERENCES

DNA Structure
http://www.accessexcellence.org/AE/AEC/CC
Explains the role of radioactive labeling and displays the structure of DNA. From National Health Museum.

Bruce Alberts, Dennis Bray, Alexander Johnson, Julian Lewis, Martin Raff, Keith Roberts, and Peter Walter, *Essential Cell Biology: An Introduction to the Molecular Biology of the Cell*, Garland Publishing, New York, 1997. Highly recommended. Stunning diagrams and readable text.

Jacob Shapiro, *Radiation Protection*, 4th Ed., Harvard University Press, Cambridge, 2002. Subtitle: A Guide for Scientists and Physicians. A very readable textbook.

Herman Cember , *Introduction to Health Physics*, McGraw-Hill, New York, 1996. Thorough and up-to-date information and instruction. Contains many illustrative calculations and tables of data. Also see Herman Cember and Thomas E. Johnson, The Health Physics Solutions Manual, PS&E Publications, 1999.

Eric J. Hall, *Radiobiology for the Radiologist*, 5th Ed., J. B. Lippincott Co., Philadelphia, 2005. An authoritative work that explains DNA damage, cell survival curves, and radiotherapy techniques.

James E. Turner, *Atoms, Radiation, and Radiation Protection*, John Wiley & Sons, New York, 2007. Shows Monte Carlo electron tracks in water to illustrate radiation effects.

Edward Pochin, *Nuclear Radiation: Risks and Benefits*, Clarendon Press, Oxford University Press, Oxford, UK, New York, 1985. Sources of radiation and biological effects, including cancer and damage to cells and genes.

Eric J. Hall, *Radiation and Life*, 2nd Ed., Pergamon Press, New York, 1984. Discusses natural background, beneficial uses of radiation, and nuclear power. The author urges greater control of medical and dental X-rays.

Fred A. Mettler, Arthur C. Upton, and Robert D. Moseley, *Medical Effects of Ionizing Radiation*, Elsevier, Netherlands, 1995.

Bernard Schleien, Lester A. Slayback, Jr., and Brian Kent Birky, Editors, *Handbook of Health Physics and Radiological Health*, 3rd Ed., Williams & Wilkins, Baltimore, 1998. A greatly expanded version of a classical document of 1970.

Health Risks from Exposure to Low Levels of Ionizing Radiation (BEIR VII), Committee to Assess Health Risks ... of National Research Council, National Academy Press, Washington DC, 2005.

Radiation and Health Physics
http://www.umich.edu/~radinfo
Many links. By University of Michigan Student Chapter of the Health Physics Society.

Radiation Dose and Biological Effects
http://www.hps.org
Select Radiation Terms and Position Statements. From Health Physics Society.

Health Effects of Low-Level Radiation
http://www.ans.org/pi/ps
American Nuclear Society Position Statement No. 41 (2001).

Radiation Effects Research Foundation (RERF)
http://www.rerf.or.jp/index_e.html
Studies of health effects of atomic bomb radiation.

Biological Effects of Low-Level Exposure (BELLE)
http://www.belleonline.com
Select Newsletters for commentary on hormesis.

Radiation, Science, and Health (RSH)
http://www.radscihealth.org/rsh
Advocates use of scientific information on radiation matters.

Zbigniew Jaworowski, "Radiation Risk and Ethics," *Physics Today*, September 1999, p. 24. The article prompted a number of Letters in the April 2000 issue, p. 11.

T. D. Luckey, *Radiation Hormesis*, CRC Press, Boca Raton, 1991.

National Council on Radiation Protection and Measurements (NCRP)
http://www.ncrp.com
Provides summaries of reports.

Fact Sheet on Biological Effects of Radiation
http://www.nrc.gov/reading-rm/doc-collections/fact-sheets/bio-effects-radiation.html
Extensive discussion. Information from NRC.

Standards for Protection Against Radiation
http://www.nrc.gov/reading-rm/doc-collections/cfr/part020
Part 20 of the Code of Federal Regulations—Energy.

Health Effects of Radiation
http://www.epa.gov/rpdweb00/understand/health_effects.html
Questions and answers from Environmental Protection Agency.

Richard Wilson, "Resource Letter EIRLD-1: Effects of Ionizing Radiation at Low Doses," tfm*American Journal of Physics* 67, pp. 372–377 (1999). Puts radiation risks into perspective.

Richard Wilson and Edmund A. C. Crouch, *Risk-Benefit Analysis*, Harvard University Press, Cambridge, MA, 2001.

Mark M. Hart, "Disabling Radiological Dispersal Terror"
http://eed.llnl.gov/ans/2002/hart/hart_ans_2002.pdf
Select Conferences/. . .Terrorism. . ./. . .2002. . .

Fact Sheet on Dirty Bombs (NRC)
http://www.nrc.gov/reading-rm/doc-collections/fact-sheets
Select Dirty Bombs.

Frequently Asked Questions (FAQs) about Dirty Bombs
http://www.bt.cdc.gov/radiation/dirtybombs.asp
Centers for Disease Control and Prevention (CDC).

Information from Isotopes 17

THE APPLICATIONS of nuclear processes can be divided into three basic classes—military, power, and radiation. In a conference[†] shortly after the end of World War II the famous physicist Enrico Fermi discussed potential applications of radio-isotopes. He then said, "It would not be very surprising if the stimulus that these new techniques will give to science were to have an outcome more spectacular than an economic and convenient energy source or the fearful destructiveness of the atomic bomb."

Perhaps Fermi would be surprised to see the extent to which radioisotopes have become a part of research, medicine, and industry, as described in the following sections.

Many important economic and social benefits are derived from the use of isotopes and radiation. The discoveries of modern nuclear physics have led to new ways to observe and measure physical, chemical, and biological processes, providing the strengthened understanding so necessary for human survival and progress. The ability to isolate and identify isotopes gives additional versatility, supplementing techniques involving electrical, optical, and mechanical devices.

Radioisotopes have become even more prominent in the wake of the tragic events of September 11, 2001. Detection of potential hazards has become a high priority of the United States Department of Homeland Security.

Special isotopes of an element are distinguishable and thus traceable by virtue of their unique weight or their radioactivity, while essentially behaving chemically as do the other isotopes of the element. Thus it is possible to measure amounts of the element or its compounds and trace movement and reactions.

When one considers the thousands of stable and radioactive isotopes available and the many fields of science and technology that require knowledge of process details, it is clear that a catalog of possible isotope uses would be voluminous. Here we will only be able to compare the merits of stable and radioactive species,

---

[†]Enrico Fermi, "Atomic Energy for Power," in *Science and Civilization, The Future of Atomic Energy*, McGraw-Hill, New York, 1946.

to describe some of the special techniques, and to mention a few interesting or important applications of isotopes.

17.1 STABLE AND RADIOACTIVE ISOTOPES

Stable isotopes, as their name suggests, do not undergo radioactive decay. Most of the isotopes found in nature are in this category and appear in the element as a mixture. The principal methods of separation, according to isotopic mass, that have been used are electromagnetic, as in the large-scale mass spectrograph; and thermal-mechanical, as in the distillation or gaseous diffusion processes. Important examples are isotopes of elements involved in biological processes (e.g., deuterium and oxygen-18). The main advantages of stable isotopes are the absence of radiation effects in the specimens under study, the availability of an isotope of a chemical for which a radioactive species would not be suitable, and freedom from necessity for speed in making measurements, because the isotope does not decay in time. Their disadvantage is the difficulty of detection.

Radioactive isotopes, or radioisotopes, are available with a great variety of half-lives, types of radiation, and energy. They come from three main sources—charged particle reactions in an accelerator, neutron bombardment in a reactor, and separated fission products. Among the principal sources of stable and longer lived isotopes are the United States Department of Energy (DOE) (see References), MDS Nordion of Canada, and Russia. A number of cyclotrons that generate radioisotopes are located at hospitals. The main advantages of the use of radioisotopes are ease of detection of their presence through the emanations and the uniqueness of identifying the half-lives and radiation properties. Potential shortage is a perennial problem for users of radioisotopes. The number of reactor sources is limited and some are being shut down. In an American Nuclear Society position statement (see References) a strong recommendation is made, stating, "There is no present U.S. policy for the purpose of maintaining reliable sources of radioisotope supplies crucial for both medical and industrial applications."

17.2 TRACER TECHNIQUES

We will now describe several special methods involving radioisotopes and illustrate their use. The tracer method consists of the introduction of a small amount of an isotope and the observation of its progress as time goes on. For instance, the best way to apply fertilizer containing phosphorus to a plant may be found by including minute amounts of the radioisotope phosphorus-32, half-life 14.28 days, emitting 1.7 MeV beta particles. Measurements of the radiation at various times and locations in the plant by a detector or photographic film provides accurate information on the rate of phosphorus intake and deposition. Similarly, circulation of blood in the human body can be traced by the injection of a harmless solution of radioactive sodium, Na-24, 14.96-h half-life. For purposes of medical diagnosis, it is desirable to administer enough radioactive material to provide the needed data, but not so much that the patient is harmed.

The flow rate of many materials can be found by watching the passage of admixed radioisotopes. The concept is the same for flows as diverse as blood in the body, oil in a pipeline, or pollution discharged into a river. As sketched in Figure 17.1, a small amount of radioactive material is injected at a point, it is carried along by the stream, and its passage at a distance d away at time t is noted. In the simplest situation, the average fluid speed is d/t. It is clear that the half-life of the tracer must be long enough for detectable amounts to be present at the point of observation but not so long that the fluid remains contaminated by radio-active material.

In many tracer measurements for biological or engineering purposes, the effect of removing the isotope by other means besides radioactive decay must be considered. Suppose, as in Figure 17.2, that liquid flows in and out of a tank of volume V (cm^3) at a rate υ (cm^3/s). A tracer of initial amount N_0 atoms is injected and assumed to be uniformly mixed with the contents. Each second, the fraction of fluid (and isotope) removed from the tank is υ/V, which serves as a flow decay constant λ_f for the isotope. If radioactive decay were small, the counting rate from a detector would decrease with time as $\exp(-\lambda_f t)$. From this

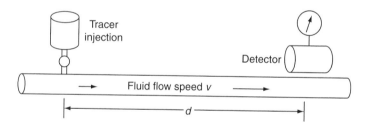

FIGURE 17.1 Tracer measurment of flow rate.

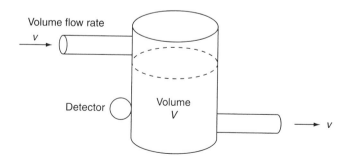

FIGURE 17.2 Flow decay.

trend, one can deduce either the speed of flow or volume of fluid, if the other quantity is known. If both radioactive decay (λ) and flow decay (λ_f) occur, the exponential formula may also be used, but with the effective decay constant $\lambda_e = \lambda + \lambda_f$. The composite effective half-life then can be found from the relationship

$$1/t_e = 1/t_H + 1/t_f.$$

This formula is seen to be of the same form as the one developed in Section 16.1 for radioactive materials in the body. Here, the flow half-life takes the place of the biological half-life.

Soon after Watson and Crick explained the structure of DNA in 1951, tracers P-13 and S-35 were used to prove that genes were associated with DNA molecules. Tritium-labeled thymidine, involved in the cell cycle, was synthesized. The field of molecular biology expanded greatly since then, leading to the Human Genome Project (see References), an international effort to map the complete genetic structure of human beings, involving chromosomes, DNA, genes, and protein molecules. Its purpose is to find which genes cause various diseases and to enable gene therapy to be applied. Part of the complex process of mapping is hybridization, in which a particular point on the DNA molecule is marked by a radioactive or fluorescent label.

An outgrowth of genetic research is DNA fingerprinting, a method of identifying individual persons, each of which (except for identical twins) has a unique DNA structure. In one of the techniques (RFLP) the procedure involves treating samples of blood, skin, or hair with an enzyme that splits DNA into fragments. The membrane containing them is exposed to a radioactive "probe" and dark bands appear on an X-ray film. The method is accurate but requires a large sample and a long exposure of film. An alternate method (PCR-STR) is more popular. It involves making multiple copies of DNA. The processes are used in crime investigation and court cases to help establish guilt or innocence and to give evidence in paternity disputes.

17.3 RADIOPHARMACEUTICALS

Radionuclides prepared for medical diagnosis and therapy are called radiopharmaceuticals. They include a great variety of chemical species and isotopes with half-lives ranging from minutes to weeks, depending on the application. They are generally beta- or gamma-ray emitters. Prominent examples are technetium-99m (6.01 h), iodine-131 (8.04 d), and phosphorus-32 (14.28 d).

A radionuclide generator is a long-lived isotope that decays into a short-lived nuclide used for diagnosis. The advantage over the use of the short-lived isotope directly is that speed or reliability of shipment is not a factor. As needed, the daughter isotope is extracted from the parent isotope. The earliest example of such a generator was radium-226 (1599 y), decaying into radon-222 (3.82 d).

Table 17.1 Radiopharmaceuticals Used in Medical Diagnosis

Radionuclide	Compound	Use
Technetium-99m	Sodium pertechnate	Brain scanning
Hydrogen-3	Tritiated water	Body water
Iodine-131	Sodium iodide	Thyroid scanning
Gold-198	Colloidal gold	Liver scanning
Chromium-51	Serum albumin	Gastrointestinal
Mercury-203	Chlormerodrin	Kidney scanning
Selenium-75	Selenomethionine	Pancreas scanning
Strontium-85	Strontium nitrate	Bone scanning

The most widely used generator is molybdenum-99 (65.9 h) decaying to technetium-99m (6.01 h). The Tc-99m is said to be "milked" from the Mo-99 "cow." Tc-99m is the most widely used radioisotope in nuclear medicine because of its favorable radiations and half-life. The parent isotope Mo-99 comes from Canada and other countries. If for any reason the United States borders were closed to imports of radioactive materials, innumerable medical tests would cease.

Several iodine isotopes are used. One produced by a cyclotron is I-123 (13.2 h). The accompanying isotopes I-124 (4.18 d) and I-126 (13.0 d) are undesirable impurities because of their excessively energetic gamma rays. Two fission products are I-125 (59.4 d) and I-131 (8.04 d).

Table 17.1 illustrates the variety of radionuclides used, their chemical forms, and the organs studied.

Specialists in radiopharmaceuticals are called radiopharmacists, who are concerned with the purity, suitability, toxicity, and radiative characteristics of the radioactive drugs they prepare.

17.4 MEDICAL IMAGING

Administering a suitable radiopharmaceutical to a patient results in a selective deposit of the radioactive material in the tissue or organ under study. The use of these radionuclides to diagnose malfunctions or disease is called "medical imaging." Approximately 20 million diagnostic nuclear medicine studies are performed each year in the United States. In imaging, a photographic screen or a detector examines the adjacent area of the body and receives an image of the organ, revealing the nature of some medical problem. A scanner consists of a sodium iodide crystal detector, movable in two directions, a collimator to define the radiation, and a recorder that registers counts in the sequence of the points it observes. In contrast, an Anger scintillation camera is stationary, with a number of photomultiplier tubes receiving gamma rays through a collimator with many holes, and an electronic data-processing circuit.

The Anger camera provides a view of activity in the form of a plane. The introduction of computer technology has made possible more sophisticated displays, including three-dimensional images. Such a process is called tomography, of which there are several types. The first is single photon emission computer tomography (SPECT), which has a rotating camera that takes a series of planar pictures of the region containing a radionuclide. A sodium iodide crystal detects uncollided photons from the radioactive source and produces electric signals. Data from 180 different angles are processed by a computer to give 2D and 3D views of the organ. SPECT is used especially for diagnosis of the heart, liver, and brain. The second is positron emission tomography (PET), in which a positron-emitting radiopharmaceutical is used. Three important examples are oxygen-15 (2 min), nitrogen-13 (10 min), and carbon-11 (20 min). They are isotopes of elements found in all organic molecules, allowing them to be used for many biological studies and medical applications, especially heart disease. A fourth, fluorine-18 (110 min), is especially important in brain studies, in which there is difficulty getting most chemicals through what is called the blood–brain barrier. In contrast, F-18 forms a compound that acts like glucose, which can penetrate brain tissue and show the location of a disease such as stroke or cancer. The isotopes are produced by a cyclotron on the hospital site, and the targets are quickly processed chemically to achieve the desired labeled compound. The gamma rays released in the annihilation of the positron and an electron are detected, taking advantage of the simultaneous emission (coincidences) of the two gammas and their motion in opposite directions. The data are analyzed by a computer to give high-resolution displays. PET scans are analogous to X-ray computerized axial tomography (CT) scans, but better for some purposes. Figure 17.3 compares the ability of CT and PET to locate a brain tumor.

An alternate diagnostic method that is very popular and does not involve radioactivity is magnetic resonance imaging (MRI). It takes advantage of the magnetic properties of atoms in cells. Formerly it was called nuclear magnetic resonance (NMR), but physicians adopted the new name to avoid the association with anything "nuclear." There are approximately 900 MRI units in the United States. References are included for the interested reader.

17.5 RADIOIMMUNOASSAY

Radioimmunoassay, discovered in 1960 by Yalow and Berson, is a chemical procedure that uses radionuclides to find the concentration of biological materials very accurately in parts per billion and less. It was developed in connection with studies of the human body's immune system. In that system a protective substance

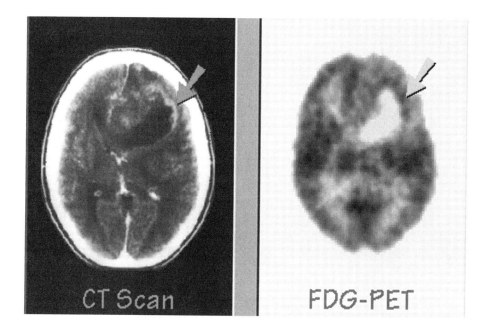

FIGURE 17.3 CT and PET scans of a brain tumor. (Courtesy of Lawrence Berkeley National Laboratory).

(antibody) is produced when a foreign protein (antigen) is introduced. The method makes use of the fact that antigens and antibodies also react. Such reactions are involved in vaccinations, immunizations, and skin tests for allergies.

The object is to measure the amount of an antigen present in a sample containing an antibody. The latter has been produced previously by repeatedly immunizing a rabbit or guinea pig and extracting the antiserum. A small amount of the radioactively labeled antigen is added to the solution. There is competition between the two antigens, known and unknown, to react with the antibody. For that reason the method is also called competitive binding assay. A chemical separation is performed, and the radioactivity in the products is compared with those in a standard reaction. The method has been extended to many other substances including hormones, enzymes, and drugs. It is said that the amounts of almost any chemical can be measured very accurately, because it can be coupled chemically to an antigen.

The method has been extended to allow medical imaging of body tissues and organs. Radiolabeled antibodies that go to specific types of body tissue provide the source of radiation. As noted in Section 18.1, the same idea applies to radiation treatment. The field has expanded to include many other diagnostic techniques not involving radioactivity (see References).

17.6 DATING

There would seem to be no relationship between nuclear energy and the humanities such as history, archaeology, and anthropology. There are, however, several interesting examples in which nuclear methods establish dates of events. The carbon dating technique is being used regularly to determine the age of ancient artifacts. The technique is based on the fact that carbon-14 is and has been produced by cosmic rays in the atmosphere (a neutron reaction with nitrogen). Plants take up CO_2 and deposit C-14, and animals eat the plants. At the death of either, the supply of radiocarbon obviously stops and the C-14 that is present decays, with half-life 5715 y. By measurement of the radioactivity, the age within approximately 50 y can be found. This method was used to determine the age of the Dead Sea Scrolls, as approximately 2000 y, making measurements on the linen made from flax; to date documents found at Stonehenge in England, by use of pieces of charcoal; and to verify that prehistoric peoples lived in the United States, as long ago as 9000 y, from the C-14 content of rope sandals discovered in an Oregon cave. Carbon dating proved that the famous Shroud of Turin was made from flax in the 14th century, not from the time of Christ.

Even greater accuracy in dating biological artifacts can be obtained by direct detection of carbon-14 atoms. Molecular ions formed from $^{14}_{6}C$ are accelerated in electric and magnetic fields and then slowed by passage through thin layers of material. This sorting process can measure three atoms of $^{14}_{6}C$ out of 10^{16} atoms of $^{12}_{6}C$. Several accelerator mass spectrometers are in operation around the world (see References).

The age of minerals in the Earth, in meteorites, or from the Moon can be obtained by a comparison of their uranium and lead contents. The method is based on the fact that Pb-206 is the final product of the decay chain starting with U-238, half-life 4.47×10^9 y. Thus the number of lead atoms now present is equal to the loss in uranium atoms, i.e.,

$$N_{Pb} = (N_U)_0 - N_U,$$

where

$$N_U = (N_U)_0 e^{-\lambda t}.$$

Elimination of the original number of uranium atoms $(N_U)_0$ from these two formulas gives a relationship between time and the ratio N_{Pb}/N_U. The latest value of the age of the Earth obtained by this method is 4.55 billion y.

For intermediate ages, thermoluminescence (heat and light) is used. Radiation shifts electrons in atoms to higher orbits (Section 2.3), whereas heating causes electrons to drop back. Thus the firing of clay in ancient pottery "starts the clock." Over the years, traces of radioactive U and Th cause a cumulative shifting, which is measured by heating and observing the light emitted. An elementary but entertaining account of the applications of this technique is provided by Jespersen and Fitz-Randolph (see References).

For the determination of ages ranging from 50,000 y to a few million y, an argon method can be used. It is based on the fact that the potassium isotope K-40 (half-life 1.25×10^9 y) crystallizes in materials of volcanic origin and decays into the stable argon isotope Ar-40. An improved technique makes use of neutron bombardment of samples to convert K-39, a stable isotope of potassium, into Ar-39. This provides a substitute for measuring the content of K. These techniques, described by Taylor and Aitken (see References), are of special interest in relation to the possible collision of an asteroid with the Earth 65 million y ago and the establishment of the date of the first appearance of man. Dating methods are used in conjunction with activation analysis described in the next section.

17.7 NEUTRON ACTIVATION ANALYSIS

This is an analytical method that will reveal the presence and amount of minute impurities. A sample of material that may contain traces of a certain element is irradiated with neutrons, as in a reactor. The gamma rays emitted by the product radioisotope have unique energies and relative intensities, in analogy to spectral lines from a luminous gas. Measurements and interpretation of the gamma-ray spectra, with data from standard samples for comparison, provide information on the amount of the original impurity.

Let us consider a practical example. Reactor design engineers may be concerned with the possibility that some stainless steel to be used in moving parts in a reactor contains traces of cobalt, which would yield undesirable long-lived activity if exposed to neutrons. To check on this possibility, a small sample of the stainless steel is irradiated in a test reactor to produce Co-60, and gamma radiation from the Co-60 is compared with that of a piece known to contain the radioactive isotope. The "unknown" is placed on a Pb-shielded large-volume lithium-drifted germanium Ge(Li) detector used in gamma-ray spectroscopy as noted in Section 10.4. Gamma rays from the decay of the 5.27-y Co-60 give rise to electrons by photoelectric absorption, Compton scattering, and pair production. The electrons produced by photoelectric absorption then give rise to electrical signals in the detector that are approximately proportional to the energy of the gamma rays. If all the pulses produced by gamma rays of a single energy were equal in height, the observed counting rate would consist of two perfectly sharp peaks at energy 1.17 MeV and 1.33 MeV. A variety of effects cause the response to be broadened somewhat as shown in Figure 17.4. The location of the peaks clearly shows the presence of the isotope Co-60 and the heights tell how much of the isotope is present in the sample. Modern electronic circuits can process a large amount of data at one time. The multichannel analyzer accepts counts caused by photons of all energy and displays the whole spectrum graphically. When neutron activation analysis is applied to a mixture of materials, it is necessary after irradiation to allow time to elapse for the decay of certain

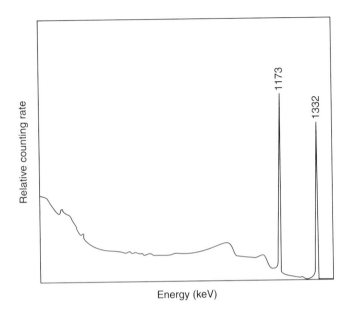

FIGURE 17.4 Analysis of gamma rays from cobalt-60. (Courtesy of Jack N. Weaver of North Carolina State University).

isotopes whose radiation would "compete" with that of the isotope of interest. In some cases, prior chemical separation is required to eliminate interfering isotope effects.

The activation analysis method is of particular value for the identification of chemical elements that have an isotope of adequate neutron absorption cross section and for which the products yield a suitable radiation type and energy. Not all elements meet these specifications, of course, which means that activation analysis supplements other techniques. For example, neutron absorption in the naturally occurring isotopes of carbon, hydrogen, oxygen, and nitrogen produces stable isotopes. This is fortunate, however, in that organic materials including biological tissue are composed of those very elements, and the absence of competing radiation makes the measurement of trace contaminants easier. The sensitivity of activation analysis is remarkably high for many elements. It is possible to detect quantities as low as a millionth of a gram in 76 elements, a billionth of a gram in 53, or even as low as a trillionth in 11.

Prompt gamma neutron activation analysis (PGNAA) is a variant on the method just described. PGNAA measures the capture gamma ray from the original (n,γ) reaction resulting from neutron absorption in the element or isotope of interest, instead of measuring gammas from new radioactive species formed in the reaction. The distinction between NAA and PGNAA is shown in Figure 17.5, which shows the series of reactions that can result from a single neutron.

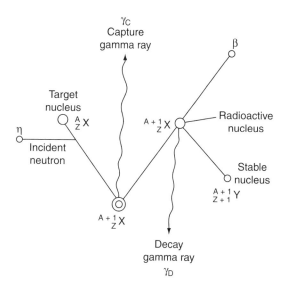

FIGURE 17.5 Nuclear reactions involved in neutron activation analysis (PGNAA) (Courtesy of Institute of Physics).

Because the reaction rate depends on the neutron cross section, only a relatively small number of elements can be detected in trace amounts. The detection limits in ppm are smallest for B, Cd, Sm, and Gd (0.01 to 0.1), and somewhat higher for Cl, Mn, In, Nd, and Hg (1 to 10). Components that can readily be measured are those often present in large quantities such as N, Na, Al, Si, Ca, K, and Fe. The method depends on the fact that each element has its unique prompt gamma-ray spectrum. The advantages of PGNAA are that it is nondestructive, it gives low residual radioactivity, and the results are immediate.

A few of the many applications of neutron activation analysis are now described.

(a) *Textile manufacturing.* In the production of synthetic fibers, certain chemicals such as fluorine are applied to improve textile characteristics, such as the ability to repel water or stains. Activation analysis is used to check on inferior imitations by comparison of the content of fluorine or other deliberately added trace elements.

(b) *Petroleum processing.* The "cracking" process for refining oil involves an expensive catalyst that is easily poisoned by small amounts of vanadium, which is a natural constituent of crude oil. Activation analysis provides a means for verifying the effectiveness of the initial distillation of the oil.

(c) *Crime investigation.* The process of connecting a suspect with a crime involves physical evidence that often can be accurately obtained by

NAA. Examples of forensic applications are the comparison of paint flakes found at the scene of an automobile accident with paint from a hit-and-run driver's car; the determination of the geographical sources of drugs by comparison of trace element content with that of soils in which plants are grown; verification of theft of copper wire by use of differences in content of wire from various manufacturers; distinguishing between murder and suicide by measurement of barium or antimony on hands; and tests for poison in a victim's body. The classic example of the latter is the verification of the hypothesis that Napoleon was poisoned, by activation analysis of arsenic in hair samples.

(d) *Authentication of art work*. The probable age of a painting can be found by testing a small speck of paint. Over the centuries the proportions of elements such as chromium and zinc used in pigment have changed, so that forgeries of the work of old masters can be detected.

An alternate method of examination involves irradiation of a painting briefly with neutrons from a reactor. The radioactivity induced produces an autoradiograph in a photographic film, so that hidden underpainting can be revealed.

It was desired to determine the authenticity of some metal medical instruments, said to be from Pompeii, the city buried by the eruption of Vesuvius in AD 79. PGNAA was applied, and by use of the fact that the zinc content of true Roman artifacts was low, the instruments were shown to be of modern origin.

(e) *Diagnosis of disease*. Medical applications (see References) include accurate measurements of the normal and abnormal amounts of trace elements in the blood and tissue as indicators of specific diseases. Other examples are the determination of sodium content of children's fingernails and the very sensitive measurement of the iodide uptake by the thyroid gland.

(f) *Pesticide investigation*. The amounts of residues of pesticides such as DDT or methyl bromide in crops, foods, and animals are found by analysis of the bromine and chlorine content.

(g) *Mercury in the environment*. The heavy element mercury is a serious poison for animals and human beings even at low concentrations. It appears in rivers as the result of certain manufacturing waste discharges. By the use of activation analysis, the Hg contamination in water or tissues of fish or land animals can be measured, thus helping to establish the ecological pathways.

(h) *Astronomical studies*. Measurement by NAA of the variation in the minute amounts of iridium (parts per billion) in geological deposits led to some startling conclusions about the extinction of the dinosaurs some 65 million years ago. A large meteorite, 6 km in diameter, is believed to have struck the Earth and to have caused atmospheric dust that reduced the sunlight needed by plants eaten by the dinosaurs. The theory is based on the fact that meteorites have a higher iridium content than the Earth. The sensitivity of NAA for Ir was vividly demonstrated by the discovery that contact of a technician's wedding ring with a sample for only 2 seconds was sufficient to invalidate results.

Evidence is mounting for the correctness of the idea. Large impact craters and buried structures have been discovered in Yucatan and Iowa. They are surrounded by geological debris whose age can be measured by the K-Ar method (see Section 17.6 and References).

(i) *Geological applications of PGNAA.* Oil and mineral exploration *in situ* of large-tonnage, low-grade deposits far below the surface has been found to yield better results than does extracting small samples. In another example, measurements were made on the ash on the ground and particles in the atmosphere from the 1980 Mount St. Helens volcano eruption. Elemental composition was found to vary with distance along the ground and with altitude. Many other examples of the use of PGNAA are found in the literature (see References).

An alternative and supplement to NAA and PGNAA is X-ray fluorescence spectrometry. It is more accurate for measuring trace amounts of some materials. The method consists of irradiating a sample with an intense X-ray beam to cause target elements to emit characteristic line spectra (i.e., to fluoresce). Identification is accomplished by either (a) measurements of the wavelengths by diffraction with a single crystal, comparison with a standard, and analysis by a computer, or (b) use of a commercial low-energy photon spectrometer, a semiconductor detector. The sensitivity of the method varies with the element irradiated, being lower than 20 ppm for all elements with atomic number above 15. The time required is much shorter than for wet chemical analyses, making the method useful when a large number of measurements are required.

17.8 RADIOGRAPHY

The oldest and most familiar beneficial use of radiation is for medical diagnosis by X-rays. These consist of high-frequency electromagnetic radiation produced by electron bombardment of a heavy-metal target. As is well known, X-rays penetrate body tissue to different degrees depending on material density, and shadows of bones and other dense materials appear on the photographic film. The term "radiography" includes the investigation of internal composition of living organisms or inanimate objects by use of X-rays, gamma rays, or neutrons.

For both medical and industrial use, the isotope cobalt-60, produced from Co-59 by neutron absorption, is an important alternative to the X-ray tube. Co-60 emits gamma rays of energy 1.17 MeV and 1.33 MeV, which are especially useful for examination of flaws in metals. Internal cracks, defects in welds, and nonmetallic inclusions are revealed by scanning with a cobalt radiographic unit. Advantages include small size and portability and freedom from the requirement of an electrical power supply. The half-life of 5.27 y permits use of the device for a long time without need for replenishing the source. On the other hand, the energy of the rays is fixed and the intensity cannot be varied, as is possible with the X-ray machine.

Other isotopes that are useful for gamma-ray radiography are: (a) iridium-192, half-life 73.8 d, photon energy approximately 0.4 MeV, for thin specimens; (b) cesium-137 (30.2 y), because of its long half-life and 0.662 MeV gamma ray; (c) thulium-170, half-life 128.6 d, emitting low-energy gammas (0.052, 0.084, 0.16 MeV), useful for thin steel and light alloys because of the high cross section of the soft radiation.

The purpose of radiography that uses neutrons is the same as that which uses X-rays, namely to examine the interior of an opaque object. There are some important differences in the mechanisms involved, however. X-rays interact principally with the electrons in atoms and molecules, and thus are scattered best by heavy high-Z elements. Neutrons interact with nuclei and are scattered according to what isotope is the target. Hydrogen atoms have a particularly large scattering cross section. Also, some isotopes have very high capture cross section (e.g., cadmium, boron, and gadolinium). Such materials are useful in detectors as well. Figure 17.6 shows the schematic arrangement of a thermal neutron radiography unit, where the source can be a nuclear reactor, a particle accelerator, or a radioisotope. Exposure times are least for the reactor source because of the large supply of neutrons; they are greatest for the isotopic source. A typical accelerator reaction that uses neutrons is the (d,n) reaction on tritium or beryllium.

Several of the radioisotopes sources use the (γ,n) reaction in beryllium-9, with gamma rays from antimony-124 (60.20 d), or the (α,n) reaction with alpha particles from americium-241 (432 y) or curium-242 (163 d). An isotope of the artificial element 98, californium-252, is especially useful as a neutron source. It

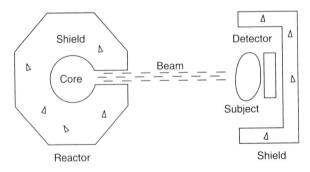

FIGURE 17.6 Schematic diagram of a thermal neutron radiography unit. Source can be an accelerator, a reactor, or a radioisotope.

decays usually (96.9%) by α-particle emission, but the other part (3.1%) undergoes spontaneous fission releasing approximately 3.5 neutrons on average. The half-lives for the two processes are 2.73 y and 85.6 y, respectively. An extremely small mass of Cf-252 serves as an abundant source of neutrons. These fast neutron sources must be surrounded by a light-element moderator to thermalize the neutrons.

Detection of transmitted neutrons is by the small number of elements that have a high thermal neutron cross section and that emit secondary radiation that readily affects a photographic film and record the images. Examples are boron, indium, dysprosium, gadolinium, and lithium. Several neutron energy ranges may be used—thermal, fast and epithermal, and "cold" neutrons, obtained by passing a beam through a guide tube with reflecting walls that select the lowest energy neutrons of a thermal distribution.

Examples of the use of neutron radiography are:

(a) Inspection of reactor fuel assemblies before operation for defects such as enrichment differences, odd-sized pellets, and cracks.

(b) Examination of used fuel rods to determine radiation and thermal damage.

(c) Inspection for flaws in explosive devices used in the United States space program. The devices served to separate booster stages and to trigger release of reentry parachutes. Items are rejected or reworked on the basis of any one of 10 different types of defects.

(d) Study of seed germination and root growth of plants in soils. The method allows continued study of the root system without disturbance. Root diameters down to $\frac{1}{3}$ mm can be discerned, but better resolution is needed to observe root hairs.

(e) "Real-time" observations of a helicopter gas turbine engine at Rolls-Royce, Ltd. Oil flow patterns that use cold neutrons are observable, and bubbles, oil droplets, and voids are distinguishable from normal density oil.

17.9 RADIATION GAUGES[†]

Some physical properties of materials are difficult to ascertain by ordinary methods but can be measured easily by observing how radiation interacts with the substance. For example, the thickness of a layer of plastic or paper can be found by measuring the transmitted number of beta particles from a radioactive source. The separated fission product isotopes strontium-90 (29.1 y, 0.546 MeV beta particle) and cesium-137 (30.2 y, 0.514 MeV beta particle) are widely used for such gauging.

The density of a liquid flowing in a pipe can be measured externally by detection of the gamma rays that pass through the substance. The liquid in the pipe

[†]Appreciation is extended to the late William Troxler for valuable information in this section.

serves as a shield for the radiation and attenuation of the beam dependent on macroscopic cross section and thus particle number density.

The level of liquid in an opaque container can be measured readily without the need for sight glasses or electric contacts. A detector outside the vessel measures the radiation from a radioactive source mounted on a float in the liquid.

Portable gauges for measurement of both moisture and density are available commercially. A rechargeable battery provides power for the electronics involving a microprocessor. Gamma rays for density measurements in materials such as soil or asphalt paving are supplied by a cesium-137 source. For operation in the direct-transmission mode, a hole is punched into the material being tested and a probe rod with radioactive source in its end is inserted. A Geiger-Mueller gamma ray detector is located at the base of the instrument, as shown in Figure 17.7. A typical calibration curve for the instrument is also shown.

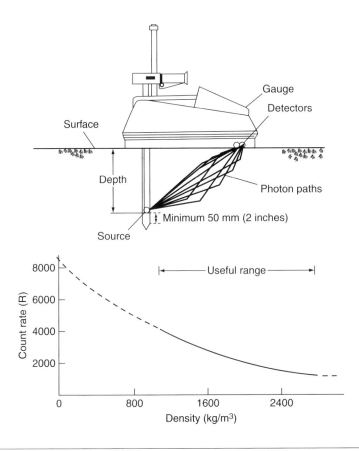

FIGURE 17.7 Direct-transmission radiation gauge to measure soil density. (Courtesy of Troxler Electronic Laboratories, Inc).

Standard blocks of test material with various amounts of magnesium and aluminum are used to determine the constants in an empirical formula that relates density to counting rate. If the source is retracted to the surface, measurements in the back-scattering mode can be made. The precision of density measurements is 0.4% or better. For moisture measurements by the instrument, neutrons of average energy 4.5 MeV are provided by an americium-beryllium source. Particles of approximately 5 MeV from americium-241, half-life 432 y, bombard beryllium-9 to produce the reaction $^9Be(\alpha,n)^{12}C$. Neutrons from the source, located in the center of the gauge base, migrate through the material and slow down, primarily with collisions with the hydrogen atoms in the contained moisture. The more water that is present, the larger is the thermal neutron flux in the vicinity of the gauge. The flux is measured by a thermal-neutron detector consisting of a helium-3 proportional counter, in which the ionization is created by the products of the reaction $^3He(n,p)^3H$ ($\sigma_a = 5330$ barns). Protons and tritons (hydrogen-3 ions) create the ionization measured in the detector. The gauge is calibrated by use of laminated sheets of the hydrocarbon polyethylene and of magnesium. The moisture content can be measured to approximately 5% in normal soil. The device requires correction if there are significant amounts of absorbers such as iron, chlorine, or boron in the ground or if there are hydrogenous materials other than water present.

A newer portable nuclear gauge[†] for measuring density and moisture content has several special features: remote control by a hand-held PDA (personal digital assistant), GPS (global positioning system) position recording, and software for transferring data to a PC. Its nuclear sources are an 8 mCi cesium-137 for gammas and a 40 mCi americium-241 plus beryllium-9 for neutrons.

Several nuclear techniques are used in the petroleum industry. In the drilling wells, the "logging" process involves the study of geological features. One method consists of the measurement of natural gamma radiation. When the detector is moved from a region of ordinary radioactive rock to one containing oil or other liquid, the signal is reduced. A neutron moisture gauge is adapted to determine the presence of oil, which contains hydrogen. Neutron activation analysis of chemical composition is performed by lowering a neutron source and a gamma ray detector into the well.

17.10 SUMMARY

Radioisotopes provide a great deal of information for human benefit. The characteristic radiations permit the tracing of processes such as fluid flow. Pharmaceuticals are radioactively tagged chemicals used in hospitals for diagnosis. Scanners detect the distribution of radioactivity in the body and form images of

---

[†]Troxler Model 3451 Enhanced RoadReader Plus Nuclear Density Gauge.

diseased tissue. Radioimmunoassay measures minute amounts of biological materials. The dates of archaeological artifacts and of rock formations can be found from carbon-14 decay data and the ratios of uranium to lead and of potassium to argon. The irradiation of materials with neutrons gives rise to unique prompt gamma rays and radioactive decay products, allowing measurement of trace elements for many applications. Radiography uses gamma rays from cobalt-60 or neutrons from a reactor, accelerator, or californium-252. Radiation gauges measure density, thickness, ground moisture, water/cement ratios, and oil deposits.

17.11 EXERCISES

17.1 A radioisotope is to be selected to provide the signal for arrival of a new grade of oil in an 800-km-long pipeline, in which the fluid speed is 1.5 m/s. Some of the candidates are:

Isotope	Half-life	Particle, Energy (MeV)
Na-24	14.96 h	β, 1.389; γ, 1.369, 2.754
S-35	87.2 d	β, 0.167
Co-60	5.27 y	β, 0.315; γ, 1.173, 1.332
Fe-59	44.5 d	β, 0.273, 0.466; γ, 1.099, 1.292

Which would you pick? On what basis did you eliminate the others?

17.2 The radioisotope F-18, half-life 1.83 h, is used for tumor diagnosis. It is produced by bombarding lithium carbonate (Li_2CO_3) with neutrons, with tritium as an intermediate particle. Deduce the two nuclear reactions.

17.3 The range of beta particles of energy 0.53 MeV in metals is 170 mg/cm^2. What is the maximum thickness of aluminum sheet, density 2.7 g/cm^3, that would be practical to measure with a Sr-90 or Cs-137 gauge?

17.4 The amount of environmental pollution by mercury is to be measured with neutron activation analysis. Neutron absorption in the mercury isotope Hg-196, present with 0.15% abundance, activation cross section 3×10^3 barns, produces the radioactive species Hg-197, half-life 2.67 days. The smallest activity for which the resulting photons can be accurately analyzed in a river water sample is 10 dps. If a reactor neutron flux of 10^{12} cm^{-2}–s^{-1} is available, how long an irradiation is required to be able to measure mercury contamination of 20 ppm (μg/g) in a 4-milliliter water test sample?

17.5 The ratio of numbers of atoms of lead and natural uranium in a certain moon rock is found to be 0.05. What is the probable age of the sample?

17.6 The activity of C-14 in a wooden figure found in a cave is only $\frac{3}{4}$ of today's value. Estimate the date the figure was carved.

17.7 Examine the possibility of adapting the uranium-lead dating analysis to the potassium-argon method. What would be the ratio of Ar-40 to K-40 if a deposit were 1 million y old? Note that only 10.72% of K-40 decay yields Ar-40, the rest going into Ca-40.

17.8 The age of minerals containing rubidium can be found from the ratio of radioactive Rb-87 to its daughter Sr-87. Develop a formula relating this ratio to time.

17.9 It has been proposed that radioactive krypton gas of 10.73 y half-life be used in conjunction with film for detecting small flaws in materials. Discuss the concept, including possible techniques, advantages, and disadvantages.

17.10 A krypton isotope $^{81m}_{36}\text{Kr}$ of half-life 13.1 sec is prepared by charged particle bombardment. It gives off a gamma ray of 0.19 MeV energy. Discuss the application of the isotope to the diagnosis of emphysema and black-lung disease. Consider production, transportation, hazards, and other factors.

17.11 Tritium (^3_1H) has a physical half-life of 12.32 y, but when taken into the human body as water, it has a biological half-life of 12.0 d. Calculate the effective half-life of tritium for purposes of radiation exposure. Comment on the result.

17.12 With the half-life relationship as given in Section 17.2, calculate the effective half-life of californium-252.

17.13 The spontaneous fission half-life of Cf-252 is 85.6 y. Assuming that it releases 3.5 neutrons per fission, how much of the isotope in micrograms is needed to provide a source of strength of 10^7 neutrons per second? What would be the diameter of the source in the form of a sphere if the Cf-252 had a density as pure metal of 20 g/cm^3?

17.14 Three different isotopic sources are to be used in radiography of steel in ships as follows:

Isotope	Half-life	Gamma Energy (MeV)
Co-60	5.27 y	1.25 (ave.)
Ir-192	73.8 d	0.4 (ave.)
Cs-137	30.2 y	0.66

Which isotope would be best for insertion in pipes of small diameter and wall thickness? For finding flaws in large castings? For more permanent installations? Explain.

17.15 The number of atoms of a parent isotope in a radionuclide generator such as Mo-Tc given by $N_p = N_{p0}E_p$, where $E_p = \exp(-\lambda_p t)$, with N_{p0} as the initial number of atoms. The number of daughter atoms for zero initially is

$$N_d = k\lambda_p N_{p0}(E_p - E_d)/(\lambda_d - \lambda_p)$$

where k is the fraction of parents that go into daughters and $E_d = \exp(-\lambda_d t)$.

   (a) Find the ratio of Tc-99m atoms to Mo-99 atoms for very long times, with $k = 0.87$.
   (b) What is the percent error in the use of the ratio found in (a) if it takes one half-life of the parent to ship the fresh isotope to a laboratory for use?

17.16 Pharmaceuticals containing carbon-14 (5715 y) and tritium (12.32 y) are both used in a biological research laboratory. To avoid an error of greater than 10% in counting beta particles, as a result of accidental contamination of C-14 by H-3, what must be the upper limit on the fraction of atoms of tritium in the sample? Assume that all betas are counted, regardless of energy.

17.17 The atom fraction of C-14 in carbon was approximately 1.2×10^{-12} before bomb tests. How many counts per minute would be expected from a 1-gram sample of carbon? Discuss the implications of that number.

Computer Exercise

17.A Recall the computer program RADIOGEN (see Computer Exercise 3.D) giving activities of parent and daughter isotopes.

   (a) Apply the radionuclide generator of Section 17.3 by use of half-lives 65.9 h for Mo-99 and 6.01 h for Tc-99 m, with $k = 0.87$. Carry the calculations out to at least 66 h in steps of 1 h.
   (b) From the formula in Computer Exercise 3.D, show that the ratio of activities of daughter to parent at very long times is

$$A_d/A_p = k/(1 - \lambda_p/\lambda_d).$$

   (c) Find out how much error there is with the formula of (b), rather than the ratio calculated by RADIOGEN, if it takes exactly one half-life of Mo-99 to ship the generator to a laboratory for use.

17.12 REFERENCES

Biology Links
http://mcb.harvard.edu/BioLinks.html
Harvard University Department of Molecular and Cellular Biology.

DOE Information Bridge
http://www.osti.gov/bridge
Search on "A Vital Legacy" for atoms in biology. 4807K pdf file.

S. James Adelstein and Frederick J. Manning, Eds., *Isotopes for Medicine and the Life Sciences*, National Academy Press, Washington, 1995. Describes sources of stable and radioactive isotopes and recommends a national dedicated accelerator.

United States Radioisotope Production
http://www.nuclear.energy.gov
Search on Radiological Facilities Management.

Radiopharmaceuticals
http://nucmedicine.com/images/Radiopharmaceuticals-1.pdf
Attractive and informative slide show by Samy Sadak.

United States Radioisotope Supply
http://www.ans.org/pi/ps/docs/ps30.pdf
ANS Position Statement 30, June 2004.

DOE Isotope Production and Distribution
http://www.ornl.gov/sci/isotopes/catalog.htm
Catalog of stable and radioactive isotopes.

Human Genome Project Information
http://www.ornl.gov/hgmis
Comprehensive information.

Access Excellence Classical Collection
http://www.accessexcellence.com/AE/AEC/CC
"A Visit with Dr. Francis Crick," "DNA Structure," and "Restriction Nucleases."

DNA and Molecular Genetics
Search Google with "DNA radioactivity Farabee"
Tutorial on DNA.

DNA Forensics
http://www.forensicmag.com/articles.asp?pid=17
Comparison of DNA fingerprinting methods RFLP and PCR-STR.

M. Krawczak and J. Schmidtke, *DNA Fingerprinting*, 2nd Ed., Springer Verlag, New York, 1998. Genetics background, applications to forensics, and legal and ethical aspects.

Guidebook on Radioisotope Tracers in Industry, International Atomic Energy Agency, Vienna, 1990. Methodology, case studies, and trends.

John Harbert and Antonio F. G. da Rocha, *Textbook of Nuclear Medicine*, Volume I: Basic Science, Volume II: Clinical Applications, 2nd Ed., Lea & Febiger, Philadelphia, 1984. Volume I contains good descriptions of radionuclide production, imaging, radionuclide generators, radiopharmaceutical chemistry, and other subjects. Volume II gives applications to organs.

Michael A. Wilson, *Textbook of Nuclear Medicine*, Lippincott Williams & Wilkins, Philadelphia, PA, 1998.

Gopal B. Saha. *Fundamentals of Nuclear Pharmacy*, 5th Ed., Springer, New York, 2005. Instruments, isotope production, and diagnostic and therapeutic uses of radiopharmaceuticals. Reflects the continued growth of the field.

Introduction to MRI
http://mritutor.org/mritutor/index.html
Elementary tutorial by Ray Ballinger.

The Basics of MRI
http://www.cis.rit.edu/htbooks/mri
Comprehensive treatment by Dr. Joseph P. Hornak. Click on skull image.

W. R. Hendee and E. R. Ritenour, *Medical Imaging Physics*, 4th Ed., Wiley-Liss, New York, 2002.

Medical Physics and Bioengineering Resources
http://www.medphys.ucl.ac.uk/inset/resource.htm
List of textbooks, videos, slides, and CD's. By University College London.

Fred A. Mettler, Jr. and Milton J. Guiberteau, *Essentials of Nuclear Medicine Imaging*, 5th Ed., W. B. Saunders Co., Philadelphia, 2005. After presenting background on radioactivity, instruments, and computers, the book describes the methods used to diagnose and treat different tissues, organs, and systems of the body.

Ramesh Chandra, *Nuclear Medicine Physics: the Basics*, Lippincott Williams & Wilkins, Philadelphia, PA, 2004. Intended for resident physicians. Covers scintillation cameras and computer tomography.

Marshall Brucer, *A Chronology of Nuclear Medicine 1600–1989*, Heritage Publications, St. Louis, 1990. Interesting and informative discussion with abundant references.

T. Chard, *An Introduction to Radioimmunoassay and Related Techniques*, Elsevier, Amsterdam, 1995. Concept, principles, and laboratory techniques. Immunoassays in general, labeling techniques, and commercial services.

Christopher P. Price and David J. Newman, Editors, *Principles and Practices of Immunoassay*, Stockton Press, New York, 1991. A great variety of immunoassay techniques.

Herman W. Knoche, *Radioisotopic Methods for Biological and Medical Research*, Oxford University Press, New York, 1991. Includes mathematics of radioimmunoassay and isotope dilution.

G. Choppin, J. Rydberg, and J. O. Liljenzin, *Radiochemistry and Nuclear Chemistry*, 3rd Ed., Butterworth-Heinemann, Oxford, 2001. Includes chapters on isotope uses in chemistry and on nuclear energy.

Radiocarbon Web-info
http://www.c14dating.com
Links to many applications of C-14 dating, by T. Higham.

Accelerator Mass Spectrometry
http://www.phys.uu.nl/ams
Principle, method, and applications.

James Jespersen and Jane Fitz-Randolph, *Mummies, Dinosaurs, Moon Rocks: How We Know How Old Things Are*, Atheneum Books for Young Readers, New York, 1996.

Sheridan Bowman, Ed., *Science and the Past*, University of Toronto Press, Toronto, 1991. Qualitative technical treatment of dating with emphasis on the artifacts.

R. E. Taylor and Martin J. Aitken, Eds., *Chronometric Dating in Archaeology*, Plenum Press, New York, 1997. Principle, history, and current research for a number of dating techniques. Articles are written by experts in the methods.

K. Heydorn, *Neutron Activation Analysis for Clinical Trace Element Research*, Vols. I & II, CRC Press, Boca Raton, 1984.

North Carolina State University's Nuclear Reactor Program
http://www.ne.ncsu.edu/NRP/reactor_program.html
Select Facilities.

INL Gamma-Ray Spectrometry Center
http://www.inl.gov/gammaray/spectrometry
Display of spectra and decay schemes.

Gabor Molnar, Editor, *Handbook of Prompt Gamma Activation Analysis with Neutron Beams*, Kluwer, Dordrecht, 2004.

Zeev Alfassi and Chien Chung, Eds., *Prompt Gamma Neutron Activation Analysis*, CRC Press, Boca Raton, 1995. Includes *in vivo* measurements of elements in the human body and *in situ* well-logging in the petroleum industry.

The Medical Radiography Home Page
http://home.earthlink.net/~terrass/radiography/medradhome.html
Internet resources by Richard Terrass of Massachusetts General.

R. Halmshaw, *Industrial Radiology, Theory and Practice*, 2nd Ed., Chapman & Hall, London, 1995. Principles and equipment using X-rays, gamma-rays, neutrons, and other particles. Computers and automation in quality control and nondestructive testing.

J. C. Domanus, Ed., *Practical Neutron Radiography*, Kluwer, Dordrecht, 1992. Sources, collimation, imaging, and applications (e.g., nuclear fuel).

A. A. Harms and D. R. Wyman, *Mathematics and Physics of Neutron Radiography*, D. Reidel, Dordrecht, Holland, 1986. Highly technical reference.

X-ray WWW Server
http://xray.uu.se
Links to synchrotron radiation facilities.

Kevin L. D'Amico, Louis J. Terminello, and David K. Shuh, *Synchrotron Radiation Techniques in Industrial, Chemical, and Materials, Science*, Plenum Press, New York, 1996. Selected papers from two conferences.

Nuclear Gauges
http://troxlerlabs.com
Details on company's products.

Useful Radiation Effects

RADIATION IN the form of gamma rays, beta particles, and neutrons is being used in science and industry to achieve desirable changes. Radiation doses control offending organisms, including cancer cells and harmful bacteria, and sterilize insects. Local energy deposition can also stimulate chemical reactions and modify the structure of plastics and semiconductors. Neutrons and X-rays are used to investigate basic physical and biological processes. In this chapter we will briefly describe some of these interesting and important applications of radiation. For additional information on the uses around the world, proceedings of international conferences can be consulted. Thanks are due to Albert L. Wiley, Jr., MD, PhD, for suggestions on the subject of nuclear medicine.

18.1 MEDICAL TREATMENT

The use of radiation for medical therapy has increased greatly in recent years, with millions of treatments given patients annually. The radiation comes from teletherapy units in which the source is at some distance from the target, from isotopes in sealed containers implanted in the body, or from ingested solutions of radionuclides.

Doses of radiation are found to be effective in the treatment of diseases such as cancer. In early times, X-rays were used, but they were supplanted by cobalt-60 gamma rays, because the high-energy (1.17 and 1.33 MeV) photons penetrated tissue better and could deliver doses deep in the body, with a minimum of skin reaction. In modern nuclear medicine, there is increasing use of accelerator-produced radiation in the range 4 to 35 MeV for cancer treatment.

Treatment of disease by implantation of a radionuclide is called interstitial brachytherapy ("brachys" is Greek for "short"). A small radioactive capsule or "seed" is imbedded in the organ, producing local gamma irradiation. The radionuclides are chosen to provide the correct dose. In earlier times, the only material available for such implantation was α-emitting radium-226 (1599 y). Most frequently used today are iridium-192 (73.8 d), iodine-125 (59.4 d), and palladium-103 (17.0 d).

Examples of tumor locations where this method is successful are the head and neck, breast, lung, and prostate gland. Other isotopes sometimes used are cobalt-60, cesium-137, tantalum-182, and gold-198. Intense fast neutron sources are provided by californium-252. For treatment of the prostate, 40 to 100 "rice-sized" seeds, (4.5 mm long and 0.81 mm diameter) containing a soft-gamma emitter, Pd-103, are implanted with thin hollow needles (see References). Computerized tomography (CT) and ultrasound aid in the implantation.

One sophisticated device for administering cancer treatment uses a pneumatically controlled string of cesium-137–impregnated glass beads encapsulated in stainless steel, of only 2.5 mm diameter. Tubes containing the beads are inserted in the bronchus, larynx, and cervix.

Success in treatment of abnormal pituitary glands is obtained by charged particles from an accelerator, and beneficial results have come from slow neutron bombardment of tumors in which a boron solution is injected. Selective absorption of chemicals makes possible the treatment of cancers of certain types by administering the proper radionuclides. Examples are iodine-125 or iodine-131 for the thyroid gland and phosphorus-32 for the bone. However, there is concern in medical circles that use of iodine-131 to treat hyperthyroidism could cause thyroid carcinoma, especially in children.

Relief from rheumatoid arthritis is obtained by irradiation with beta particles. The radionuclide dysprosium-165 (2.33 h) is mixed with ferric hydroxide, which serves as a carrier. The radiation from the injected radionuclide reduces the inflammation of the lining of joints.

Table 18.1 shows some of the radionuclides used in treatment.

A great deal of medical research is an outgrowth of radioimmunoassay (see Section 17.5). It involves monoclonal antibodies (MAbs), which are radiolabeled substances that have an affinity for particular types of cancer, such as those of the skin and lymph glands. The diseased cells are irradiated without damage to

Table 18.1 Radionuclides Used in Therapy

Radionuclide	Disease Treated
P-32	Leukemia
Y-90	Cancer
I-131	Hyperthyroidism
	Thyroid cancer
Sm-153 and Re-186	Bone cancer pain
Re-188 and Au-198	Ovarian cancer

neighboring normal tissue. The steps in this complex procedure start with the injection into mice of human cancer cells as antigens. The mouse spleen, a part of the immune system, produces antibodies through the lymphocyte cells. These cells are removed and blended with myeloma cancer cells to form new cells called hybridoma. In a culture, the hybridoma clones itself to produce the MAb. Finally, a beta-emitting radionuclide such as yttrium-90 is chemically bonded to the antibody.

A promising treatment for cancer is boron neutron capture therapy (BNCT). A boron compound that has an affinity for diseased tissue is injected, and the patient is irradiated with neutrons from a reactor. Boron-10, with abundance of 20% in natural boron, strongly absorbs thermal neutrons to release lithium-7 and helium-4 ions. An energy of more than 2 MeV is deposited locally because of the short range of the particles. The technique was pioneered in the 1950s by Brookhaven National Laboratory, but the program was suspended from 1961 to 1994 and terminated in 1999. Research is continuing at other locations, however (see References). A compound, Bisphenol A (BPA), was found that localized boron better, and thermal neutrons were replaced by intermediate energy neutrons, with favorable results. A single treatment with BNCT is as effective as many conventional radiation-chemotherapy sessions. The method has been found to be effective in treatment of malignancies such as melanoma (skin) and glioblastoma multiforme (brain). The discovery of monoclonal antibodies opens up new possibilities for large-scale use of BNCT.

The mechanism of the effects of radiation is known qualitatively. Abnormal cells that divide and multiply rapidly are more sensitive to radiation than normal cells. Although both types are damaged by radiation, the abnormal cells recover less effectively. Radiation is more effective if the dosage is fractionated (i.e., split into parts and administered at different times, allowing recovery of normal tissue to proceed).

Use of excess oxygen is helpful. Combinations of radiation, chemotherapy, and surgery are applied as appropriate to the particular organ or system affected. The ability to control cancer has improved over the years, but a cure on the basis of better knowledge of cell biology is yet to come.

18.2 RADIATION PRESERVATION OF FOOD

The ability of radiation treatment to eliminate insects and microorganisms from food has been known for many years. Significant benefits to the world's food supply are beginning to be realized, as a number of countries built irradiation facilities.[†] Such application in the United States has been slow because of fears related to anything involving radiation.

[†]Thanks are due Food Technology Service in Mulberry, Florida, for extensive literature on food irradiation.

Spoilage of food before it reaches the table is due to a variety of effects: sprouting as in potatoes, rotting caused by bacteria as in fruit, and insect infestation as in wheat and flour. Certain diseases stem from microorganisms that contaminate food. Examples are the bacteria Salmonella, found in much of poultry products, and the parasite trichinae that infest some pork. The National Centers for Disease Control and Prevention state that foodborne illnesses affect millions of people in the United States each year, with thousands of deaths.

Various treatments are conventionally applied to preserve food, including drying, pickling, salting, freezing, canning, pasteurization, sterilization, the use of food additives such as nitrites, and until they were banned, the application of fumigants such as ethylene dibromide (EDB). Each treatment method has its advantages, but nitrites and EDB are believed to have harmful physiological effects.

On the other hand, research has shown that gamma radiation processing can serve as an economical, safe, and effective substitute and supplement for existing treatments. The shelf life of certain foods can be extended from days to weeks, allowing adequate time for transportation and distribution. It has been estimated that 20% to 50% of the food supplied to certain countries is wasted by spoilage that could be prevented by radiation treatment. The principal sources of ionizing radiation suitable for food processing are X-rays, electrons from an accelerator, and gamma rays from a radionuclide. Much experience has been gained from the use of cobalt-60, half-life 5.27 y, with its two gamma rays of energy 1.17 MeV and 1.33 MeV. The largest supplier of cobalt-60 is a Canadian firm, MDS Nordion, formerly part of Atomic Energy of Canada, Ltd. The isotope is prepared by irradiating pure cobalt-59 target pellets with neutrons in the CANDU reactors of Ontario Hydro. The targets are disassembled and shipped for processing into double-layer capsules of approximately 10 Ci each. Another attractive isotope is cesium-137, gamma ray 0.662 MeV, because of its longer half-life of 30.2 y and its potential availability as a fission product. A considerable amount of cesium-137 has been separated at Hanford, Washington, as a part of the radioactive waste management strategy. Arrangements for loans of capsules from the DOE to industrial firms have been made. Additional cesium-137 could be obtained through limited reprocessing of spent reactor fuel.

Many people are concerned about the use of irradiated products because of the association with nuclear processes. The first worry is that the food might become radioactive. The concern is unfounded, because there is no detectable increase in radioactivity at the dosages and particle energies of the electrons, X-rays, or gamma rays used. Even at higher dosages than are planned, the induced radioactivity would be less than that from natural amounts of potassium-40 or carbon-14 in foods. Another fear is that hazardous chemicals may be produced. Research shows that the amounts of unique radiolytic products (URP) are small, less than those produced by cooking or canning, and similar to natural food constituents. No indication of health hazard has been found, but scientists recommend continuing monitoring of the process. A third concern is that there would be a loss in nutritional value. Some loss in vitamin content occurs, just as it does in ordinary cooking. Research is continuing on the effects of radiation on nutritional value. It seems that the loss is minor at the low dose levels used. On various food products, there are certain organoleptic effects (taste, smell, color, texture), but these are a matter of personal reaction, not of health. Even these effects can be eliminated by operating the targets at reduced temperatures. The astronauts of the Apollo missions and the space shuttle have dined regularly on treated foods while in orbit. They were enthusiastic about the irradiated bread and meats. Many years ago, some scientists in India reported that consumption of irradiated wheat caused polyploidy, an increase in cell chromosomes. Extensive studies elsewhere disproved the finding.

Finally, it has been suggested that radiation might induce resistance of organisms, just as with pesticides and antibiotics, but the effect seems not to occur. The difference is attributed to the fact that there is a broad effect on enzymes and compounds.

The radiation dosages required to achieve certain goals are listed in Table 18.2. Note that 1 gray (Gy) is 100 rads.

The main components of a multiproduct irradiation facility that can be used for food irradiation on a commercial basis are shown in Figure 18.1. Important parts are: (a) transfer equipment, involving conveyors for pallets, which are portable platforms on which boxes of food can be loaded; (b) an intense gamma ray source, of approximately 1 million curies strength, consisting of doubly encapsulated pellets of cobalt-60; (c) water tanks for storage of the source, with a cooling and purification system; and (d) a concrete biological shield, approximately 2 meters thick. In the operation of the facility, a rack of cobalt rods is pulled up out of the water pool, and the food boxes are exposed as they pass by the gamma source. Commercial firms providing irradiation equipment and carrying out irradiations, mainly for sterilization of medical supplies, are MDS Nordion of Kanata Ontario, Canada; Food Technology Service, Inc. of Mulberry, FL; Isomedix, Inc. of Whippany, NJ; and SteriGenics International of Oak Brook,

Table **18.2** Doses to Achieve Beneficial Effects	
Effect	**Dose (Gy)**
Inhibit sprouting of potatoes and onions	60–150
Eliminate trichinae in pork	200–300
Kill insects and eggs in fruits	200–500
Disinfect grain, prolong berry life	200–1000
Delay ripening of fruit	250–350
Eliminate salmonella from poultry	1000–3000

IL, which has facilities around the world (see References). Among services provided is mold remediation in books, documents, and records with gamma rays.

A number of experimental facilities and irradiation pilot plants have been built and used in some 70 countries. Some of the items irradiated have been grain, onions, potatoes, fish, fruit, and spices. The most active countries in the development of large-scale irradiators have been the United States, Canada, Japan, and the former U.S.S.R.

Table 18.3 shows the approvals for irradiation as issued by the United States Food and Drug Administration (FDA). Limitations are typically set on dosages to foodstuffs of 1 kilogray (100 kilorad) except for dried spices, not to exceed 30 kGy (3 Mrad).

Labeling of the packages to indicate special treatment is required by use of a phrase such as "treated with radiation." In addition, packages will exhibit the international logo, called a radura, shown in Figure 18.2. The symbol's solid circle represents an energy source; the two petals signify food; the breaks in the outer circle mean rays from the energy source. The radura label is required for shipment to the first purchaser, not for a consumer in a restaurant. Labeling of radiation-treated food is obviously a factor in acceptance by the public. The FDA is planning to allow the word "pasteurized" for products that are unchanged except for shelf life. The FDA also seeks a label other than "irradiated."

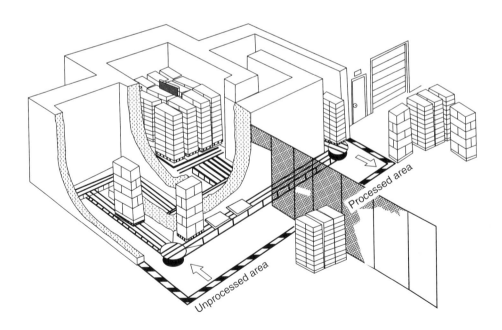

FIGURE 18.1 Gamma irradiation of Sterigenics International, at Haw River, NC. Pallets containing boxes of products move on a computer-controlled conveyor through a concrete maze past a gamma-emitting screen.

Table 18.3 Approvals by the Food and Drug Administration for Use of Irradiated Substances

Product	Purpose of Irradiation	Dose (krad)	Date
Wheat and powder	Disinfest insects	20–50	1963
White potatoes	Extend shelf life	5–15	1965
Spices, seasonings	Decontaminate	3000	1983
Food enzymes	Control insects	1000	1985
Pork products	Control trichinae	30–100	1985
Fresh fruits	Delay spoilage	100	1986
Enzymes	Decontaminate	1000	1986
Dried vegetables	Decontaminate	3000	1986
Poultry	Control salmonella	300	1990
Beef, lamb, pork	Control pathogens	450	1997
Shell eggs	Control salmonella	300	2000
Shellfish	Control bacteria	550	2005
Seeds	Control pathogens	800	2005

Note: 1 krad = 10 Gy.
Source: FDA, (see References).

Final rules on red meat irradiation as a food additive were issued by the FDA in December 1997 and by United States Department of Agriculture (USDA) in December 1999 (see References). The action was prompted in part by the discovery of the bacterium *E. coli* contamination of hamburger by an Arkansas supplier. Some 25 tons of meat were recalled and destroyed. The new rule cites statistics on outbreaks of disease and numbers of deaths related to beef. Maximum permitted doses for meat are 4.5 kGy (450 krad) as refrigerated and 7.0 kGy (700 krad) as frozen. More than 80 technical references are cited on all aspects of the subject.

Irradiated fresh ground beef prepared by SureBeam Corporation of San Diego is available at thousands of groceries in the United States. Electron beams are used to process beef at prices comparable to those of nonirradiated meat.

Approval to irradiate does not guarantee that it actually will be done, however. Many large food processors and grocery chains tend to shy away from the use of irradiated food products, believing that the public will be afraid of all of their products. Obviously, people will not have much opportunity to find treated foods acceptable if there are few products on the market. Anti-irradiation activists, who claim that the nuclear irradiation process is unsafe, have taken advantage of that reluctance. In contrast, enthusiastic endorsement of food irradiation is provided

FIGURE 18.2 International logo to appear on irradiated food.

by organizations such as World Health Organization, American Medical Association, American Dietetic Association, International Atomic Energy Agency, Grocery Manufacturers of America, and many others (see References).

At the First World Congress on Food Irradiation, held in 2003, a number of facts were reported. The 2002 Food Bill specified that irradiated food should be made available to the National School Lunch Program; regulation of facilities was developed by an International Consultative Group on Food Irradiation (ICGFI) and adopted worldwide; consumers are generally aware of beef irradiation (68%) and favor marketing irradiated ground beef (78%).

Plans were laid in 2005 by a fruit company, Pa'ina Hawaii, to install a cobalt-60 irradiator at the Honolulu airport to be used for treatment of exotic fruits such as papayas. Approval by the Nuclear Regulatory Commission (NRC) was received. *E. coli* contamination in 2006 of spinach and lettuce have stimulated new interest in food irradiation.

18.3 STERILIZATION OF MEDICAL SUPPLIES

Ever since the germ theory of disease was discovered, effective methods of sterilizing medical products have been sought. Example items are medical instruments, plastic gloves, sutures, dressings, needles, and syringes. Methods of killing bacteria in the past include dry heat, steam under pressure, and strong chemicals such as carbolic acid and gaseous ethylene oxide. Some of the chemicals are too harsh for equipment that is to be reused, and often the substances themselves are hazardous. Most of the previous methods are batch processes, difficult to scale up to handle the production needed. More recently, accelerator-produced electron beams have been introduced and preferred for some applications.

The special virtue of cobalt-60 gamma-ray sterilization is that the rays penetrate matter very well. The item can be sealed in plastic and then irradiated, assuring freedom from microbes until the time it is needed in the hospital. Although the radioactive material is expensive, the system is simple and reliable, consisting principally of the source, the shield, and the conveyor. A typical auto-mated plant requires a source of approximately 1 MCi.

18.4 PATHOGEN REDUCTION

In the operation of public sewage treatment systems, enormous amounts of solid residues are produced. In the United States alone the sewage sludge amounts to 6 million tons a year. Typical methods of disposal are by incineration, burial at sea, placement in landfills, and application to cropland. In all of these there is some hazard caused by pathogens—disease-causing organisms such a parasites, fungi, bacteria, and viruses. Experimental tests of pathogen reduction by cobalt-60 or cesium-137 gamma irradiation have been made in Germany and in the United States. The program in the United States was part of the Department of Energy's (DOE) studies of beneficial uses of fission product wastes and was carried out at Sandia Laboratories and the University of New Mexico. Tests of the effectiveness of radiation were made, and the treated sludge was found to be suitable as a feed supplement for livestock, with favorable economics. However, no use of those results was made in the United States. Apparently, the only large-scale application of sewage sludge irradiation is in Argentina, in the large city of Tucuman (see References). It is conceivable that the time is not yet ripe in the United States and Europe for such application of radiation. It took a number of years to adopt recycling of household wastes.

18.5 CROP MUTATIONS

Beneficial changes in agricultural products are obtained through mutations caused by radiation. Seeds or cuttings from plants are irradiated with charged par-ticles, X-rays, gamma rays, or neutrons; or chemical mutagens are applied. Genetic effects have been created in a large number of crops in many countries. The science of crop breeding has been practiced for many years. Unusual plants are selected and crossed with others to obtain permanent and reproducible hybrids. However, a wider choice of stock to work with is provided by mutant species. In biological terms, genetic variability is required.

Features that can be enhanced are larger yield, higher nutritional content, bet-ter resistance to disease, and adaptability to new environments, including higher or lower temperature of climate. New species can be brought into cultivation, opening up sources of income and improving health.

The leading numbers of mutant varieties of food plants that have been devel-oped are as follows: rice, 28; barley, 25; bread wheat, 12; sugar cane, 8; and soy-beans, 6. Many mutations of ornamental plants and flowers have also been produced, improving the income of small farmers and horticulturists in developing countries. For example, there are 98 varieties of chrysanthemum. The International Atomic Energy Agency (IAEA), since its creation in 1957, has fostered mutation breeding through training, research support, and information transfer. The improvement of food is regarded by the IAEA as a high-priority endeavor in light of the expanding population of the world.

More recently, the application of genetic engineering to improve crops and foodstuffs has drawn a great deal of criticism, especially in Europe, and a deep-seated conflict with the United States over use of biotechnology will be difficult to resolve.

18.6 INSECT CONTROL

To suppress the population of certain insect pests the sterile insect technique (SIT) has been applied successfully. The standard method is to breed large numbers of male insects in the laboratory, sterilize them with gamma rays, and release them for mating in the infested area. Competition of sterile males with native males results in a rapid reduction in the population. The classic case was the eradication of the screwworm fly from Curaçao, Puerto Rico, and the southwestern United States. The flies lay eggs in wounds of animals and the larvae feed on living flesh and can kill the animal if untreated. After the numbers were reduced in the early 1960s, flies came up from Mexico, requiring a repeat operation. As many as 350 million sterile flies were released each week, bringing the infestations from 100,000 to zero. The annual savings to the livestock industry was approximately $100 million.

The rearing of large numbers of flies is a complex process, involving choice of food, egg treatment, and control of the irradiation process to provide sterilization without causing body damage. Cobalt-60 gamma rays are typically used to give doses that are several times the amounts that would kill a human being.

SIT has been used against several species of mosquito in the United States and India and stopped the infestation of the Mediterranean fruit fly in California in 1980.

The discovery of a screwworm infestation in Libya in 1988 prompted international emergency action by the United Nations Food and Agriculture Organization, the International Atomic Energy Agency, and others (see References). Arrangements were made for the fly factory in Mexico to supply millions of radiation-sterilized males to Libya. There, light aircraft dropped them in a grid pattern, starting in 1990. Within 5 months the screwworm was eradicated, thus protecting Libyan wildlife as a whole.

The technique was effective on the island of Zanzibar, part of Tanzania, in combating the tsetse fly (see References). The insect is a carrier of trypanosomiasis, a livestock disease, and of sleeping sickness, which affects humans. Prior pesticide use made SIT feasible, and within 4 years, by 1996, there were no flies left.

Unfortunately, vast areas of Africa are infested with tsetse flies, and the fly-free zones are overloaded.

SIT can potentially control *Heliothis* (American bollworm, tobacco budworm, and corn earworm) and other pests such as ticks and the gypsy moth. Other related techniques include genetic breeding that will automatically yield sterile males.

Some of the organizations providing gamma-ray insect irradiation are ceasing operations for fear of terrorist action. An alternative is the use of X-rays.

18.7 APPLICATIONS IN CHEMISTRY

Radiation chemistry refers to the effect of high-energy radiation on matter, with particular emphasis on chemical reactions. Examples are ion-molecule reactions, capture of an electron that leads to dissociation, and charge transfer without a chemical reaction when an ion strikes a molecule. Many reactions have been studied in the laboratory, and a few have been used on a commercial scale. For a number of years, Dow Chemical used cobalt-60 radiation in the production of ethyl bromide (CH_3CH_2Br), a volatile organic liquid used as an intermediate compound in the synthesis of organic materials. The application terminated for reasons of cost and safety. As catalysts, gamma rays have been found to be superior in many cases to chemicals, to the application of ultraviolet light, and to electron bombardment.

Various properties of polymers such as polyethylene are changed by electron or gamma ray irradiation. The original material consists of long parallel chains of molecules, and radiation damage causes chains to be connected in a process called cross-linking. Irradiated polyethylene has better resistance to heat and serves as a good insulating coating for electrical wires. Fabrics can be made soil-resistant by radiation bonding of a suitable polymer to a fiber base.

Highly wear-resistant wood flooring is produced commercially by gamma irradiation (see References). Wood is soaked in a monomer plastic, encased in aluminum, and placed in a water pool containing a cesium-137 source of 661 keV photons. The process of polymerization takes place throughout the wood. The molecular structure is changed so that the surface cannot be scratched or burned.

A related process has been applied in France to the preservation of artistic or historic objects of wood or stone. The artifact is soaked in a liquid monomer and transferred to a cobalt-60 gamma cell where the monomer is polymerized into a solid resin.

18.8 TRANSMUTATION DOPING OF SEMICONDUCTORS

Semiconductor materials are used in a host of modern electrical and electronic devices. Their functioning depends on the presence of small amounts of impurities such as phosphorus in the basic crystal element silicon. The process of adding impurities is called "doping." For some semiconductors, impurities can be introduced in the amounts and locations needed by use of neutron irradiation to create an isotope that decays into the desired material.

The process is relatively simple. A pure silicon monocrystal is placed in a research or experimental reactor of several megawatts power level. The sample is irradiated with a previously calibrated thermal neutron flux for a specified time. This converts one of the silicon isotopes into a stable phosphorus isotope by the reactions

$$^{30}_{14}\text{Si} + ^{1}_{0}\text{n} \rightarrow ^{31}_{14}\text{Si} + \gamma$$

$$^{31}_{14}\text{Si} \rightarrow ^{31}_{15}\text{P} + ^{0}_{-1}\text{e},$$

where the abundance of Si-30 is 3.1% and the half-life of Si-31 is 2.62 h. After irradiation, the silicon resistivity is too high because of radiation damage caused by the fast neutron component of the flux. Heat treatment is required before fabrication to anneal out the defects.

The principal application of neutron transmutation doping (NTD) has been in the manufacture of power thyristors, which are high-voltage, high-current semiconductor rectifiers (see References), so named because they replaced the thyratron, a vacuum tube. The virtue of NTD compared with other methods is that it provides a uniform resistivity over the large area of the device. Annual yields of the product material are more than 50 tons, with a considerable income to the reactor facilities involved in the work. NTD is expected to become even more important in the future for household and automotive devices. The doping method is also applicable to other substances besides silicon (e.g., germanium and gallium arsenide).

18.9 NEUTRONS IN FUNDAMENTAL PHYSICS

Intense neutron beams produced in a research reactor serve as powerful tools for investigation in physics. Three properties of the neutron are important in this work: (a) the lack of electrical charge, which allows a neutron to penetrate atomic matter readily until it collides with a nucleus; (b) a magnetic moment, resulting in special interaction with magnetic materials; and (c) its wave character, causing beams to exhibit diffraction and interference effects.

Measurements of neutron cross section of nuclei for scattering, capture, and fission are necessary for reactor analysis, design, and operation. An area of study that goes beyond those needs is called inelastic neutron scattering. It is based on the fact that the energy of thermal neutrons, 0.0253 eV, is comparable to the energy of lattice vibrations in a solid or liquid. Observations of changes in the energy of bombarding neutrons provide information on the interatomic forces in materials, including the effects of impurities in a crystal, of interest in semiconductor research. Also, inelastic scattering yields understanding of microscopic magnetic phenomena and the properties of molecular gases.

We recall that the magnetic moment of a bar magnet is the product of its length s and the pole strength p. For charges moving in a circle of radius r, the magnetic moment is the product of the area πr^2 and the current i. Circulating and spinning electrons in atoms and molecules also give rise to magnetic moments. Even though the neutron is uncharged, it has an intrinsic magnetic moment. Thus the neutron interacts differently with materials according to their magnetic properties. If the materials are paramagnetic, with randomly oriented atomic moments, no special effect occurs. Ferromagnetic materials such as iron and manganese have unpaired electrons, and moments are all aligned in one direction. Antiferromagnetic materials have aligned moments in each of two directions. Observations of scattered neutrons lead to understanding of the microscopic structure of such materials.

The wavelength of a particle of mass m and speed v according to the theory of wave mechanics is

$$\lambda = h/mv$$

where h is Planck's constant, 6.64×10^{-34} J–s. For neutrons of mass 1.67×10^{-27} kg, at the thermal energy, 0.0253 eV, speed 2200 m/s, the wavelength is readily calculated to be $\lambda = 1.8 \times 10^{-10}$ m. This is fairly close to d, the spacing of atoms in a lattice; for example, in silicon d is 3.135×10^{-10} m. The wave property is involved in the process of neutron diffraction, in analogy to X-ray and optical diffraction, but the properties of the materials that are seen by the rays differ considerably. Whereas X-rays interact with atomic electrons and thus diffraction depends strongly on atomic number Z, neutrons interact with nuclei according to their scattering lengths, which are unique to the isotope, and are rather independent of Z. Scattering lengths, labeled a, resemble radii of nuclei but have both magnitude and sign. For nearby isotopes, a values and the corresponding cross sections $\sigma = \pi r^2$ differ greatly. For example, the approximate σ values of three nickel isotopes are: Ni-58, 26; Ni-59, 1; Ni-60, 10. In neutron diffraction one applies the Bragg formula $\lambda = 2d \sin \theta$, where d is the lattice spacing and θ is the scattering angle. A host of isotopes, elements, and compounds have been investigated by neutron diffraction, as discussed by Bacon (see References).

A still more modern and sophisticated application of neutrons is interferometry in which neutron waves from a nuclear reactor source are split and then recombined. We can describe the essential equipment needed. A perfect silicon crystal is machined very accurately in the form of the letter E, making sure the planes are parallel. A neutron beam entering the splitter passes through a mirror plate and analyzer. Reflection, refraction, and interference take place, giving rise to a periodic variation of observed intensity. Insertion of a test sample causes changes in the pattern. The method has been used to measure accurately the scattering lengths of many materials. Images of objects are obtained in phase topography, so named because the introduction of the sample causes a change in phase in the neutron waves in an amount dependent on thickness, allowing observation of surface features. Interference fringes have been observed for neutrons passing through slightly different paths in the Earth's magnetic field. This suggests the possibility of studying the relationship of gravity, relativity, and cosmology.

In the Spallation Neutron Source (SNS) (in Section 8.6), wavelengths and energies of neutrons produced match the size and energy scales of many materials of interest. Its enhanced neutron beams allow higher resolution images of biological materials. Of special benefit in the SNS is the ability to locate hydrogen atoms in complex molecules. Crystallography studies will lead to more effective drugs.

The broad scope of research with SNS can be appreciated by a listing of areas adapted from the Oak Ridge National Laboratory (ORNL) Web site (see References):

Chemistry. Use of neutron scattering to study microstructures in chemical products.

Complex fluids. Investigation of new time-release drug-delivery systems targeting specific parts of the human body.

Crystalline materials. Research on ways to tailor structures and properties of new materials.

Disordered materials. Study of proteins of interest to biological industries.

Engineering. Knowledge on material failures and substitutes.

Magnetism and superconductivity. Understanding leading to improved devices.

Polymers. Small-angle scattering to reveal behavior of molecular chains.

Structural biology. Neutrons as complement to X-rays for studying vitally important chemicals.

18.10 NEUTRONS IN BIOLOGICAL STUDIES

One of the purposes of research in molecular biology is to describe living organisms by physical and chemical laws. Thus, finding sizes, shapes and locations of components of biological structures is the first step in understanding. Neutron scattering provides a useful tool for this purpose. The radiation does not destroy the specimen; cross sections of materials of interest are of the same order for all nuclei so that heavier elements are not favored as in the case of X-rays; long wavelength neutrons needed to study the large biological entities are readily obtained from a reactor. Of special importance is the fact that scattering lengths for hydrogen (3.8×10^{-15} m) and deuterium (6.5×10^{-15} m), are quite different, so that the neutron scattering patterns from the two isotopes can be readily distinguished.

An example is the investigation of the ribosome. It is a particle approximately 25 nanometers in diameter that is part of a cell and helps manufacture proteins. The *E. coli* ribosome is composed of two subunits, one with 34 protein molecules and two RNA molecules, the other with 21 proteins and one RNA. The proteins are quite large, with molecular weight as high as 65,000. Study with X-rays or an electron microscope is difficult because of the size of the ribosome. For the neutron experiment, two of the 21 proteins are "stained" with deuterium (i.e., they are prepared by growing bacteria in D_2O rather than H_2O).

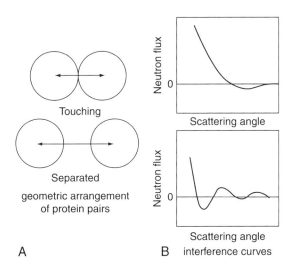

FIGURE 18.3 Interference patterns for the ribosome, a particle in the cell. Estimates of size and spacing are a start toward understanding biological structures.

Early research on ribosomes was performed at the Brookhaven National Laboratory High Flux Beam Reactor (now shut down). A beam of neutrons was scattered from a graphite crystal that selected neutrons of a narrow energy range at wavelength 2.37×10^{-10} m. The specimen to be studied was placed in the beam in front of a helium-3 detector, which counted the number of neutrons as a function of scattering angle. The neutron wave, when scattered by a protein molecule, exhibited interference patterns similar to those of ordinary light. A distinct difference in pattern would be expected depending on whether the two molecules are touching or separated, as shown in Figure 18.3. For the ribosome, the distance between centers of molecules was deduced to be 35×10^{-10} m. Tentative "maps" of the ribosome subunit were developed, as well.

18.11 RESEARCH WITH SYNCHROTRON X-RAYS

Knowledge of the structure of molecules is made possible by the use of synchrotron X-rays because of their high intensity and sharp focus. Studies are faster and less damaging than those with conventional X-rays (see References). Materials in crystalline form are bombarded with photons, and the diffraction patterns are produced on a sensitive screen. The patterns are analyzed by computer by use of the Fourier transform to determine electron densities and thus atom locations. Suitable manipulations yield 3D data. Knowing molecular structure provides information on how chemicals work and helps find better drugs and treatment for disease. A classic example of a synchrotron X-ray study result was the determination of the structure of the rhinovirus HRV14, the cause of the common cold. The crystals were very sensitive to radiation and would cease to diffract before data were obtained by ordinary X-rays. Many other macromolecular proteins, enzymes, hormones, and viruses have been investigated. It is possible to observe chemical processes as they occur (e.g., photodissociation of hydrocarbons and of ozone). Information for improvement of industrial processes and products is also made available.

18.12 SUMMARY

Many examples of the use of radiation for beneficial purposes can be cited. Diseases such as cancer can be treated by gamma rays. Food spoilage is reduced greatly by irradiation. Medical supplies are rendered sterile within plastic containers. Sewage sludge can be disinfected by irradiation. New and improved crops are produced by radiation mutations. The sterile insect technique has controlled insect pests in many areas of the world. Radiation serves as a catalyst in the production of certain chemicals. Properties of fibers and wood are enhanced by radiation treatment. Desirable impurities can be induced in semiconductor materials by neutron bombardment. The scattering by neutrons provides information on magnetic materials, and interference of neutron beams is used to examine surfaces. Scattered neutrons yield estimations of location and size of minute biological structures. Synchrotron X-rays are required for detailed study of biological molecules.

18.13 EXERCISES

18.1 Thyroid cancer is treated successfully by the use of iodine-131, half-life 8.04 d, energy release approximately 0.5 MeV. The biological half-life of I-131 for the thyroid is 4 d. Estimate the number of millicuries of the isotope that should be administered to obtain a dose of 25,000 rads to the 20-grams thyroid gland.

18.2 The disease polycythemia vera (PV) is characterized by an excess of red blood cells. Treatment by chemotherapy and radiation is often successful. In the latter, the patient is injected with a solution of sodium phosphate containing phosphorus-32, half-life 14.28 d, average beta energy 0.69 MeV. Estimate the dose in rads resulting from the administration of an initial 10 mCi of P-32, of which 10% goes to the 3-kg bone marrow. Recall 1 rad $= 10^{-5}$ J/g and 1 mCi $= 3.7 \times 10^7$ dps. Suggestion: Neglect biological elimination of the isotope.

18.3 A company supplying cobalt-60 to build and replenish radiation sources for food processing uses a reactor with thermal flux $10^{14}/cm^2-s$. To meet the demand of a megacurie a month, how many kilograms of cobalt-59 must be inserted in the reactor? Note that the density of Co-59 is 8.9 g/cm^3 and the neutron cross section is 37 barns.

18.4 A cobalt source is to be used for irradiation of potatoes to inhibit sprouting. What strength in curies is needed to process 250,000 kg of potatoes per day, providing a dose of 10,000 rads? Note that the two gammas from Co-60 total approximately 2.5 MeV energy. What is the amount of isotopic power? Discuss the practicality of absorbing all of the gamma energy in the potatoes.

18.5 Transmutation of silicon to phosphorus is to be achieved in a research reactor. The capture cross section of silicon-30, abundance 3.1%, is 0.108 barns. How large must the thermal flux be to produce an impurity content of 10 parts per billion in a day's irradiation?

Computer Exercises

18.A The classic "predator–prey" balance equations simulate interacting populations such as foxes and rabbits. Run the program PREDPREY to see trends with time.

18.B An adaptation of the predator–prey equations can be used to analyze the control of the screwworm fly by the sterile male technique. Study the trend in population under different assumptions and initial conditions with the program ERADIC (eradicate/irradiate). In particular, find the time required to reduce the population to less than one female fly.

18.14 REFERENCES

Interstitial Collaborative Working Group: Lowell L. Anderson, et al., *Interstitial Brachytherapy: Physical, Biological, and Clinical Considerations*, Raven Press, New York, 1990.

IAEA Bulletin, Vol. 33, No. 1, 1991. The issue features nuclear medicine.

Nuclear Medicine Resource Manual
http://www-pub.iaea.org/MTCD/publications/PDF/Pub1198_web.pdf
An IAEA guide for establishing nuclear medicine service. 532 pages.

Major Advances in Nuclear Medicine, Diagnosis and Treatment (American Nuclear Society)
http://www.ans.org/pi/np/diagnosis
Data on remission rates with cell-targeted therapy.

The Basis of Boron Neutron Capture Therapy
http://web.mit.edu/nrl/www/bnct/info/description/description.html
Explanation of effect on tumor.

Boron Neutron Capture Therapy of Cancer: Current Status and Future Prospects
Reprint of paper by Rolf F. Barth, et al., 2005.
Search on Rolf F. Barth in Google and Select title of paper.

"The Present Status of Boron-neutron Capture Therapy for Tumors,"
H. Hatanaka, et al. *Pure & Appl. Chem.* 63, 373 (1991)
http://www.iupac.org/publications/pac/1991/pdf/6303x0373.pdf

Nanotech News, National Cancer Institute
Research in Denmark using boron carbide nanoparticles
http://nano.cancer.gov/news_center/nanotech_news_2006-03-06e.asp

United States-Argentina BNCT Program
http://www.inl.gov/featurestories/2007-02-13.shmtl
Collaboration on research.

Mark B. Garnick and William R. Fair, "Combatting Prostate Cancer," *Scientific American*,
    December 1998, p. 74 ff.

Prostate Cancer Treatment
http://www.theragenics.com
A commercial supplier of radioactive particles. Select TheraSeed.

Foods permitted to be irradiated
http://www.cfsan.fda.gov/~dms/irrafood.html
FDA approval history.

John Henkel, "Irradiation: A Safe Measure for Safer Food," *FDA Consumer*, May–June 1998,
    Cambridge Press.
http://www.fda.gov/fdac/features/1998/398_rad.html
Excellent article, not being updated.

Background and Status of Labeling of Irradiated Food
http://www.organicconsumers.org/Irrad/LabelingStatus.cfm

"FDA seeks to ease labeling requirements," *Nuclear News*, May 2007, p. 61.

"Rick Michael Irradiated food, good; foodborne pathogens, bad," *Nuclear News*, July 2003, p. 62

Irradiation of Food and Packaging: An Overview
http://www.cfan.gov/~dms/irraover.html
Article by Morehouse and Komolprasert, 2004.

Iowa State Food Safety Project
http://www.extension.iastate.edu/foodsafety
Information and links. Select Food Irradiation.

Food Irradiation Information from Food Safety and Inspection Service
http://www.fsis.usda.gov/Fact_Sheets/Irradiation_Resources/index.asp
Links to many documents.

IAEA Report.
Google "IAEA Thematic Planning Food Irradiation"
Select Summary Report.

Radiation Information Network: Food Irradiation
http://www.physics.isu.edu/radinf/food.htm
Extensive discussion, references, and links by Idaho State University.

V. M. Wilkinson and G. W. Gould, *Food Irradiation: A Reference Guide*, Butterworth-
    Heinemann, Oxford, 1996. Topics, definitions, and discussion with references to all
    relevant terms.

E. A. Murano (Ed.), *Food Irradiation: A Sourcebook*, Iowa State University Press, Ames, IA, 1995. Processing, microbiology, food quality, consumer acceptance, and economics.

Argentina Irradiates Urban Sludge
http://www.iaea.org/Publications/Magazines/Bulletin/Bull391/argentina.html
IAEA publication, *INSIDE Technical Cooperation*, March 1997.

D. A. Lindquist and M. Abusowa, "Eradicating the New World Screwworm from the Libyan Arab Jamahiriya," *IAEA Bulletin*, Vol. 34, No. 4, 1992, p. 9. Describes the success of the international program. With 12,000 cases in 1990, the number dropped to zero by May 1991.

History of Screwworm Eradication
http://www.nal.usda.gov/speccoll/screwworm/history.htm
Discovery and application of Sterile Male Technique. From National Agricultural Library.

New World Screwworm Eradication
http://www-tc.iaea.org/tcweb/publications/factsheets/jamaica.pdf
IAEA plans for Jamaica project, still underway.

Eradication of Tsetse Fly in Zanzibar
http://www.fao.org/NEWS/1998/980505-e.htm
http://www-tc.iaea.org/tcweb/Publications/factsheets/tsetse2.pdf
News items from sponsoring agencies.

Simulation Exercise on SIT
http://ipmworld.umn.edu/chapters/SirSimul.htm
Program "Curaçao" by Phil A. Arneson.

Gammapar Impregnated Flooring
http://www.gammapar.com/faqs.jsp
Commercial production of irradiated acrylic impregnated hardwood.

G. E. Bacon, *Neutron Diffraction*, 3rd Ed., Clarendon Press, Oxford, 1975. A classical reference on the subject.

G. Foldiak, Editor, *Industrial Applications of Radioisotopes*, Elsevier, Amsterdam, 1986.

SNS and Biological Research
http://www.ornl.gov/info/ornlreview/v34_1_01/sns.htm
Drug studies.

William D. Ehmann and Diane E. Vance, *Radiochemistry and Nuclear Methods of Analysis*, John Wiley & Sons, New York, 1991. Covers many of the topics of this chapter.

Robert J. Woods and Alexei K. Pikaev, *Applied Radiation Chemistry: Radiation Processing*, John Wiley & Sons, New York, 1994. Includes synthesis, polymerization, sterilization, and food irradiation.

Spallation Neutron Source
http://neutrons.ornl.gov

T. E. Mason, et al., "The Spallation Neutron Source: A Powerful Tool for Materials Research."
http://arxiv.org/abs/physics/0007068
A frequently cited article describing the equipment used in research. Select Download PDF.

The World of Synchrotron Radiation
http://www.srs.dl.ac.uk/SRWORLD
Locations of facilities.

Massimo Altarelli, Fred Schlacter, and Jane Cross, "Ultrabright X-ray Machines," *Scientific American*, December 1998, p. 66 ff.

Kevin L. D'Amico, Louis J. Terminello, and David K. Shuh, *Synchrotron Radiation Techniques in Industrial, Chemical, and Materials Science*, Plenum Press, New York, 1996. Emphasis on structural biology and environmental science.

Albert Hofmann, *The Physics of Synchrotron Radiation*, Cambridge University Press, New York, 2004.

J. R. Helliwell and P. M. Rentzepis, (Eds.), *Time-resolved Diffraction*, Oxford University Press, Oxford, UK, 1997. Research on time-dependent structural changes by use of X-rays (including synchrotron radiation), electrons, and neutrons.

Reactor Safety and Security

It is well known that the accumulated fission products in a reactor that has been operating for some time constitute a potential source of radiation hazard. Assurance is needed that the integrity of the fuel is maintained throughout the operating cycle, with negligible release of radioactive materials. This implies limitations on power level and temperature and adequacy of cooling under all conditions. Fortunately, inherent safety is provided by physical features of the fission chain reaction. In addition, the choice of materials, their arrangement, and restrictions on modes of operation give a second level of protection. Devices and structures that minimize the chance of accident and the extent of radiation release in the event of accident are a third line of defense. Finally, nuclear plant location at a distance from centers of high population density results in further protection.

We will now describe the dependence of numbers of neutrons and reactor power on the multiplication factor, which is in turn affected by temperature and control rod absorbers. Then we will examine the precautions taken to prevent release of radioactive materials to the surroundings and discuss the philosophy of safety.

Thanks are due to Earl M. Page for suggestions on reactor safety and Robert M. Koehler on reactor design and operation.

19.1 NEUTRON POPULATION GROWTH

The multiplication of neutrons in a reactor can be described by the effective multiplication factor k, as discussed in Chapter 11. The introduction of one neutron produces k neutrons; they in turn produce k^2, and so on. Such a behavior tends to be analogous to the increase in principal with compound interest or the exponential growth of the human population. The fact that k can be less than, equal to, or greater than 1 results in significant differences, however.

The total number of reactor neutrons is the sum of the geometric series $1 + k + k^2 + \ldots$. For $k < 1$ this is finite, equal to $1/(1 - k)$. For $k > 1$ the sum is infinite (i.e., neutrons multiply indefinitely). We thus see that knowledge of the effective multiplication factor of any arrangement of fuel and other material is needed to assure safety. Accidental criticality is prevented in a number of situations: (a) chemical processing of enriched uranium or plutonium, (b) storage of fuel in arrays of containers or of fuel assemblies, and (c) initial loading of fuel assemblies at time of startup of a reactor. A classic measurement involves the stepwise addition of small amounts of fuel with a neutron source present. The thermal neutron flux without fuel ϕ_0 and with fuel ϕ is measured at each stage. Ideally, for a subcritical system with a nonfission source of neutrons in place, in a steady-state condition, the multiplication factor k appears in the relation

$$\phi/\phi_0 = 1/(1 - k).$$

As k gets closer to 1, the critical condition, the flux increases greatly. On the other hand, the reciprocal ratio

$$\phi_0/\phi = 1 - k$$

goes to zero as k goes to 1. Plotting the measured flux ratio as it depends on the mass of uranium or the number of fuel assemblies allows increasingly accurate predictions of the point at which criticality occurs, as shown in Figure 19.1. Fuel additions are always intended to be less than the amount expected to bring the system to criticality.

Let us now examine the time-dependent response of a reactor to changes in multiplication. For each neutron, the gain in number during a cycle of time length ℓ is $\delta k = k - 1$. Thus for n neutrons in an infinitesimal time dt the gain

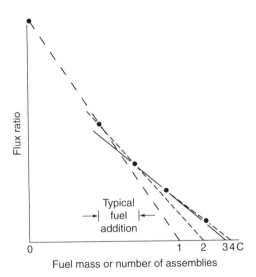

FIGURE 19.1 Critical experiment.

is $dn = \delta k \, n \, dt/\ell$. This can be treated as a differential equation. For constant δk, the solution is

$$n = n_0 \exp(t/T),$$

where T is the period, the time for the population to increase by a factor $e = 2.718\ldots$, given by $T = \ell/\delta k$. When applied to people, the formula states that the population grows more rapidly the more frequently reproduction occurs and the more abundant the progeny.

A typical cycle time ℓ for neutrons in a thermal reactor is very short, approximately 10^{-5} s, so that a δk as small as 0.02 would give a very short period of 0.0005 s. The growth according to the formula would be exceedingly rapid, and, if sustained, would consume all of the atoms of fuel in a fraction of a second.

A peculiar and fortunate fact of nature provides an inherent reactor control for values of δk in the range 0 to approximately 0.0065. Recall that approximately 2.5 neutrons are released from fission. Of these, some 0.65% appear later as the result of radioactive decay of certain fission products and are thus called *delayed neutrons*. Quite a few different radionuclides contribute these, but usually six groups are identified by their different fractions and half-lives (see Exercise 19.12). The average half-life of the isotopes from which they come, taking account of their yields, is approximately 8.8 s. This corresponds to a mean life $\tau = t_H/0.693 = 12.7$ s, as the average length of time required for a radioactive isotope to decay. Although there are very few delayed neutrons, their presence extends the cycle time greatly and slows the rate of growth of the neutron population. The effect of delayed neutrons on reactor transients has an analogy to the growth of principal in an investment, say at a bank. Imagine that the daily interest were mailed out to a client, who had to reinvest by sending the interest back. This "checks in the mail" process would cause principal to increase more slowly.

Then, to understand the mathematics of this effect, let β be the fraction of all neutrons that are delayed, a value 0.0065 for U-235; $1 - \beta$ is the fraction of those emitted instantly as "prompt neutrons." They take only a very short time ℓ to appear, whereas the delayed neutrons take a time $\ell + \tau$. The average delay is thus

$$\bar{\ell} = (1 - \beta)\ell + \beta(\ell + \tau) = \ell + \beta\tau.$$

Now because $\beta = 0.0065$ and $\tau = 12.7$ s, the product is 0.083 s, greatly exceeding the multiplication cycle time, which is only 10^{-5} s. The delay time can thus be regarded as the effective generation time, $\bar{\ell} \cong \beta\tau$. This approximation holds for values of δk much less than β. For example, let $\delta k = 0.001$, and use $\bar{\ell} = 0.083$ s in the exponential formula. In 1 second $n/n_0 = e^{0.012} = 1.01$, a very slight increase.

On the other hand, if δk is greater than β, we still find very rapid responses, even with delayed neutrons. If all neutrons were prompt, one neutron would give a gain of δk, but because the delayed neutrons actually appear much later, they cannot contribute to the immediate response. The apparent δk is then $\delta k - \beta$, and the cycle time is ℓ. We can summarize by listing the period T for the two regions.

$$\delta k << \beta, \ T \cong \frac{\beta\tau}{\delta k}$$

$$\delta k >> \beta, \ T \cong \frac{\ell}{\delta k - \beta},$$

Even though β is a small number, it is conventional to consider δk small only if it is less than 0.0065 but large if it is greater. Figure 19.2 shows the growth in reactor power for several different values of reactivity ρ, defined as $\delta k/k$. These curves were generated with the full set of delayed neutron emitters. Because k is close to 1, $\rho \cong \delta k$. We conclude that the rate of growth of the neutron population or reactor power is very much smaller than expected, so long as δk is kept well below the value β, but rapid growth will take place if δk is larger than β.

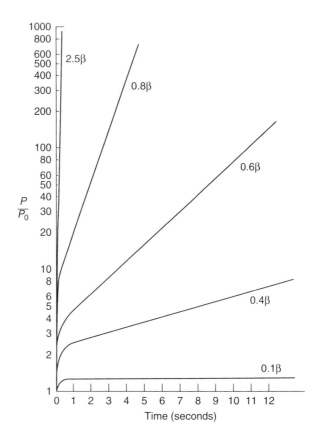

FIGURE 19.2 Effect of delayed neutrons.

We have used the value of β for U-235 for illustration but should note that its effective value depends on reactor size and type of fuel (e.g., β for Pu-239 is only 0.0021). Also, the value of the neutron cycle time depends on the energy of the predominant neutrons. The ℓ for a fast reactor is much shorter than that for a thermal reactor.

In the many hundreds of critical experiments and manipulations of nuclear fuel in processing plants, there have been serious criticality accidents, involving radiation exposure and several deaths. In the early days, fewer precautions were taken (see References for summary information). Even as late as 1999, an accident in Japan resulted from the addition of an excess of enriched uranium to a process vessel.

In Computer Exercises 19.A and 19.B we demonstrate the growth with time of the neutron population as it depends on reactivity. One equivalent delayed neutron group is used in 19.A; six groups are used in 19.B.

19.2 **ASSURANCE OF SAFETY**

The inherent nuclear control provided by delayed neutrons is aided by proper design of the reactor to favor certain negative feedback effects. These are reductions in the neutron multiplication factor resulting from increases in reactor power. With additional heat input the temperature increases, and the negative reactivity tends to shut the reactor down. Design choices include the size and spacing of fuel rods and the soluble boron content of the cooling water. One of the temperature effects is simple thermal expansion. The moderator heats up, it expands, the number density of atoms is reduced, and neutron mean free paths and leakage increase, whereas thermal absorption goes down. In early homogeneous aqueous reactors this was a dominant effect to provide shutdown safety. In heterogeneous reactors it tends to have the opposite effect in that reductions in boron concentration accompany reduction in water density. Thus there must be some other effect to override moderator expansion effects. The process of Doppler broadening of resonances provides the needed feedback. An increase in the temperature of the fuel causes greater motion of the uranium atoms, which effectively broadens the neutron resonance cross section curves for uranium shown in Figure 4.6. For fuel containing a high fraction of uranium-238 the multiplication decreases as the temperature increases. The Doppler effect is "prompt" in that it responds to the fuel temperature, whereas the moderator effect is "delayed" as heat is transferred from fuel to coolant. The use of the term "Doppler" comes from the analogy with frequency changes in sound or light when there is relative motion of the source and observer.

The amounts of these effects can be expressed by formulas such as

$$\rho = \alpha \Delta T$$

in which the reactivity ρ is proportional to the temperature change ΔT, with a temperature coefficient α that is a negative number. For example, if the value of α is $-10^{-5}/\,°C$, a temperature rise of 20°C would give a reactivity of -0.0002. Another relationship is

$$\rho = a\,\Delta P/P$$

with a negative power coefficient a and fractional change in power $\Delta P/P$. For example, in a pressurized water reactor (PWR) if $a = -0.012$, a 2% change in power would give a reactivity of -0.00024.

Temperature effects cause significant differences in the response of a reactor to disturbances. The effects were ignored in Figure 19.2, and the population grew exponentially, but if effects are included, as in Figure 19.3, the power flattens out and becomes constant.

Even though a reactor is relatively insensitive to increases in multiplication in the region $\delta k < \beta$, and temperature rises provide stability, additional protection is provided in reactor design and operating practices. Part of the control of a reactor of the PWR type is provided by the boron solution (see Section 11.5). This "chemical shim" balances the excess fuel loading and is adjusted gradually as fuel is consumed during reactor life. In addition, reactors are provided with several groups of movable rods of neutron-absorbing material, as shown in Figure 19.4. The rods serve three main purposes: (a) to permit temporary increases in multiplication that brings the reactor up to the desired power level or to make adjustments in power; (b) to cause changes in the flux and power shape in the core, usually striving for uniformity; and (c) to shut down the reactor manually or automatically in the event

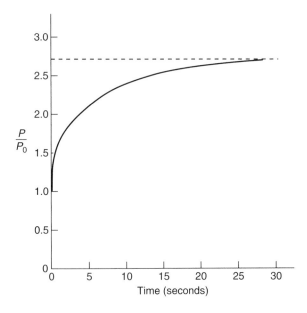

FIGURE 19.3 Effect of temperture of power.

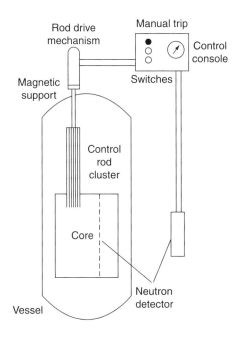

FIGURE 19.4 Reactor control.

of unusual behavior. To ensure effectiveness of the shutdown role, several groups of safety rods are kept withdrawn from the reactor at all times during operation. In the PWR they are supported by electromagnets that release the rods on interruption of current, whereas in the boiling water reactor (BWR) they are driven in from the bottom of the vessel by hydraulic means.

The reactivity worth of control and safety rods as a function of depth of insertion into the core can be measured by a comparison technique. Suppose a control rod in a critical reactor is lifted slightly by a distance δz and a measurement is made of the resulting period T of the rise in neutron population. By use of the approximate formula from Section 19.1,

$$T \cong \beta \tau / \delta k,$$

we deduce the relation of δk to δz. The reactor is brought back to critical by an adjustment of the soluble boron concentration. Then the operation is repeated with an additional shift in rod position. The experiment serves to find both the reactivity worth of the rod as a function of position and, by summation, the total worth of the rod. Figure 19.5 shows the calibration curves of a control rod in an idealized case of a core without end reflectors. It is noted that the effect of a rod movement in a reactor depends strongly on the location of the tip. The basis for the S-shaped curves of Figure 19.5 is found in reactor theory, which tells us that the reactivity effect of an added absorber sample to a reactor is approximately

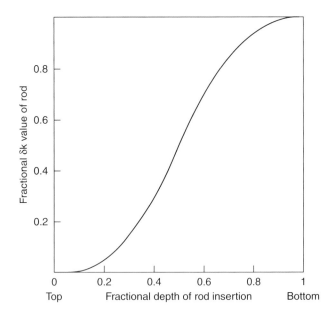

FIGURE 19.5 Control rod worth as it depends on depth of insertion in an unreflected reactor core.

dependent on the square of the thermal flux that is disturbed. Thus if a rod is fully inserted or fully removed, such that the tip moves in a region of low flux, the change in multiplication is practically zero. At the center of the reactor, movement makes a large effect. The slope of the curve of reactivity vs. rod position when the tip is near the center of the core is twice the average slope in this simple case.

Estimates of total reactivity worth can also be made by the rod-drop technique. A control rod is allowed to fall from a position outside the core to a full-in position. The very rapid change of neutron flux from an initial value ϕ_0 to a final value ϕ_1 is shown in Figure 19.6. Then the reactivity worth is calculated from the formula

$$\rho/\beta = (\phi_0/\phi_1) - 1.$$

The result somewhat depends on the location of the detector.

An instrumentation system is provided to detect an excessive neutron flux and thus power level to provide signals calling for a "trip" of the reactor. As sketched in Figure 19.4, independent detectors are located both inside the core and outside the reactor vessel. Data from core detectors are processed by a computer to determine whether or not power distributions are acceptable.

Because almost all of the radioactivity generated by a reactor appears in the fuel elements, great precautions are taken to ensure the integrity of the fuel. Care

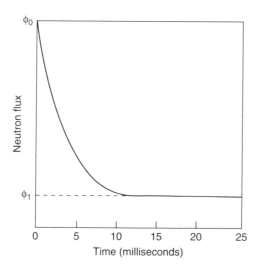

FIGURE 19.6 Neutron flux variation with time in the rod-drop method of measuring reactivity.

is taken in fuel fabrication plants to produce fuel pellets that are identical chemically, of the same size and shape, and of common U-235 concentration. If one or more pellets of unusually high fissile material content were used in a reactor, excessive local power production and temperature would result. The metal tubes that contain the fuel pellets are made sufficiently thick to stop the fission fragments, to provide the necessary mechanical strength to support the column of pellets, and to withstand erosion by water flow or corrosion by water at high temperatures. Also, the tube must sustain a variable pressure difference caused by moderator-coolant outside and fission product gases inside. The cladding material usually selected for low neutron absorption and for resistance to chemical action, melting, and radiation damage in thermal reactors is zircaloy, an alloy that is approximately 98% zirconium with small amounts of tin, iron, nickel, and chromium. The tube is formed by an extrusion process that eliminates seams, and special fabrication and inspection techniques are used to ensure that there are no defects such as deposits, scratches, holes, or cracks.

Each reactor has a set of specified limits on operating parameters to ensure protection against events that could cause hazard. Typical of these is the upper limit on total reactor power, which determines temperatures throughout the core. Another is the ratio of peak power to average power that is related to hot spots and fuel integrity. Protection is provided by limiting the allowed control rod position, reactor imbalance (the difference between power in the bottom half of the core and the top half), reactor tilt (departure from symmetry of power across the core), maximum reactor coolant temperature, minimum coolant flow, and maximum and minimum primary system pressure. Any deviation causes the safety rods to be inserted to trip the reactor. Maintenance of chemical purity of the coolant to minimize corrosion, limitation on allowed leakage rate from the primary cooling system, and continual observations on the level of radioactivity in the coolant serve as further precautions against release of radioactive materials.

Protection of fuel against failure that would release fission products into the coolant is thus an important constraint in the operation of a reactor. Correct choices must be made of the enrichment of U-235, the operating power level, the length of time between refuelings, and the arrangement of new and partially burned fuel, all with an eye on cost.

The term "burnup" is widely used. Take a typical cubic centimeter of fuel and let all fission be due to U-235. The macroscopic fission cross section is Σ_f, the fission rate in a neutron flux ϕ is $f = \phi\Sigma_f$, and the power density is $p = fw$, where w is the energy per fission. The energy produced in a time t is $W = pt$. Now the density of uranium is $d = N_U m_U$, where N_U and m_U are the number density and mass of a uranium atom, respectively. The burnup in watt-seconds per gram is then $B = W/d$. A numerical factor allows easy conversion to MWd/tonne. As shown in Exercise 19.13, B can be shown to depend on the enrichment in U-235, expressed as the ratio N_{235}/N_U.

In Section 11.5 we examined the trends in fuel and control boron for a reactor visualized as a single region. Modern power reactor cores consist of several regions. At the start of an operating cycle, it will contain fresh and partially burned fuel; at the end, partially and fully burned fuel.

For a reactor core with n zones, let k_i be the multiplication constant of fuel in zone i and assume nearly equal power over the core. Then the average k is

$$k = \sum_{i=1}^{i=n} k_i.$$

It has been found that k_i varies with burnup according to

$$k_i = k_0 - aB$$

where B is the burnup in megawatt-days per metric ton (MWd/tonne), k_0 is the initial multiplication constant, and a is a constant.

The amount of control absorber required to keep the reactor critical is a measure of the average k of the core. Figure 19.7 shows its variation with time for different numbers of zones. As noted, the larger is n, the smaller is the initial control absorber.

A little algebra shows us (see Exercise 19.14) that the discharge burnup of fuel depends on the number of zones. Letting $B(1)$ be that for one zone, the burnup for n zones is

$$B(n) = (2n/(n+1))B(1).$$

Thus $B(2) = (4/3)B(1)$, $B(3) = (3/2)B(1)$, etc. For very large n, corresponding to continuous refueling as in the Canadian reactors, the burnup turns out to be twice $B(1)$.

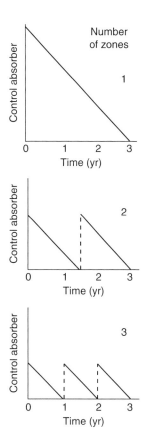

FIGURE 19.7 Reactor operation with different numbers of fuel zones. The initial control absorber varies inversely with the number of regions.

In the foregoing paragraphs we have alluded to a few of the physical features and procedures used in the interests of safety. These have evolved from experience over a number of years, and much of the design and operating experience has been translated into widely used standards, which are descriptions of acceptable practice. Professional technical societies, industrial organizations, and the federal government cooperate in the development of these useful documents.

In addition, requirements related to safety have a legal status, because all safety aspects of nuclear systems are rigorously regulated by federal law and administered by the United States Nuclear Regulatory Commission (NRC). Before a prospective owner of a nuclear plant can receive a permit to start construction, he or she must submit a comprehensive preliminary safety analysis report (PSAR) and an environmental impact statement. On approval of these, a final safety analysis report (FSAR), technical specifications, and operating procedures must be developed in parallel with the manufacture and construction. An exhaustive testing program of components and systems is carried out at the plant. The documents and test results form the basis for an operating license.

Throughout the analysis, design, fabrication, construction, testing, and operation of a nuclear facility, adequate *quality control* (QC) is required. This consists of a careful documented inspection of all steps in the sequence. In addition, a *quality assurance* (QA) program that verifies that quality control is being exercised properly is imposed. Licensing by the NRC is possible only if the QA program has satisfactorily performed its function. During the life of the plant, periodic inspections of the operation are made by the NRC to ascertain whether or not the owner is in compliance with safety regulations, including commitments made in Technical Specifications and the FSAR.

19.3 EMERGENCY CORE COOLING AND CONTAINMENT

The design features and operating procedures for a reactor are such that under normal conditions a negligible amount of radioactivity will get into the coolant and find its way out of the primary loop. Knowing that abnormal conditions can exist, the worst possible event, called a design basis accident, is postulated. Backup protection equipment, called engineered safety features, is provided to render the effect of an accident negligible. A loss of coolant accident (LOCA) is the condition typically assumed, in which the main coolant piping somehow breaks and thus the pumps cannot circulate coolant through the core. Although in such a situation the reactor power would be reduced immediately by use of safety rods, there is a continuing supply of heat from the decaying fission products that would tend to increase temperatures above the melting point of the fuel and cladding. In a severe situation, the fuel tubes would be damaged and a considerable amount of fission products released. To prevent melting, an emergency core cooling system (ECCS) is provided in water-moderated reactors, consisting of auxiliary pumps that inject and circulate cooling water to keep temperatures down. Detailed analysis of heat generation and transfer is required in an application to the NRC for a license to operate a nuclear power plant (see References). The operation of a typical ECCS can be understood by study of some schematic diagrams.

The basic PWR system (Figure 19.8) includes the reactor vessel, the primary coolant pump, and the steam generator, all located within the containment building. The system actually may have more than one steam generator and pump—these are not shown for ease in visualization. We show in Figure 19.9 the auxiliary equipment that constitutes the engineered safety (ES) system. First is the *high-pressure injection system*, which goes into operation if the vessel pressure, expressed in pounds per square inch (psi), drops from a normal value of approximately 2250 psi to approximately 1500 psi as the result of a small leak. Water is taken from the borated water storage tank and introduced to the reactor through the inlet cooling line. Next is the *core flooding tank*, which delivers borated

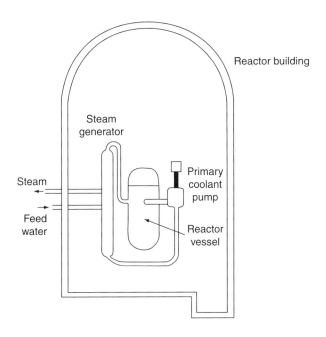

FIGURE 19.8 Reactor containment.

water to the reactor through separate nozzles in the event a large pipe break occurs. Such a rupture would cause a reduction in vessel pressure and an increase in building pressure. When the vessel pressure becomes approximately 600 psi, the water enters the core through nitrogen pressure in the tank. Then if the primary loop pressure falls to approximately 500 psi, the *low-pressure injection* pumps start to transfer water from the borated water storage tank to the reactor. When this tank is nearly empty, the pumps take spilled water from the building sump as a reservoir and continue the flow through coolers that remove the decay heat from fission products. Another feature, the building spray system, also goes into operation if the building pressure increases above approximately 4 psi. It takes water from the borated water storage tank or the sump and discharges it from a set of nozzles located above the reactor to provide a means for condensing steam. At the same time, the emergency cooling units of the reactor building are operated to reduce the temperature and pressure of any released vapor, and reactor building isolation valves are closed on unnecessary piping to prevent the spread of radioactive materials outside the building.

We can estimate the magnitude of the problem of removing fission product heat. For a reactor fueled with U-235, operated for a long time at power P_0 and then shut down, the power associated with the decay of accumulated fission products is $P_f(t)$, given by an empirical formula such as

$$P_f(t) = P_0 A t^{-a}.$$

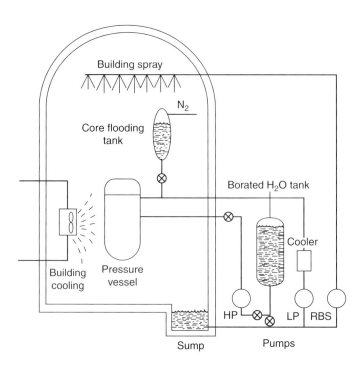

FIGURE 19.9 Emergency core cooling system. Pumps: HP, high pressure; LP, low pressure; RBS, reactor building spray.

For times greater than 10 s after reactor shutdown, the decay is represented approximately by use of $A = 0.066$ and $a = 0.2$. We find that at 10 s the fission power is 4.2% of the reactor power. By the end of a day, it has dropped to 0.68%, which still corresponds to a sizable power, viz., 20 MW for a 3000-MWt reactor. The ECCS must be capable of limiting the surface temperature of the zircaloy cladding to specified values (e.g., 2200 °F), of preventing significant chemical reaction, and of maintaining cooling over the long term after the postulated accident.

The role of the steel-reinforced concrete reactor building is to provide containment of fission products that might be released from the reactor. It is designed to withstand internal pressures and to have a very small leak rate. The reactor building is located within a zone called an exclusion area, of radius of the order of half a kilometer, and the nuclear plant site is several kilometers from any population center.

A series of experiments called Loss of Flow Tests (LOFT) were done at Idaho Falls to check the adequacy of mathematical models and computer codes related to LOCA/ECCS. A double-ended coolant pipe break was introduced and the ability to inject water against flow reversal and water vapor determined. Tests showed that peak temperatures reached were lower than predicted, indicating conservatism in the calculation methods.

19.4 PROBABILISTIC RISK ASSESSMENT

The results of an extensive investigation of reactor safety were published in 1975. The document is variously called "Reactor Safety Study," "WASH-1400," or "Rasmussen Report," after its principal author. The study (see References) involved 60 scientists and cost several million dollars. The technique used was probabilistic risk analysis (PRA), a formal method of analyzing reactor systems. The objective is to find the chance of an undesired event such as core damage, breach of containment, or release of radioactivity, and to determine potential causes. The first step is to investigate all of the possible faults in the equipment or processes. Flow diagrams of fluid systems and circuit diagrams of electrical systems serve as reference. *Event trees* are logic diagrams relating an initiating event to either successful mitigation or failure. Figure 19.10 shows a simple event tree. Probabilities of success and failure at each branch are applied. The principal logic

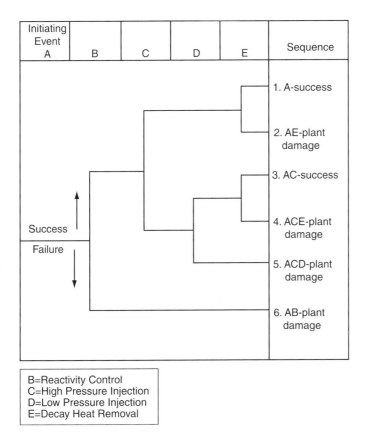

FIGURE 19.10 Simple event tree (After NUREG/CR-4350, Vol.1).

diagrams are the *fault trees*, which trace causes and effects mathematically by use of Boolean algebra, a form of set theory. Figure 19.11(A) shows a simple high-pressure injection system to which we can apply the concept for illustration. The failure of both pumps and/or the valve prevents cooling water to reach the reactor. In Figure 19.11(B) the fault tree diagram shows two types of "gate," the AND ($\cap$) that requires two or more events to result in failure, and the OR ($\cup$) that requires only one event. We have attached symbols A, B, C, F, and T to the various events for use in the mathematical manipulation. Note that F occurs if both A and B occur, expressed in Boolean algebra as

$$F = A \cap B,$$

an intersection. Also, T occurs if either C or F occurs, expressed as

$$T = C \cup F,$$

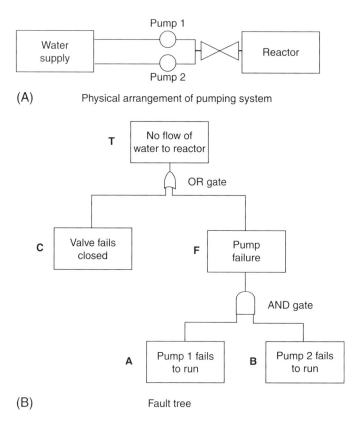

(A) Physical arrangement of pumping system

(B) Fault tree

FIGURE 19.11 Simple example of PRA diagrams (after NUREG-0492).

a union. Theory (e.g., Lewis in References) tells us what the probability of T is in terms of C and F, viz.,

$$P(T) = P(C) + P(F) - P(C \cap F).$$

Insert the formula for F and note that because A, B, and C are independent events, the probabilities $P(A \cap B)$ and $P(C \cap A \cap B)$ are simply products of the separate probabilities. Thus,

$$P(T) = P(C) + P(A) \, P(B) - P(A) \, P(B) \, P(C).$$

The virtue of Boolean algebra is seen by comparison of this formula with the statement in words that the probability of failure of the high-pressure injection system is the sum of the probabilities of individual failures of the valve and the pumps less the probability of failure of both valves and pumps, which was included already. To illustrate numerically, let event probabilities $P(A)$ and $P(B)$ be 10^{-3} and $P(C)$ be 10^{-4}. Inserting numbers,

$$P(T) = 10^{-4} + 10^{-6} - 10^{-10} \cong 1.01 \times 10^{-4},$$

which shows that the top event is dominated by the possibility of valve failure. The product of three probabilities can be neglected assuming rare events. The numerical result illustrates two ideas—that fault trees can reveal potential vulnerabilities and that redundancy in safety equipment is beneficial. The figure calculated can be included in the simple event tree of Figure 19.10.

Several good books on fault trees are listed in References. Among important topics discussed in those references are: Venn diagrams, used to visualize relationships of intersections and unions; conditional probability, related to sequences of events; the Bayes theorem, a technique for updating failure probability data; and common cause failures, those where several components can fail from a single cause.

The ultimate objective of PRA is to calculate risks to people calculated by use of a principle most simply stated as

$$\text{Risk} = \text{Frequency} \times \text{Consequences}.$$

For reactors, *frequency* means the number of times per year of operation of a reactor that the incident is expected to occur, and *consequences* means the number of fatalities, either immediate or latent. The technique of PRA is used to determine which changes in equipment or operation are most important to ensure safety and also give guidance on emergency plans.

In recent years the regulation of nuclear activities including reactor operation and handling of radioisotopes has changed. Currently, regulations are risk-informed and performance-based, in contrast with previous prescriptive approaches. As discussed in an American Nuclear Society position statement (see References), risk-informed implies use of probability in prioritizing challenges to safety, and performance-based makes use of measurable safety parameters. Fuller explanations are found in a publication of the NRC (see References).

If an incident occurring at a nuclear plant has the potential of releasing radio-activity to the atmosphere, a chain of reactions to alert or warn the public is set in motion. The NRC and the Federal Emergency Management Agency (FEMA) cooperate in providing requirements and in monitoring tests of readiness. Each nuclear station and the state in which it is located are required to have emergency plans in place and to hold drills periodically, resembling action to be taken in a real accident situation. In such exercises, state and local officials are notified, and an emergency team made up of many organizations makes a coordinated response. Included are radiation protection staff, police and fire departments, highway patrol, public health officers, and medical response personnel. Command posts are set up; weather observations are correlated with radiation conditions to evaluate the possible radiation exposure of the public. Advisories are sent out by radio, sirens are sounded, and the public is advised to take shelter in homes or other buildings. In extreme cases people would be urged to evacuate the affected area.

In case of actual accident involving reactors or transportation of fuel or waste, members of the public who suffer a loss can be compensated. The Price-Anderson Act was passed by Congress in 1957 to provide rules about nuclear insurance that were favorable to the development of the nuclear industry. The Act was renewed in 2005 for 20 y. Nuclear plants are required to take out insurance from private companies in the amount of $300 million. In the event of an accident, all reactors would be assessed to bring the total liability to approximately $10 billion. The Act has been criticized as unfairly benefiting the nuclear industry because any excess cost would be borne by government and thus taxpayers.

19.5 THE THREE MILE ISLAND ACCIDENT AND LESSONS LEARNED

On March 28, 1979, an accident occurred in the reactor Three Mile Island (TMI) Unit 2 near Harrisburg, Pennsylvania. A small amount of radioactivity was released, causing great alarm throughout the region. We briefly review with Figure 12.7 what happened at TMI and the resultant improvements in reactor safety.

The steam generator's feedwater system malfunctioned, causing the turbine generator to trip and control rods to go into the core to reduce power. Backup feedwater pumps failed to operate because a valve to the steam generator had been left closed by mistake. The steam generator dried causing the primary water coolant temperature and pressure to increase. This caused a relief valve on the pressurizer to open and stay stuck open. Core water escaped to a quench tank that in turn released radioactive water to the containment building and then to an auxiliary building. The ECCS actuated. Operators thought the pressurizer was full of water and shut off the ECCS and later the coolant pumps. The core heated up and became uncovered. Decay heat caused major damage to the fuel.

Radioactive gases were detected outside the containment building. The radiation dosage to anyone was estimated to be less than 100 mrem. This was based on assumed continuous exposure outdoors at the site boundary for 11 days. The average exposure to people within 50 miles was estimated to be only 11 mrems, noted to be less than that caused by a medical X-ray. As a result of a warning by the governor of Pennsylvania, many people, especially pregnant women, left the area for several days. Estimates published by the Department of Health, Education, and Welfare indicate that the exposure over the lifetimes of the 2 million people in the region there would be statistically only one additional cancer death (of 325,000 from other causes).

The TMI accident was due to a combination of (a) design deficiency—inadequate control of water and insufficient instrumentation, (b) equipment failure—the stuck pressurizer valve, and (c) operator error—especially turning off the ECCS and the pumps. Some would view the event as proof that reactors are unsafe; others would note that even with core damage little radioactivity was released. Before the TMI-2 accident, the movie "China Syndrome" had been released. It focused on a hypothetical accident in which the whole core is assumed to melt its way through the reactor vessel and go on in the earth toward China. No such scenario is valid, but public fears were aroused.

A recovery program for TMI was initiated. The interior of the reactor pressure vessel was examined by use of miniature TV cameras attached to the ends of long cables inserted from the top. The damage was greater than originally thought. The upper 5 feet of the core was missing, having slumped into the portion below, and solidified molten fuel was found in the lower part of the vessel. Special handling tools were devised to extract the damaged fuel. Care was taken by measurements and analysis to ensure that the debris would not go critical during recovery. The fuel was transferred to a series of always-safe canisters for storage and shipment.

President Jimmy Carter took a keen interest in the accident. He created The President's Commission on the Accident at Three Mile Island (called the Kemeny Commission after its chairman, John Kemeny, president of Dartmouth College). It was composed of qualified people without nuclear industry connections. A number of recommendations were made in its report, including the need for the nuclear industry to enhance operator and supervisor training, to set its own standards of excellence, and to conduct performance evaluations. Insights on the Kemeny Commission are found in References. In cooperation with utility leaders such as Duke Power's Bill Lee, many of the recommendations were implemented. One of the most significant outcomes of the TMI accident was the formation by the industry of the Institute of Nuclear Power Operations (INPO). This organization reviews all aspects of the performance of United States nuclear power plants and provides recommendations for improvements. Details of the function of INPO are found in Section 23.6. Shortly after the TMI-2 accident the NRC requested that utilities take a number of corrective actions to improve safety. In anticipation of NRC's expectations the industry conducted a study called Industry Degraded Core Rulemaking (IDCOR). Its purpose was to provide well-documented databases on reactor safety. It was concluded that fission product releases would be much lower than those predicted by the Safety Study (Section 19.4). This discrepancy prompted new studies of the "source term," the radioactive release in case of accident.

A second study sponsored by NRC was described in report BMI-2104. New computer codes by Battelle Memorial Institute were applied to all processes, giving an improvement over the Safety Study. Risks were found to be dependent on containment design. Other studies were made by the American Nuclear Society and by the American Physical Society. The general conclusion was that source terms were lower because of particle retention at containment wall.

Figure 19.12 illustrates the improvement in going from WASH-1400 to BMI-2104 for one example reactor, the Surry Nuclear Station of Virginia Electric Power Company. The interpretation of the lower curve is as follows: the chance for as

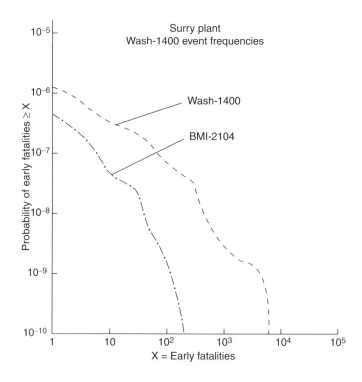

FIGURE 19.12 Distribution function for the Surry, Virginia, plant. The probabilities for various numbers of early fatalities are shown (Adapted from NUREG-0956).

many as *one* early fatality is seen to be less than 10^{-6} per reactor year[†]. If one selects a larger number of fatalities, for example 200, the chance drops by a factor of approximately 5000. However, the chance of latent cancer fatalities is quoted to be larger, 3.4×10^{-3} per reactor year. This still corresponds to a prediction of less than one death per year for the more than 100 United States reactors.

When the findings on source term are used in establishing emergency plans, evacuation of people from a large area surrounding a damaged plant would be an inappropriate action. Over the period 1985–2001 the NRC carried out a program called Individual Plant Examination (IPE) to seek out vulnerabilities and report them. PRA was the only way to accomplish that. No significant problems were uncovered.

19.6 THE CHERNOBYL ACCIDENT

On April 26, 1986, a very serious reactor accident occurred at the Chernobyl[‡] reactor near Kiev in the U.S.S.R. Ukraine. An explosion took place that blew a hole in the roof of the building housing the reactor, the graphite moderator caught fire, and a large amount of radioactive material from the damaged nuclear fuel was released into the atmosphere. The amount of radiation exposure to workers and the public is not precisely known, but the doses exceeded the fallout from earlier weapons tests. A number of workers were killed, nearby towns were contaminated, and it is estimated that the collective dose to the public increased the cancer risk. A number of people were evacuated from the town of Pripyat. Agriculture was disrupted in the Soviet Union, and a ban on food imports was imposed by several European countries.

The Chernobyl reactor Unit 3 was of the type labeled RBMK. The core was cylindrical, 7 m high, 12 m diameter. Moderator graphite blocks were pierced by vertical holes to hold pressure tubes. These contained slightly enriched UO_2 fuel rods and ordinary water coolant. Figure 19.13 shows the reactor and its building.

An experiment related to the electrical supply in case of emergency was being performed. A separate group had planned the experiment. Operators were under pressure to complete the test because the next available maintenance period was over a year away. The power was to be reduced from 3200 MWt to approximately 700 MWt, but the operators allowed it to drop to 30 MWt. There, the neutron flux was too low to burn out accumulated xenon-135. The buildup of absorber made it very difficult to raise the power, and only 200 MWt could be reached.

[†]Many do not understand the terminology used by reactor safety analysts (e.g., "a 10^{-6} chance per year of harm"). Norman Rasmussen, after which the Report (Section 19.4) was named, remarks, "For most people, a rare event is one that occurs once in a lifetime, like Halley's comet, frequency 10^{-2}. Once in 100 lifetimes is 10^{-4}, that's getting hard to believe."

[‡]Ukraine prefers the spelling "Chornobyl" but we will use the more familiar form.

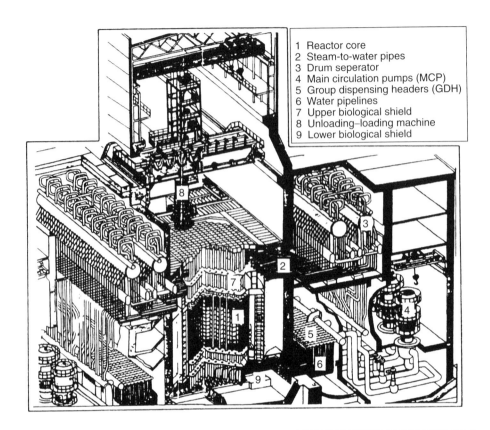

1 Reactor core
2 Steam-to-water pipes
3 Drum seperator
4 Main circulation pumps (MCP)
5 Group dispensing headers (GDH)
6 Water pipelines
7 Upper biological shield
8 Unloading–loading machine
9 Lower biological shield

FIGURE 19.13 The Chernobyl reactor and building before the 1986 accident.

In violation of rules, most of the control rods were pulled out. Coolant flow was reduced to a point where steam voids were created. The RBMK had a positive void coefficient of reactivity, in contrast with that of light water reactors (LWRs). The reactivity caused by steam caused the power to flash up to 30,000 MWt, 10 times the operating level. The power could not be reduced because the rods were too far out to have an effect.

The excess energy pulverized fuel, caused steam pressure to build, and ruptured coolant tubes. Chemical reactions involved steam, graphite, zirconium, and fuel, creating heat that vaporized fuel and set the graphite afire. The explosion blew off the roof of the building housing the reactor and released some 80 million curies of activity, including cesium-137, iodine-131, noble gases, and fission products. A radioactive cloud passed over several countries of Europe, but activity was detected worldwide. Food crops had to be discarded because of contamination.

A total of 203 operating personnel, firefighters, and emergency workers were hospitalized with radiation sickness, 31 of whom died. Their exposures ranged from 100 rems to as high as 1500 rems. Thousands of people were evacuated, many of whom were permanently relocated, with great cost and undoubtedly much distress. A total of 135,000 people were evacuated from a 30-km zone, including 45,000 from the town of Pripyat. Most of those in the evacuation zone received less than 25 rem. With the total estimated dose of 1.6 million person-rems, an increase of up to 2% in cancer deaths over the next 70 y would be predicted. The exposure outside the U.S.S.R. was considerably less, being only several times natural background radiation.

A structure called a sarcophagus was erected around the damaged reactor in an effort to prevent future releases of radioactivity. There is evidence of deterioration, with the possibility of in-leakage of rainwater.

Several implications of the accident were noted: (a) The RBMK reactor type is inherently unsafe and should be phased out. (b) Reactor safety philosophy and practice needs to be revised with greater attention to human factors and safety systems. (c) International cooperation and information exchange should be strengthened. (d) Reactor accidents have global significance. (e) Users of LWRs need to examine equipment and practices, even though the reactors are far safer with their negative power coefficients and strong containment buildings. (f) The consequences of the Chernobyl accident should continue to be monitored.

One consequence of the accident was the formation of a set of joint research projects between the United States and the Russian Federation. These emphasized databases, computer codes, and the development of a plan for Russian nuclear safety research.

At the tenth, fifteenth, and twentieth anniversaries of the Chernobyl accident, reviews were made to assess consequences. It was noted in 1999 that radioactivity persisted, with damage to plants and animals. People were seriously afflicted with anxiety. In 2001 a series of 20 lessons learned was presented, with emphasis on financial and social effects (see References). In 2006 note was made of the thousands of childhood thyroid cancers, with treatment generally successful. Total cancer mortality in the population was estimated to be a few percent. Decommissioning of Unit 4 was deemed of high priority (see References).

19.7 PHILOSOPHY OF SAFETY

The subject of safety is a subtle combination of technical and psychological factors. Regardless of the precautions that are provided in the design, construction, and operation of any device or process, the question can be raised "Is it safe?" The answer cannot be a categorical "yes" or "no" but must be expressed in more ambiguous terms related to the chance of malfunction or accident, the nature of protective systems, and the consequences of failure. This leads to more philosophical questions such as "How safe is safe?" and "How safe do we want to be?"

In an attempt to answer such questions, the NRC adopted in 1986 what are called *safety goals*. These are intended to free neighbors of nuclear plants from worry. Regulations are " . . . to provide reasonable assurance . . ." that a severe core accident will not occur at a United States nuclear plant." Design and operation are to be such that risks of death from a nuclear accident are no greater than a thousandth of known and accepted risks. The comparison is to be made with other common accidents for those people living within a mile of the plant and with cancer from all causes for those living within 10 miles.

Every human endeavor is accompanied by a certain risk of loss, damage, or hazard to individuals. In the act of driving an automobile on the highways, or in turning on an electrical appliance in the home, or even in the process of taking a bath, one is subject to a certain danger. Everybody agrees that the consumer deserves protection against hazard outside his personal control, but it is not at all clear as to what lengths it is necessary to go. In the absurd limit, for instance, a complete ban on all mechanical conveyances would ensure that no one would be killed in accidents involving cars, trains, airplanes, boats, or spacecraft. Few would accept the restrictions thus implied. It is easy to say that reasonable protection should be provided, but the word "reasonable" has different meanings among people. The concept that the benefit must outweigh the risk is appealing, except that it is very difficult to assess the risk of an innovation for which no experience or statistical data are available or for which the number of accidents is so low that many years would be required for adequate statistics to be accumulated. Nor can the benefit be clearly defined. A classic example is the use of a pesticide that ensures protection of the food supply for many, with finite danger to certain sensitive individuals. To the person affected adversely, the risk completely overshadows the benefit. The addition of safety measures is inevitably accompanied by increased cost of the device or product and the ability or willingness to pay for the increased protection varies widely among people.

It is thus clear that the subject of safety falls within the scope of the social-economic-political structure and processes and is intimately related to the fundamental conflict of individual freedoms and public protection by control measures. It is presumptuous to demand that every action possible should be taken to provide safety, just as it is negligent to contend that because of evident utility, no effort to improve safety is required. Between these extreme views, there remains an opportunity to arrive at satisfactory solutions, applying technical skill accompanied by responsibility to assess consequences. It is most important to provide understandable information on which the public and its representatives can base judgments and make wise decisions as to the proper level of investment of effort and funds.

19.8 NUCLEAR SECURITY

Protection of nuclear facilities from adverse actions has always been regarded as important, but the terrorist attacks on the United States on September 11, 2001, prompted a need for significant enhancement of security.

The NRC defined the "design basis threat" (DBT), which involves the number of attackers, their weapons capability, and probable mode of action on the basis of intelligence gathered. A ground-based threat is assumed to be by a well-armed, well-trained, suicidal group that uses vehicles and explosives.

The nuclear industry spent well over a billion dollars to extend security at power plants. Included were improved training and arming of security guards, additional physical barriers, better intrusion surveillance and detection, stronger access controls, institution of protection of plant computer systems, and improved background checks of plant employees.

An innovation called "force-on-force" exercises was required by NRC. At a plant, the roles of attacker and responder are played, with reviews of performance. At nuclear plants in the United States some 8000 security officers are available to counter a ground threat. Physical barriers are erected between three zones labeled "owner controlled," "protected," and "vital."

After 9/11, concern was expressed about the possibility of an aircraft attack on nuclear facilities. A report on the effects was prepared in 2002 by the Electric Power Research Institute (EPRI). The selected aircraft was the Boeing 767 with wing span 170 feet, larger than the diameter of a typical containment building of 140 feet. A low speed of 350 mph was assumed because of the difficulty in precision flying near the ground. The containment was composed of steel-reinforced concrete approximately 4 feet thick, designed to be impervious to natural disasters such as hurricanes, tornadoes, earthquakes, and floods. Its curved surface prevents a full impact of the airplane. Conservative assumptions were made to give the maximum force. The result of the analyses was that the containment was not breached, so no parts of the plane entered the building. The report also concluded that fuel storage pools, dry storage units, and shipping casks would be safe against air attack. The targets would be very near the ground, requiring a sharp dive of the airplane, with only a glancing blow.

The subject was addressed by the 9/11 Commission (National Commission of Terrorists Attacks Upon the United States). The July 2004 report concluded that there was adequate protection against airplanes.

19.9 SUMMARY

Prevention of release of radioactive fission products and fuel isotopes is the ultimate purpose of safety features. Inherent reactor safety is provided by delayed neutrons and temperature effects. Control rods permit rapid shutdown, and reactor components are designed and constructed to minimize the chance of failure. Emergency core cooling equipment is installed to reduce the hazard in the event of an accident. Licensing is administered by the Nuclear Regulatory Commission, which expects plants to use probabilistic safety risk analysis (PRA).

An accident at Three Mile Island Unit 2 in 1979 resulted in considerable damage to the reactor core but little radioactive material was released. The event stimulated the nuclear industry to make many changes that enhance reactor safety.

A serious accident occurred in 1986 at Chernobyl, U.S.S.R. As a result of an unauthorized experiment, there was an explosion and fire, accompanied by the release of a great deal of radioactivity. Nearby cities were evacuated, a number of people were killed, and many received significant dosage.

Security at nuclear plants was greatly enhanced after 9/11. A study indicates that terrorist aircraft do not pose a problem.

19.10 EXERCISES

19.1 (a) If the total number of neutrons from fission by thermal neutrons absorbed in U-235 is 2.42, how many are delayed and how many are prompt? (b) A reactor is said to be "prompt critical" if it has a positive reactivity of β or more. Explain the meaning of the phrase. (c) What is the period for a reactor with neutron cycle time 5×10^{-6} s if the reactivity is 0.013? (d) What is the period if instead the reactivity is 0.0013?

19.2 A reactor is operating at a power level of 250 MWe. Control rods are removed to give a reactivity of 0.0005. Noting that this is much less than β, calculate the time required to go to a power of 300 MWe, neglecting any temperature feedback.

19.3 Measurements of the fast neutron cycle time ℓ were made on EBR-I, the first reactor to produce electricity. Calculate its value in two different ways: (a) With the ratio β/ℓ, called the Rossi-α , with a value of 1.74×10^5/s and β of 0.0068; (b) With a rough formula $\ell = 1/(v\,\Sigma_a)$ with an average energy of 500 keV neutrons. At that energy, $\sigma_c = 0.1$ barn and $\sigma_f = 0.62$ barn. Note $N_U = 0.054$ (in units of 10^{24}), $v = 2200$ m/s for $E = 0.0253$ eV. (Thanks are due Professor Robert Busch for this exercise and its answers.)

19.4 During a critical experiment in which fuel is initially loaded into a reactor, a fuel element of reactivity worth 0.0036 is suddenly dropped into a core that is already critical. If the temperature coefficient is -9×10^{-5}/°C, how high will the temperature of the system go above room temperature before the positive reactivity is canceled out?

19.5 How long will it take for a fully withdrawn control rod in a reactor of height 4 m to drop into a reactor core neglecting all friction and buoyancy effects? (Recall $s = \frac{1}{2}gt^2$ with $g = 9.8$ m/s^2.)

19.6 Calculate the ratio of fission product power to reactor power for four times after shutdown—1 day, 1 week, 1 month, and 1 year, with the approximation $A = 0.066$, $a = 0.2$.

19.7 A reactivity of -0.0025 caused by Doppler effect results when the thermal power goes from 2500 MW to 2800 MW. Estimate the contribution of this effect on the power coefficient for the reactor.

19.8 Assuming a probability of reactor core meltdown of 3×10^{-4} per reactor year, calculate the chance of one meltdown for 100 reactors in a period of 20 y.

19.9 Counting rates for several fuel addition steps in a critical experiment are listed below.

Number of Fuel Assemblies	Counting Rate (Counts/min)
0	200
50	350
100	800
125	1,600
140	6,600
150	20,000

At the end of each fuel addition, what is the estimated critical number of assemblies? Was the addition always less than the amount expected to make the array critical?

19.10 When a control rod is raised 4 cm from its position with tip at the center of a critical reactor, the power rises on a period of 200 s. With a value $\beta = 0.008$ and $\tau = 13$ s, estimate the δk produced by the rod shift and the slope of the calibration curve $\triangle k/\triangle z$. Estimate the rod worth if the core height is 300 cm.

19.11 Measurements are made of the periods of power rise in a research reactor of height 24 inches for shifts in control rod position. From the periods, values are obtained for the slope of the reactivity $\triangle \rho_i/\triangle z_i$, with units percent per inch, as listed below:

i	z_i	$\triangle\rho_i/\triangle z_i$	i	z_i	$\triangle\rho_i/\triangle z_i$
1	0		10	12.5	
		0.02			1.03
2	3		11	13	
		0.16			1.08
3	5.5		12	14	
		0.38			1.02
4	7.5		13	15	
		0.68			0.95
5	9		14	16.5	
		0.83			0.77

(Continued)

i	z_i	$\Delta\rho_i/\Delta z_i$	i	z_i	$\Delta\rho_i/\Delta z_i$
6	10		15	18.5	
		0.89			0.40
7	11		16	21	
		0.96			0.11
8	11.5		17	24	
		0.98			
9	12				
		1.02			

Plot the slope against average position $\bar{z}_i = (z_{i+1} + z_i)/2$. Pass a smooth curve through the points, then find the area under the curve as a function of z. Estimate the rod worth when the tip is 16 inches up from the bottom.

19.12 Commonly used fractions and half-lives of the nuclides that are delayed neutron emitters for thermal neutron fission in uranium-235 are as follows:

Group I	Fraction β_i	Half-life $(t_H)_i$
1	0.000247	54.51
2	0.001385	21.84
3	0.001222	6.00
4	0.002645	2.23
5	0.000832	0.496
6	0.000169	0.179

Verify that the total fraction is 0.0065 and that the average half-life is approximately 8.8 s.

19.13 (a) Show that a megawatt per tonne is the same as a watt per gram. (b) Show that the burnup in MWd/tonne is given by the formula

$$B = cE\phi\ \sigma_{f235}wt/m_U$$

where the enrichment is

$$E = N_{235}/N_U$$

and

$$c = 1.157 \times 10^{-5}.$$

(c) Calculate B for a flux of 2×10^{13}/cm²-s for three years with enrichment 0.03. Note $m_U = 395 \times 10^{-24}$ grams, $\sigma_{f235} = 586 \times 10^{-24}$ cm², and $w = 3.04 \times 10^{-11}$ W-s/fission.

19.14 To remain critical at the end of a cycle of operation, a power reactor must have an average multiplication factor k_F. For a one-zone core, this is related to the burnup B by

$$k_F = k_0 - aB$$

where a is a constant, so that the discharge burnup is

$$B(1) = (k_0 - k_F)/a.$$

For a two-zone core, we have

$$k_F = (k_0 - aB)/2 + (k_0 - 2aB)/2 = k_0 - (3/2)aB.$$

The discharge burnup is $2B$ or

$$B(2) = (4/3)B(1).$$

Continue the analysis to find $B(3)$ and $B(4)$. Check the results against the formulas quoted in the text.

Computer Exercises

19.A A simplified version of the analysis of neutron population growth is called the one-delayed-group model. The six emitters listed in Exercise 19.12 are replaced by a single emitter with mean life $\tau = 12.7$ s, effective neutron lifetime $\bar{\ell} = 0.083$ s, decay constant $\lambda = 0.0785$ s^{-1}, total fraction $\beta = 0.0065$. Differential equations for the neutron population n and the delayed emitter concentration c are written:

$$dn/dt = n(r - \beta)/\bar{\ell} - \lambda c$$

$$dc/dt = n\beta/\bar{\ell} + \lambda c.$$

To solve, the program OGRE (One Group Reactor Kinetics) is used.
(a) Try various reactivity values such as 0.0001, 0.0005, and 0.001, with a time step of 0.01 s. (b) Plot the time responses of neutron population. (c) Change the time step for $\rho = 0.001$ from 0.01 s to 0.1 s. Explain the results and discuss actions required.

19.B The program KINETICS solves the time-dependent equations for neutrons and delayed emitters, yielding the neutron population as a function of time. Six emitters are used, and feedback is neglected. (a) Run the program with the menus, observing symbols, equations, and input data. (b) Try various input reactivity values–positive, negative and zero; small and large with respect to $\beta = 0.0065$.

19.C The effect of temperature feedback on the time response of a reactor can be estimated by use of the program RTF (Reactor Transient with

Feedback). RTF solves simple differential equations that express the rates of change with time of power and temperature. There is a negative temperature coefficient of reactivity and power is extracted according to a temperature difference. (a) Load the program RTF and scan the tables to see how the power varies with time for the sample problem. (b) Draw a graph of power P versus time t. (c) Examine the effect of changing the reactor fuel from uranium to plutonium. Pu has an effective neutron lifetime of only 0.04 s compared with the value for U of 0.083 s. Let all other factors be the same as in (a) above.

19.D A typical PWR core contains approximately 200 fuel assemblies, arranged to optimize production and safety. The computer program COREFUEL shows top views of cores with different fuel patterns, including that of Three Mile Island Unit 2 before its accident. Run the program to inspect the cores with the menus.

19.E The power excursion without cooling in the Three Mile Island Unit 2 reactor (TMI-2) turned fuel assemblies into a mass of broken and melted material. Load and run the program RUBBLE, which sketches the cavity formed by the slumping of damaged fuel.

19.F Features of the Chernobyl reactor before its accident are sketched in three computer programs: CIRCLE6, which shows the array of 19 fuel rods within an aluminum tube, forming their assembly; SQRCIR6, which shows the array of holes in the graphite core for insertion of fuel or rods; and CORODS, which illustrates the arrangement of control rods that led to the accident. Load and run the programs.

19.11 REFERENCES

Ronald Allen Knief, *Nuclear Criticality Safety: Theory and Practice*, American Nuclear Society, La Grange Park, IL, 1985. Methods of protection against inadvertent criticality, with an appendix on the Three Mile Island recovery program.

Brian L. Koponen, Ed., Nuclear Criticality Safety Experiments, Calculations, and Analyses—1958 to 1998: Compilation of Papers from the *Transactions of the American Nuclear Society*, Golden Valley Publications, Livermore, CA, 1999.

Criticality Accidents
http://www.cddc.vt.edu/host/atomic/accident/critical.html
Atomic Energy Commission, 1943–1970.

A Review of Criticality Accidents 2000 Revision
http://www.orau.org/ptp/Library/accidents/la-13638.pdf
Original edition (1967) by William R. Stratton. Includes Russian information and more recent Japanese accident.

The Virtual Nuclear Tourist
http://www.nucleartourist.com
All about nuclear power. By Joseph Gonyeau.

E. E. Lewis, *Nuclear Power Reactor Safety*, John Wiley & Sons, New York, 1977.

F. R. Farmer, Editor, *Nuclear Reactor Safety*, Academic Press, New York, 1977.

History of Nuclear Power Plant Safety
http://www.nuclearsafetyhistory.org
Select decade time line, 1940–2000.

ECCS Evaluation Models
http://www.nrc.gov/reading-rm/doc-collections/cfr/part050
Select Appendix K for regulation on emergency core cooling system.

Dennis D. Wackerly, William Mendenhall III, and Richard L. Scheaffer *Mathematical Statistics with Applications*, 5th Ed., Wadsworth Publishing Co., Belmont, CA, 1996.
A popular textbook on basic statistics.

Reactor Safety Study: An Assessment of Accident Risks in U.S. Commercial Nuclear Power, WASH-1400 (NUREG-75/014), United States Nuclear Regulatory Commission, 1975. Often called the Rasmussen Report after the director of the project, Norman Rasmussen. The first extensive use of probabilistic risk assessment in the nuclear field.

PRA Procedures Guide: A Guide to the Performance of Probabilistic Risk Assessments for Nuclear Power Plants, NUREG/CR-2300, Vols. 1 and 2, Nuclear Regulatory Commission, January 1983. A comprehensive 936-page report prepared by the American Nuclear Society and the Institute of Electrical and Electronic Engineers. Issued after the Three Mile Island accident, it serves as a primary source of training and practice.

Norman J. McCormick, *Reliability and Risk Analysis: Method and Nuclear Power Applications*, Academic Press, New York, 1981.

N. H. Roberts, W. E. Vesely, D. F. Haasl, and F. F. Goldberg *Fault Tree Handbook*, NUREG-0492, Nuclear Regulatory Commission, Washington, DC, 1981.
http://www.nrc.gov/reading-rm/doc-collections/nuregs/staff/sr0492
A frequently cited tutorial containing fundamentals and many sample analyses.

Ralph R. Fullwood and Robert E. Hall, *Probabilistic Risk Assessment in the Nuclear Power Industry: Fundamentals and Applications*, Pergamon Press, Oxford, 1988.

Ernest J. Henley and Hiromitsu Kumamoto, *Probabilistic Risk Assessment*, IEEE Press, New York, 1992.

E. E. Lewis, *Introduction to Reliability Engineering*, John Wiley & Sons, New York, 1996.

Society for Risk Analysis
http://www.sra.org
Select Resources/Glossary.

Risk-Informed and Performance-Based Regulations for Nuclear Power Plants
American Nuclear Society Position Statement 46, June 2004
http://www.ans.org/pi/ps/docs/ps46.pdf

Understanding Risk Analysis: A Short Guide for Health, Safety, and Environmental Policy Making, American Chemical Society and Resources For the Future, Washington, DC, 1998.
http://www.rff.org/rff/Publications/upload/14418_1.pdf
The 39 pages can be downloaded. A good qualitative discussion of risk, including history, perceptions, methodology, and limitations.

Ronald M. Eytchison, "Memories of the Kemeny Commission," *Nuclear News*, March 2004, p. 61.

John Kemeny, et al., *The Need for Change: The Legacy of TMI*, Pergamon Press, Elmsford, NY, 1979. Subtitle: Report of the President's Commission on the Accident at Three Mile Island.
http://www.threemileisland.org

M. Rogovin and G. T. Frampton, Jr., Three Mile Island: A Report to the Commissioners and the
Public, NUREG/CR-1250, Vols. I and II, Parts 1-3, Nuclear Regulatory Commission, 1980.
http://www.threemileisland.org

Three Mile Island 2 Accident
http://www.nrc.gov/reading-rm/doc-collections/fact-sheets/3mile-isle.html
Account of the accident with diagram, references, and glossary by NRC.

TMI-2 Recovery and Decontamination Collection
http://www.libraries.psu.edu/tmi
TMI-2 Cleanup Highlights Program (24-min QuickTime video).
Pennsylvania State University Library.

M. Sandra Wood and Suzanne M. Schultz, *Three Mile Island: A Selectively Annotated Bibliography*, Greenwood Press, New York, 1988.

J. Samuel Walker, *Three Mile Island: A Nuclear Crisis in Historical Perspective*, University of
California Press, Berkeley, CA, 2004. The most authoritative book on the accident. A scholarly but gripping account, it details the accident, the crisis situation, the heroic role of
Harold Denton, the aftermath, and the implications.

IDCOR Nuclear Power Plant Response to Serious Accidents, Technology for Energy Corp.,
Knoxville, TN, November 1984.

Report on the Accident at the Chernobyl Nuclear Station, NUREG-1250, United States Nuclear
Regulatory Commission, January 1987. A compilation of information obtained by DOE, EPRI,
EPA, FEMA, INPO, and NRC.

David R. Marples , *Chernobyl and Nuclear Power in the USSR*, St. Martin's Press, New York,
1986. A comprehensive examination of the nuclear power industry in the U.S.S.R., the
Chernobyl accident, and the significance of the event in the U.S.S.R. and elsewhere, by a
Canadian specialist in Ukraine studies.

Consequences of the Chernobyl Accident
http://www-ns.iaea.org/appraisals/chernobyl.htm
Select link to "Fifteen Years after the Chernobyl Accident."

Revisiting Chernobyl: 20 years later
http://www.iaea.org/NewsCenter/Focus/Chernobyl
Status and recommended actions.

IAEA Department of Safety and Security
http://www-ns.iaea.org/publications/default.htm
Links to publications that can be downloaded.

Darryl. Randerson (Ed.), *Atmospheric Science and Power Production*, Department of Energy,
1984. Third version of book on estimation of concentrations of radioactivity from accidental
releases.

WANO—World Association of Nuclear Operators
http://www.wano.org.uk

"Security Effectiveness: Independent Studies and Drills"
http://www.nei.org
Select safety and security. From Nuclear Energy Institute.

"Nuclear plant damage from air attacks not likely," *Nuclear News*, August 2002, p. 21.

"Aircraft Crash Impact Analyses Demonstrate Nuclear Power Plant's Structural Strength," EPRI, December 2002.
http://evacuationplans.org/epri-crash-study.pdf
Conclusion that fuel is protected.

Richard Wolfson, *Nuclear Choices: A Citizen's Guide to Nuclear Technology*, MIT Press, Cambridge, 1993. Nuclear power and nuclear weapons, with comments on the accidents at TMI and Chernobyl.

Dominic Golding, Jeanne X. Kasperson, and Roger E. Kasperson, Eds., *Preparing for Nuclear Power Plant Accidents*, Westview Press, Boulder, 1995. Critical of existing emergency plans, use of PRA, source term analysis, and warning systems. Sponsored by Three Mile Island Public Health Fund.

Nuclear Propulsion

NUCLEAR PROCESSES are logical choices for compact energy sources in vehicles that must travel long distances without refueling. The most successful application is in the propulsion of naval vessels, especially submarines and aircraft carriers. Thermoelectric generators that use the isotope plutonium-238 provide reliable electric power for interplanetary spacecraft. Research and development has been done on reactors for aircraft and rockets, and reactors may be used in future missions.

20.1 REACTORS FOR NAVAL PROPULSION†

The discovery of fission stimulated interest on the part of the United States Navy in the possibility of the use of nuclear power for submarine propulsion. The development of the present fleet of nuclear ships was due largely to Admiral H. G. Rickover, a legendary figure because of his reputation for determination, insistence on quality, and personalized management methods. The team that he brought to Oak Ridge in 1946 to learn nuclear technology supervised the building of the land-based prototype at Idaho Falls and the first nuclear submarine, *Nautilus*. As noted by historians for the project (see References), the name had been used for submarines before, including Jules Verne's fictional ship.

The principal virtue of a nuclear-powered submarine is its ability to travel long distances at high speed without refueling. It can remain submerged because the reactor power plant does not require oxygen. Research on the Submarine Thermal Reactor was conducted by Argonne National Laboratory, and the development was carried out at the Bettis Laboratory of Westinghouse Electric Corporation.

The power plant for the *Nautilus* was a water-moderated, highly enriched uranium core with zirconium-clad plates. The submarine's first sea trials were made

---

†Thanks are due Commander (Ret.) Marshall R. Murray, USN, for some of the information in this section.

in 1955. Some of its feats were a 1,400-mile trip with an average speed of 20 knots, the first underwater crossing of the Arctic ice cap, and traveling a distance of more than 62,000 miles on its first core loading. Subsequently the *Triton* reproduced Magellan's trip around the world, but completely submerged. The Nautilus was decommissioned in 1980 and is now in a museum at Groton, CT.

Over the years of the Cold War, the United States nuclear fleet was built up, with more than 100 nuclear-powered submarines and a number of aircraft carriers. The first of the latter was the *Enterprise*, deployed in 1961. It has eight reactors, 85 aircraft, and 5,830 men. Figure 20.1 shows the carrier with Einstein's familiar formula spelled out on the deck by members of the crew. Since then, several additional carriers have been built, some of which saw service in the Gulf War and subsequent military activities.

Attack submarines are designed to seek and destroy enemy submarines and surface ships. In 1995 the *Seawolf* was launched. It was powered with one reactor and armed with *Tomahawk* cruise missiles. Ballistic missile submarines are designed as deterrents to international conflict. Examples are the *Ohio*-class, which carry 24 long-range *Trident* strategic missiles (see References). These weapons can be ejected by compressed air while the vessel is under water, with the rocket motors started when the missile clears the surface. The number of United States nuclear-powered naval vessels is gradually being reduced by obsolescence and decision, and by international agreement, as part of the START program (see Section 26.3).

FIGURE 20.1 The nuclear-powered aircraft carrier USS *Enterprise*. Sailors in formation on the flight deck spell out Einstein's formula. The accompanying ships are the USS *Long Beach* and the USS *Bainbridge* (Courtesy United States Navy).

Commercial nuclear power has benefited in two ways from the Navy's nuclear program. First, industry received a demonstration of the effectiveness of the pressurized water reactor. Second, utilities and vendors have obtained the talents of a large number of highly skilled professionals who are retired officers and enlisted men.

The United States built only one commercial nuclear vessel, the merchant ship N.S. *Savannah*. Its reactor was designed by Babcock & Wilcox Company Carrying both cargo and passengers, it was successfully operated for several years in the 1960s, making a goodwill voyage to many countries (see References). After being on display at a naval museum in South Carolina, the N.S. *Savannah* was moved in 1994 to Virginia as a national landmark.

Several icebreakers powered by nuclear reactors were built by the U.S.S.R. and continue to be used in the far north for expedition cruises (see References). The newest and most powerful Russian icebreaker is the *50 Years Since Victory*.

Japan launched an experimental nuclear-powered merchant ship *Mutsu* in 1962. It successfully passed several rigorous sea trials, performing well in rough seas caused by a typhoon. Decommissioned in 1995 and placed in a museum, its experience served as the basis for the design of two other vessels (see References).

20.2 SPACE REACTORS

Many years before the advent of the space program, an attempt was made to develop an aircraft reactor. A project with the acronym NEPA (Nuclear Energy for the Propulsion of Aircraft) was started at Oak Ridge in 1946 by the United States Air Force. The basis for the program was that nuclear weapon delivery would require supersonic long-range (12,000 miles) bombers not needing refueling. An important technical question that still exists is how to shield a crew without incurring excessive weight. As described by Hewlett and Duncan (see References), the program suffered from much uncertainty, changes of management, and frequent redirection. It was transferred from Oak Ridge to Cincinnati under General Electric as the Aircraft Nuclear Propulsion (ANP) program. The effort was terminated for several reasons: (a) the need for a much larger airplane than expected, (b) improvements in performance of chemically fueled jet engines, and (c) the selection of intercontinental ballistic missiles to carry nuclear weapons. Some useful technical information had been gained, but the project never came close to its objective.

The space program was given new impetus in 1961 with President Kennedy's goal of a manned lunar landing. Other missions visualized were manned exploration of the planets and ultimately colonization of space. For such long voyages requiring high power, the light weight of nuclear fuel made reactors a logical choice for both electrical power and propulsion. One concept that was studied extensively was ion propulsion, with a reactor supplying the energy needed to accelerate the ions that give thrust. A second approach involved a gaseous core reactor, in which a mixture of uranium and a gas would be heated by the fission reaction and be expelled as propellant. Another more exotic idea was to explode a number of small nuclear weapons next to a plate mounted on the space vehicle, with the reaction to the explosion giving a repetitive thrust.

Fission reactors with thermoelectric conversion systems were developed in the period 1955–1970 by the Atomic Energy Commission. Its contractor, Atomics International, conducted the Systems for Auxiliary Nuclear Power (SNAP) program. The most successful of these was SNAP-10A, which was the first and only United States reactor to be flown in space. Two systems were built—one tested on Earth, the other put in orbit. Their fuel was an alloy of enriched uranium and zirconium hydride to operate at high temperatures (810 K). The coolant was liquid sodium-potassium (NaK) for efficient heat transfer. The NaK was circulated through the reactor and a thermoelectric converter system that produced 580 watts of electrical power. The total weight of one system was 435 kg. The space version was launched in 1965 by an Agena rocket and started up by remote control. It operated smoothly in orbit for 43 days until it was accidentally shut down by an electric failure in the spacecraft. The ground version operated satisfactorily for 10,000 h. Further details are provided by Bennett (see References). Another successful reactor SNAP-8 used mercury as coolant, with conversion to 50 kW of electric power in a Rankine cycle. Further details of these reactors appear in the book by Angelo and Buden (see References).

The nuclear system that received the most attention in the space program was the solid core nuclear rocket. Liquid hydrogen would be heated to a high temperature as gas on passing through holes in a reactor with graphite moderator and highly enriched uranium fuel. In the proposed vehicle the hydrogen would be exhausted as propellant through a nozzle.

The nuclear thermal rocket sketched in Figure 20.2 is a relatively simple device. Hydrogen propellant is stored in a tank as a liquid. The reason that space

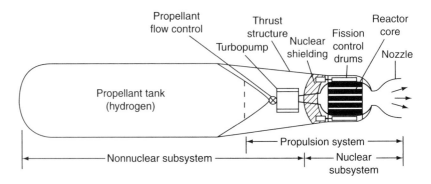

FIGURE 20.2 Nuclear-thermal rocket system. Hydrogen stored in liquid form is heated in the solid core and expelled as propellant. (Courtesy Gary Bennett).

travel by nuclear rocket is advantageous can be seen from the mechanics of propulsion. The basic rocket equation relating spacecraft velocity v, fuel exhaust velocity v_f, and the masses of the full and empty rocket m_0 and m, is

$$v = v_f \log_e(m_0/m)$$

or the inverse relation

$$m/m_0 = \exp(-v/v_f),$$

with the mass of vehicle plus payload being $m_0 - m$. The burning products of a chemical system are relatively heavy molecules, whereas a nuclear reactor can heat light hydrogen gas. Thus for a given temperature, v_f is much larger for nuclear and m is closer to m_0 (i.e., less fuel is needed).

To escape from the Earth or from an orbit around the Earth requires work to be done on the spacecraft against the force of gravity. The escape velocity v_e for vertical flight is

$$v_e = \sqrt{2g_0 r_E}$$

where g_0 is the acceleration of gravity at the Earth's surface, 32.174 ft/s^2 or 9.80665 m/s^2, and r_E is the radius of the Earth, approximately 3,959 miles or 6,371 km. Inserting numbers we find the escape velocity to be approximately 36,700 ft/s, 25,000 mi/h, or 11.2 km/s.

The Rover project at Los Alamos was initiated with a manned mission to Mars in mind. Flight time would be minimized with hydrogen as propellant, because its specific impulse would be approximately twice that of typical chemical fuels. A series of reactors named Kiwi, NRX, Pewee, Phoebus, and XE' were built and tested at the Nuclear Rocket Development Station located in Nevada. The systems used uranium carbide fuel, graphite moderator, and once-through hydrogen coolant, entering as a liquid and leaving as a gas. The best performance obtained in the Nuclear Engine for Rocket Vehicle Application (NERVA) program was a power of 4,000 MW for 12 min. The program was a technical success but was terminated in 1973 because of a change in NASA plans. After the lunar landing in the Apollo program, a decision was made not to have a manned Mars flight. It was judged that radioisotope generators and solar power would be adequate for all future space needs.

Various R&D programs on space reactors to provide electric power were initiated subsequently (e.g., the SP-100), which was to be a reactor in the 100 kW to 1 MW range. Most of the projects were eventually canceled.

20.3 SPACE ISOTOPIC POWER

Chemical fuels serve to launch and return space vehicles such as the shuttle. For long missions such as interplanetary exploration, where it is necessary to supply electric power to control and communication equipment for years, nuclear power is needed. The radioisotope thermoelectric generator (RTG) has been developed and used successfully for many missions. It uses a long-lived radionuclide to supply heat that is converted into electricity. The power source has many desirable features: (a) lightness and compactness, to fit within the spacecraft readily; (b) long service life; (c) continuous power production; (d) resistance to environmental effects such as the cold of space, radiation, and meteorites; and (e) independence from the sun, permitting visits to distant planets.

The isotope used to power the RTGs is plutonium-238, half-life 87.74 y, which emits alpha particles of 5.5 MeV. The isotope is produced by reactor neutron irradiation of the almost-stable isotope neptunium-237, half-life 2.14×10^6 y. The latter is a decay product of uranium-237, a 6.75-d beta emitter that arises from neutron capture in uranium-236 or by (n,2n) and (γ,n) reactions with uranium-238. The high-energy alpha particles and the relatively short half-life of Pu-238 give the isotope the high specific activity of 17 Ci/g and the favorable power to weight ratio quoted to be 0.57 W/g.

The earliest use was in Pioneer 10, launched in 1972 and Pioneer 11, launched in 1973. The missions were to explore Jupiter. The last radio signal from Pioneer 10 was received on January 22, 2003. The spacecraft was then 7.6 billion miles from Earth, after more than 30 y in space. It was powered by an RTG of initial power 160 W along with several one-watt radioisotope heater units (RHUs). The two spacecraft are at the edge of the solar system and experiencing an unexplained slowing, the "Pioneer Anomaly." Reasons advanced for the effect include thermal recoil, gravity from the Kuiper Belt, "dark matter," and new physics.

Typical of the RTGs is the one sent to the Moon in the Apollo-12 mission. It powered a group of scientific instruments called Apollo Lunar Surface Experimental Package (ALSEP), which measured magnetic fields, dust, the solar wind, ions, and earthquake activity. The generator is shown schematically in Figure 20.3.

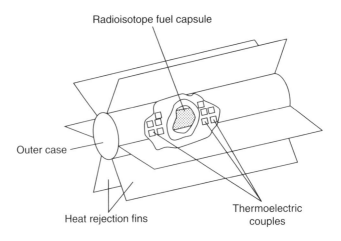

FIGURE 20.3 Isotopic electrical power generator. (SNAP-27 used in Apollo-12 mission).

Lead-telluride thermoelectric couples are placed between the PuO_2 and the beryllium case. Data on the generator are shown in Table 20.1.

This generator, called SNAP-27, was also used in several other Apollo missions, and data were returned to Earth for the period 1969–1977. For the 1975 Viking mission, the somewhat smaller SNAP-19 powered the Mars landers, which sent back pictures of the surface of that planet.

An advanced model, called multi-hundred watt (MHW), provided all electrical power for the two Voyager spacecraft (Figure 20.4), designed and operated by the Jet Propulsion Laboratory of NASA. They were launched in the summer of 1977 and reached Jupiter in 1979 and Saturn in late 1980 and early 1981, sending back pictures of Saturn's moons and rings. Voyager 1 was then sent out of the solar system to deep space. Taking advantage of a rare alignment of three planets, Voyager 2 was redirected to visit Uranus in January 1986. The reliability of the power source

Table 20.1 Radioisotope Thermoelectric Generator SNAP-27	
System weight 20 kg	Thermal power 1480 W
Pu-238 weight 2.6 kg	Electrical power 74 W
Activity 44,500 Ci	Electrical voltage 16 V
Capsule temperature 732°C	Operating range −173°C to 121°C

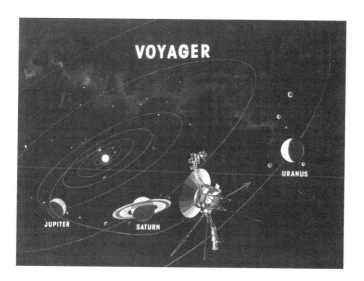

FIGURE 20.4 Voyager 2 spacecraft as it passed Saturn in 1981. (Courtesy National Aeronautics and Space Administration).

after 9 y in space was crucial to the mission. Because of limited light at 1.8 billion miles from the Sun, long exposure times of photographs and thus great stability of the spacecraft were needed. By sending radio signals to Voyager 2, the onboard computers were reprogrammed to allow very small corrective thrusts (see References). Several new moons of Uranus were discovered, including some whose gravity stabilizes the planet's rings. Voyager 2 arrived at Neptune in 1989; it then went on to outer space. The MHW generator used silicon-germanium as thermoelectric material rather than lead-telluride; each generator was heavier and more powerful than SNAP-27. Similar power supplies are used for the Lincoln Experimental Satellites (LES 8/9), which can communicate with each other and with ships and aircraft.

A still larger supply, called General Purpose Heat Source (GPHS), was used in the Galileo spacecraft sent out in October 1990 toward Jupiter. On its way it photographed the asteroids Gaspra and Ida, viewed the impacts of the Shoemaker-Levy 9 comet on the surface of Jupiter, and made flybys of moons Io and Europa. A battery-powered instrumented probe was sent down through Jupiter's atmosphere. Photographs and further information have been provided by NASA (see References). Studies of this distant planet complement those made of nearby Venus by the solar-powered Magellan. Pictures of the moon Europa indicate an icy surface of a liquid ocean. The mission ended when the spacecraft entered Jupiter's heavy atmosphere.

The spacecraft Ulysses, launched in November 1990, is also powered by a GPHS. It is a cooperative mission between the United States and Europe to study the solar wind—a stream of particles from the Sun—and the star's magnetic field. Ulysses will rendezvous with Jupiter to use the planet's gravity to take the spacecraft out of the ecliptic (the plane in which the planets move).

The Cassini spacecraft, launched in 1997 toward Saturn and its moon Titan, contains three RTGs to power instruments and computers, each with approximately 10.9 kg of PuO_2. The total power was initially 888 W. In addition to radio power, RTGs were used to keep the electronic components warm. Unusual views of the planet's rings were obtained. The accompanying Huygens probe landed on Titan. Details on the spacecraft and its instrumentation are found on the Web site of NASA (see References).

The spacecraft New Horizons was launched in 2006 to go past Jupiter to Pluto in 2015 and the Kuiper Belt in 2016–2020. There, many planetary objects will be encountered (see References).

To help maintain proper temperatures for sensitive electronic components, small (2.7 g, 1 W) Pu-238 sources—radioisotope heating units (RHUs)—are provided. These were used in the missions to distant planets and also in the Sojourner minirover that explored the surface of Mars (see References).

Power supplies planned for missions of the more distant future will be in the multikilowatt range, have high efficiency, and make use of a different principle. In the dynamic isotope power system (DIPS), the isotopic source heats the organic fluid Dowtherm A, the working fluid for a Rankine thermodynamic cycle, with the vapor driving a turbine connected to an electric generator. In a ground test the DIPS operated continuously for 2000 h without failure. Details of all of these RTGs are given in the book by Angelo and Buden (see References).

Long-range missions for the 21st century planned by NASA include the recovery of resources at a lunar base and from an asteroid, a space station orbiting the Earth, and eventually a manned Mars mission. Such activities will require nuclear power supplies in the multimegawatt range.

Other isotopes that can be used for remote unattended heat sources are the fission products strontium-90 in the form of SrF_2 and cesium-137 as CsCl. When the use of an oil-fired power unit is not possible because of problems in fuel delivery or operability, an isotopic source is very practical, despite the high cost. If the two isotopes were extracted by fuel reprocessing to reduce the heat and radiation in radioactive waste, many applications would surely materialize.

Success with power sources for space applications prompted a program to develop a nuclear-powered artificial heart. It involved a Pu-238 heat source, a piston engine, and a mechanical pump. The research program was suspended and is unlikely to be revived, with the advent of battery-powered implantable artificial hearts (see References).

20.4 FUTURE NUCLEAR SPACE APPLICATIONS

The extent to which nuclear processes are used in space depends on the degree of commitment to a space program. Over the years, United States enthusiasm for space programs has varied greatly. The Russian Sputnik of 1957 prompted a flurry of activity; President Kennedy's proposal to put a man on the Moon gave the space effort new impetus. Public support has waned, as launches became more routine and new national social problems gained prominence. The Challenger tragedy of 1986 resulted in a loss of confidence in NASA and was a setback to plans for new missions.

In 1989, President George H. W. Bush announced a new program called Space Exploration Initiative (SEI), involving return to the Moon and establishing a base there, then to make a manned trip to the planet Mars. A report by the Synthesis Group (see References) discussed justification and strategies. One nuclear aspect of the project was the possibility of mining helium-3 from the surface of the Moon for use in fusion reactors, as discussed in Section 14.5. The proposed SEI program was not accepted by Congress, and more modest NASA activities involving unmanned spacecraft such as Mars Pathfinder took its place. The exploration of the surface of Mars by the remotely controlled Sojourner minirover was viewed on television by millions of people.

President George W. Bush revived prospects for a Mars mission. The first step would be to return to the Moon. The objective would be exploration for scientific knowledge, to prepare for a base there, and to gain experience relevant to a human visit to Mars. The method by which travel would be made has not been firmed up. Originally, a nuclear rocket of the NERVA type (see Section 20.2) was considered. Later, electric propulsion with ions providing thrust was proposed, and still later matter–antimatter annihilation energy generation was suggested. However, NASA has indicated that a Mars mission would only be around 2037. Some believe that a manned trip to an asteroid would be easier and less expensive.

Whatever the mode of transport to Mars, it is presumed that an initial unmanned vehicle would carry cargo, including a habitat and a reactor for power on the planet's surface. Both timing and duration are important factors for a manned mission. Travel to Mars would take approximately 160 d, allowing 550 d for exploration, until the planets are in correct position for return, which would take 160 d again. Only nuclear power is available for such a schedule. For the descent and ascent between Mars orbit and the planet's surface, chemical rockets would be needed. Studies of geology and microbiology would be carried out, investigating further the possibility of life forms. The fuel produced on Mars— methane (CH_4) and liquid oxygen—would come from the thin CO_2 atmosphere and the supply of H_2 brought from Earth.

For power to process materials on the Moon, a small reactor might be used. Potential resources include water, oxygen, and hydrogen. With adequate heat, the fine dust (regolith) that covers the surface could be formed into solid for construction and radiation shielding. Helium-3 as a fuel for a fusion reactor could be recovered for return to Earth.

Computer Exercise 20.B describes two simple programs that simulate planetary motion.

Calculations of trajectory can be made with the program ORBIT1, described in Computer Exercise 20.A.

The Challenger shuttle accident in 1986 resulted in increased attention to safety. It also raised the question as to the desirability of the use of robots for missions instead of human beings. The benefit is protection of people from harm; the disadvantage is loss of capability to cope with unusual situations. Among the hazards experienced by astronauts are high levels of cosmic radiation outside the Earth's atmosphere, possible impacts of small meteorites on the spacecraft, debilitating effects of long weightlessness, and in the case of a nuclear-powered vehicle, radiation from the reactor. If a reactor were used for transport, to avoid the possibility of contamination of the atmosphere with fission products if the mission is aborted, the reactor would be started only when it is safely in Earth orbit.

For power supplies that use radioisotopes, encapsulation of the Pu-238 with iridium and enclosing the system with graphite fiber reduces the possibility of release of radioactivity. For space missions, risk analyses analogous to those for power reactors are carried out.

Some time in the distant future, electric propulsion may be used. Charged particles are discharged backwards to give a forward thrust. Its virtue is the low mass of propellant that is needed to permit a larger payload or a shorter travel time.

Several possible technologies exist: (a) electrothermal, including arcjets and resistojets (in which a propellant is heated electrically), (b) electrostatic, which uses an ion accelerator, or (c) electromagnetic, such as a coaxial magnetic plasma device. The distinction between electric propulsion and thermal propulsion is in the ratio of thrust and flow rate of propellant, which is the specific impulse, I_{sp}. For example, the shuttle launcher has a high thrust but also a high flow rate, and its I_{sp} is approximately 450 s. Electric propulsion has a low thrust but a very low flow rate, giving an I_{sp} of some 4000 s.

Research on the Hall Thruster, an ion engine, is in progress at Princeton Plasma Physics Laboratory. Electrons are injected to neutralize space charge and permit heavy ion flow to provide thrust (see References).

Prospects for nuclear power for space propulsion have waxed and waned over many decades. A recent development is Project Prometheus, which was designed to use a nuclear reactor as a power source, driving an ion engine that expels xenon ions for its thrust. The engine was successfully tested on the ground in 2003. Speeds were expected to be up to 200,000 miles per hour, 10 times that of the space shuttle. The system was originally intended for an exploration of Jupiter's icy moons, but the mission was shifted to the Earth's moon. Note that present Web sites mainly refer to the original plan. For a 2005 account, search for "NASA's Prometheus." A decision was then made to suspend or possibly abandon Prometheus in favor of some other approach. The ion engine concept will be retained, however. The spacecraft Dawn, launched in 2007 toward asteroids Vesta and Ceres, has solar-powered ion engines rather than an RTG, but its detectors are unique. The 21 sensors measure cosmic ray gammas and neutrons that bounce off the asteroids, providing information about composition.

The asteroid *Apophis*, 390 m wide, is predicted to come very close to Earth in 2029 and 2036. Various proposed techniques to prevent a catastrophic collision include nudging it into a new path with a nuclear-powered spacecraft. Destruction by a nuclear weapon is not favored because the fragments would still strike the Earth.

Looking into the very distant future, some scientists contemplate the "terraforming" of Mars by the introduction of chemicals that change the atmosphere and ultimately permit the normal existence of life forms. Finally, the vision is always present of manned interstellar travel, paving the way for colonization of planets outside our solar system. The discovery of a number of stars with planets has given encouragement to that idea.

What the future of nuclear applications in space will be depends on the accomplishments and aspirations of mankind in space. The urge to investigate and understand is a strong and natural aspect of the human psyche, and some say it is desirable or necessary to plan for interplanetary colonization. Supporters of space exploration cite its many spinoff benefits. Others remind us that there are many serious problems on Earth that need attention and money. How to balance these views remains an issue to be resolved by the political process.

20.5 SUMMARY

Nuclear reactors serve as the power source for the propulsion of submarines and aircraft carriers. Tests of reactors for aircraft and for rockets have been made, and reactors are being considered for future space missions. Thermoelectric generators that use plutonium-238 provided electric power for lunar exploration in the Apollo program and for interplanetary travel of the spacecrafts Voyager, Galileo, Ulysses, and Cassini.

20.6 EXERCISES

20.1 (a) Verify that plutonium-238, half-life 87.7 y, α-particle energy 5.5 MeV, yields an activity of 17 Ci/g and a specific power of 0.57 W/g. (b) How much plutonium would be needed for a 200 microwatt heart pacemaker?

20.2 Note that the force of gravity varies inversely with r^2 and that centrifugal acceleration balances gravitational attraction for an object in orbit. (a) Show that the velocity of a satellite at height h above the Earth is

$$v_s = r_E \sqrt{g_0/(r_E + h)}$$

where g_0 is the acceleration of gravity at the surface of the Earth, of radius r_E. (b) Calculate the velocity of a shuttle in orbit at 100 miles above the Earth. (c) Derive a formula and calculate h in miles and kilometers for a geosynchronous (24 h) communications satellite.

20.3 If the exhaust velocity of rocket propellant is 11,000 ft/s (3.3528 km/s), what percent of the initial mass must be fuel for vertical escape from the Earth?

Computer Exercises

20.A The initial velocity of a rocket ship determines whether it falls back to Earth, goes into orbit about the Earth, or escapes into outer space. The program ORBIT1 calculates the position of a spacecraft and its distance from the center of the Earth for various input values of the starting point and velocity.
(a) Try 100 miles and 290 miles per minute.
(b) Explore various starting points and velocities. Comment on the results.

20.B (a) To view the motions of Earth and Mars about the Sun, run the program PLANETS.

(b) To see numerical features of the relative motion of the planets over the years, run the program PLANETS1. Verify that the phase differences are $0°$ and $180°$ when the planets are in conjunction and in opposition, respectively. Find out how many years it takes to return to the initial phase difference of $44.3°$.

20.C A trip to Mars will probably be made in a spacecraft assembled in orbit around the Earth at altitude of, say, 100 miles (160.9 km). Find its initial speed with the formula for v_s in Exercise 20.2. What is its period, as the time for one revolution? With computer program ALBERT from Chapter 1, find the fractional increase of mass of the ship (and the astronauts) at that speed. Recall that the radius of the Earth is 6378 km.

20.7 REFERENCES

Richard G. Hewlett and Francis Duncan, *Nuclear Navy 1946–1962*, University of Chicago Press, Chicago, 1974.

Commander William R. Anderson, U.S.N., and Clay Blair, Jr., *Nautilus-90-North*, World Publishing Co., Cleveland, 1959. An account by the chief officer of the nuclear submarine *Nautilus* of the trip to the North Pole.

Nautilus Museum
http://www.ussnautilus.org
History, tour, and links.

Nuclear-powered ships
http://www.uic.com.au/nip32.htm
Submarines, naval fleets, civil vessels, and power plants.

United States Submarine Classes
http://www.milnet.com/pentagon/subclass.htm
Descriptions of vessels.

United States Attack Submarines
http://www.navsource.org/archives/08/05idx.htm
Specifications, histories, and photographs.

Center for Defense Information (CDI)
http://www.cdi.org/issues/naval/seawolf.html
http://www.cdi.org/issues/nukef&f/database/usnukes.html#ohio
Information on nuclear naval vessels.

John W. Simpson, *Nuclear Power from Underseas to Outer Space*, American Nuclear Society, La Grange Park, IL, 1994. History of the development of the nuclear submarine *Nautilus*, the first commercial nuclear power plant at Shippingport, and the nuclear thermal rocket engine, NERVA.

Theodore Rockwell, *The Rickover Effect: How One Man Made a Difference*, Naval Institute Press, Annapolis, 1992. Describes the key role Admiral Rickover had in the United States Navy's nuclear submarine program and the first nuclear power plant.

Adams Atomic Insights on N. S. Savannah
http://www.atomicinsights.com/AEI_home.html
Select Topical Index/Nuclear Ships. Additional information by Rod Adams.

N. S. Savannah
https://voa.marad.dot.gov/programs/ns_savannah/index.asp
History, decommissioning, and preservation.

Nuclear-powered Icebreaker
Google "nuclear icebreaker Yamal"
Specs on ship and cruise to North Pole.

Japanese Nuclear Ship Mutsu
http://jolisfukyu.tokai-sc.jaea.go.jp/fukyu/tayu/ACT95E/06/0601.htm
Specifications and history.

RTG History
http://www.osti.gov/accomplishments/rtg.html
Links to many documents.

Nuclear Space
http://www.nuclearspace.com
Articles, opinion, and interviews.

Gary Bennett, "Space Nuclear Power: Opening the Final Frontier"
http://www.fas.org/nuke/space/bennett0706.pdf
A spectacular paper, 2006.

Nuclear Power in Space
http://www.ne.doe.gov/pdfFiles/NPSPACE.PDF
Review of all missions by Department of Energy.

Mohamed S. El-Genk, Editor, *A Critical Review of Space Nuclear Power and Propulsion, 1984–1993*, AIP Press, New York, 1994. Includes papers on radioisotope generators and nuclear thermal propulsion.

Joseph A. Angelo, Jr. and David Buden, *Space Nuclear Power*, Orbit Book Co., Malabar, FL, 1985. A definitive textbook.

Joseph A. Angelo, Jr., *Encyclopedia of Space and Astronomy*, Facts on File, New York, 2001.

Richard P. Laeser, William I. McLaughlin, and Donna M. Wolff, "Engineering Voyager 2's Encounter with Uranus," *Scientific American*, November 1986, p. 36. Discusses the power problem caused by decay of Pu-238 in RTGs.

Ellis D. Miner, "Voyager 2's Encounter with the Gas Giants," *Physics Today*, July 1990, p. 40.

Pioneer Anomaly
http://www.planetary.org
Unexplained slowing of the spacecraft. Select The Pioneer Anomaly.

Galileo Mission
http://www2.jpl.nasa.gov/galileo
14-year odyssey.

Radioisotope Thermoelectric Generators (RTG)
http://www2.jpl.nasa.gov/galileo/messenger/oldmess/RTGs.html
Description of power units in the spacecraft Galileo.

Cassini
http://www.nasa.gov/cassini
Details on spacecraft and voyage.

Guy Gugliotta, "Can We Survive On The Moon?" *Discover*, March 2007, p. 32.

New Horizons Mission
http://pluto.jhuapl.edu
Trip to Pluto and the Kuiper Belt.

Rover Sojourner to Mars
http://mars.jpl.nasa.gov/MPF/mpf/rover.html
Equipped with RTGs.

Artificial Heart
http://www.fda.gov/bbs/topics/NEWS/2006/NEW01443.html
Mechanical implantable heart, approved by FDA.

America at the Threshold: Report of the Synthesis Group on America's Space Exploration Initiative, NASA, Arlington, VA, 1991. An illustrated description of the proposed goals and plans for the space trips to the Moon and Mars.

Whole Mars Catalog
http://www.marstoday.com
News articles.

The Mars Society
http://www.marssociety.org
Activities, news, publications, and conventions.

Hall Thruster
http://htx.pppl.gov/ht.html
High thrust plasma propulsion system. Research by Princeton Lab.

Radiation Protection

PROTECTION OF biological entities from the hazard of radiation exposure is a fundamental requirement in the application of nuclear energy. Safety is provided by the use of one or more general methods that involve control of the source of radiation or its ability to affect living organisms. We will identify these methods and describe the role of calculations in the field of radiation protection. Thanks are due to Dr. James E. Watson, Jr. for his excellent suggestions on this chapter.

21.1 PROTECTIVE MEASURES

Radiation and radioactive materials are the link between a device or process as a source and the living being to be protected from hazard. We can try to eliminate the source, remove the individual, or insert some barrier between the two. Several means are thus available to help ensure safety.

The first is to avoid the generation of radiation or isotopes that emit radiation. For example, the production of undesirable emitters from reactor operation can be minimized by the control of impurities in materials of construction and in the cooling agent. The second is to be sure that any radioactive substances are kept within containers or multiple barriers to prevent dispersal. Isotope sources and waste products are frequently sealed within one or more independent layers of metal or other impermeable substance, and nuclear reactors and chemical processing equipment are housed within leak-tight buildings. The third is to provide layers of shielding material between the source of radiation and the individual and to select favorable characteristics of geological media in which radioactive wastes are buried. The fourth is to restrict access to the region where the radiation level is hazardous and take advantage of the reduction of intensity with distance. The fifth is to dilute a radioactive substance with very large volumes of air or water on release to lower the concentration of harmful material. The sixth is to limit the time that a person remains within a radiation zone to reduce the dose received. We thus see that radioactive materials may be treated in several different ways: *retention, isolation,* and *dispersal*; whereas exposure to radiation can be avoided by methods involving *distance, shielding,* and *time.*

The analysis of radiation hazard and protection and the establishment of safe practices is part of the function of the science of radiological protection or health physics. Every user of radiation must follow accepted procedures, and health physicists should provide specialized technical advice and monitor the user's methods. In the planning of research involving radiation or in the design and operation of a process, calculations must be made that relate the radiation source to the biological entity by use of exposure limits provided by regulatory bodies. Included in the evaluation are necessary protective measures for known sources, or limits that must be imposed on the radiation source, the rate of release of radioactive substances, or the concentration of radioisotopes in air, water, and other materials.

The detailed calculations of radiological protection are very involved for several reasons. A great variety of situations should be considered, including reactor operations and uses of isotopes. Many scientific and engineering disciplines are needed—physics, chemistry, biology, geology, meteorology, and several engineering fields. Increased use of computers favors the development of more sophisticated calculation methods while providing increasing convenience. The collection of new experimental data on the interaction of radiation and matter and the relationship of dose and effect results in evolving recommendations and regulations. Finally, the enhanced awareness of radiation and concern for safety on the part of the public have prompted increased conservatism, which entails refinement in methods and a requirement for fuller justification of methods and results.

In the operation of nuclear power plants and uses of radioisotopes, adherence to government regulations is mandatory to maintain a license. The principal document of the United States Nuclear Regulatory Commission (NRC) is *Code of Federal Regulations: 10 Energy* (see References). Part 20 "Standards for Protection Against Radiation," has an abbreviated designation 10CFR20.

The establishment of regulations is a slow process, starting with the study of research information by advisory bodies such as International Committee on Radiological Protection (ICRP) and National Council on Radiation Protection and Measurements (NCRP), recommendations for dose limits and protection policies, review by the regulatory body with input from the public, institutions, and industry, with final issuance of mandatory requirements, along with guidance documents. As a consequence, the limits and method for different situations may be inconsistent but fundamentally safe. A case in point is the older use of a "critical organ" and maximum permissible concentrations of radionuclides and the newer use of "committed effective dose equivalent" referring to the summation of all effects on the body. The old and the new are contrasted in the NRC's discussion of regulation 10CFR20 in the *Federal Register* of January 9, 1986. Some of the earlier regulations are still applicable. We will present examples of both methods for two reasons: (a) to help the reader make use of all pertinent literature of radiological protection, and (b) to illustrate the trend toward greater precision and realism in radiation protection.

We now discuss the relationship of dose to flux, the effect of distance and shielding materials, internal exposure, environmental assessment, and dose limits for workers and the public.

21.2 CALCULATION OF DOSE

Some simple idealized situations will help the reader understand concepts without becoming involved in intricate calculations. The estimation of radiation dose or dose rate is central to radiation protection. The dose is an energy absorbed per unit mass as discussed in Section 16.2. It depends on the type, energy, and intensity of the radiation, as well as on the physical features of the target. Let us imagine a situation in which the radiation field consists of a stream of gamma rays of a single energy. The beam of photons might be coming from a piece of radioactive equipment in a nuclear plant. The stream passes through a substance such as tissue with negligible attenuation. We use the principles of Chapter 4 to calculate the energy deposition. Flux and current are the same for this beam (i.e., j and ϕ are both equal to nv). With a flux ϕ cm^{-2}–s^{-1}, and cross section Σ cm^{-1}, the reaction rate is $\phi\Sigma$ cm^{-3}–s^{-1}. If the gamma ray energy is E joules, then the energy deposition rate per unit volume is $\phi\Sigma E$ J cm^{-3}–s^{-1}. If the target density is ρ g-cm^{-3}, the dose in joules per gram with exposure for a time t seconds is

$$H = \phi\Sigma Et/\rho.$$

This relationship can be used to calculate a dose for given conditions or to find limits on flux or on time.

For example, let us find the gamma ray flux that yields an external dose of 0.1 rem in 1 y with continuous exposure. This is the dose limit to members of the public according to 10CFR20 (Section 16.2). Suppose that the gamma rays have an energy of 1 MeV and that the cross section for energy absorption with soft tissue of density 1.0 g/cm^3 is 0.0308 cm^{-1}. With a quality factor of 1 for this radiation, the numerical values of the dose and the dose equivalent are the same, so

$$H = (0.1\,\text{rad})\left(1 \times 10^{-5}\text{J/g-rad}\right) = 1 \times 10^{-6}\text{J/g}.$$

Also $E = 1$ MeV $= 1.60 \times 10^{-13}$ J. Solving for the flux,

$$\phi = \frac{H\rho}{\Sigma Et} = \frac{\left(1.0 \times 10^{-6}\,\text{J/g}\right)(1\text{g/cm}^3)}{(0.0308\,\text{cm}^{-1})(1.60 \times 10^{-13}\,\text{J})(3.16 \times 10^7\,\text{s})}$$

or

$$\phi = 6.42\,\text{cm}^{-2}\text{s}^{-1}.$$

This value of the gamma ray flux may be scaled up or down if another dose limit is specified. The fluxes of various particles corresponding to 0.1 rem/y are shown in Table 21.1.

Table 21.1 Radiation Fluxes (0.1 rem/y)

Radiation Type	Flux ($\text{cm}^{-2} - \text{s}^{-1}$)
X- or gamma rays	6.4
Beta particles	0.10
Thermal neutrons	3.1
Fast neutrons	0.085
Alpha particles	10^{-5}

Another situation is the exposure of a person to air containing a radioactive contaminant, for example the noble gas krypton-85, half-life 10.73 y, an emitter of beta particles of average energy 0.251 MeV. Let us derive and apply a formula for the case of continuous exposure during working hours. We wish to relate dose H in rems to activity A in µCi, with an exposure time of t seconds. A rough estimate comes from a simple assumption—that the person is immersed in a large radioactive cloud, and that the energy absorption in air, E_a, is the same as in the human body and the same as that released by decay of the radionuclide, E_r. Write expressions for each of these,

$$E_a = H(\text{rems})(1 \text{ rad/rem})(10^{-5} \text{ J/g-rad})(1.293 \times 10^{-3} \text{ g/cm}^3)$$

$$E_r = A(\mu\text{Ci})(3.7 \times 10^4 \text{ dps-}\mu\text{Ci})(E \text{ MeV})(1.60 \times 10^{-13} \text{ J/MeV})(t \text{ s}).$$

Equate these and solve for the dose, but reducing the figure by a factor of 2 if the person is on the ground and the cloud occupies only half of space. The result is

$$H = 0.229 \, AEt.$$

Assume continuous exposure for 40 h/w, 50 w/y, 3600 s/h, so that $t = 7.2 \times 10^6$ s. Insert $E = 0.251$ MeV and $H = 5$ rems, the annual dose limit for plant workers. Solve for the activity

$$A = 5/[(0.229)(0.251)(7.2 \times 10^6)]$$
$$= 1.2 \times 10^{-5} \mu\text{Ci}.$$

This agrees fairly well with the figure of 1×10^{-5} listed in the 1993 edition of the old NRC 10CFR20. We will see in Section 21.7 that the latest method yields a larger dose limit.

21.3 EFFECTS OF DISTANCE AND SHIELDING

For protection, advantage can be taken of the fact that radiation intensities decrease with distance from the source, varying as the *inverse square of the distance*. Let us illustrate by an idealized case of a small source, regarded as a mathematical point, emitting S particles per second, the source "strength." As in Figure 21.1, let the rate of flow through each unit of area of a sphere of radius R about the point be labeled ϕ $(cm^{-2} - s^{-1})$. The flow through the whole sphere surface of area $4\pi R^2$ is then $\phi\, 4\pi R^2$, and if there is no intervening material, it can be equated to the source strength S. Then

$$\phi = \frac{S}{4\pi R^2}.$$

This relation expresses the inverse square spreading effect. If we have a surface covered with radioactive material or an object that emits radiation throughout its volume, the flux at a point of measurement can be found by addition of elementary contributions.

Let us consider the neutron radiation at a large distance from an unreflected and unshielded reactor operating at a power level of 1 MW. Because 1 W gives 3.3×10^{10} fissions per second (Section 6.4) and the number of neutrons per fission is 2.42 (Section 6.3), the reactor produces 8.0×10^{16} neutrons per second. Suppose that 20% of these escape the core as fast neutrons, so that S is $1.6 \times 10^{16}\,s^{-1}$. Apply the inverse square relation, neglecting attenuation in air, an assumption that would only be correct if the reactor were in a spacecraft. Let us find the closest distance of approach to the reactor surface to keep the dose below 100 mrems/y as in Table 21.1. The limiting fast flux is 0.085 cm^{-2}–s^{-1}. Solving the inverse-square formula, we obtain

$$R = \sqrt{S/(4\pi\phi)} = \sqrt{(1.6 \times 10^{16})/(4\pi 0.085)} = 1.22 \times 10^8 \text{ cm.}$$

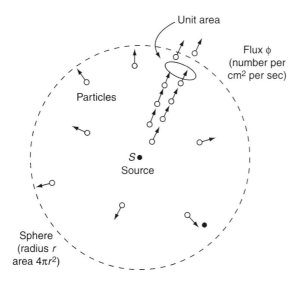

FIGURE 21.1 Inverse square spreading of radiation.

This is a surprisingly large distance, approximately 760 miles. If the same reactor were on the Earth, neutron attenuation in air would reduce this figure greatly, but the necessity for shielding by solid or liquid materials is clearly revealed by this calculation.

As another example, let us find how much radiation is received at a distance of 1 mile from a nuclear power plant, if the dose rate at the plant boundary, $\frac{1}{4}$-mile radius, is 5 mrems/y. Neglecting attenuation in air, the inverse-square reduction factor is $\frac{1}{16}$ giving 0.31 mrems/y. Attenuation would reduce the dose to a negligible value.

The evaluation of necessary protective shielding from radiation makes use of the basic concepts and facts of radiation interaction with matter described in Chapters 4 and 5. Let us consider the particles with which we must deal. Because charged particles—electrons, alpha particles, protons, etc.—have a very short range in matter, attention needs to be given only to the penetrating radiation—gamma rays (or X-rays) and neutrons. The attenuation factor with distance of penetration for photons and neutrons may be expressed in exponential form $\exp(-\Sigma r)$, where r is the distance from source to observer and Σ is an appropriate macroscopic cross section. In shielding analysis; this is called the linear attenuation coefficient, μ, with units cm^{-1}. Now Σ or μ depends on the number of target atoms, and through the microscopic cross section σ also depends on the type of radiation, its energy, and the chemical and nuclear properties of the target.

For fast neutron shielding, a light element is preferred because of the large neutron energy loss per collision. Thus hydrogenous materials such as water, concrete, or earth are effective shields. The objective is to slow neutrons within a small distance from their origin and to allow them to be absorbed at thermal energy. Thermal neutrons are readily captured by many materials, but boron is preferred, because accompanying gamma rays are very weak.

Let us compute the effect of a water shield on the fast neutrons from the example reactor used earlier. The macroscopic cross section appearing in the exponential formula $\exp(-\Sigma r)$ is now called a "removal cross section," because many fast neutrons are removed from the high-energy region by one collision with hydrogen and eventually are absorbed as thermal neutrons. Its value for fission neutrons in water is approximately $0.10\ cm^{-1}$. A shield of thickness $2.5\ m = 250\ cm$ would provide an attenuation factor of $\exp(-25) = 10^{-10.9} = 1.39 \times 10^{-11}$. The inverse-square reduction with distance is

$$\frac{1}{4\pi R^2} = \frac{1}{4\pi(250)^2} = 1.27 \times 10^{-6}.$$

The combined reduction factor is 1.77×10^{-17}; and with a source of 1.6×10^{16} neutrons/s, the flux is down to 0.28 neutrons/cm^2 – s, which is somewhat higher than the safe level of 0.085 as in Table 21.1. The addition of a few centimeters of water shield would provide adequate protection, for steady reactor operation at least. Computer Exercise 21.B describes a program NEUTSHLD that finds fast flux from a fission source as a point or a plane.

For gamma ray shielding, in which the main interaction takes place with atomic electrons, a substance of high atomic number is desired. Compton scattering varies as Z, pair production as Z^2, and the photoelectric effect as Z^5. Elements such as iron and lead are particularly useful for gamma shielding. The amount of attenuation depends on the material of the shield, its thickness, and the photon energy. The literature gives values of the mass attenuation coefficient μ/ρ, which is the ratio of the linear attenuation coefficient μ (macroscopic cross section Σ) and the material density ρ, thus it has units cm^2/g. Typical values for a few elements at different energies are shown in Table 21.2. For 1 MeV gammas in iron, for example, density 7.86 g/cm^3, we calculate $\Sigma = (0.0596)(7.86) = 0.468$ cm^{-1}. In contrast, for water H_2O, molecular weight $2(1.008) + 16.00 = 18.016$, the average value of μ with numbers from Table 21.2 with weight fractions is

$$(0.112)(0.126) + (0.888)(0.0636) = 0.0706 \text{ cm}^{-1}.$$

This is also the value of Σ because $\rho = 1$. Thus to achieve the same reduction in gamma flux in iron as in water, the thickness only need be 15% as much.

As an example of gamma shielding calculations, let us find the flux of 1-MeV gamma rays that have made no collision in arriving from a point source. This *uncollided flux* is a product of a source strength S, an exponential attenuation factor $\exp(-\Sigma r)$, and an inverse square spreading factor $1/(4\pi r^2)$, i.e.,

$$\phi_u = S\exp(-\Sigma r)/(4\pi r^2).$$

For example, find the uncollided flux at 10 cm from a 1 millicurie source ($S = 3.7 \times 10^7$/s). We readily calculate Σ for lead, $\mu/\rho = 0.0684$ cm^2/g, density 11.3 g/cm^3, to be 0.773 cm^{-1}, and $\Sigma r = 7.73$. Inserting numbers,

$$\phi_u = (3.7 \times 10^7)(4.39 \times 10^{-4})/(4\pi 100) = 12.9 \text{ cm}^{-2}\text{s}^{-1}.$$

This is not the complete flux that strikes a receptor at the point, because those scattered by the Compton effect can return to the stream and contribute as

Table 21.2 Mass Attenuation Coefficients (cm^2/g). (NUREG/CR-5740, 1991)

Energy (MeV)	H	O	A1	Fe	Pb	U
0.01	0.385	5.76	2.58	169.6	125.7	173.7
0.1	0.294	0.151	0.161	0.342	5.35	1.72
1	0.126	0.0636	0.0613	0.0596	0.0684	0.0757
2	0.0876	0.0445	0.0432	0.0425	0.0454	0.0479
10	0.0324	0.0208	0.0231	0.0299	0.0496	0.0519

sketched in Figure 21.2. To account for this "buildup" of radiation a multiplying *buildup factor B* depending on Σr is introduced. Figure 21.3 shows B for 1 MeV gammas in the most common shielding materials—lead, iron, and water. The total flux is then

$$\phi = B\phi_u,$$

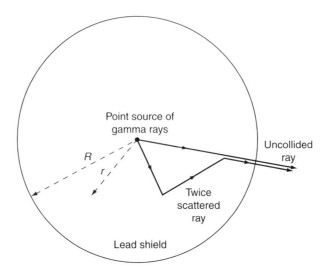

FIGURE 21.2 Buildup effect.

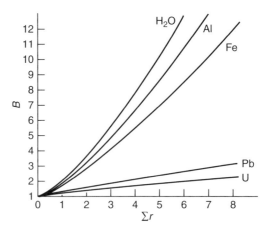

FIGURE 21.3 Buildup factors, 1 MeV gammas. B for concrete and aluminum are about the same.

which shows that the buildup factor is the ratio of the actual flux to the uncollided flux. It remains to find B from the graph or tables, as 3.04, so that the flux is

$$\phi = (3.04)(12.9) = 39.2 \, \text{cm}^{-2} - \text{s}^{-1}.$$

This calculation was rather straightforward, but it is more difficult if the flux is known and one wants to find the distance. Note that r appears in three places in the flux formula, so trial-and-error methods are needed. This tedious process is greatly assisted by use of the computer program EXPOSO, see Computer Exercise 21.A. To bring the exposure down to 5 mrems/y, the value of r is approximately 15 cm.

Although calculations are performed in the design of equipment or experiments involving radiation, protection is ultimately assured by the measurement of radiation. Portable detectors used as "survey meters" are available commercially. They use the various detector principles described in Chapter 10, with the Geiger-Mueller counter having the greatest general usefulness. Special detectors are installed to monitor general radiation levels or the amount of radioactivity in effluents.

The possibility of accidental exposure to radiation always exists in a laboratory or plant despite all precautions. To have information immediately, personnel wear dosimeters, which are pen-size self-reading ionization chambers that detect and measure dose. For a more permanent record, film badges are worn. These consist of several photographic films of different sensitivity, with shields to select radiation types. They are developed periodically, and if significant exposure is noted, individuals are relieved of future work in areas with potential radiation hazards for a suitable length of time.

Newer devices include the thermoluminescent detector reader (TLD), discussed in Section 10.3, which when heated releases photons in proportion to dose received. Solid-state meters have liquid crystal display (LCD) that involves transparent polarizing sheets with electric potential between them. Audible warning signals are available on some instruments.

Operation, maintenance, and repair of nuclear equipment involve some possible exposure to radiation. Even though it is assumed that any radiation is undesirable, it is necessary on practical grounds to allow a certain amount of exposure. It would be prohibitively expensive to reduce the level to zero. A basis for what action to take is the philosophy expressed in the phrase, "as low as is reasonably achievable," with the acronym ALARA. Planning, design, and operation are done with the ALARA principle in mind. For example, a repair job on contaminated equipment is planned after making careful surveys of radiation levels. The repair is to be carried out by a small crew of well-trained people who will do the work quickly and with minimum contact with the radiation sources. Temporary shielding, special clothing, and respirators are used as needed to minimize doses. Factors considered are: (a) the maximum exposure both to individuals and to the group of workers as a whole, (b) other nonradiological risks, (c) the state of technology, and (d) the economic importance of the operation being performed.

If the expected total dosage to the group is more than a fraction of the allowed quarterly dose, a formal ALARA evaluation is made, accounting for both the dollar costs and the dose costs. For a complete discussion by the NRC of the regulatory implications of ALARA, see References.

21.4 INTERNAL EXPOSURE

We now turn to the exposure of internal parts of an organism as a result of having taken in radioactive substances. Special attention will be given to the human body, but similar methods will apply to other animals and even to plants. Radioactive materials can enter the body by drinking, breathing, or eating, and to a certain extent can be absorbed through pores or wounds. The resulting dosage depends on many factors: (a) the amount that enters, which in turn depends on the rate of intake and elapsed time; (b) the chemical nature of the substance, which affects affinity with molecules of particular types of body tissue and which determines the rate of elimination (the term biological half-life is used in this connection, being the time for half of an initial amount to be removed); (c) the particle size, which relates to progress of the material through the body; (d) the radioactive half-life, the energy, and kind of radiation, which determine the activity and energy deposition rate, and the length of time the radiation exposure persists; and (e) the radiosensitivity of the tissue, with the gastrointestinal tract, reproductive organs, and bone marrow as the most important.

In the older regulatory framework, limiting concentrations of radionuclides in air or water are calculated with the concept of "critical organ," the one receiving the greatest effective dose from a certain ingested radionuclide. The organ selected thus dominates the hazard to the body, and effects on other organs are neglected. We apply the method to calculate the maximum permissible concentration (MPC) in units μCi/cm^3 of iodine-131 in water consumed by plant workers. I-131 has a half-life of 8.0 d and releases 0.23 MeV of beta-gamma energy per decay. The thyroid gland, of mass 20 g, will be taken as the critical organ because of the affinity of the thyroid for iodine. According to ICRP 2 (see References), the allowed annual dose is 30 rad. We first find the activity A that will yield that dose. The method of Section 21.2 is applied again. The energy absorbed is

$$(30\,\text{rad})\left(10^{-5}\,\text{J/g-rad}\right)(20\,\text{g}) = 0.0060\,\text{J}.$$

The energy released is

$$(A\,\mu\text{Ci})(0.23\,\text{MeV/dis})\left(3.7 \times 10^4\,\text{dis/sec-}\mu\text{Ci}\right) \times \left(1.60 \times 10^{-13}\,\text{J/MeV}\right)\left(3.16 \times 10^7\,\text{s}\right).$$

Equate and solve for $A = 0.139$ μCi.

Now we find the rates of supply and elimination of I-131 to the organ, assumed to be in balance in steady state. With the formula of Section 16.1, with biological half-life of 138 days, the effective half-life t_E is 7.56 days and the decay constant λ_E is 0.0917 d^{-1}. Thus the elimination rate is proportional to

$\lambda_E A = (0.0917)(0.139) = 0.0127$. The consumption rate of water for the standard man is 2200 cm^3 per day, but it is assumed that workers drink 1.5 times the average during their 8-h day, and they work only 50 wk at 40 h/wk. The rate of intake of contaminated water is thus 755 cm^3/d, and if 30% of the iodine goes to the thyroid, the supply rate of I-131 is (755)(0.3)(MPC). Equate rates and solve for MPC = 5.65×10^{-5} μCi/cm^3, which rounds off to 6×10^{-5} μCi/cm^3, the figure appearing in the older (1993) version of 10CFR20.

When there is more than one radioisotope present, the allowed concentrations must be limited. The criterion used is

$$\Sigma_i \frac{C_i}{(MPC)_i} \leq 1$$

where i is an index of the isotope. This equation says the sum of quotients of actual concentrations and maximum permissible concentrations must be no greater than 1.

21.5 THE RADON PROBLEM

The hazard of breathing air in a poorly ventilated uranium mine has long been recognized. The death rate of miners has historically been higher than that for the general population. The suspected source is the radiation from radioactive isotopes in the decay chain of uranium-238, which by emission of a series of α particles eventually becomes lead-206. The data are clouded by the fact that uranium miners tend to be heavy smokers.

Well down the chain is radium-226, half-life 1599 y. It decays into radon-222, half-life 3.82 d. Although radon-222 is an alpha emitter, its shorter-lived daughters provide most of the dosage. Radon as a noble gas along with its suspended particulate decay products is breathed in with air. Some radioactive particles deposit on the lung surfaces. Decay of the radon and its daughters releases ionizing radiation.

The problem of radon near piles of residue from uranium mining, the mill tailings, has been known, and rules adopted about earth covers to inhibit radon release and about use of the tailings for fill or construction. More recently it has been discovered that a large number of United States homes have higher than normal concentrations of radon. Such excessive levels are due to the particular type of rock on which houses are built. Many homes have a concentration of 20 picocuries per liter, in contrast with the average of approximately 1.5 pCi/L and in excess of the EPA limit of 4 pCi/L. In recent years, EPA has given the subject a great deal of attention.

Application of dose-effect relationships yields estimates of a large number of cancer deaths from the radon effect, as high as 20,000 per year in the United States. Such numbers depend on the validity of the linear relationship of dose and effect discussed in Section 16.3. If there were a threshold or if there were a hormesis effect, the hazard would be very much smaller and mitigation costs greatly reduced. See References for the history of radon in mines, spas, and homes.

It was originally believed that the radon concentrations in houses were high because of conservation measures that reduced ventilation. Investigations revealed that the radon comes out of the ground and is brought into the home by drafts, similar to chimney action. Temperature differences between the air in the house and in the ground beneath cause pressure differences that cause the flow. One might think that covering the earth under a house with plastic would solve the problem, but even slight leaks let the radon through. In areas known to have significant radon levels, it is considered wise for homeowners to obtain radon test kits, which are rather inexpensive. If levels well above 4 pCi/L are found, action is recommended. The best solution is to ventilate a crawl space or to provide a basement with a small blower that raises the pressure and prevents radon from entering.

The dimensions of the problem are yet not fully appreciated nationally; continued study is required to determine the proper course of action at the national level.

21.6 ENVIRONMENTAL RADIOLOGICAL ASSESSMENT[†]

The NRC requires that the ALARA principle, discussed in Section 21.3, be applied to the releases of radioactive materials from a nuclear power plant. A deliberate effort is to be made to stay below the specified limits. These refer to any person in the unrestricted area outside the plant. According to 10CFR50, Appendix I, the annual dose resulting from a liquid effluent must be less than 3 millirems to the individual's total body or 10 millirems to any organ. The dose from air release must be less than 10 millirems from gamma rays and 20 millirems from beta particles. To comply with ALARA, it is necessary for the plant to correlate a release of contaminated water or air to the maximum effect on the most sensitive person. An acceptable method to calculate releases and doses is found in NRC's Regulatory Guide 1.109, October 1977 (see References). This "Reg. Guide" discusses the factors to be considered, gives useful formulas, and provides basic data. Older health physics methods are used, but because the dose limit sought is very small, the results are conservative. Among the important factors are:

1. The amounts of each radioisotope in the effluent, with special attention to cesium-137, carbon-14, tritium, iodine, and noble gases.
2. The mode of transfer of material. The medium by which radioactivity is received may be drinking water, aquatic food, shoreline deposits, or irrigated food. For the latter, pathways include meat and milk. If the medium

---

[†]Appreciation is extended to Mary Birch for helpful discussions.

is air, human beings may be immersed in a contaminated cloud or breathe the air, or material may be deposited on vegetables.

3. The distance between the source of radioactivity and person affected and how much dilution by spreading takes place.

4. The time of transport, to account for decay during flow through air or by streams, or in the case of foodstuffs, during harvesting, processing, and shipment.

5. The age group at risk: infant (0 to 1 y), child (1 to 11 y), teenager (11 to 17 y), and adult (17 and older). Sensitivities to radiation vary considerably with age.

6. The dose factor, which relates dose in millirems to the activity in picocuries. These numbers are tabulated according to isotope, age group, inhalation or ingestion, and organ (bone, liver, total body, thyroid, kidney, lung, and GI tract).

As an example, let us make an approximate calculation of the dose resulting from a continuous release of radioactive water from a nuclear power station into a nearby river. Assume that each day there is a release of 1000 gallons of water contaminated with a single radioisotope cesium-137, half-life 30.2 y. Also assume an activity in the water of 10^5 pCi/L. The activity in the discharge of (1000 gal/d) (1440 min/d) $= 0.694$ gal/min is diluted by a stream flow of 2×10^4 gal/min, down to $(10^5)(0.694)/(2 \times 10^4) = 3.47$ pCi/L. The potential radiation hazard to the population downstream is by two types of ingestion: drinking the water or eating fish that live in the water. The age groups at risk are infants (I), children (C), teenagers (T), and adults (A). Consumption data are as shown in Table 21.3.

The row in the table that refers to fish must be multiplied by a bioaccumulation factor of 2,000 (its units are pCi/liter per pCi/kg). Consider the dose to an adult. To the consumption rate of water of 730 liters/y must be added the effect of eating fish, (2,000) (21) $= 42,000$, giving a total of 4.27×10^4 liter/y. Now apply a dose conversion in mrems per pCi for cesium-137 as in Table 21.4. Each number should be multiplied by 10^{-5}. The adult total body dose conversion factor is 7.14×10^{-5} mrems/pCi. Thus, the yearly dose is

$$D = (3.47)(4.27 \times 10^4)(7.14 \times 10^{-5}) = 10.6 \text{ mrems}.$$

Because this is well above the limit of 3 mrems, a reduction in rate of release will be required.

The general environmental effect of supporting parts of the nuclear fuel cycle must be described in an application for a construction permit for a power reactor.

Table 21.3 Consumption by Age Group (Table E-5, Reg. Guide 1.109)

	I	C	T	A
Water (liters/y)	330	510	510	730
Fish (kg/y)	0	6.9	16	21

Table 21.4 Ingestion Dose Conversion Factors in units of 10^{-5} (Table E-12, Reg. Guide 1.109)

Group	Bone	Liver	Total body	Kidney	Lungs	GI tract
I	52.2	61.1	4.33	16.4	6.64	0.191
C	32.7	31.3	4.62	10.2	3.67	0.196
T	11.2	14.9	5.19	5.07	1.97	0.212
A	7.97	10.9	7.14	3.70	1.23	0.211

Data acceptable to the NRC for that purpose appear in the *Code of Federal Regulations,* Part 51.51, as "Table of Uranium Fuel Cycle Environmental Data."

21.7 NEWER RADIATION STANDARDS

A major revision of regulations on radiation exposure was proposed by the NRC in 1986, published as a Final Rule in 1991, and required for use from January 1, 1994. The newer version of the rule 10CFR20,[†] intended to provide greater protection for both workers and the public, was based on recommendations of the International Committee on Radiological Protection (ICRP).

The improved regulations are more realistic in terms of hazards and bring to bear accumulated knowledge about radiation risk. The complicated task of deducing doses is accomplished by computer methods. Whereas the traditional limits on dosage are based on the critical organ, the new 10CFR20 considers the dosage to the whole body from whatever sources of radiation are affecting organs and tissues. Radiations from external and internal sources are summed to obtain the total dose. Also, long-term effects of radionuclides fixed in the body are added to any short-term irradiation effects. The bases for the limits selected are the risk of cancer in the case of most organs and tissues and the risk of hereditary diseases in offspring in the case of the gonads.

A new concept called "committed effective dose equivalent" is introduced. Recall from Section 16.2 that dose equivalent is the product of absorbed dose and the quality factor. The word "committed" implies taking account of future exposure after ingestion of radioactive material. The time span is taken to be a typical working life of 50 y (e.g., between ages 20 and 70). Suppose that a certain

---

[†]Federal Register, Vol. 56, No. 98, Tuesday, May 21, 1991, p. 23360 ff. The Introduction contains useful reading on the history of dose regulations in the United States.

radionuclide is deposited in an organ of the human body. Over time thereafter the nuclide decays and is eliminated but provides a dose to that organ. The total dose, labeled H_{50}, is called a committed dose equivalent. It is assumed that the dose is experienced within the year the nuclide is deposited, which will be more nearly true the shorter the effective life in the body.

To calculate H_{50}, suppose that N_0 atoms are deposited in a gram of an organ or tissue. The number left after a time t is

$$N = N_0 (1/2)^{t/te},$$

where t_e is the effective half-life, as discussed in Section 16.1. The number that have been lost is $N_L = N_0 - N$, and the fraction of these that decay is t_e/t_H, as shown in Exercise 16.6. Thus the number that decay is

$$N_D = N_L(t_e/t_H).$$

As each nucleus decays, it delivers energy E, and thus the committed dose equivalent is

$$H_{50} = N_D E.$$

Let us apply these relations to some radionuclides. The half-life of tritium of 12.3 y is a fairly large fraction of 50 y but the biological half-life is only $t_b = 10$ d, so t_e is also approximately 10 d. The fraction that decays within the organ is $10/(4.5 \times 10^3)$ and the fraction lost is almost exactly 1. In contrast, for plutonium-239, $t_H = 2.4 \times 10^4$ y, $t_b = 100$ y for bone, and $t_e = 99.6$ y. The fraction left after 50 y is $(1/2)^{50/100} = 0.707$, whereas the fraction lost is 0.293. Of these, decay accounts for only $99.6/(2.4 \times 10^4) = 0.0042$.

Finally, the word "effective" takes account of the relative risk associated with different organs and tissues by forming a weighted sum by use of weighting factors w_T as listed in Table 21.5. If $(H_{50})_T$ represents the committed dose to organ or tissue T, the effective dose is a sum over T,

$$(H_{50})_E = \Sigma_T \, w_T (H_{50})_T.$$

If only one organ were important, as in the case of iodine-131 in the thyroid, the effective dose to the whole body would only be 3% of what it would be if the same dose were delivered throughout the body.

From the factors in Table 21.5 and from the knowledge of chemical properties, half-life, radiations, and organ and tissue data, the NRC has deduced the limits on concentration of specific radionuclides. Dose restrictions are for an annual limit of intake (ALI) by inhalation or ingestion of 5 rems/y (or a 50-year dose of 50 rems) for a plant worker. The derived air concentration (DAC) would give one ALI in a working year through breathing contaminated air. Extensive tables of ALI and DAC for hundreds of radioisotopes are provided in the new 10CFR20. They allow the calculation of exposure to mixtures of isotopes.

The two quantities are related by

$$DAC \, (\mu Ci/ml) = ALI \, (\mu Ci)/(2.4 \times 10^9)$$

Table 21.5 Organ and Tissue Radiation Weighting Factors (10CFR20)

Organ or Tissue	Weighting Factor
Gonads	0.25
Breast	0.15
Red bone marrow	0.12
Lung	0.12
Thyroid	0.03
Bone surfaces	0.03
Remainder[†]	0.30
Whole body	**1.00**

[†]*0.06 each for five organs.*

where the numerical factor is a product of four things: 50 wk/y; 40 h/wk; 60 min/h; and 2×10^4 ml (air breathed per minute).

A distinction is made between two types of dose: The first is "stochastic," which is the same as "probabilistic," defined as dosages related to the chance of cancer or hereditary effect, with the number of health effects proportional to the dose. The worker dose limit for stochastic effects is 5 rems/y. The second is "nonstochastic" or "deterministic," which are doses to tissues for which there is a threshold dose for an effect, so that a definite limit can be set on an annual dose (e.g., 50 rems). The skin and the eye lens are examples.

We can revisit the situation of a cloud of radioactive krypton-85 as in Section 21.2. Detailed calculation on all organs lead to the conclusion that only the skin will be significantly affected and thus the nonstochastic limit applies. The ALI and DAC values are correspondingly higher, the latter being 1×10^{-4} μCi/cm^3, 10 times the value in the old 10CFR20. For other radionuclides and modes of exposure, the new calculated concentrations can be smaller, the same, or larger than the old.

An example adapted from NRC material will be helpful in understanding the new rule. Suppose that a worker in a nuclear plant receives 1 rem of external radiation and also is exposed over 10 working days to concentrations in air of iodine-131 of 9×10^{-9} μCi/ml and of cesium-137 of 6×10^{-8} μCi/ml (these correspond to the older MPCs). What is the fraction (or multiple) of the annual effective dose equivalent limit? We sum the fractions that each exposure is of the annual limit of 5 rems. The external exposure contributes $\frac{1}{5} = 0.2$. The ALI figures, taking account of the ICRP weighting factors for the various organs for the two isotopes, are 50 µCi for I-131 and 200 µCi for Cs-137. We need to find the actual activities taken in. With the standard breathing rate of 1.2 m^3/h, in 80 h the air intake is 96 m^3. The activities received are thus 0.86 µCi for I and 5.8 µCi for Cs. The corresponding fractions are 0.86/50 = 0.017 and 5.8/200 = 0.029, giving a total of external and internal fractions of

$$0.2 + 0.017 + 0.029 = 0.246$$

or approximately $\frac{1}{4}$ of the limit. In this particular case, the expected hazard is lower than by the older method.

Other features of the new rule are separate limits on exposures (a) of body extremities—hands, forearms, feet, lower legs; (b) of the lens of the eye; and (c) of an embryo and fetus. The risk to the whole body per rem of dosage is 1 in 6000. For the limit of 5 rems the annual risk is 8×10^{-4}, which is approximately eight times acceptable rates in "safe" industries. The figure is to be compared with the lifetime risk of cancer from all causes of approximately 1 in 6.

Dose limits for individual members of the public (0.1 rem/y) are quite a bit lower than those working with radionuclides (5 rems/y). In calculating concentrations of radionuclides in air released to an unrestricted area, differences in time of exposure, breathing rate, and average age are accounted for by dividing worker DAC values by 300 for inhalation or 219 for submersion. Examples (in µCi/ml) are Cs-137 $(6 \times 10^{-8})/300 = 2 \times 10^{-10}$ and gaseous Xe-133 $(1 \times 10^{-4})/219 = 5 \times 10^{-7}$.

21.8 SUMMARY

Radiation protection of living organisms requires control of sources, barriers between source and living being, or removal of the target entity. Calculations required to evaluate external hazard include the dose as it depends on flux and energy, material, and time; the inverse square geometric spreading effect; and the exponential attenuation in shielding materials. Internal hazard depends on many physical and biological factors. Maximum permissible concentrations of radioisotopes in air and water can be deduced from the properties of the emitter and the dose limits. Application of the principle of ALARA is designed to reduce exposure to levels that are as low as reasonably achievable. There are many biological pathways that transport radioactive materials. New dose limit rules are based on the total effects of radiation—external and internal—on all parts of the body.

21.9 EXERCISES

21.1 What is the rate of exposure in mrems/y corresponding to a continuous 1-MeV gamma ray flux of 100 cm^{-2} – s^{-1}? What dose equivalent would be received by a person who worked 40 h/w throughout the year in such a flux?

21.2 A Co-60 source is to be selected to test radiation detectors for operability. Assuming that the source can be kept at least 1 m from the body, what is the largest strength acceptable (in μCi) to assure an exposure rate of less than 500 mrems/y? (Note that two gammas of energy 1.17 and 1.33 MeV are emitted.)

21.3 By comparison with the Kr-85 analysis, estimate the MPC in air for tritium, average beta particle energy 0.006 MeV.

21.4 The nuclear reactions resulting from thermal neutron absorption in boron and cadmium are

$$^{10}_{5}B + ^{1}_{0}n \rightarrow ^{7}_{3}Li + ^{4}_{2}He,$$

$$^{113}_{48}Cd + ^{1}_{0}n \rightarrow ^{114}_{48}Cd + \gamma(5MeV).$$

Which material would you select for a radiation shield? Explain.

21.5 Find the uncollided gamma ray flux at the surface of a spherical lead shield of radius 12 cm surrounding a very small source of 200 mCi of 1 MeV gammas.

21.6 Concentration limits of some radionuclides in water released to the public, according to 10CFR20 in the old and new versions are listed:

Radionuclide	Concentration Limits (μCi/ml)	
	Old	New
Tritium	3×10^{-3}	1E-3
Cobalt-60	3×10^{-5}	3E-6
Strontium	3×10^{-7}	5E-7
Iodine-131	3×10^{-7}	1E-6
Cesium-137	2×10^{-5}	1E-6

Calculate the ratio new/old for each radionuclide.

21.7 Water discharged from a nuclear plant contains in solution traces of strontium-90, cerium-144, and cesium-137. Assuming that the concentrations of each isotope are proportional to their fission yields, find the allowed activities per ml of each. Note the following data:

Isotope	Half-life	Yield	Limit (μCi/ml)[†]
^{90}Sr	29.1 y	0.0575	5E-7
^{144}Ce	284.6 d	0.0545	8E-6
^{137}Cs	30.2 y	0.0611	1E-6

[†]According to 10CFR20 (1993 version).

21.8 A 50-year exposure time is assumed in deriving the dose factors listed in Section 21.6. These take account of the radioisotope's physical half-life t_p and also its biological half-life t_b, which is the time it takes the chemical to be eliminated from the body. The effective half-life t_e can be calculated from the formula

$$1/t_e = 1/t_p + 1/t_b.$$

Find t_e for these three cases cited by Eichholz (see References):

Radionuclide	t_p	t_b
Iodine-131	8.04 d	138 d
Cobalt-60	5.27 y	99.5 d
Cesium-137	30.2 y	70 d

If t_p and t_b are greatly different from each other, what can be said about the size of t_e?

21.9 The activities of U-238, Ra-226, and Rn-222 in a closed system are approximately equal, in accord with the principle of secular equilibrium. Assuming that the natural uranium content of soil is 10 ppm, calculate the specific activities of the isotopes in microcuries per gram of soil (Table 3.1 gives half-lives needed to calculate).

Computer Exercises

21.A Program EXPOSO looks up gamma ray attenuation coefficients and buildup factors on data tables and finds the radiation exposure at a distance from a point source.
(a) Run the program and explore its menus.
(b) Verify that the flux at 10 cm from a point millicurie 1 MeV gamma ray source in lead is $39.2/cm^2$-s.
(c) Use the program to find the lead distance from a millicurie 1 MeV source that yields 5 mrems/y, to within one millimeter.
(d) Check the figures for a reactor in space (Section 21.3) with the shield option 7 (none).

21.B A small research reactor core is located near the bottom of a deep pool of water. The water serves as moderator, coolant, and shield. (a) With a power of 10 MW and a fission neutron leakage fraction of 0.3, estimate, with the point source version of the computer program NEUTSHLD, the uncollided flux of fast neutrons at a distance of 20 ft from the core, treated as a point source. (b) Samples to be irradiated are placed near the core, the dimensions of which are 30 cm $\times$ 30 cm $\times$ 60 cm high. Assuming that the neutron source strength per unit area is uniform, calculate, with the plane version of NEUTSHLD, the fast neutron flux at 10 cm from the center of a large face of the core.

21.C   A study is made of leukemia incidence over a 100-km^2 area in the vicinity of a nuclear power plant. Some apparent clustering of cases is observed that might be attributed to proximity or wind direction. Run computer program CLUSTER to see how small samples of completely random statistical data normally are clustered. Then edit line 410 of the program from 100 to 1,000 and then to 10,000 to see the population becoming more uniform.

21.D   To improve the uniformity of irradiation of large objects in a water pool, a set of five "point" cobalt-60 sources (average gamma ray energy 1.25 MeV) are arranged in a plane at coordinates in centimeters (0, 0), (20, 20), (20, −20), (−20, −20), and (−20, 20). Explore the variation of total gamma flux over a parallel plane 10 cm away with computer program EXPOSO to calculate contributions of each source. Compare with results in a case where all five sources are concentrated near the point (0,0).

21.10 REFERENCES

Radiation Information Network
http://www.physics.isu.edu/radinf/index1.html
Numerous links to sources. Created by Bruce Busby, Idaho State University.

Bernard Schleien, Lester A. Slayback, Jr., and Brian Kent Birky, Editors, *Handbook of Health Physics and Radiological Health*, 3rd Ed., Williams & Wilkins, Baltimore, 1998. A greatly expanded version of a classical document of 1970.

Herman Cember and Thomas E. Johnson, *Introduction to Health Physics*, 4th Ed., McGraw-Hill, New York, 2008. A thorough and easily understood textbook.

Herman Cember and Thomas E. Johnson, *The Health Physics Solutions Manual*, PS&E Publications, Silver Spring, MD, 1999. Recommended by the second author.

Joseph John Bevelacqua, *Contemporary Health Physics: Problems and Solutions*, John Wiley & Sons, New York, 1995.

Joseph John Bevelacqua, *Basic Health Physics: Problems and Solutions*, John Wiley & Sons, New York, 1999. Helpful in preparing for CHP exam.

James E. Turner, *Atoms, Radiation, and Radiation Protection*, John Wiley & Sons, New York, 1995.

Steven B. Dowd and Elwin R. Tilson, Eds., *Practical Radiation Protection and Applied Radiobiology*, W. B. Saunders, Philadelphia, 1999. All about radiation, its effects, and protection, from a nuclear medicine viewpoint. An appendix of Web sites maintained also at http://www.radscice.com/dowd.html.

J. Kenneth Shultis and Richard E. Faw, *Radiation Shielding*, Prentice-Hall, Upper Saddle River, NJ, 1996. Includes transport theory and Monte Carlo methods.

James Wood, *Computational Methods in Reactor Shielding*, Pergamon Press, Oxford, 1982.

Theodore Rockwell III, *Reactor Shielding Design Manual*, McGraw-Hill, New York, 1956. A classic book on shielding calculations that remains a valuable reference.

Geoffrey G. Eichholz, *Environmental Aspects of Nuclear Power*, Lewis Publishers, Chelsea, MI, 1985.

Richard E. Faw and J. Kenneth Shultis, *Radiological Assessment: Sources and Doses*, American Nuclear Society, La Grange Park, IL, 1999. Fundamentals and extensive data. A reprint with a few changes of a 1991 book published by Prentice-Hall.

John E. Till and H. Robert Meyer, Editors, *Radiological Assessment, A Textbook on Environmental Dose Analysis*, Nuclear Regulatory Commission, Washington, DC, NUREG/CR-3332, 1983.

Kenneth L. Miller and William A. Weidner, Editors, *CRC Handbook of Management of Radiation Protection Programs*, CRC Press, Boca Raton, 1986. An assortment of material not found conveniently elsewhere, including radiation lawsuit history, the responsibilities of health physics professionals, information about state radiological protection agencies, and emergency planning. More than half of the book is a copy of regulations of the Department of Transportation.

Radon Exposure of the United States Population—Status of the Problem, NCRP Commentary No. 6, National Council of Radiation Protection and Measurements, Bethesda, MD 1991.

International Basic Safety Standards for Protection Against Ionizing Radiation and for the Safety of Radiation Sources. International Atomic Energy Agency, Vienna, Safety Series No. 115, CD-ROM 2004. Document can be downloaded from IAEA website.http://www-ns.iaea.org/standards/documents/default.asp?sub=160

Code of Federal Regulations, Energy 10, Office of the Federal Register, National Archives and Records Administration, United States Government Printing Office, Washington, DC (annual issuance).

10CFR Part 20—Standards for Protection Against Radiation
http://www.nrc.gov/reading-rm/doc-collections/cfr/part020
Links to each subpart (e.g., C: Occupational Dose Limits and D: Radiation Dose Limits for Individual Members of the Public).

NRC Update: New Reactor Licensing
http://hps.ne.uiuc.edu/numug/archive/2006/presentations/harvey_ppt.pdf
PowerPoint presentation by R. Brad Harvey of NRC Office of Nuclear Reactor Regulation, including present activities and future plans, 2006.

NRC Regulatory Guides
http://www.nrc.gov/reading-rm/doc-collections/reg-guides
Issued in ten broad divisions.

Select division 1...

1.109 Calculation of Annual Doses to Man from Routine Releases of Reactor Effluents for the Purpose of Evaluating Compliance with 10CFR Part 50 Appendix I, October 1977 (pdf 2.71 MB).

1.111 Methods for Estimation of Atmospheric Transport and Dispersion of Gaseous Effluents in Routine Releases from Light-Water-Cooled Reactors, July 1977 (pdf 1.25 MB).

Select division 8. Occupational Health.

8.8 Information Relevant to Ensuring that Occupational Radiation Exposures at Nuclear Power Stations Will be as low as Reasonably Achievable, June 1978 (pdf 1.19 MB).

Committee on the Biological Effects of Ionizing Radiations, National Research Council, *Health Risks of Radon and Other Internally Deposited Alpha-Emitters, BEIR IV*, National Academy Press, Washington DC, 1988.
http://books.nap.edu/openbook.php?isbn=0309037972.
Select sections from Web site. Emphasizes lung cancer and the relationship of smoking and radon.

Committee on Health Effects of Exposure to Radon (BEIR VI), National Research Council, *Health Effects of Exposure to Radon*, National Academy Press, Washington DC, 1999.
http://books.nap.edu/openbook.php?isbn=0309056454.
Select sections from Web site. Conclusion: new information needs to be considered and improved models developed.

Radon Update
http://www.physics.isu.edu/radinf/radon.htm#top
Based on an article by Dr. A. B. Brill in *Journal of Nuclear Medicine*, February 1994, provided by Physics Department, Idaho State University.

Radon and hormesis
http://www.belleonline.com/newsletters/volume7/vol7-1/riskmanagement.html
Article by K. T. Bogen and D. W. Layton.

Radiation, Science, and Health
http://www.radscihealth.org/rsh
Select Documents. Organization criticizes conservatism of standards advisory bodies and government regulators.

NIST Physics Laboratory
http://physics.nist.gov
Select Physical Reference Data, X-ray and Gamma-Ray Data. Figures differ slightly from those in NUREG/CR-5740.

ICRP Publication 2, "Recommendations of the International Commission on Radiological Protection," *Health Physics 3*, Pergamon Press, Oxford, 1960.

Radioactive Waste Disposal 22

MATERIALS THAT contain radioactive atoms and that are deemed to be of no value are classed as radioactive wastes. They may be natural substances, such as uranium ore residues with isotopes of radium and radon, products of neutron capture, with isotopes such as those of cobalt and plutonium, or fission products, with a great variety of radionuclides. Wastes may be generated as byproducts of national defense efforts, of the operation of commercial electric power plants and their supporting fuel cycle, or of research and medical application at various institutions. The radioactive components of the waste may emit alpha particles, beta particles, gamma rays, and in some cases neutrons, with half-lives of concern from the standpoint of storage and disposal ranging from several days to thousands of years.

Because it is very difficult to render the radioactive atoms inert, we face the fact that the use of nuclear processes must be accompanied by continuing safe management of materials that are potentially hazardous to workers and the public. The means by which this essential task is accomplished is the subject of this chapter.

22.1 THE NUCLEAR FUEL CYCLE

Radioactive wastes are produced throughout the nuclear fuel cycle sketched in Figure 22.1. This diagram is a flowchart of the processes that start with mining and end with disposal of wastes. Two alternative modes are shown—once-through and recycle.

Uranium ore contains very little of the element uranium, approximately 0.1% by weight. The ore is treated at processing plants known as mills, where mechanical and chemical treatment gives "yellowcake," which is mainly U_3O_8, and large residues called mill tailings. These still have the daughter products of the uranium decay chain, especially radium-226 (1599 y), radon-222 (3.82 d), and some polonium isotopes. Tailings are disposed of in large piles near the mills, with an earth cover to reduce the rate of release of the noble radon gas and thus prevent

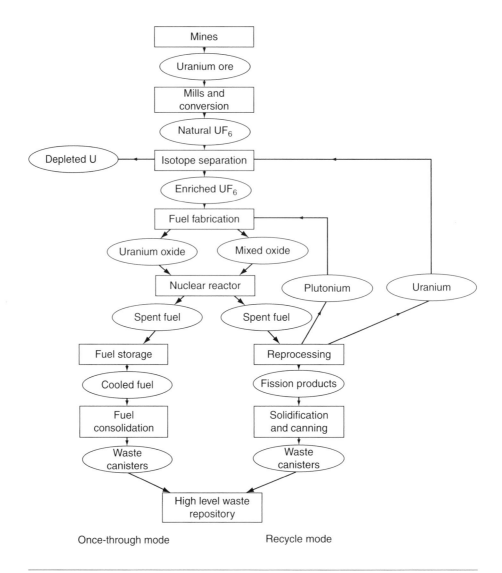

FIGURE 22.1 Nuclear fuel cycles. The once-through shown on the left is used in the United States; the recycle shown on the right is used in other countries.

excessive air contamination. Strictly speaking the tailings are waste, but they are treated separately.

Conversion of U_3O_8 into uranium hexafluoride, UF_6, for use in isotope enrichment plants produces relatively small amounts of slightly radioactive material. The separation process, which brings the uranium-235 concentration from 0.7 wt% to 3 to 5%, also has little waste. It does generate large amounts of depleted uranium ("tails") at approximately 0.3% U-235. Depleted uranium is stored and could be used as fertile material for future breeder reactors. The fuel fabrication operation, involving the conversion of UF_6 to UO_2 and the manufacture of fuel assemblies, yields considerable waste despite recycling practices. Because U-235 has a shorter half-life than U-238, the slightly enriched fuel is more radioactive than natural uranium.

The operation of reactors gives rise to liquids and solids that contain radioactive materials from two sources. One is activation of metals by neutrons to produce isotopes of iron, cobalt, and nickel. The other is fission products that escape from the fuel tubes or are produced from uranium residue on their surfaces.

Spent fuel, resulting from neutron irradiation in the reactor, contains the highly radioactive fission products and various plutonium isotopes, along with the sizeable residue of uranium that is near natural concentration. As shown on the left side of Figure 22.1, the fuel will be stored, packaged, and disposed of by burial according to current United States practice.

In some other countries the spent fuel is being reprocessed. As sketched in the right side of Figure 22.1, uranium is returned to the isotope separation facility for re-enrichment, and the plutonium is added to the slightly enriched fuel to produce "mixed-oxide" fuel. Only the fission products are subject to disposal.

22.2 WASTE CLASSIFICATION

For purposes of management and regulation, classification schemes for radioactive wastes have evolved. The first contrasts defense and nondefense wastes. The original wastes were from the Hanford reactors used in World War II to produce weapons material. The wastes were stored in moist form in large underground tanks. Over subsequent years part of these defense wastes have been processed for two reasons: (a) to fix the wastes in stable form; and (b) to separate out the two intermediate half-life isotopes, strontium-90 (29.1 y) and cesium-137 (30.2 y), leaving a relatively inert residue. Additional defense wastes were generated by reactor operation over the years for the stockpile of plutonium and tritium for nuclear weapons, and the spent fuel from submarine reactors was reprocessed.

Nondefense wastes include those produced in the commercial nuclear fuel cycle as described previously by industry and by institutions. Industrial wastes come from manufacturers who use isotopes and from pharmaceutical companies. Institutions include universities, hospitals, and research laboratories.

Another way to classify wastes is according to the type of material and the level of radioactivity. The first class is high-level waste (HLW) from reactor operations. These are the fission products that have been separated from other materials in spent fuel by reprocessing. They are characterized by their very high radioactivity; hence the name.

A second category is spent fuel, which really should not be called a waste, because of its residual fissile isotopes. However, in common use, because spent fuel in the United States is to be disposed of in a high-level waste repository, it is often thought of as HLW.

A third category is transuranic wastes, abbreviated TRU, which are wastes that contain plutonium and heavier artificial isotopes. Any material that has an activity caused by transuranic materials of as much as 100 nanocuries per gram is classed as TRU. The main source is nuclear weapons fabrication plants.

Mill tailings are the residue from processing uranium ore. The main radioactive elements other than residual uranium are radium (1599 y) and thorium (7.54×10^4 y). Nuclear Regulatory Commission (NRC) regulations call for covers of tailings piles to prevent the release of radon (3.82 d).

Another important category is low-level waste (LLW), which officially is defined as material that does not fall into any other class. LLW has a small amount of radioactivity in a large volume of inert material and generally is subject to placement in a near-surface disposal site. The name "low-level waste" is misleading in that some LLW can have a curie content comparable to that of some old high-level waste.

Two other categories are naturally occurring radioactive materials (NORM) such as byproducts of phosphate mining and accelerator-produced materials (NARM). Both have slight radioactivity.

Still other categories are used for certain purposes e.g., remedial action wastes, coming from the cleanup of formerly used facilities of the Department of Energy (DOE). A category called mixed LLW has the characteristics of LLW but also contains hazardous organic chemicals or heavy metals such as lead or mercury. An EPA Web site gives a full description of this type of waste (see References).

For a number of years the category "below regulatory concern" (BRC) referred to wastes having trivial amounts of activity and subject to unrestricted release. The NRC was unsuccessful in obtaining consensus on the subject and abandoned the category around 1990.

Some perspective of the fuel cycle and nuclear wastes can be gained from Table 22.1, adapted from an Australian Web site (see References).

Table 22.1 Typical Annual Weights (Tonnes) 1000 MWe Reactor, 3.5% Enriched Fuel

Fuel supplied and discharged	25.0
Enriched UF_6	32.5
Depleted UF_6	217.5
U_3O_8 from uranium mill	200
Uranium ore and tailings	50,000

The 25 tonnes of spent fuel is to be compared with the burning of 3.2 million tonnes of coal with a release of 7 million tons of CO_2.

22.3 SPENT FUEL STORAGE

The management of spent fuel at a reactor involves a great deal of care in mechanical handling to avoid physical damage to the assemblies and to minimize exposure of personnel to radiation. At the end of a typical operating period of 1 y for a Pressurized Water Reactor (PWR), the head of the reactor vessel is removed and set aside. The whole space above the vessel is filled with borated water to allow fuel assemblies to be removed while immersed. The radiation levels at the surface of an unshielded assembly are millions of rems per hour. By use of movable hoists, the individual assemblies weighing approximately 600 kg (1320 lb) are extracted from the core and transferred to a water-filled storage pool in an adjacent building. Computer Exercise 22.A shows the arrangement of fuel assemblies in racks of a water storage pool. Approximately a third of the core is removed; fuel remaining in the core is rearranged to achieve the desired power distribution in the next cycle; and fresh fuel assemblies are inserted in the vacant spaces. The water in the 40-ft-deep storage pool serves as a shielding and cooling medium to remove the fission product residual heat. We may apply the decay heat formula from Section 19.3 to estimate the energy release and source strength of the fuel. At a time after shutdown of 3 months (7.9×10^6 s) the decay power from all the fuel of a 3000 MWt reactor is

$$P = 3000(0.066)(7.9 \times 10^6)^{-0.2} = 8.26 \text{ MW}.$$

If we assume that the typical particles released have an energy of 1 MeV, this corresponds to 1.4 billion curies (5.2×10^{19} Bq). To ensure integrity of the fuel, the purity of the water in the pool is controlled by filters and demineralizers, and the temperature of the water is maintained by use of coolers.

The storage facilities consist of vertical stainless steel racks that support and separate fuel assemblies to prevent criticality, because the multiplication factor k of one assembly is rather close to 1. When most reactors were designed, it was expected that fuel would be held for radioactive "cooling" for only a few months, after which time the assemblies would be shipped to a reprocessing plant. Capacity was provided for only about two full cores, with the possibility of having to unload all fuel from the reactor for repairs. The abandonment of reprocessing by the United States required utilities to store all spent fuel on site, awaiting acceptance of fuel for disposal by the federal government in accordance with the Nuclear Waste Policy Act of 1982 (NWPA). Re-racking of the storage pool was the first action taken. Spacing between assemblies was reduced, and neutron-absorbing materials were added to inhibit neutron multiplication. For some reactors this was not an adequate solution of the problem of fuel accumulation, and thus alternate storage methods were investigated. There were several choices. The first was to ship spent fuel to a pool of a newer plant in the utility's system. The second was for the plant to add more water basins or for a commercial organization to build basins at another central location. The third was to use storage at government facilities, a limited amount of which was promised in NWPA. The fourth was rod consolidation, in which the bundle of fuel rods is collapsed and put in a container, again to go in a pool. A volume reduction of about two can be achieved. A fifth was to store a number of dry assemblies in large casks sealed to prevent access by water. A variant is the storage of intact assemblies in dry form in a large vault. Dry storage is the favored alternative. An ideal solution would be to use the same container for storage, shipment, and disposal. A combination of methods may instead be adopted as DOE accepts spent fuel.

The amount of material in spent fuel to be disposed of annually can be shown to be surprisingly small. Dimensions in meters of a typical PWR fuel assembly are $0.214 \times 0.214 \times 4.06$, giving a volume of 0.186 m^3. If 60 assemblies are discharged from a typical reactor, the annual volume of spent fuel is 11.2 m^3 or 394 cubic feet. For 100 United States reactors this would be 39,400 ft^3, which would fill a standard football field (300 ft $\times$ 160 ft) to a depth of less than 10 inches, assuming that the fuel assemblies could be packed closely.

The amount of fission products can be estimated by letting their weight be equal to the weight of fuel fissioned, which is 1.1 g per MWd of thermal energy. For a reactor operating at 3000 MW this implies 3.3 kg/d or approximately 1200 kg/y. If the specific gravity is taken to be 10 (i.e., 10^4 kg/m^3), the annual volume is 0.12 m^3, corresponding to a cube 50 cm on a side. This figure is the origin of the claim that the wastes from a year's operation of a reactor would fit under an office desk. Even with reprocessing the actual volume would be considerably larger than this.

The detailed composition of a spent fuel assembly is determined by the number of Mwd/tonne of exposure it has received. A burnup of 33,000 MWd/tonne corresponds to a 3-year operation in an average thermal neutron flux of 3×10^{13}/cm^2−s. Figure 22.2 shows the composition of fuel before and after. The fissile material content has only been changed from 3.3% to 1.43%, and the U-238 content is reduced only slightly.

22.4 TRANSPORTATION

Regulations on radioactive material transportation are provided by the federal Department of Transportation and the NRC. Container construction, records, and radiation limits are among the specifications. Three principles used are: (a) packaging is to provide protection; (b) the greater the hazard, the stronger the package must be; and (c) design analysis and performance tests assure safety. A classification scheme for containers has been developed to span levels of radioactivity from exempt amounts to that of spent nuclear fuel. For LLW coming from processing reactor water, the cask consists of an outer steel cylinder, a lead lining,

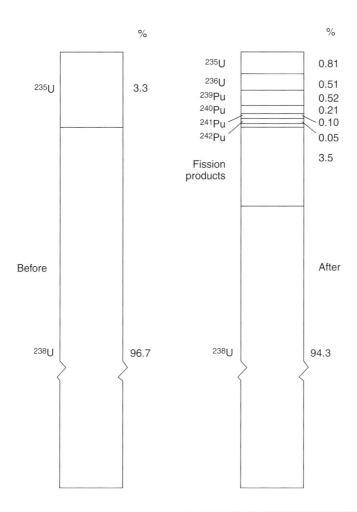

FIGURE 22.2 Composition of nuclear fuel before and after irradiation with neutrons in a reactor. (From Raymond L. Murray, *Understanding Radioactive Waste*, 2003, courtesy of Battelle Press, Columbus, OH.)

and an inner sealed container. For spent fuel, protection is required against (a) direct radiation exposure of workers and the public, (b) release of radioactive fluids, (c) excessive heating of internals, and (d) criticality. The shipping cask shown in Figure 22.3(A) consists of a steel tank of length 5 m (16.5 ft) and diameter 1.5 m (5 ft). When fully loaded with 7 PWR assemblies, the cask weighs up to 64,000 kg (70 tons). The casks contain boron tubes to prevent criticality, heavy metal to shield against gamma rays, and water as needed to keep the fuel cool and to provide additional shielding. A portable air-cooling system is attached when the cask is loaded on a railroad car as in Figure 22.3(B). The cask

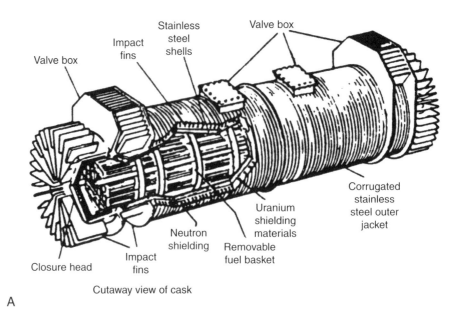

Cutaway view of cask

A

Cask loaded on railroad car

B

FIGURE 22.3 Spent fuel shipping cask. (Courtesy of General Electric Company.)

is designed to withstand normal conditions related to temperature, wetting, vibration, and shocks. In addition, the cask is designed to meet four performance specifications that simulate real conditions in road accidents. The cask must withstand a 30 ft (~10 m) free fall onto an unyielding surface, a 40 in. (~1 m) fall to strike a 6-in. (~15 cm) diameter pin, a 30-min exposure to a fire at temperature 1475°F (~800°C), and complete immersion in water for a period of 8 h. Some extreme tests have been conducted to supplement the design specifications. In one test a trailer rig carrying a cask was made to collide with a solid concrete wall at 84 mph. Only the cooling fins were damaged; the cask would not have leaked if radioactivity had been present.

Public concern has been expressed about the possibility of accident, severe damage, and a lack of response capability. The agencies responsible for regulation do not assume that accidents can be prevented but expect all containers to withstand an incident. In addition, efforts have been made to make sure that police and fire departments are familiar with the practice of shipping radioactive materials and with resources available in the form of state radiological offices and emergency response programs with backup by national laboratories.

22.5 REPROCESSING

The physical and chemical treatment of spent nuclear fuel to separate the components—uranium, fission products, and plutonium—is given the name reprocessing. The fuel from the Hanford and Savannah River Plant weapons production reactors and the naval reactors has been reprocessed in the defense program at the federal government national laboratories. Commercial experience with reprocessing in the United States has been limited. In the period 1966–1972, Nuclear Fuel Services (NFS) operated a facility at West Valley, NY. Another was built by Allied General Nuclear Service (AGNS) at Barnwell, SC, but it never operated on radioactive material as a matter of national policy. To understand that political decision it is necessary to review the technical aspects of reprocessing.

On receipt of a shipping cask of the type shown in Figure 22.3, the spent fuel is unloaded and stored for further decay in a water pool. The assemblies are then fed into a mechanical shear that cuts them into pieces approximately 3 cm long to expose the fuel pellets. The pieces fall into baskets that are immersed in nitric acid to dissolve the uranium dioxide and leave zircaloy "hulls." The aqueous solution from this chop-leach operation then proceeds to a solvent extraction (Purex) process. Visualize an analogous experiment. Add oil to a vessel containing salt water. Shake to mix. When the mixture settles and the liquids separate, some salt has gone with the oil (i.e., it has been extracted from the water). In the Purex process the solvent is the organic compound tributyl phosphate (TBP) diluted with kerosene. Countercurrent flow of the aqueous and organic materials is maintained in a packed column as sketched in Figure 22.4. Mechanical vibration assists contact.

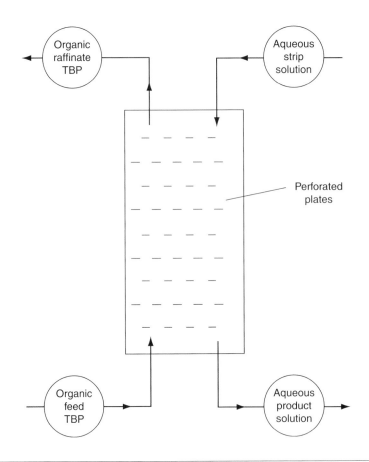

FIGURE 22.4 Solvent extraction by the Purex method.

A flow diagram of the separation of components of spent fuel is shown in Figure 22.5. The amount of neptunium-239, half-life 2.355 d, depends on how fresh the spent fuel is. After a month of holding, the isotope will be practically gone. The three nitrate solution streams contain uranium, plutonium, and an array of fission product chemical elements. The uranium has a U-235 content slightly higher than natural uranium. It can either be set aside or re-enriched in an isotope separation process. The plutonium is converted into an oxide that is suitable for combining with uranium oxide to form a mixed oxide (MOX) that can form part or all of the fuel of a reactor. Precautions are taken in the fuel fabrication plant to protect workers from exposure to plutonium.

In the reprocessing operations, special attention is given to certain radioactive gases. Among them are 8.04-d iodine-131, 10.73-y krypton-85, and 12.32-y tritium, which is the product of the occasional fission into three particles. The iodine

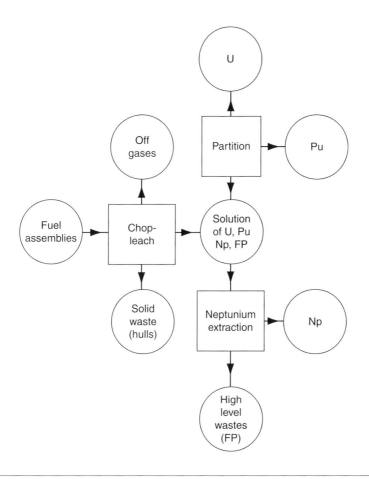

FIGURE 22.5 Simplified flowchart of nuclear fuel reprocessing.

concentration is greatly reduced by reasonable holding periods. The long-lived krypton poses a problem because it is a noble gas that resists chemical combination for storage. It may be disposed of in two ways: (a) release to the atmosphere from tall stacks with subsequent dilution, or (b) absorption on porous media such as charcoal maintained at very low temperatures. The hazard of tritium is relatively small, but water containing it behaves as ordinary water.

Reprocessing has merit in several ways other than making uranium and plutonium available for recycling:

(a) The isolation of some of the long-lived transuranic materials (other than plutonium) would permit them to be irradiated with neutrons, achieving additional energy and transmuting them into useful species or innocuous forms for purposes of waste disposal.

(b) Numerous valuable fission products such as krypton-85, strontium-90, and cesium-137 have industrial applications or may be used as sources for food irradiation.

(c) The removal of radionuclides with intermediate half-lives allows canisters of wastes to be placed closer together in the ground because the heat load is lower.

(d) Several rare elements of economic and strategic national value can be reclaimed from fission products. Availability from reprocessing could avoid interruption of supply from abroad for political reasons. Examples are rhodium, palladium, and ruthenium.

(e) The volume of wastes to be disposed of would be lower because the uranium has been extracted.

(f) Even if it were not recycled, the recovered uranium could be saved for future use in breeder reactor blankets.

Several countries abroad—France, the United Kingdom, Germany, Japan, and the former U.S.S.R.—have working reprocessing facilities and benefit from some of the preceding virtues.

An important aspect of reprocessing is that the plutonium made available for recycling can be visualized as a nuclear weapons material. Concern about international proliferation of nuclear weapons prompted President Carter in 1977 to issue a ban on reprocessing. It was believed that if the United States refrained from reprocessing, it would set an example to other countries. The action had no effect, because the United States had made no real sacrifice, having abundant uranium and coal reserves, and countries lacking resources saw full utilization of uranium in their best interests. It was recognized that plutonium from nuclear reactor operation was unsuitable for weapons because of the high content of Pu-240, which emits neutrons in spontaneous fission. Finally, it is possible to achieve weapons capability through the completely different route of isotope separation yielding highly enriched uranium. The ban prevented the AGNS plant from operating. President Reagan lifted the ban in 1981, but industry was wary of attempting to adopt reprocessing because of uncertainty in government policy and lack of evidence that there was a significant immediate economic benefit. However, the DOE is anticipating a revival of commercial reprocessing to ensure a sustainable fuel supply for the expected increased number of nuclear plants and to facilitate waste disposal, as discussed in Chapter 27.

22.6 HIGH-LEVEL WASTE DISPOSAL

The treatment given wastes containing large amounts of fission products depends on the cycle chosen. If the fuel is reprocessed, as described in the previous section, the first step is to immobilize the radioactive residue. One popular method is to mix the moist waste chemicals with pulverized glass similar to Pyrex, heat the mixture in a furnace to molten form, and pour the liquid into metal containers called canisters. The solidified waste form can be stored conveniently, shipped, and disposed of. The glass-waste is expected to resist leaching by water for hundreds of years.

If the fuel is not reprocessed, there are several choices. One is to place intact fuel assemblies in a canister. Another is to consolidate the rods (i.e., bundle them closely together in a container). A molten metal such as lead could be used as a filler if needed. What would be done subsequently with waste canisters has been the subject of a great deal of investigation concerning feasibility, economics, and social-environmental effects. Some of the concepts that have been proposed and studied are the following:

1. Send nuclear waste packages into space by shuttle and spacecraft. The weight of protection against vaporization in accidental re-entry to the Earth's atmosphere would make costs prohibitive.

2. Place canisters on the Antarctic ice cap, either held in place or allowed to melt their way down to the base rock. Costs and environmental uncertainty rule out this method.

3. Deposit canisters in mile-deep holes in the Earth. The method is impractical with available drilling technology.

4. Drop canisters from a ship, to penetrate the layer of sediment at the bottom of the ocean. Although considered as a backstop, there are evident environmental concerns.

5. Sink vertical shafts a few thousand feet deep, and excavate horizontal corridors radiating out. In the floors of these tunnels, drill holes in which to place the canisters, as sketched in Figure 22.6, or place waste packages on the floor of the corridor itself. The latter is the currently preferred technology in the United States high-level waste disposal program.

The design of a repository for high-level radioactive waste or spent fuel uses a multibarrier approach. The first level of protection is the waste form, which may be glass-waste or an artificial substance, or uranium oxide fuel, which itself inhibits diffusion of fission products and is resistant to chemical attack. The second level is the container, which can be chosen to be compatible with the surrounding materials. Choices of metal for the canister include steel, stainless steel, copper, and nickel alloys. The third level is a layer of clay or other packing that tends to prevent access of water to the canister. The fourth is a backfill of concrete or rock. The fifth and final level is the geological medium. It is chosen for its stability under heat as generated by the decaying fission products. The medium will have a pore structure and chemical properties that produce a small water flow rate and a strong filtering action.

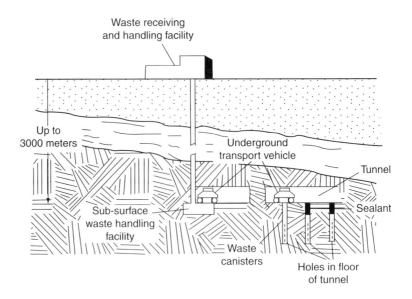

FIGURE 22.6 Nuclear waste isolation by geologic emplacement.

The system must remain secure for thousands of years. It must be designed to prevent contamination of water supplies that would give significant doses of radiation to members of the public. The radionuclides found in fission products can be divided into several classes as follows:

1. Nuclides of short half-life, up to about a month. Examples are xenon-133 (5.24 d) and iodine-131 (8.04 d). These would pose a problem in case of accident and give rise to heat and radiation that affect handling of fuel but are not important to waste disposal. The storage time for fuel is long enough that they decay to negligible levels.

2. Materials of intermediate half-life, up to 50 y, which determine the heating in the disposal medium. Examples are: cerium-144 (284.6 d), ruthenium-106 (1.020 y), cesium-134 (2.065 y), promethium-147 (2.62 y), krypton-85 (10.73 y), tritium (12.32 y), plutonium-241 (14.4 y), strontium-90 (29.1 y), and cesium-137 (30.2 y).

3. Isotopes that are still present after many thousands of years and that ultimately determine the performance of the waste repository. Important examples are radium-226 (1599 y), carbon-14 (5715 y), selenium-79 (2.9×10^5 y), technetium-99 (2.13×10^5 y), neptunium-237 (2.14×10^6 y), cesium-135 (2.3×10^6 y), and iodine-129 (1.7×10^7 y). Radiological hazard is contributed by some of the daughter products of these isotopes; for example, lead-210 (22.6 y) comes from radium-226, which in turn came from almost-stable uranium-238.

Several candidate types of geologic media are found in various parts of the United States. One is rock salt, identified many years ago as a suitable medium because its very existence implies stability against water intrusion. It has the ability to self-seal through heat and pressure. Another is the dense volcanic rock basalt. Third is tuff, a compressed and fused volcanic dust. Extensive deposits of these three rocks as candidates for repositories are found in the states of Texas, Washington, and Nevada, respectively. Still another is crystalline rock, an example of which is granite as found in the eastern United States.

A simplified model of the effect of a repository is as follows. It is known that there is a small but continued flow of water past the emplaced waste. The container will be leached away in a few hundred years and the waste form released slowly over perhaps 1,000 y. The chemicals migrate much more slowly than the water flows, making the effective time of transfer tens of thousands of years. All of the short and intermediate half-life substances will have decayed by this time. The concentration of the long half-life radionuclides is greatly reduced by the filtering action of the geological medium. For additional details on the process of performance assessment, see References.

A pair of Computer Exercises provide an introduction to the mathematical modeling of the behavior of radioactive waste in a repository or disposal facility. A simple moving pulse with decay is studied in 22.B, and the spreading of a pulse by dispersion is shown in 22.C.

A plan and a timetable for establishment of an HLW repository in the United States was set by Congress. The NWPA called for a search of the country for possible sites, the selection of a small number for further investigation, and characterization of one or more sites, taking account of geology, hydrology, chemistry, meteorology, earthquake potential, and accessibility.

In 1987, Congress decreed that site studies in Texas and Washington State should cease and mandated that Nevada would be the host state. The location would be Yucca Mountain, near the Nevada Test Site for nuclear weapons. The project was delayed for several years by legal challenges from the State of Nevada, but characterization was begun in 1991, with cognizance by DOE's Office of Civilian Radioactive Waste Management. To test suitability of the site, an Exploratory Studies Facility was dug consisting of a corridor 10 m in diameter and 5 miles long. Among features investigated were the effect of heating to 300°C and the flow of water down through the rock. As reported in the Viability Assessment document, the Yucca Mountain site is favorable because of the desert climate (only approximately 7 inches of water per year), the unsaturated zone with deep water table (2,000 ft), the stability of the geological formation, and a very low population density nearby. A Reference Design Document (RDD) was issued in January 1999. Some of the features cited are the following:

100 mi (160 km) northeast of Las Vegas, NV
70,000 tonnes of spent fuel and other wastes in 10,200 packages
Underground horizontal tunnels (drifts)

Diameter of drifts 18 ft (5.5 m); spacing 92 ft (28 m)
Emplacement level approximately 1,000 ft (305 m) below the surface
Waste packages hold 21 PWR or 44 BWR fuel assemblies

Multiple engineered barriers include the solid waste form (UO_2), the metal fuel rod cladding, a container of special corrosion-resistant nickel alloy C-22,[†] a "drip shield" to deflect water, a V-shaped trough for support, and underneath, the invert composed of stainless steel and volcanic rock to slow water flow. Figure 22.7 shows the proposed design.

Dedicated trains to carry spent fuel and high-level waste to Yucca Mountain are proposed by DOE. The choice as an alternative to trucks leads to fewer shipments, with 3,500 estimated.

The project was brought under question by the revelation in March 2005 of some e-mail messages in 1998 suggesting falsification of quality assurance data related to water infiltration. Excerpts of the messages are found in References. Investigations were made by Congress, DOE, United States Geological Survey (USGS), and the Federal Bureau of Investigation (FBI), and certain measurements were repeated as a corrective action needed to verify repository safety. The investigations were completed in 2005 as described in References.

Safety standards developed by the Environmental Protection Agency (EPA) (40CFR191) are to be used in licensing and regulation by the NRC (10CFR60).

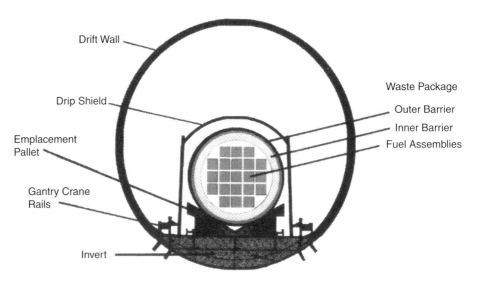

FIGURE 22.7 Spent fuel at Yucca Mountain.

[†]Nominal percentages in Hastelloy C-22: 56 Ni, 22 Cr, 13 Mo, 3 Fe, 2.5 Co, 3 W, 0.5 other.

The EPA placed limits on the maximum additional radiation dosage to members of the public because of the release of radioactive material. Two time frames were established: (a) up to 10,000 y, with 15 mrem per year; and (b) to 1 million y, with 350 mrem per year, slightly under the United States average (see Section 16.2). The selection by EPA of a time span of 10,000 y for protection against hazard from waste deposits was based on a logical analysis. A comparison was made between two radioactivities. The first was that of natural uranium as found in the ground, a figure that remains constant. The second was the declining activity of spent fuel, as the fission products and activation products decay. It was assumed that when the two figures are equal, the radiation dose caused by the waste is no greater than that caused by the original uranium. Calculations led to a time of approximately 1,000 y, and a safety factor of 10 was applied. For further details, see an article in *Nuclear News*, February 2006. The longer time frame was set on the basis that the highest radiation from waste may occur beyond the 10,000-y period.

The EPA was reluctant to establish regulations that applied to a million years, stating in 2001, "It is not possible to make reliable estimates over such a long time frame." (see References). The American Nuclear Society in a 2006 position statement with background information (see References) concurred, stating, "... extrapolating beyond 10,000 years is not scientifically sensible..."

The Yucca Mountain license application was delivered in 2008 by the Department of Energy to the NRC. DOE will certify that the repository will meet standards set by the EPA. A plan will be prepared for transportation of spent fuel by rail from reactors to the final destination in Nevada. The cap in loading of the repository of 70,000 tonnes might be raised by Congress. Money from the Nuclear Waste Fund will have to be authorized and approval for use of federal land obtained from the Department of the Interior.

In October 2005 DOE announced a change in the plan for handling spent fuel for disposal at Yucca Mountain. Instead of requiring several stages involving storage, packaging, shipping, and disposal, fuel is to be loaded into containers that can go directly into disposal. This is said to provide a "clean" (noncontaminated) repository. Benefits of the new method include elimination of expensive handling facilities and the avoidance of damage of fuel. An undesired consequence, however, is the necessary delay in submission of a license application by DOE to NRC. The new plan is ridiculed by representatives of Nevada.

The projected date for the start of burial is 2017. According to the Nuclear Waste Policy Act, DOE was required to accept spent fuel by 1998, but has not complied, to the concern of the nuclear industry.

The law called for a study of a monitored retrieval storage (MRS) system to serve as a staging center before disposal in a repository. Efforts to find a host were unsuccessful. Use of the Nevada Weapons Testing Grounds as a storage area for spent fuel has been promoted as a stopgap.

Financing for the waste disposal program being carried out by the federal government is provided by a Nuclear Waste Fund. The consumers of electricity generated by nuclear reactors pay a fee of 1/10 cent per kilowatt hour collected by the power companies. This adds only approximately 2% to the cost of nuclear electric power. Concern has been expressed about the fact that Congress has used some of the Fund for other purposes.

Progress in establishing the repository at Yucca Mountain has been slow, and the completion date has been repeatedly extended. The difficulties and uncertainties of the project have prompted consideration of alternatives. One is to irradiate certain radioisotopes in the spent fuel to destroy problem isotopes such as cesium-137 and strontium-90 that contribute to heating in the early period and neptunium-237, technetium-99, and iodine-129 that dominate the hazard at long times. These constitute only about a percent of the waste stream. If they are removed, the remaining waste needs to be secure for only approximately 100 y rather than the 10,000 y for spent fuel.

An R&D program titled Accelerator Transmutation of Wastes (ATW) was conducted over a number of years at Los Alamos National Laboratory. The concept first involves reprocessing spent fuel by pyroprocessing, which uses the IFR technology (see Section 13.3). The key fission products, actinides, and possibly plutonium would then be irradiated in a subcritical system with an accelerator that causes spallation (see Section 8.6). A beam of protons of 100 MW was to be directed to a molten lead target. A surrounding liquid would contain the isotopes to be burned, with heat removed by liquid lead. Electricity would be produced. Some 15 of such burners were estimated to be able to handle United States spent fuel. A roadmap for further development of the concept was prepared in 1999 (see References). However, spallation research on ATW at LASL was suspended and the ATW program was merged by DOE with the program Accelerator Production of Tritium (APT, Section 26.6) and essentially abandoned. There is a current lack of interest in ATW, but it remains a possibility for the more distant future.

22.7 LOW-LEVEL WASTE GENERATION, TREATMENT, AND DISPOSAL

The nuclear fuel cycle, including nuclear power stations and fuel fabrication plants, produces approximately two thirds of the annual volume of LLW. The rest comes from companies that use or supply isotopes and from institutions such as hospitals and research centers.

In this section we look at the method by which low-level radioactive materials are produced, the physical and chemical processes that yield wastes, the amounts to be handled, the treatments that are given, and the methods of disposal.

In the primary circuit of the nuclear reactor the flowing high-temperature coolant erodes and corrodes internal metal surfaces. The resultant suspended or dissolved materials are bombarded by neutrons in the core. Similarly, core metal structures absorb neutrons, and some of the surface is washed away. Activation products as listed in Table 22.2 are created, usually through an (n,γ) reaction.

Table 22.2 Activation Products in Reactor Coolant

Isotope	Half-life (years)	Radiation Emitted	Parent Isotope
C-14	5715	β	N-14*
Fe-55	2.73	x	Fe-54
Co-60	5.27	β, γ	Co-59
Ni-59	7.6×10^4	x	Ni-58
Ni-63	100	β	Ni-62
Nb-94	2.4×10^4	β, γ	Nb-93
Tc-99	2.13×10^5	β	Mo-98, Mo-99†

*(n,p) reaction.
†Beta decay.

Computer Exercise 22.D displays a series of radionuclides involved in the activation process and decay with release of radiation. In addition, small amounts of fission products and transuranic elements appear in the water as the result of small leaks in cladding and the irradiation of uranium deposits left on fuel rods during fabrication. The isotopes involved are similar to those of concern for HLW.

Leaks of radioactive water from the primary coolant are inevitable and result in contamination of work areas. Also, radioactive equipment must be removed for repair. For such reasons, workers are required to wear elaborate protective clothing and use a variety of materials to prevent spread of contamination. Much of it cannot be cleaned and reused. Contaminated dry trash includes paper, rags, plastics, rubber, wood, glass, and metal. These may be combustible or noncombustible, compactible or noncompactible. Avoidance of contamination of inert materials by radioactive materials is an important technique in waste reduction. The modern trend in nuclear plants is to try to reduce the volume of waste by whatever method is appropriate. Over the period 1980–1998, by a combination of methods, the nuclear industry reduced the LLW volume by a factor of more than 15. Costs of disposal have not decreased proportionately, however, because capital costs tend to be independent of waste volume.

One popular technique is incineration, in which the escaping gases are filtered, and the ash contains most of the radioactivity in a greatly reduced volume. Another method is compaction, with a large press to give a reduced volume and also to make the waste more stable against further disturbance after disposal. "Supercompactors" that reduce the volume greatly are popular. A third approach is grinding or shredding, then mixing the waste with a binder such as concrete or asphalt to form a stable solid.

Purification of the water in the plant, required for re-use or safe release to the environment, gives rise to a variety of wet wastes. They are in the form of solutions, emulsions, slurries, and sludges of both inorganic and organic materials. Two important physical processes that are used are filtration and evaporation. Filters are porous media that take out particles suspended in a liquid. The solid residue collects in the filter that may be a disposable cartridge or may be reusable if backwashed. Figure 22.8 shows the schematic arrangement of a filter in a nuclear plant. The evaporator is simply a vessel with a heated surface over which liquid flows. The vapor is drawn off leaving a sludge in the bottom. Figure 22.9 shows a typical arrangement.

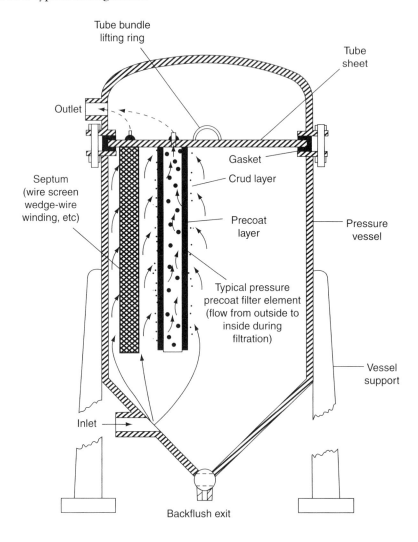

FIGURE 22.8 Disposal-cartridge filter unit used to purify water and collect LLW. (Courtesy ORNL.)

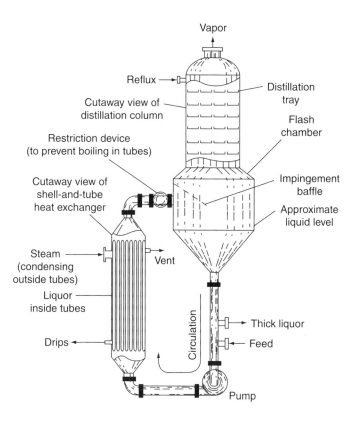

FIGURE 22.9 Natural-circulation evaporator used to concentrate LLW. (Courtesy ORNL.)

The principal chemical treatment of wet LLW is ion exchange. A solution containing ions of waste products contacts a solid such as zeolite (aluminosilicate) or synthetic organic polymer. In the mixed-bed system, the liquid flows down through mixed anion and cation resins. As discussed by Benedict, Pigford, and Levi (see References), ions collected at the top move down until the whole resin bed is saturated, and some ions appear in the effluent, a situation called "breakthrough." Decontamination factors may be as large as 10^5. The resin may be reused by application of an elution process, in which a solution of Na_2SO_4 is passed through the bed to extract the ions from the resin. The resulting waste solution will be smaller than before but will probably be larger than the exchanger. Whether to discard or elute depends on the cost of the ion-exchanger material.

The variety of types of LLW from institutions and industry is indicated in Table 22.3. The institutions include hospitals, medical schools, universities, and research centers. As discussed in Chapter 17, labeled pharmaceuticals and biochemicals are used in medicine for diagnosis and therapy and in biological research to study the physiology of humans, other animals, and plants. Radioactive

Table 22.3 Institutional and Industrial Low-Level Waste Streams (Adapted from the Environmental Impact Statement for NRC 10CFR61)

Fuel fabrication plant	Industrial
Trash	Trash
Process wastes	Source and special nuclear materials[†]
Institutions	Special
Liquid scintillation vials	Isotope production facilities
Liquid wastes	Tritium manufacturing
Biowastes	Accelerator targets
	Sealed sources, e.g., radium

[†]*SNM = Pu, U-233, etc.*

materials are used in schools for studies in physics, chemistry, biology, and engineering and are produced by research reactors and particle accelerators. The industries make various products: (a) radiography sources; (b) irradiation sources; (c) radioisotope thermoelectric generators; (d) radioactive gages; (e) self-illuminating dials, clocks, and signs; (f) static eliminators; (g) smoke detectors; and (h) lightning rods. Radionuclides that often appear in LLW from manufacturing include carbon-14, tritium, radium-226, americium-241, polonium-210, californium-252, and cobalt-60. LLW disposal from the decommissioning of nuclear power reactors is of considerable future importance and is discussed separately in Section 22.8.

Although defined by exclusion, as noted in Section 22.2, low-level radioactive waste generally has low enough activity to be given near-surface disposal. There are a few examples of very small contaminations that can be disregarded for disposal purposes and some highly radioactive materials that cannot be given shallow-land burial.

The method of disposal of low-level radioactive wastes for many years was similar to a landfill practice. Wastes were transported to the disposal site in various containers such as cardboard or wooden boxes and 55-gallon drums and were placed in trenches and covered with earth without much attention to long-term stability.

A total of six commercial and 14 government sites around the United States operated for a number of years until leaks were discovered, and three sites at West Valley, NY, Sheffield, IL, and Maxey Flats, KY, were closed. One problem was subsidence, in which deterioration of the package and contents by entrance of water would cause local holes in the surface of the disposal site. These would fill with water and aggravate the situation. Another difficulty was the "bathtub effect," in which water would enter a trench and not be able to escape rapidly, causing the contents to float and be exposed.

Three remaining sites at Richland, WA, Beatty, NV, and Barnwell, SC, handled all of the LLWs of the country. These sites were more successful, in part because trenches had been designed to allow ample drainage. Managers of the sites, however, became concerned with the waste generators' practices and attempted to reduce the amount of waste accepted. This situation prompted Congress to pass in 1980 the Low-Level Radioactive Waste Policy Act (LLRWPA), followed by the Low-Level Radioactive Waste Policy Amendments Act of 1985. These laws placed responsibility on states for wastes generated within their boundaries but recommended regional disposal. Accordingly, a number of interstate compacts were formed, with several states remaining independent. Figure 22.10 shows the division of the United States into states and compacts. The alignment of states has tended to change over the years.

At the same time, the NRC developed a new rule governing LLW management. Title 10 of the Code of Federal Regulations Part 61 (10CFR61) calls for packaging

FIGURE 22.10 United States interstate compacts for disposal of low-level radioactive wastes. States shaded are unaffiliated. (Courtesy of Afton Associates and LLW Forum.)

of wastes by the generator according to isotope type and specific activity (Ci/m^3). Waste classes A, B, and C are defined in 10CFR61 and increasing levels of security prescribed. Greater-than-Class-C wastes are unsuitable for near-surface disposal and are managed by DOE as equivalent to high-level waste.

Computer Exercise 22.E describes an elementary "expert system" that determines the proper class of a given waste on the basis of half-life and specific activity.

The required degree of waste stability increases with the radioactive content. Limits are placed on the amount of liquid present with the waste, and the use of stronger and more resistant containers is recommended in the interest of protecting the public during the operating period and after closure of the facility.

Regulation 10CFR61 calls for a careful choice of the characteristics of the geology, hydrology, and meteorology of the site to reduce the potential radiation hazard to workers, the public, and the environment. Special efforts are to be made to prevent water from contacting the waste. Performance specifications include a limit of 25 mrems per year whole-body dose of radiation to any member of the public. Monitoring is to be carried out over an institutional surveillance period of 100 y after closure. Measures are to be taken to protect the inadvertent intruder for an additional 500 y. This is a person who might build a house or dig a well on the land. One method is to bury the more highly radioactive material deep in the trench; another is to put a layer of concrete over the wastes.

The use of an alternate technology designed to improve confinement stems from one or more public viewpoints. First is the belief that the limiting dose should be nearer zero or even should be actually zero. Second is the concern that some unexpected event might change the system from the one analyzed. Third is the idea that the knowledge of underground flow is inadequate and not capable of being modeled to the accuracy needed. Fourth is the expectation that there may be human error in the analysis, design, construction, and operation of the facility. It is difficult to refute such opinions, and in some states and interstate compact regions, legislation on additional protection has been passed to make a waste disposal facility acceptable to the public. Some of the concepts being considered as substitutes for shallow land burial are listed.

Belowground vault disposal involves a barrier to migration in the form of a wall such as concrete. It has a drainage channel, a clay top layer and a concrete roof to keep water out, a porous backfill, and a drainage pad for the concrete structure. *Aboveground vault disposal* makes use of slopes on the roof and surrounding earth to assist runoff. The roof substitutes for an earthen cover. *Shaft disposal* uses concrete for a cap and walls and is a variant on the belowground vault that conceivably could be easier to build. *Modular concrete canister disposal* involves a double container, the outer one of concrete, with disposal in a shallow-land site. *Mined-cavity disposal* consists of a vertical shaft going deep in the ground, with radiating corridors at the bottom, similar to the planned disposal system for spent fuel and high-level wastes from reprocessing. It is only applicable to the most active LLWs. *Intermediate-depth disposal* is similar to shallow-land disposal except for the greater trench depth and thickness of cover. *Earth-mounded concrete bunker disposal*, used in France, combines several favorable features. Wastes of higher activity are encased in concrete below grade and those of lower activity are placed in a mound with concrete and clay cap, covered with rock or vegetation to prevent erosion by rainfall.

Each of the interstate compacts embarked on investigations in accord with LLRWPA and 10CFR61. These involved site selection processes, geological assessments, and designs of facilities. The nature of the facilities proposed depended on the location, with shallow land burial deemed adequate for the California desert at Ward Valley, but additional barriers and containers planned for North Carolina in the humid Southeast. However, as the result of concerted opposition taking the form of protests, lawsuits, political action and inaction, and occasional violence, progress was very slow in establishing LLW disposal capability. Thus despite excellent planning and vigorous efforts, and the expenditure of millions of dollars in preparation, political and regulatory factors prevented most of the programs in the United States from coming to fruition. The only sites receiving low-level wastes as of the year 2008 were the Northwestern at Hanford and Barnwell in South Carolina, with certain materials accepted by Envirocare in Utah. Some 36 states have no outlet for LLW and must store it until new disposal facilities are available. A comprehensive review of the situation appears in a 2004 report by the General Accounting Office (GAO). Although noting the scarcity of disposal facilities, GAO was not very concerned. That view was strongly disputed by Alan Pasternak of the Cal Rad Forum (see References). He predicts higher costs for services and curtailment of biomedical research and medical use.

An elementary analysis by use of a spreadsheet of the behavior of a selected set of radionuclides in low-level radioactive waste is described in Computer Exercise 22.F. The effects of storage, decay, and retardation are displayed.

22.8 ENVIRONMENTAL RESTORATION OF DEFENSE SITES

The legacy of World War II and the Cold War includes large amounts of radioactive waste and contamination of many defense sites. Priority was given to weapons production rather than environmental protection, leaving a cleanup task that will take several decades to carry out and cost many billions of dollars.

One of the most pressing problems to solve is the degraded condition of underground tanks at Hanford used to store the waste residue from reprocessing to extract plutonium. The single-wall tanks have leaked, and there is concern for the contamination of the nearby Columbia River. Some of the wastes have been processed to extract the valuable Cs-137 and Sr-90, and the contents of some tanks have been successfully stabilized to prevent hydrogen explosion. Ideally, all of the waste would be transferred to double-layered tanks or immobilized in glass. Similar tanks are located at the Savannah River Plant in South Carolina, where plutonium and tritium were produced.

Transuranic wastes (TRU) consist of materials and equipment contaminated by small amounts of plutonium. They have been stored or temporarily buried over the years, especially at Hanford, Idaho Falls, Los Alamos, Oak Ridge, and Savannah River. These wastes are scheduled to be buried in the Waste Isolation Pilot Plant (WIPP), a repository near Carlsbad, NM, that opened in 1999. The geological medium is salt, which has several advantages—its presence demonstrates the absence of water and it is plastic, self-sealing under pressure. The TRU is packaged in 55-gallon drums and shipped to WIPP in a cylindrical cask called TRUPAC II, which contains seven drums in each of two layers. The waste is buried approximately 2,160 ft (658 m) below the surface. Construction of WIPP was under the supervision of DOE, with advice by the National Research Council, and regulation by the EPA. Performance assessment was done by Sandia. For details of the roles of the various organizations, see References.

The monumental challenge of environmental restoration of sites used in the United States defense program is being addressed by the DOE. It has been recognized that it is not feasible to completely decontaminate the sites. Instead, cleanup to an extent practical is followed by "stewardship," involving isolation, monitoring, and maintenance of certain locations for a very long period. To achieve the goal of protection of the public and the environment, the DOE Environmental Management (EM) program has initiated research on new efficient technologies to handle radioactive materials. As described in the DOE document Five Year Plan (see References), the mission of EM is the safe and successful cleanup of facilities, including stabilization of tank waste for treatment, disposal of all types of waste, and remediation of sites.

22.9 NUCLEAR POWER PLANT DECOMMISSIONING

"Decommissioning," a naval term meaning to remove from service e.g., a ship, is applied to actions taken at the end of the useful life of a nuclear power plant (30 to 40 y). The process begins at shutdown of the reactor and ends with disposal of radioactive components in a way that protects the public. LLW disposal from dismantled reactors will be a major problem in decades to come.

The first action is to remove and dispose of the spent nuclear fuel. Several choices of what to do with the remainder of the plant are available. The options as identified formally by the NRC are (a) SAFSTOR or mothballing, in which some decontamination is effected, the plant is closed up, and then monitored and guarded for a very long period, perhaps indefinitely; (b) ENTOMB or entombment, in which concrete and steel protective barriers are placed around the most radioactive equipment, sealing it to prevent release of radioactivity, again with some surveillance; (c) DECON or immediate dismantlement, in which decontamination is followed by destruction, with all material sent to a LLW disposal site; (d) delayed dismantlement, the same as the previous case, but with a time lapse of a number of years to reduce personnel exposure. The distinction among these various options is blurred if it is assumed that the facility must eventually be disassembled. It becomes more a question of "when." Aside from the aesthetic impact of an essentially abandoned facility, there is a potential environment problem related to the finite life of structural materials.

Operation of the reactor over a long period of time will have resulted in neutron activation, particularly of the reactor vessel and its stainless steel internal parts. Contamination of other equipment in the system will include the same isotopes that are of concern in LLW disposal. Various techniques are used to decontaminate—washing with chemicals, brushing, sand blasting, and ultrasonic vibration. To cut components down to manageable size, acetylene torches and plasma arcs are used. Because such operations involve radiation exposure to workers, a great deal of preplanning, special protective devices, and extra manpower are required. A very large volume of waste is generated. Some of it may be too active to put into a LLW disposal site but will not qualify for disposal in a high-level waste repository. Cobalt-60 dominates for the first 50 y, after which the isotopes of concern are 76,000-y nickel-59 and 24,000-y niobium-94.

The NRC requires nuclear plant owners to provide a License Termination Plan and to set aside funds for decommissioning. A standardized cost-estimation procedure has been developed. Costs vary with units but are $300 million or more. This cost, a small fraction of the value of electricity generated over the reactor life, is borne by the consumers of electrical power. Data on reactors that have been decommissioned are provided by NRC (see References).

An option that has not yet been fully explored is "intact" decommissioning, in which the highly radioactive region of the system would be sealed off, making surveillance unnecessary. The virtues claimed are low cost and low exposure. Ultimately, renewal of the license after replacing all of the worn-out components may be the best solution.

A number of reactors will need to be decommissioned in the first quarter of the 21st century. Factors that will determine action include the degree of success in reactor life extension, license renewal, and the general attitude of the public about the disposal of material from nuclear stations as low-level radioactive waste.

22.10 SUMMARY

Radioactive wastes arise from a great variety of sources, including the nuclear fuel cycle, and from beneficial uses of isotopes and radiation by institutions.

Spent fuel contains uranium, plutonium, and highly radioactive fission products. In the United States spent fuel is accumulating, awaiting the development of a high-level waste repository. A multibarrier system involving packaging and geological media will provide protection of the public over the centuries the waste must be isolated. The favored method of disposal is in a mined cavity deep underground. In other countries, reprocessing the fuel assemblies permits recycling of materials and disposal of smaller volumes of solidified waste. Transportation of wastes is by casks and containers designed to withstand severe accidents.

LLWs come from research and medical procedures and from a variety of activation and fission sources at a reactor site. They generally can be given near-surface burial. Isotopes of special interest are cobalt-60 and cesium-137. Transuranic wastes are being disposed of in the Waste Isolation Pilot Plant. Establishment of regional disposal sites by interstate compacts has generally been unsuccessful in the United States. Decontamination of defense sites will be long and costly. Decommissioning of reactors in the future will contribute a great deal of low-level radioactive waste.

22.11 EXERCISES

22.1 Compare the specific activities (dps/g) of natural uranium and slightly enriched fuel, including the effect of uranium-234. Note the natural uranium density of 18.9 g/cm^3 and the half-lives and atom abundances in percent for the three isotopes:

Isotope	Half-life (y)	Natural	Enriched
U-235	7.04×10^8	0.720	3.0
U-238	4.47×10^9	99.2745	96.964
U.234	2.45×10^5	0.0055	0.036

What fraction of the activity is due to uranium-234 in each case?

22.2 With the data below (a) calculate the power capacity of all United States PWRs, Boiling Water Reactors (BWRs), and the Light Water Reactor (LWR) total, and (b) estimate the total annual amount of solid radioactive waste produced by United States power reactors.

	PWR			BWR	
No.	Average Power (MWe)	Waste (m^3/GWe-y)	No.	Average Power (MWe)	Waste (m^3/GWe-y)
69	949.33	23.2	35	929.31	91.5

22.3 A batch of radioactive waste from a processing plant contains the following isotopes:

Isotope	Half-life	Fission Yield, %
I-131	8.04 d	2.9
Ce-141	32.50 d	6
Ce-144	284.6 d	6.1
Cs-137	30.2 y	5.9
I-129	1.7×10^7 y	1

Form the products of the decay constants (in s^{-1}) and fission yields (in %) to serve as relative initial activities of the isotopes. Find the times where successive semilog graphs of activity would intersect by use of equality of activities (e.g., $A_n = A_{n+1}$).

22.4 Traces of plutonium remain in certain waste solutions. If the initial concentration of Pu-239 in water were 100 parts per million ($\mu g/g$), find how much of the water would have to be evaporated to make the solution critical, neglecting neutron leakage as if the container were very large. Note: for H, $\sigma_a = 0.332$; for Pu, $\sigma_f = 752$, $\sigma_a = 1022$, $v = 2.88$.

22.5 If the maximum permissible concentration of Kr-85 in air is 1.5×10^{-9} $\mu Ci/cm^3$, and the yearly reactor production rate is 5×10^5 Ci, what is a safe diluent air volume flow rate (in cm^3/s and ft^3/min) at the exit of the stack? Discuss the implications of these numbers in terms of protection of the public.

22.6 Calculate the decay heat from a single fuel assembly of the total of 180 in a 3,000-MWt reactor at 1 day after shutdown of the reactor. How much longer is required for the heat generation rate to go down an additional factor of 2?

22.7 Data on fission products (in %) to accompany numbers in Figure 22.2 are as follows: U-238, 0.16; U-235, 1.98, Pu-239, 1.21; and Pu-241, 0.15.
(a) Calculate the percentages of total power caused by each fissionable isotope.
(b) Assuming that one third of the 180 fuel assemblies in the reactor are removed each year and that each contains 470 kg of U, find what weight of fission products the 60 assemblies contain.
(c) What mass of fission products would be produced annually in the whole reactor if operated at its full rating of 3,000 MWt, knowing that 1.1 grams of fuel fissions per MWd?
(d) Deduce a capacity factor (actual energy divided by rated energy) from the results of (b) and (c) above.

22.8 Assume that high-level wastes should be secured for a time sufficient for decay to reduce the concentrations by a factor of 10^{10}. How many half-lives does this require? How long is this in years for strontium-90? For cesium-137? For plutonium-239?

22.9 A 55-gallon drum contains an isotope with 1 MeV gamma ray, distributed uniformly with activity 100 $\mu Ci/cm^3$. For purposes of radiation protection planning, estimate the radiation flux at the surface, treating the container as a sphere of equal volume of water, and neglecting buildup. (a) Show that the gamma flux at the surface, radius R, is given by $SRP_e/3$, where

P_e is the escape probability and S is the source strength in dps/cm^3.
(b) With attenuation coefficient Σ and $x = \Sigma R$

$$P_e = \left[3/(8x^3)\right]\left[2x^2 - 1 + (1 + 2x)(\exp(-2x))\right]$$

calculate the gamma flux at R.

Computer Exercises

22.A   For a computer display of a stylized water pool for the storage of spent fuel at a nuclear plant, load and run the program FUELPOOL.

22.B   If buried radioactive waste is dissolved at a constant rate by water infiltration, it will be released as a square pulse. As the pulse migrates in an aquifer with some effective speed, the number of nuclei decreases because of decay. Program WASTPULS displays the motion in time. Load and run the program, trying a variety of combinations of distances, speeds, and half-lives.

22.C   The transport of a waste radionuclide by groundwater involves the flow with retardation because of holdup in pores. A process called dispersion causes an initial square pulse to be rounded as it moves along. Computer program WTT gives numerical values of the contaminant concentration observed at a point in space for various times. Run the program with the default values, then change individual parameters such as dispersivity to observe effects.

22.D   The sequence of products resulting from neutron capture in a nonradioactive nucleus is displayed in the program ACTIVE. Included are the activation product and the residual nucleus after decay. Load and run the program to observe the sequence. Suggest a set of specific nuclear species for which the diagram is appropriate, giving cross sections and half-lives wherever possible.

22.E   The NRC specifies in the *Code of Federal Regulations* 10 *Energy Part* 61 *Section* .55 (10CFR61.55) a classification scheme for low-level radioactive waste. The radionuclides present and their concentrations determine whether a shipment is Class A, B, or C. Computer program LLWES (LLW expert system) provides an easy way to classify a given waste. The program also illustrates an expert system, which yields answers by a specialist to questions by a worker. Load and run the program, then use the menus to learn about the NRC's rule and to test the expert's knowledge. Select some isotope or combination of isotopes and assign specific activity values to find out the classification. Note the effect of increasing or decreasing the concentration significantly.

22.12 **REFERENCES**

Robert G. Cochran and Nicholas Tsoulfanidis, *The Nuclear Fuel Cycle: Analysis and Management*, 2nd Ed., American Nuclear Society, La Grange Park, IL, 1999.

The Nuclear Waste Primer , by The League of Women Voters Education Fund, Lyons & Burford, New York, 1993. Brief and elementary information.

Raymond L. Murray, "Radioactive Waste Storage and Disposal," *Proceedings of the IEEE*, Vol. 74, No. 4, April 1986, p. 552. A survey article that covers all aspects.

Raymond L. Murray, *Understanding Radioactive Waste*, 5th Ed., Battelle Press, Columbus, OH, 2003. An elementary survey intended to answer typical questions by the student or the public.

A. M. Platt, J. V. Robinson, and O. F. Hill, *The Nuclear Fact Book*, Harwood Academic Publishers, New York, 1985. A large amount of useful data on wastes is included.

Donald C. Stewart, *Data for Radioactive Waste Management and Nuclear Applications*, John Wiley & Sons, New York, 1985.

Roy G. Post, Ed., Proceedings of the Symposium on Waste Management, Tucson, AZ, WM Symposia, Inc. An annual event with a collection of papers on all aspects of radioactive waste around the world.

Alan Moghissi, Herschel W. Godbee, and Sue A. Hobart, Editors, *Radioactive Waste Technology*, The American Society of Mechanical Engineers, New York, 1986. Sponsored by ASME and the American Nuclear Society.

Mill tailings (Nuclear Regulatory Commission)
http://www.nrc.gov/reading-rm/doc-collections/nuregs/brochures/br0216/#mill_tailings

Mixed Low-Level Waste (Environmental Protection Agency)
http://www.epa.gov/radiation/mixed-waste

Status of "Below Regulatory Concern" (DOE)
http://homer.ornl.gov/nuclearsafety/nsea/oepa/guidance/radwaste/brc.pdf

Radioactive Waste Management
http://www.uic.com.au/wast.htm
From Australian Uranium Association. Source of data for Table 22.1.

Nuclear Waste Policy Act As Amended
http://ymp.gov/documents/nwpa/css/nwpa_2004.pdf
High-level waste and spent fuel.

Nuclear Waste Policy Act of 1982
http://ocrwm.doe.gov/documents/nwpa/css/nwpa.htm
Text of law on HLW and spent fuel. Includes update to 1987 policy.

Office of Civilian Radioactive Waste Management
http://www.ocrwm.doe.gov
Comprehensive discussion of repository. Select Yucca Mountain Repository, Transportation of Nuclear Wastes, or License Application, with links.

"Investigation closes on project's questionable QA records," *Nuclear News*, June 2006, pp. 52–53.

A. Alan Moghissi, "The origin of the EPA's 10,000-year Time Frame for the High-level Waste Repository," *Nuclear News,* February 2006, p. 41.

About Yucca Mountain and the Standards
http://www.epa.gov/radiation/yucca/about.html
Answers to questions and discussion of legal aspects.

The EPA Radiation Standard for Spent-Fuel Storage in a Geological Repository
ANS position statement and background information, November 2006
http://www.ans.org/pi/ps/docs/ps81.pdf
http://www.ans.org/pi/ps/docs/ps81-bi.pdf

Yucca Mountain Site Suitability Evaluation
http://www.ocrwm.doe.gov/documents/sse_a/index.htm
DOE report of 2002.

Robert E. Berlin and Catherine C. Stanton, *Radioactive Waste Management*, John Wiley & Sons,
New York, 1989. Many tables, charts, and diagrams on all types of nuclear wastes.

William Bebbington, "The Reprocessing of Nuclear Fuels," *Scientific American*, December 1976,
p. 30.

Manson Benedict, Thomas H. Pigford, and Hans Wolfgang Levi, *Nuclear Chemical Engineering*,
2nd Ed., McGraw-Hill, New York, 1981.

Nuclear Wastes: Technologies for Separation and Transmutation, National Academy Press,
Washington, DC, 1996. Study by a National Research Council committee on the feasibility of
reducing long-lived isotopes by reprocessing followed by neutron irradiation.

A Roadmap for Development of ATW Technology
http://www.osti.gov/bridge/product.biblio.jsp?osti_id=750787
Outdated plans.

DOE's Advanced Accelerator Applications Program
http://www.ne.doe.gov/pdfFiles/AAARptConMarch2001.pdf
The combination of ATW and APT into AAA.

United States Nuclear Regulatory Commission, Final Environmental Impact Statement on 10 CFR
Part 61 "Licensing Requirements for Land Disposal of Radioactive Waste," NUREG-0945,
Vols. 1–3, November 1982.

J. Howard Kittel, Editor, *Near-Surface Land Disposal*, *Radioactive Waste Management Handbook*, Vol. 1, Harwood, Chur, Switzerland, 1989.

Performance Assessment for LLW Disposal Facilities
http://www.osti.gov/bridge
Search on 563204 for article by S. M. Birk of INEEL.

Status of Low-Level Radioactive Waste Disposal
http://www.gao.gov
Search on RCED-99-238 for 1999 status.

Low-Level Radioactive Waste Disposal
http://www.gao.gov/new.items/d04604.pdf
Report GAO-04-604 on the status of disposal facilities (2004).

"Cal Rod Forum's Alan Pasternak: Time is Running Out for a Permanent LLW solution,"
Nuclear News, December 2004, p. 22. Interview on disposal facilities.

Improving the Regulation and Management of Low-Activity Radioactive Wastes
http://books.nap.edu/openbook.php?record_id=11595&page=R1

Nancy J. Zacha, "Low-level Radioactive Waste Disposal: Are we Having a Crisis Yet?"
Nuclear News, August 2007, p. 29.

Thomas E. Carleson, Nathan A. Chipman, and Chien M. Wai, Editors, *Separation Techniques in Nuclear Waste Management*, CRC Press, Boca Raton, FL, 1995. Describes many processes in various areas of the world.

Robert Noyes, *Nuclear Waste Cleanup Technology and Opportunities*, Noyes Publications, Park Ridge, NJ, 1995. Survey of the national problem and available technologies based on government reports.

Roy E. Gephart and Regina E. Lundgren, *Hanford Tank Cleanup: A Guide to Understanding the Technical Issues*, Battelle Press, Columbus, OH, 1998.

The Waste Isolation Pilot Plant, National Academy Press, Washington, DC 1996. A review by a committee of the National Research Council.

Improving Operations and Long-Term Safety of the Waste Isolation Pilot Plant
http://books.nap.edu/openbook.php?record_id=10143&page=R1
Online book of National Academies Press, 2001.

EPA's WIPP Program
http://www.epa.gov/radiation/wipp
Select Frequent Questions.

Science @ WIPP
http://www.wipp.energy.gov/science/index.htm
Physics and biology research underground.

DOE Environmental Management (EM)
http://www.em.doe.gov/Pages/EMHome.aspx
Select Mission, History, or EM Five-Year Plan (2008-2012).

Methodology and Technology of Decommissioning Nuclear Facilities, IAEA Technical Reports Series, No. 267, Vienna, 1986.

Decommissioning of Nuclear Reactors
http://www.nrc.gov/reading-rm/doc-collections/fact-sheets/decommissioning.html
Regulations and history from NRC.

Standard Review Plan for License Termination
http://www.nrc.gov/reading-rm/doc-collections/nuregs/staff/sr1700
NRC requirements for decommissioning, 2003.

Decommissioning Nuclear Facilities
http://www.uic.com.au/nip13.htm
Information from Europe, Japan, and the United States.

Bernard L. Cohen, *Before It's Too Late: A Scientist's Case for Nuclear Energy*, Plenum Press, New York, 1983. Includes data and discussion of radiation, risk, and radioactive waste. A powerful statement by a strong advocate of nuclear power.

Marvin Resnikoff, *The Next Nuclear Gamble: Transportation and Storage of Nuclear Waste*, Council on Economic Priorities, New York, 1983. Written by an opponent and critic of nuclear power.

Frederick Frankena and Joann Koelln Frankena, *Radioactive Waste as a Social and Political Issue: A Bibliography*, AMS Press, New York, 1991. Inclusive of pro- and anti-nuclear references on several technical aspects.

Laws, Regulations, and Organizations[†]

23

After World War II, Congress addressed the problem of exploiting the new source of energy for peaceful purposes. This led to the Atomic Energy Act of 1946, which was expanded in 1954. The Atomic Energy Commission (AEC) had functions of promotion and regulation for 28 y. Compliance with licensing rules plays an important role in the operation of any nuclear facility. A number of other organizations have evolved to provide technical information, develop standards, protect against diversion of nuclear materials, improve nuclear power operations, and perform private research and development.

23.1 THE ATOMIC ENERGY ACTS

The first law in the United States dealing with control of nuclear energy was the Atomic Energy Act of 1946. Issues of the times were involvement of the military, security of information, and freedom of scientists to do research (see References).

In the declaration of policy, the Act says, ". . . the development and utilization of atomic energy shall, so far as practicable, be directed toward improving the public welfare, increasing the standard of living, strengthening free competition in private enterprise, and promoting world peace." The stated purposes of the Act were to carry out that policy through both private and federal research and development, to control information and fissionable material, and to provide regular reports to Congress. Special mention was given to the distribution of "byproduct material," which includes the radioactive substances used for medical therapy and for research. The act created the United States Atomic Energy Commission (AEC), consisting of five commissioners and a general manager. The AEC was given broad powers to preserve national security while advancing

---

[†]Thanks are due Angelina Howard of the Nuclear Energy Institute for helpful information.

the nuclear field. A Joint Committee on Atomic Energy (JCAE) provided oversight for the new AEC. It included nine members each from the Senate and the House. Advice to the AEC was provided by the civilian General Advisory Committee and the Military Liaison Committee.

The Atomic Energy Act of 1954 revised and liberalized the previous legislation and expanded the AEC's role in disseminating unclassified information while retaining control of restricted weapons data. The groundwork was laid for a national program of reactor research and development with cooperation between the AEC and industry, including some degree of private ownership. The act authorized sharing of atomic technology with other countries, spelled out licensing procedures for the use of nuclear materials, and clarified the status of patents and inventions.

The powerful AEC carried out its missions of supplying material for defense, promoting beneficial applications, and regulating uses in the interests of public health and safety. It managed some 50 sites around the United States. Seven of the sites were labeled "national laboratories," each with many R&D projects under way. The AEC owned the facilities, but contractors operated them. For example, Union Carbide Corporation had charge of Oak Ridge National Laboratory. During the Cold War of the late 1940s and early 1950s new plutonium and enriched uranium plants were built, weapons tests were conducted in the South Pacific, and a major uranium exploration effort was begun. Under AEC sponsorship a successful power reactor research and development program was carried out. Both the United States and the U.S.S.R. developed the hydrogen bomb, and the nuclear arms race escalated.

Critics pointed out that the promotional and regulatory functions of the AEC were in conflict, despite an attempt to separate them administratively. Eventually, in 1974, the activities of the AEC were divided between two new agencies, the Energy Research and Development Administration and the Nuclear Regulatory Commission (NRC).

23.2 THE ENVIRONMENTAL PROTECTION AGENCY

The National Environmental Policy Act of 1969 (NEPA) included a Council on Environmental Quality in the executive branch and required environmental impact statements on all federal projects. The Environmental Protection Agency (EPA) was then proposed and accepted. A prominent part of EPA is the administration of the Superfund to clean up old waste sites. EPA has responsibility for standards on hazardous, solid, and radioactive wastes. EPA also sets standards for radiation protection that are used by the NRC in its licensing and regulation.

The topics of EPA as listed in its 2008 budget (see References) are clean air and global climate change, clean and safe water, land preservation and restoration, healthy communities and ecosystems, and compliance and environmental stewardship. The activity of EPA that is most relevant to nuclear energy is Radiation Protection (see References). EPA also has a research program on the causes and effects of acid rain. More recently it has initiated a program related to terrorism, assuring capabilities for detection and recovery.

The EPA provides key nuclear regulations, appearing in Code of Federal Regulations. In Title 40 Part 61 the release of radionuclides is limited to a value calculated to be less than 10 mrem/y. The computer code to use is specified. In 40CFR191 dealing with disposal of high-level wastes and spent fuel it is required that during the first 10,000 y no member of the public will receive an annual dose larger than 15 mrems (150 µSv). In addition, release limits are given for several radionuclides, expressed as curies per thousand metric tons of heavy metal (U, Pu, etc.). The lowest figure, 10 Ci, is for Th-230 or Th-232; most isotopes are at 100 Ci; the highest figure 10,000 Ci is for Tc-99. Complete copies of the regulations are available online (see References).

23.3 THE NUCLEAR REGULATORY COMMISSION

The federal government through the NRC has the authority to license and regulate nuclear facilities of all types, from a multireactor power station down to isotope research in an individual laboratory. The Office of Nuclear Reactor Regulation of the NRC requires applicants for a reactor license to submit a voluminous and detailed Safety Analysis Report and an Environmental Report. These documents provide the basis for issuance of a construction permit, and later when the plant is completed, an operating license. The process involves several steps: review of the application by the NRC staff; an independent safety evaluation by the Advisory Committee on Reactor Safeguards (ACRS); the holding of public hearings in the vicinity of the proposed plant by an Atomic Safety and Licensing Board (ASLB); and the testing of qualifications of the people who will operate the plant. In addition to completing a written examination, operators are tested on the plant's simulator and on their knowledge of the location and operation of equipment. The NRC and the Federal Emergency Management Agency (FEMA) collaborate in setting criteria for emergency response programs that are developed by the utilities, state government, and local government. The five NRC commissioners make the final decision on low-power operation and full-power operation.

Once a plant is licensed, the Office of Inspection and Enforcement has oversight. The nuclear operations are subject to continual scrutiny by the resident inspector and periodic inspection by teams from the regional NRC office. Training of operating personnel goes on continuously, with one shift in training while other shifts run the plant. Periodic exercises of the emergency plan for the 10-mile radius zone about the plant are conducted. Nuclear stations are required to report unusual events to the NRC promptly. The NRC maintains a nuclear engineer on duty at all times to receive calls and take action as needed.

The staff routinely reviews all incidents. For a number of years NRC administered a program called Systematic Assessment of Licensee Performance (SALP). A new substitute is the Reactor Oversight Process (see References), which involves monitoring performance in three areas—reactor safety, radiation safety, and safeguards (against security threats). The process gives attention to human performance, safety culture, and corrective actions. Plants provide reports to the NRC on a set of performance indicators. Companies are subject to fines for lack of compliance with regulations, and if necessary, NRC can shut a plant down. The principal reference is the *Code of Federal Regulations Title 10, Energy*. Key sections of that annually updated book are: Part 20, Standards for Protection Against Radiation; Part 50, Domestic Licensing of Production and Utilization Facilities; Part 60, Disposal of High Level Radioactive Wastes in Geological Repositories; Part 61, Licensing Requirements for Land Disposal of Radioactive Waste; Part 63, Disposal of High-Level Radioactive Wastes in a Geologic Repository at Yucca Mountain, Nevada; Part 71, Packaging and Transportation of Radioactive Material; and Part 100, Reactor Site Criteria. Part 50 has a number of appendices covering criteria for general design, quality assurance, emergency plans, emergency core cooling system, and fire protection. For Web access to the complete document, see References. Standards for NRC licensed facilities and DOE sites in 10CFR61 include annual dose limits of 25 mrems to the whole body or 75 mrems to the thyroid or 25 mrems to any other organ of members of the public.

The NRC reviewed its security programs and made enhancements at many facilities. The actions are documented in a report titled "Protecting Our Nation—Since 9-11-01."

Other NRC references are the Regulatory Guides ("Reg. Guides"), each consisting of many pages of instructions. Titles appear on NRC's Web site (see References), with ability to download key Guides.

NRC's policies and practices underwent a transition. Traditionally, evaluation of compliance was based on deterministic design information that involved engineering data and analysis. It also was prescriptive in nature, in which specific instructions to nuclear facilities were provided (e.g., Appendix A of 10CFR50 that covers general design criteria). In 1995, NRC adopted risk-informed regulation. Probabilistic risk assessment (PRA, see Section 19.4) was to be used to decide the most important areas for attention in terms of safety. NRC also endorsed the idea of performance-based regulation, in which goals of performance are provided, but the utilities are able to decide how to achieve the goals. The combination of approaches is designated as Risk-Informed Performance-Based regulation. Definitions and discussion of the various approaches to regulation appear in a white paper on the Internet (see References).

An example of regulation that required much effort to implement was the Maintenance Rule, a brief statement by NRC in 1996 of expectations on monitoring the performance of structures, systems, and components (SSC) with respect to maintenance. PRA was not mandated, but needed to define the scope of safety significance. The nuclear industry responded with detailed guidance documents.

The NRC can delegate some of its authority to individual states by negotiation. An Agreement State can develop its own regulations for users of radiation and radioactive material (i.e., facilities other than those of the nuclear fuel cycle). However, the regulations must be compatible with, and no less strict than, those of the NRC.

In addition to its licensing and regulatory activities, the NRC carries out an extensive research program related to radiation protection, nuclear safety, and radioactive waste disposal. Part of the research is "in-house" and part is through contractors to the NRC. The workload of NRC has increased greatly as many nuclear power plants seek license extension and as applications for licenses of new reactors have been received.

The Office of Nuclear Material Safety and Safeguards has responsibility for interaction with, and reporting to, the International Atomic Energy Agency on fissionable material for safeguards purposes.

23.4 **THE DEPARTMENT OF ENERGY**

The federal government has legal responsibility for assuring adequate energy supply through the Department of Energy (DOE). This cabinet-level department was formed in 1977 from several other groups and is headed by the Secretary of Energy.

The agency supports basic research in science and engineering and engages in energy technology development. It also manages national defense programs such as nuclear weapons design, development, and testing. DOE operates several multiprogram laboratories[†] and many smaller facilities around the United States. The Office of Civilian Radioactive Waste Management has responsibility for carrying out the Nuclear Waste Policy Act of 1982, which involves management of the Waste Fund, repository site selection, and the design of a storage facility. The broader scope of DOE activities can be seen from some of the sections in the Annual Performance and Accountability Report (to Congress): Environmental Cleanup, Security, Safety and Health, Project Management, Oversight of Contractors, Information Technology Management, Human Capital, and Stockpile Surveillance & Testing. The document highlights the need for a balanced and diversified mix of energy sources, including conservation, coal, and nuclear power. The Report describes accomplishments in 59 program goals.

A report that describes United States energy policy is *National Energy Strategy* (see References). Goals of this plan are listed as: (I) Improve the efficiency of the energy system; (II) Ensure against energy disruption; (III) Promote energy

---

[†]Argonne National Laboratory, Brookhaven National Laboratory, Idaho National Laboratory, Lawrence Berkeley Laboratory, Lawrence Livermore National Laboratory, Los Alamos National Laboratory, Oak Ridge National Laboratory, Pacific Northwest Laboratory, and Sandia National Laboratories.

production and use in ways that respect health and environmental values; (IV) Expand future energy choices; and (V) Cooperate internationally on global issues.

The Energy Independence and Security Act of 2007 has these objectives: improve vehicle fuel economy, reduce dependence on foreign oil, and reduce projected CO_2 emissions through "... greater use of cleaner coal technology, solar and wind energy, and clean, safe nuclear power."

23.5 INTERNATIONAL ATOMIC ENERGY AGENCY

In 1953, President Dwight Eisenhower gave a speech titled "Atoms for Peace" that had an important influence on all aspects of nuclear energy. After describing the danger of nuclear war, he proposed the formation of an Atomic Energy Agency that would be responsible for receiving contributed fissionable materials, storing them, and making them available for peaceful purposes. He hoped to thus prevent the proliferation of nuclear weapons. For a copy of the speech, (see References).

In response to the speech, the United Nations established the International Atomic Energy Agency (IAEA), through a statute ratified by the necessary number of countries in 1957. More than 130 nations support and participate in the programs administered by headquarters in Vienna. The objective of the IAEA is "to accelerate and enlarge the contribution of atomic energy to peace, health and prosperity throughout the world."

One of the actions of the United States was to supply enriched uranium to other countries for use in research reactors. Universities (Pennsylvania State and North Carolina State) and a national laboratory (Argonne) provided formal training in nuclear technology for scientists and engineers of many nations.

International conferences sponsored by IAEA were held in 1955, 1958, 1964, and 1971 at Geneva, with all countries of the world invited to participate. The first of these revealed the progress made by the U.S.S.R. in nuclear R&D.

The main functions of IAEA are as follows:

(a) To help its members develop nuclear applications to agriculture, medicine, science, and industry. Mechanisms are conferences, expert advisor visits, publications, fellowships, and the supply of nuclear materials and equipment. Special emphasis is placed on isotopes and radiation. Local research on the country's problems is encouraged. Nuclear programs sponsored by IAEA often help strengthen basic science in developing countries, even if they are not yet ready for nuclear power.

(b) To administer a system of international safeguards to prevent diversion of nuclear materials to military purposes. This involves the review by the IAEA of reports by individual countries on their fissionable material inventories and on-the-spot inspections of facilities. Included are reactors, fuel fabrication plants, and reprocessing facilities. Such monitoring is done for countries that signed the Non-Proliferation Treaty of 1968 and do not have nuclear weapons. The form of the monitoring is set by agreement. If a serious violation is found, the offending nation could lose its benefits from the IAEA.

IAEA is one of the largest science publishers in the world because it sponsors a number of symposia on nuclear subjects each year and publishes the proceedings of each. The outlet in the United States is Bernan UNIPUB. IAEA also promotes international rules, for example, in the area of transportation safety.

Recent initiatives of the IAEA include the establishment of agreements with countries on the application of safeguards. A large number of safeguards seminars are given each year. Annual reports on nuclear-related information are available online (see References).

Unfortunately, IAEA has had difficulties in making inspections in certain countries. Over several years Iraq either refused entry or limited access to its facilities. A similar situation existed in North Korea and Iran.

23.6 INSTITUTE OF NUCLEAR POWER OPERATIONS

Many organizations contribute to the safety and effectiveness of nuclear power generation, not the least of which are the operating companies themselves. One organization, however, provides a broad stimulus to excellence that warrants special attention. The Institute of Nuclear Power Operations (INPO) is the industry's self-regulation organization. Its objective is to promote the highest level of safety and reliability in the operation of nuclear electric generating plants.

Based in Atlanta, GA, INPO has approximately 350 employees, a number of whom are on loan from industry. It was founded by the electric utilities operating nuclear plants in 1979 shortly after the Three Mile Island accident. Corporate leaders saw the need for the utilities to be actively responsible for safety rather than merely complying with NRC regulations. The Kemeny Commission, in its report on the accident, recognized the need for forming INPO. All utilities that have nuclear plants are members of INPO as a private nonprofit organization. Its programs embrace nuclear systems vendors and utilities outside the United States. In its work to promote excellence in safety and reliability of operation of nuclear electric generating plants it has four cornerstone programs. It evaluates the operational performance of utilities, analyzes plant events and distributes lessons-learned information, evaluates training and provides accreditation, and assists member companies. More than 100 reactors operated by more than 40 utilities are influenced by INPO's activities. INPO has no role as an advocate of nuclear power but recognizes that excellent performance is vital to public confidence.

Evaluations are performed regularly by teams of INPO staff members and personnel from other utilities. They visit a facility for 2 weeks—reviewing, observing, and discussing activities. Day-to-day operations and maintenance programs are examined, along with management practices. Candid interactions lead to an evaluation report that identifies both strengths and areas needing improvement. Such evaluations are shared only with the utility for its use in improving performance. This ability to communicate freely is regarded as very important.

Data on operational events are obtained by the INPO program called Significant Event Evaluation and Information Network (SEE-IN), established in 1980. It is designed to share experiences. INPO receives reports from the utilities and other sources, studies them for possible precursors of severe problems, and sends out information on a computer-based communication system NUCLEAR NETWORK. INPO also prepares formal documents including Significant Event Reports (SERs) that describe the most important occurrences, and Significant Operating Experience Reports (SOERs) that are comprehensive reviews of key topics. The latter documents provide recommendations for solutions in areas such as radiological protection, training, and maintenance practices.

An enormous amount of information on nuclear power plant equipment has been collected and put into INPO's database Equipment Performance Information Exchange (EPIX). Events and incidents involving equipment failure are reported and analyzed for root causes and ways to prevent future problems. A continuous flow of information to and from INPO keeps the industry up to date on equipment performance. Of special value is the ability of a utility to quickly obtain information on the solution of an equipment problem by access to EPIX.

In the area of training of personnel, INPO administers the National Academy for Nuclear Training. The Academy's objective is to assure ample knowledge and skill on the part of nuclear personnel and to promote professionalism among nuclear workers. INPO issues guidelines on training in classes and on simulators. It reviews the training programs set up by utilities for supervisors, shift technical advisors, operators, maintenance personnel, and technicians. It also manages the accreditation done by the independent National Nuclear Accrediting Board. The Academy provides workshops, meetings, training courses, and reports, all aimed at the improvement of performance by workers, supervisors, and management.

Assistance programs that continually evolve to meet the changing needs of the nuclear industry help member utilities improve nuclear operations. Through assistance visits, working meetings, workshops, technical documents, and loan of personnel, INPO fosters comparison and exchange of successful methods among members. INPO carefully watches a set of performance indicators, which are quantified trends that measure success in achieving excellence. Examples are plant availability to produce electricity, industrial safety, safety system performance, fuel reliability, unplanned automatic scrams, radiation exposure, and volume of radioactive waste. With input from its Board of Directors and Advisory Council, INPO assists in setting target goals for the industry, with distinctions between PWRs and BWRs as appropriate. Figure 23.1 shows trends over the years of two of the key performance indicators, unplanned scrams and radiation exposure, as a composite for the two types of reactor.

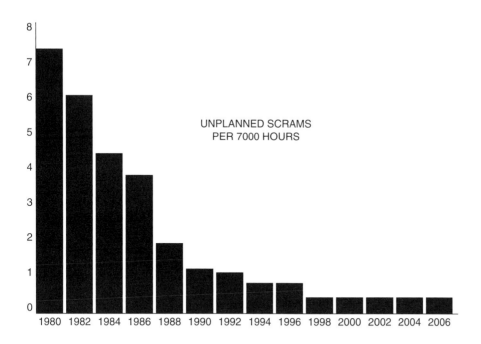

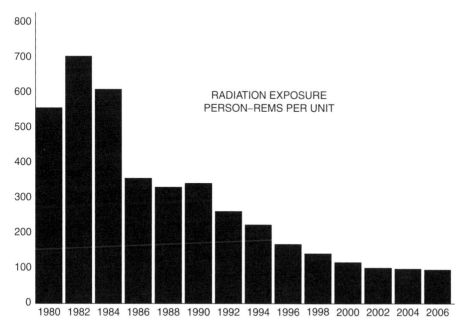

FIGURE 23.1 Trends in two nuclear utility key performance indicators. (Adapted from INPO data).

The organization welcomes utilities from other countries as participants who receive benefits of information exchange but are not subject to evaluations or accreditations. Other countries often assign liaison engineers to the INPO staff. International cooperation on nuclear power is stimulated by an allied organization called the World Association of Nuclear Operators (WANO), with centers in Atlanta, Paris, Moscow, and Tokyo, and a coordinating center in London (see References). It establishes the performance indicators and facilitates communication, comparison, and emulation among organizations in many countries. INPO is the United States representative to WANO and makes its information capabilities available worldwide. The WANO-Atlanta Center is co-located with INPO. Whenever possible, WANO helps maintain stable nuclear power operations in countries that have economic and social problems.

INPO's activities are recognized as independent and supplementary to those of the NRC. The industry supports and oversees INPO but gives it authority to enforce its recommendations, thus providing self-regulation by peer review. It is widely accepted that the activities of the INPO have significantly contributed to the improvement in the level of safety in the United States and abroad.

Thanks are due to Philip McCullough for helpful information on INPO.

23.7 DEPARTMENT OF HOMELAND SECURITY

This federal organization was created after the terrorist attack of September 11, 2001. It was authorized by the Homeland Security Act of 2002. The Domestic Nuclear Detection Office of Department of Homeland Security (DHS) has the goal of improving the nation's capability to thwart attempts to use nuclear or radiological materials against the United States. The principal emphasis is on detection that reduces vulnerability. Section 101 of the Act states the primary mission of DHS. Included is the assignment as focal point for natural and manmade crises and emergency planning. Section 504 identifies a Nuclear Incident Response Team (NIRT). A part of that Team according to Section 506 is DOE's Oak Ridge–based Radiation Emergency Assistance Center/Training Site (REAC/TS) that is prepared to go into action on short notice.

The document National Strategy for Homeland Security was issued in 2007. It identified four main purposes: to prevent terrorist attacks, to protect people, infrastructure, and resources, to respond to and recover from incidents, and to strengthen the security system. A companion document of 2006 is titled National Strategy for Combating Terrorism. All of the preceding references are found online.

23.8 OTHER ORGANIZATIONS

The following brief descriptions of organizations that supply information and assistance to the nuclear industry do not do justice to their importance in electrical power generation.

The Electric Power Research Institute (EPRI) is a private nonprofit organization in Palo Alto, CA. It was founded in 1973 to carry out the major research program needed to meet the expected electric power demand. Its stated mission is "to discover, develop, and deliver advances in science and technology for the benefit of member utilities, their customers, and society." It supports studies by its contractors in the general energy field, in coal combustion, nuclear power, and electrical systems. Its product is in the form of research and development reports, distributed widely for use by the industry. EPRI has sponsored the development of computer codes to be used by utilities in managing their fuel cycle and reactor safety analysis programs. EPRI has major initiatives in reactor safety as related both to operations and maintenance, in reliability methodology aimed at reducing operating and maintenance costs while assuring safety, planning for operating license renewal, and contributions to the industry's plan to install advanced light water reactors. Its Nuclear Safety Department staff makes in-depth analyses of potential accidents and recommends ways to avoid them. The process involves the study of event reports, setting priorities, and proposing remedies. A few of the topics covered are probabilistic risk assessment, pressurized thermal shock of reactor vessels, steam generator tube rupture, fuel failure, control of hydrogen, seismic protection, station blackout, the effect of the fission product source term on emergency planning, decay heat removal capability, diesel generator reliability, and reduction in reactor trips. EPRI's mission relates to all electricity generation, but it works in close cooperation with INPO on nuclear power generation.

The Edison Electric Institute (EEI), named for inventor Thomas Edison, was formed in 1933 to represent investor-owned electric utilities. It consists of more than 300 companies, affiliates, and associates, and its staff draws on thousands of experts in the industry to serve on the organization's many committees. Examples are the Policy Committee on Energy Resources and the Nuclear Power Executive Advisory Committee. EEI deals with broad issues of interest to the electric industry, such as management, economics, legislation, regulation, and environmental matters. Subjects of concern to EEI are the future of the nuclear option and maintenance of reliable transmission capability in a changing regulatory environment. Links to its organizations and related sites are found in References.

The Nuclear Energy Institute (NEI) is the Washington-based policy organization of the nuclear energy industry. NEI, formed in 1994, has 300 corporate members in 15 countries. They include companies that operate nuclear power plants, design and engineering firms, fuel suppliers, companies involved in nuclear medicine and nuclear industrial applications, universities, and labor unions. NEI, with member participation, develops policy on key legislative and regulatory issues affecting the industry. It then serves as a unified industry voice before Congress, Executive Branch agencies, and federal regulators. NEI also provides a forum to resolve technical and business issues for the industry through a number of committees and task forces. Finally, NEI provides accurate and timely information through speeches, print publications, and its Web site (see References).

The wide range of information on nuclear energy and technologies is provided for the general public, students and teachers, journalists, financial analysts, and congressional staff members. Featured topics include reliability and efficiency, environmental preservation, transportation safety, education and careers, as well as the basics of nuclear plant operations, nuclear policy issues, and facts and statistics. The organization is committed to maintaining the nuclear option as an environmentally friendly, emission-free source for the United States and the world, and to promoting the values of safety, reliability, and efficiency. More than 6000 industry professionals participate in NEI activities and programs. NEI maintains close relations with other industry organizations, such as INPO, WANO, EEI, ANS, National Rural Electric Cooperative Association (NRECA), and American Public Power Association (APPA).

The American Nuclear Society is the principal professional organization of those working in the nuclear field in industry, government, and universities. Founded in 1954, it has approximately 11,000 members. Its stated objective is "to advance science and engineering related to the atomic nucleus." This is achieved by providing objective technical evaluation of nuclear issues and educating the public, particularly students and teachers, about nuclear matters. ANS emphasizes the importance to its members of professionalism and responsibility. The society is very active in generating and updating nuclear standards. It has a committee of a large number of scientists and engineers. A thorough discussion of ANS standards is available on the Web (see References). ANS publishes journals including *Nuclear Science and Engineering*, *Nuclear Technology*, *Radwaste Solutions*, and *Nuclear News*. ANS also coordinates the publication of technical books and conference reports, including *Transactions of the American Nuclear Society*. Its divisions represent major subject areas such as Reactor Physics, Nuclear Criticality Safety, and Isotopes and Radiation. Its committees serve functions such as public information, planning, and standards. Local sections and student chapters throughout the country hold regular technical meetings in behalf of members and the nuclear field.

The Institute of Electrical and Electronics Engineers (IEEE) has two major nuclear groups—the Nuclear Power Engineering Committee and the Power Generation Committee. These have subcommittees on topics such as Operations, Surveillance, and Testing; Energy Development; Nuclear Power; Quality Assurance; and Human Factors and Control Facilities. The monthly publication *Proceedings of the IEEE* often contains survey articles on nuclear topics.

Several other journals provide technical information on nuclear energy. Examples are *Annals of Nuclear Energy, Waste Management*, and *Nuclear Engineering International*, a British publication that covers world nuclear activities.

Nuclear utility groups on various subjects are informal working associations of experts with common technical or administrative problems. Of the more than 30 topics, examples are PWR steam generators, nuclear waste management, seismic qualification, degraded core rule making, and plant life extension. Nuclear owners groups are composed of people from companies owning equipment supplied by one of the four vendors—Westinghouse, General Electric, Babcock and Wilcox, and ABB Combustion Engineering—and having a common technical problem.

The APPA represents and provides services to 1,750 community-owned electrical utilities. The NRECA supports rural electrification and development. It embraces a variety of other cooperative organizations.

Related organizations are the National Association of Regulatory Utility Commissioners (NARUC), whose principal function is to improve the quality and effectiveness of public utility regulation in the United States; and the Nuclear Non-Operating Owners Group (NNOG), which as its name implies is an association of organizations that own nuclear facilities operated by others. It is principally a forum for exchange of information and ideas.

Standards are descriptions of acceptable engineering practice. Professional technical societies, industrial organizations, and the federal government cooperate in the development of these useful documents. They represent general agreement, arrived at by careful study, writing, review, and discussion by qualified practitioners. Many hundreds of scientists and engineers participate in standards development.

The American National Standards Institute (ANSI) provides an umbrella under which standards are written and published for use by reactor designers, manufacturers, constructors, utilities, and regulators (see References). Some of the societies that are active in standards development are the American Nuclear Society (ANS), the Health Physics Society (HPS), the American Association of Mechanical Engineers (ASME), the Institute of Electrical and Electronics Engineers (IEEE), and the American Society for Testing and Materials (ASTM).

The first nuclear engineering education program in the United States was initiated in 1950 at then North Carolina State College. Subsequently, some 80 programs were established, a number with research and training reactors. Graduates assumed positions of leadership in the development of nuclear applications. In the course of time, however, some departments were merged with others or terminated, and many of the reactors were shut down (see References). Many companies and government agencies provide assistance to students of nuclear science and engineering. An example involving NEI and INPO is found in References.

The Nuclear Energy Research Initiative (NERI) consists of grant awards by the DOE for research at universities, national laboratories, and industry. Three areas are emphasized: new reactors (Gen IV), advanced fuel cycle, and hydrogen production. A part of the NERI funding is in a program called Innovation in Nuclear Infrastructure and Engineering (INIE) that is specifically for universities. The programs have been very helpful in increasing enrollments in nuclear programs, as discussed in References. There will be a continuing need for new manpower in the nuclear renaissance, not only to design and build new systems but also to replace the aging workforce.

23.9 ENERGY POLICY ACTS

Efforts were underway for several years in the United States to develop a comprehensive energy program that would integrate the activities of the DOE, the NRC, the EPA, and other federal agencies, with contributions by the private sector. These initiatives culminated in the passage by Congress of the legislation titled Energy Policy Act of 1992 (Public Law 102-486). It provided energy efficiency goals and standards, promoted alternative fuels, prescribed new R & D on electric vehicles, restructured the production of electricity, addressed radioactive waste disposal, established a uranium enrichment corporation, and simplified nuclear plant licensing. In essence, the law affirmed the nation's commitment to preserve and extend the nuclear option as part of a broad energy mix. From the more than 350 pages of the Act, we can highlight the features that are related primarily to nuclear energy.

> *Energy efficiency.* Standards, guidelines, and incentives were provided for conservation efforts.
>
> *Electric cars.* DOE was to work with manufacturers and the electric utility industry to develop practical vehicles.
>
> *Electrical generation.* Other organizations besides utilities were permitted to generate power.
>
> *High-level radioactive waste.* EPA was to provide safety standards for Yucca Mountain, taking cognizance of recommendations of the National Academy of Science. NRC rules were to be consistent with those of EPA.
>
> *United States Enrichment Corporation.* This was created to operate as a business enterprise on a profitable and efficient basis. It was allowed to lease facilities and sell enrichment services.
>
> *Fusion energy.* United States was to participate in ITER and pursue various fusion concepts.
>
> *Advanced nuclear reactors.* Designs alternative to light water reactors (LWRs) were to be designed and certified.
>
> *Nuclear plant licensing.* NRC was to use combined construction and operating licenses.

The Energy Policy Act of 2005 had a number of provisions related to nuclear energy. Included are the following:

(a) Amendment of the nuclear insurance legislation, the Price-Anderson Act.
(b) Decommissioning of the SEFOR fast breeder reactor.
(c) Permission to send U-235 abroad for medical isotope production (with adequate protection).
(d) Arrangement for a study by the National Academy of Science (NAS) of medical isotope production.
(e) Management of Greater-than-Class-C low-level radioactive wastes.

(f) Prohibition of shipment of nuclear materials, equipment, or technology to countries sponsoring terrorism.
(g) Demonstration of hydrogen production by a nuclear power reactor.
(h) NRC user fees and annual charges.
(i) Standby support for six new reactors.
(j) Establishment of a Next Generation Nuclear Power Plant Project for production of electricity and hydrogen.
(k) Security evaluations of nuclear plants by NRC.
(l) Determination of design basis threat to nuclear facilities by NRC.
(m) Partnerships with minority colleges.
(n) Arrangement with NAS for a study of uses of radiation sources.
(o) Assurance of security of radiation sources.
(p) Use of firearms to protect nuclear facilities or materials.
(q) Security of transfer of nuclear material.

Criticism has been leveled at Congress by the organization Public Citizen for "nuclear giveaways" in the Energy Policy Act of 2005 (see References). In contrast, President Bush in signing the Act strongly supported nuclear power as follows:

"Nuclear power is another of America's most important sources of electricity. Of all our nation's energy sources, only nuclear power plants can generate massive amounts of electricity without emitting an ounce of air pollution or greenhouse gases. And thanks to the advances in science and technology, nuclear plants are far safer than ever before. Yet America has not ordered a nuclear plant since the 1970s. To coordinate the ordering of new plants, the bill I sign today continues the Nuclear Power 2010 Partnership between government and industry. It also offers a new form of federal risk insurance for the first six builders of new nuclear power plants. With the practical steps in this bill, America is moving closer to a vital national goal. We will start building nuclear power plants again by the end of this decade."

23.10 SUMMARY

Congress passed the Atomic Energy Act of 1946, amended in 1954, to further peaceful purposes, as well as to maintain defense. The AEC was formed to administer the programs. Later, the AEC was split. Currently the DOE is responsible for development of nuclear energy and the NRC enforces rules on radiation set by the EPA. The IAEA helps developing countries and monitors nuclear inventories. The DHS emphasizes detection of nuclear materials. Among other influential organizations are the INPO, the EPRI, and the NEI. The ANSI and the ANS are active in developing standards for processes and procedures in the nuclear industry. Education and training in nuclear technology are provided by a number of universities. The Energy Policy Act of 1992 provided comprehensive national goals and requirements on energy efficiency and generation. The Energy Policy Act of 2005 has many provisions favoring nuclear and deals with protection against terrorism.

23.11 REFERENCES

Atomic Energy Act of 1946
http://www.osti.gov/atomicenergyact.pdf
Excerpts from legislative history.

List of United States Federal Government Agencies
http://www.lib.lsu.edu/govdocs/federal/list.html
A comprehensive set of links maintained by Louisiana State University Libraries.

History of Nuclear Energy
http://www.ne.doe.gov/pdfFiles/History.pdf
Brief (28 pages) pamphlet. Chronology to 1992, references, and glossary.

Environmental Protection Agency
http://www.epa.gov
Select About EPA/EPA's Budget or Policy Statements and Strategy Documents/Strategic Plan.

Environmental Protection Agency activities
http://www.gpoaccess.gov/usbudget/fy07/pdf/budget/epa.pdf
Description of all programs.

EPA Radiation Program
http://www.epa.gov/radiation
Select from a variety of links (e.g., Laws and Regulations/List of Federal and State Regulations/
    EPA/Part 191).

George T. Mazuzan and J. Samuel Walker, *Controlling the Atom: The Beginnings of Nuclear
    Regulation 1946–1962*, California University Press, Berkeley and Los Angeles, 1984. Written
    by historians of the Nuclear Regulatory Commission, the book provides a detailed regulatory
    history of the Atomic Energy Commission.

A Short History of Nuclear Regulation, 1946–1999
http://www.nrc.gov/about-nrc/short-history.html
The first chapter is drawn from the book by Mazuzan and Walker.

Code of Federal Regulations, Energy, Title 10, Parts 0–199, United States Government Printing
    Office, Washington, DC (annual revision). All rules of the Nuclear Regulatory Commission
    appear in this book, published by the Office of the Federal Register, National Archives and
    Records Service, General Services Administration. Available to be downloaded by sections in
    pdf at the NRC Web site.

Maintenance Rule
http://www.nrc.gov
Search on 10CFR 50.65.

Nuclear Regulatory Commission
http://www.nrc.gov
Select About NRC/Organization and Functions or /How We Regulate.

NRC Regulatory Guides
http://www.nrc.gov/reading-rm/doc-collections/reg-guides
Select from list of ten divisions.

NRC New Reactor Licensing
http://www.nrc.gov/reactors/new-reactors.html
Select Combined License Application.

NRC Reactor License Renewal
http://www.nrc.gov/reactors/operating/licensing/renewal.html
Select Status of Current Applications.

Risk-Informed Performance-Based Regulation
http://www.nrc.gov/about-nrc/regulatory/risk-informed.html
Explore links.

Nuclear Security and Safeguards
http://www.nrc.gov/security.html
Select Reports Protecting Our Nation and Force-on-Force Exercises at Nuclear Plants.

United States Department of Energy
http://www.doe.gov
Select About DOE or National Energy Policy.

Comprehensive National Energy Strategy
http://www.pi.energy.gov/documents/cnes.pdf
A 1998 discussion of goals.

DOE Performance and Accountability Report
http://www.cfo.doe.gov/progliaison/2005pr.pdf
Goals and progress of the Department of Energy.

DOE Information Bridge
http://www.osti.gov/bridge
A large collection of online reports.

Energy Security for the 21st Century
http://www.whitehouse.gov/infocus/energy
Select Fact Sheet: Energy Independence and Security Act of 2007.

Dwight D. Eisenhower, "Atoms for Peace" *Nuclear News*, November 2003, p. 38.

"Atoms for Peace—Updating the Vision," *Nuclear News*, January 2004, p.55. Report on a topical
    meeting assessing the realization of Eisenhower's vision.

International Atomic Energy Agency
http://www.iaea.org
Select About IAEA or Our Work.

IAEA Safeguards Overview
http://www.world-nuclear.org/info/iaeasafeguards.htm
Protocol and inspections. by World Nuclear Association.

IAEA 2007 Year in Review
http://www.iaea.org/NewsCenter/News/2007/year_in_review.html
Events and developments (e.g., Iran's nuclear ambitions).

Lawrence Scheinman, *The International Atomic Energy Agency and World Nuclear Order*,
    Resources for the Future, Washington, DC, 1987. History of origin of IAEA, structure and
    activities, and relationship to safeguards.

World Association of Nuclear Operators (WANO)
http://www.wano.org.uk
Select Programmes.

"Zack Pate: His WANO career," *Nuclear News*, August 2002, p. 25.

Department of Homeland Security
http://www.dhs.gov/index.shtm
Select About the Department/Prevention and Protection/Domestic Nuclear Detection Office/
    Laws and Regulations.

Electric Power Research Institute
http://www.epri.com
Select About EPRI or Global Climate Change.

Edison Electric Institute
http://www.eei.org
Select About EEI.

Nuclear Energy Institute
http://www.nei.org
A wealth of nuclear information.

American Nuclear Society
http://www.ans.org
Select Public Information.

Institute of Electrical and Electronics Engineers
http://www.ieee.org/portal/site
Select "journal or magazine."

American National Standards Institute
http://www.ansi.org
Select About ANSI.

American Nuclear Society Standards Committee Report of Activities 2007
http://www.ans.org/standards/resources/downloads/docs/comactivitiesreport2007.pdf
A comprehensive report that lists all ANS standards.

U.S. Nuclear Engineering Education: Status and Prospects, National Academy Press,
    Washington, DC, 1990. A study by a committee of the National Research Council.

Educational Assistance Program
http://www.nei.org
Select Search/INPO for information on National Academy for Nuclear Training.

Nuclear Engineering Education Sourcebook
http://www.ne.ncsu.edu/NE%20sourcebook/index.html
Includes links to lists of faculty in all institutions. Editors: Man-Sung Yim and John Gilligan.

Nuclear Engineering Education Research (NEER)
http://www.ne.doe.gov/universityPrograms/neUniversity2d.html
Select DOE program to support universities.

NERI Nuclear Energy Research Initiative
http://nuclear.energy.gov/neri/neNERIresearch.html
Descriptions of the three areas of investigation.

Rick Michal, "INIE Funding: University Consortia Put Over $9 Million to Good Use," *Nuclear
    News*, February 2007, p. 22. Included is an informative diagram of infrastructure.

Energy Policy Act of 2005
http://www.epa.gov/oust/fedlaws/publ_109-058.pdf
Full text of the law. Nuclear Matters in Sections 621-657.

Nuclear Giveaways
http://www.citizen.org/documents/NuclearEnergyBillFinal.pdf

Federal Energy Regulatory Commission
http://www.ferc.gov
Select About FERC.

Jacek Jedruch, *Nuclear Engineering Data Bases, Standards, and Numerical Analysis*, Van Nostrand Reinhold, New York, 1985. A useful collection of information that would be difficult to find elsewhere.

References for the full text of laws related to nuclear power and radioactive waste as originally written appearing in *United States Statutes at Large*, United States Government Printing Office. Most of the laws can be accessed by entering their title in Google.

"Atomic Energy Act of 1946," Public Law 585, 79th Congress, August 1, 1946.

"Atomic Energy Act of 1954," Public Law 703, 83rd Congress, August 30, 1954.

"Price-Anderson Act", Public Law 85-256, 85th Congress, September 2, 1957.

"National Environmental Policy Act of 1969," Public Law 91-190, 91st Congress, January 1, 1970.

"Energy Reorganization Act of 1974," Public Law 93-438, 93rd Congress, October 11, 1974.

"Department of Energy Organization Act of 1977," Public Law 95-91, 95th Congress, August 4, 1977.

"Low-Level Radioactive Waste Policy Act (of 1980)," Public Law 99-240, 96th Congress, December 22, 1980.

"Nuclear Waste Policy Act of 1982," Public Law 97-425, 97th Congress, January 7, 1983.

"Low-Level Radioactive Waste Amendments Act of 1985," Public Law 99-240, 99th Congress, January 15, 1986.

"Energy Policy Act of 1992," Public Law 102-486, 102nd Congress, October 24, 1992.

"Energy Policy Act of 2005," Public Law 109-58, 109th Congress, August 10, 2005.

CHAPTER

Energy Economics

24

THE DEFINITION of economics appearing in a popular textbook[†] is as follows:

Economics is the study of how societies use scarce resources to produce valuable commodities and distribute them among different people.

The definition is relevant in that we seek answers to questions such as these:

(a) What are the comparative costs of electricity from nuclear plants and from coal or oil plants?
(b) What is the expected use for nuclear power in the future?
(c) What choices of nuclear power research and development must be made?

In this chapter we will consider the first of these questions, examining the origin of costs of electricity and reviewing past events and trends. In a later chapter we study the long-range role of nuclear power.

As background for the discussion of electric power and nuclear's role, it is instructive to examine the energy flow diagram of Figure 24.1. Several points to note are: (a) nuclear accounts for more than 8% of the total supply; (b) oil imports are more than 72% of the total oil supply; (c) the total of all renewables, including hydro, solar, wind, biofuels, and geothermal, is comparable to nuclear; (d) energy for transportation is more than 28% of the total consumption. The chart strongly suggests that conservation of natural gas and oil should have high priority. The amounts of electricity generated by the various energy sources can be seen in Table 24.1.

24.1 COMPONENTS OF ELECTRICAL POWER COST

The consumer's interest lies in the unit cost of electricity delivered to the home. The author's light bill, in a region of the country that uses nuclear for more than one third of electricity generation, gives a figure of close to 8 cents/kWh.

[†]Paul A. Samuelson and William D. Nordhaus, *Economics*, 18th Ed., Irwin/McGraw-Hill, Inc., New York, 2006.

415

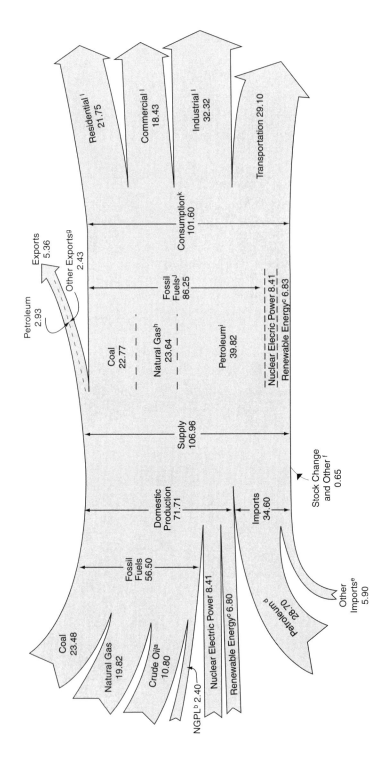

FIGURE 24.1 United States energy sources and uses. Units are quads (quadrillion Btu). Source: Annual Energy Review 2007, DOE/EIA-0384 (2007).

Table 24.1 Electric Energy Amounts and Percentages from Various Sources (DOE/EIA Web site)

Source	Amount (× 10⁹ kWh)	Percent
Coal	1991	49.0
Nuclear	787	19.4
Hydro	283	7.0
Gas	829	20.4
Oil	64	1.6
Other renewables[†]	96	2.4
Other	14	0.3
Total	**4064**	**100.0**

[†]*Geothermal, solar, wind, biomass, etc.*

This cost includes three typical components—generation (55%), transmission (32%), and administration (13%). This says that the generation or "bus bar" cost of electricity for this particular area would be approximately 4 cents/kWh or 40 mills/kWh.

The comparison between costs of nuclear and its main competitor, coal, varies in several ways. On the average, the two have about that same cost, but there are large variations among countries, with the ratio coal/nuclear between 0.8 and 1.7. Electrical generation costs are dominated by fuel costs for coal and by capital costs for nuclear. Thus differences on a global, national, or regional basis depend on the distance from coal fields and on the discount rate. For example, nuclear electricity is relatively inexpensive in Japan and many European countries because of the cost of importing coal. Another factor is the regulatory environment of the country and the degree of emphasis on clean air or nuclear safety. Operating and maintenance (O&M) costs for nuclear plants are generally high because of the great complexity of the equipment and the stringent safety requirements of the regulators. However, O&M costs can vary widely among utilities with comparable facilities because of differing degrees of management effectiveness. Capital costs of both fossil and nuclear plants were high during the decade of high interest rates and high inflation, but the increase in cost was greater for nuclear plants, because they are basically more expensive and the time to construct was excessive. Table 24.2 gives the trend of plant costs for half the United States reactors over four time periods in which commercial operation began.

The capital costs of nuclear plants vary greatly, but the figure is $2-3 billion. This represents the money required to construct the plant, including interest.

Table 24.2 Construction Costs for Nuclear Units. (Energy Information Administration, United States Department of Energy, DOE/EIA-0439(84))

Period[†]	Number of Units	Average Cost ($/kWe)
1971-1974	13	313
1975-1976	12	460
1977-1980	13	576
1982-1984	13	1229

[†]*During which units entered commercial operation.*

Nuclear power has long been regarded as "capital-intensive," because equipment costs are high, whereas fuel costs are low. Typically, the main parts of the nuclear plant itself and percentages of the cost are reactor and steam system (50%), turbine generator (30%), and balance of plant (20%). Additional costs include land, site development, plant licensing and regulation, operator training, interest and taxes during construction, and an allowance for contingencies.

Further perspective is needed on the capital cost component. Utilities that are not affected by deregulation serve an assigned region without competition. In exchange, the price that they can charge for electricity is regulated by public commissions of state governments. When a utility decides to add a plant to its system, it raises capital by the sale of bonds, with a certain interest rate, and by the sale of stock, with a dividend payment to the investor. These payments can be combined with income tax and depreciation to give a charge rate that may be as high as 20%. The interest charge on the capital invested must be paid throughout the construction period. This is an important matter, because the average total time required to put a plant into operation by 1985 was approximately 13 y, in contrast with a figure of less than 6 y in 1972. Figure 24.2 shows the trend in construction periods for the recent past. Several reasons have been advanced for the long time between receipt of a construction permit and commercial operation. In some cases plants were well along when new regulations were imposed, requiring extensive modifications. Others have been involved in extended licensing delays resulting from intervention by public interest groups. Others suffered badly from lack of competent management.

24.2 FORECASTS AND REALITY

The demand for electrical power varies on a daily basis as a result of the activities of individuals, businesses, and factories. It also varies with the season of the year, showing peaks when either heating or air conditioning is used extensively.

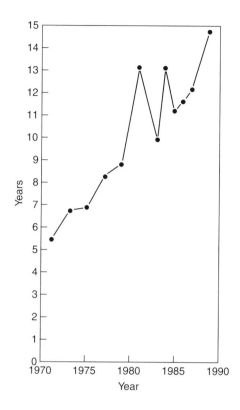

FIGURE 24.2 Construction times for nuclear power plants. Adapted from DOE/EIA-0439(84).

The utility must be prepared to meet the peak demand, avoiding the need for voltage reduction or rotating blackouts. The existing megawatts capacity must include a margin or reserve, prudently a figure such as 20%. Finally, the state of the national economy and the rate of development of new manufacturing determine the longer range trends in electrical demand. Utilities must continually be looking ahead and predicting when new plants are required to meet power demand or to replace older obsolete units.

Such forecasts have to be made well into the future because of the long time required to build a new power plant. But forecasts can readily turn out to be wrong because of unforeseen events or trends, including the interruption of energy supplies from abroad, shifts in the state of the economy, and major changes in the regulatory climate. If an estimate of power demand is too low and stations are not ready when needed, customers face the problem of shortages; but if the estimate is too high, and excessive capacity is built, customers and shareholders must bear the effects of added expense.

The history of nuclear growth and eventual stagnation over the last several decades of the 20th century serves well to illustrate this situation. It is not possible to identify any single cause for the situation, but we can indicate many of the factors that had an effect. Optimism about nuclear energy was based on the successful development of military applications and the belief that translation into peaceful uses was relatively easy and straightforward. After studying and testing several reactor concepts, the United States chose the light water reactor. Hindsight indicates that safety might have been assured with far less complexity and resultant cost by adoption of heavy water reactors or gas-cooled reactors.

There was a post-World War II economic boom in which the demand for electric power was approximately 7% per year. New coal-fired plants provided most of the growth. In 1957 the first commercial power reactor was started at Shippingport, Pennsylvania, and new designs of larger units were developed by two concerns. Some of these were attractive to utilities because they were turnkey plants, priced very favorably. A large number of orders were placed in the 1960s to the main vendors—Westinghouse, General Electric (GE), Babcock & Wilcox, and Combustion Engineering. These orders were placed on the basis of sustained electric power demand growth well to the end of the century and an expected construction time of approximately six years.

Predictions were optimistic in that period. For example, in 1962 and later in 1967, the Atomic Energy Commission (AEC) predicted[†] the following installed nuclear capacities:

Year	GWe
1970	10
1980	95
2000	734

The reasons for the optimism were expectation that the United States economy would continue to expand, that electricity would substitute for other fuels, and that nuclear would fill a large fraction of the demand, reaching 56% by the year 2000. As it turned out, the level of 95 GWe was actually reached late in the 1980s, but the figure only reached 100 GWe after an additional decade.

What is the reason for the great discrepancy between forecasts and reality? The first is that it took longer and longer to build nuclear plants, adding large interest costs to the basic capital cost. Inflation in the 1970s drove costs of construction up dramatically, as we saw in Table 24.2. The effect on nuclear plants was especially, severe because of their complexity and the requirement of quality assurance at every stage from material selection to final testing.

[†]Civilian Nuclear Power—A Report to the President-1962 (and 1967 Supplement), United States Atomic Energy Commission.

The Middle East oil boycott of 1973 caused an increase in the cost of energy in general, accentuated a national recession, and prompted conservation practices by the public. The growth rate of electrical demand fell to 1% per year. As a consequence, many orders for reactors were canceled. However, by this time, a large number of reactors were in various stages of completion—reactors that would not be needed for many years, if ever. Some that were approximately 80% completed were finished, but work on others of 50% or less was stopped completely. The hard fact was that it was cheaper to abandon a facility on which half a billion dollars had been invested rather than to complete it.

Nuclear power was barely getting started when the environmental movement began and consumers' interests became more vocal and influential. Opposition to nuclear power was composed of many elements. Early activists expressed themselves as opposed to the power of the entity called the military-industrial complex. Because nuclear energy is involved in both weapons and commercial power, it became a ready target for attack. Those philosophically inclined toward decentralized authority, the return to a simpler lifestyle, and the use of renewable energy were enlisted into the antinuclear cause. Those fearful of radiation hazard and those concerned about the growth of nuclear weapons were willing recruits also. Well-organized opposition forces set about to obstruct or delay reactor construction through intervention wherever possible in the licensing process. The Nuclear Regulatory Commission (NRC) had a liberal attitude toward intervenors in the interests of fairness. The net effect in many cases was to delay construction and thus increase the cost. The high costs then served as an additional argument against nuclear power. The general public has tended to be swayed by statements of the organized opposition and to become doubtful or concerned. Traditional distrust of government was accentuated in the 1970s by the pains of the war in Viet Nam. The aftermath of the Watergate affair was a loss in confidence in national leadership. The public was further sensitized by the revelation that industrial chemicals were affecting plant and animal life and that wastes had been mismanaged, as at Love Canal. Because of accompanying radiation, wastes from nuclear power were regarded as more dangerous than ordinary industrial wastes. Concerns were aggravated by the apparent inability of government and industry to deal effectively with nuclear wastes. Changes in policy and plans between national administrations on the basis of differences in approach were ascribed to ignorance.

In the 1980s the demand for electricity began to increase again, but by that time, other factors had developed that discouraged utility management from resuming a building program. In earlier years, the utilities in a state were regulated monopolies that could readily pass on costs to the consumers and could show continued decreases in the cost of electricity. When the recession occurred, costs increased, and customers adopted conservation measures, reducing income from the sale of electricity.

In the interests of improved protection of the public, the NRC increased the number and detail of its rules and guidelines, often requiring that changes in equipment be made or additional equipment be installed. Examples of mistakes in design, installation, testing, cost overruns, shoddy workmanship, and inept management received a great deal of media attention, further eroding confidence among investors and the general public.

In this period, the role of the Public Utilities Commissions (PUCs) became more important. These state regulatory organizations are committed to protect the consumers' interest. They became alarmed at the rising costs of nuclear plants and were reluctant to allow utilities to pass costs on to consumers, thus reducing the margin of profit to the company and its stockholders. The practice of prudent review is applied to the construction of facilities after construction is complete. Questions are asked, "Would a reasonable person have incurred those costs, or canceled the project?" A related "used and useful" test asks, "Were those facilities actually needed?" or "Should a cheaper power source have been built?" Expenditure by utilities was disallowed for many reasons. Some costs were unreasonable, such as cost overruns that could have been avoided with better project management. Others were in the category of errors in judgment but only in hindsight. An example is a decision to build a generating plant that turned out to be larger than necessary. In many cases, expenditures by the utility were disallowed even though they were outside the control of the management. As a consequence of unhappy experience, utility executives became increasingly wary of any new large-scale long-term commitment. The prospect of fiscal disaster outweighed that of criticism for failing to anticipate and meet electricity needs.

On a very positive note, more than 100 reactors were in operation by the turn of the century, contributing approximately 20% of the total United States electricity, with no harm to the public, and at a cost that was well below that of oil-fired units and many coal-burning plants. Realistically, however, it is a fact that the cost of nuclear plants had increased dramatically. Utilities found little sympathy for their requests for rate increases to meet costs of operation. The last new orders for reactors were in 1978.

The Three Mile Island accident of 1979 (Section 19.5) dealt a severe blow to the nuclear power industry in the United States. Although releases of radioactivity were minimal and no one was hurt, the image of nuclear power was seriously tarnished. Media attention was disproportionate to the significance of the event and greatly increased the fears of local residents. The apparent confusion that existed immediately after the incident and the revelation of errors in design, construction, and operation caused national concern over the safety of all reactors.

The Chernobyl accident of 1986 (Section 19.6) commanded international attention. The effect on public opinion may have been greater in Europe than in America, in part because of the geographic proximity to the event. It is generally appreciated in the United States that the Chernobyl reactor was operated by the U.S.S.R. without adequate precautions, was basically more unstable than LWRs, and lacked a full containment. Nonetheless, the spectre of Chernobyl remained over the United States nuclear industry.

The environment surrounding modern nuclear utilities is complex and demanding. Competition among utilities and between utilities and other independent generators has increased, because consumers are anxious to obtain the lowest cost electricity. To attract and keep customers, the margin of profit must be reduced or production costs must be minimized, or both. Among the steps taken by utilities were removal of excess layers of management and reductions in staffing. Recognizing that operations and maintenance (O&M) costs are a major part of the cost of producing electricity, utilities increased attention to efficiency in the maintenance and repair process. They concentrated on reduction in the time required for refueling outages and eliminated unscheduled reactor trips to enhance capacity factors. Such actions had to be taken with care that safety was not jeopardized. It is clear that every additional monitoring device or safety equipment or special procedure intended to enhance safety adds to the product cost. As performance improves, it becomes more difficult to find areas of further improvement. But it is difficult for a regulator, either from government or industry, to refrain from recommending new safety initiatives. In the limit, the industry could be put out of business by escalating costs. From another point of view, the addition of excessive complexity to a facility can be counterproductive to safety. This suggests that greater attention be given to establishing priorities and to reducing costs in other areas besides those that are safety sensitive. A more important goal still is the achievement of a uniform level of excellence in every nuclear unit in the country.

A recent trend toward consolidation of management is notable. Companies such as Entergy bought nuclear plants for market value, which is much lower than initial cost. As a result, their electric power from nuclear is generated at cost competitive with natural gas or coal.

24.3 TECHNICAL AND INSTITUTIONAL IMPROVEMENTS

The nuclear industry has made great efforts to improve efficiency and safety. Some of the activities are cited in the following.

Computer assistance. Word processing capability for generation of reports and correspondence is standard, and plant computers provide status information displays on the basis of measurements by a host of instruments. The computer plays a vital role in scheduling actions during planned outages. The down times have been reduced to well under 1 month.

Expert systems. These capture a large body of human knowledge and make it available for decisions and problem solving. It is one facet of the broader subject of artificial intelligence (AI), which has been researched extensively. Some other relevant components of AI are (a) simulation, in which a computer program imitates a reactor control room, providing a variety of experiences for operators in training, (b) robotics, involving computer, circuits, and mechanical devices to simulate movement of human beings and perform tasks in hazardous environments, (c) neural networks, computer programs that model the processing of neurons in the brain, and (d) virtual reality, which provides visual and tactile experience for workers in preparation for complex operations.

Digital control of nuclear plants. The instrumentation and control (I&C) system of a power reactor provides (a) continuous information on status, including neutron flux, power level, power distribution, temperatures, water level, and control rod position; (b) provides commands to trip the reactor if preset limits are exceeded; and (c) reports deviations from normal or failures of components. Traditionally, the I&C systems were of the analog type, involving a sensor, a feedback circuit, and a display device. Such systems tended to become unreliable with time. The industry has converted to digital I&C, which consists of computer software and microprocessor-based hardware. The NRC recommended and supported the conversion in the interests of reliability and safety. Large expenditures were incurred to achieve the transition.

Reactor life extension. The life of a nuclear reactor system was set by Congress on economic grounds as 40 y, with prior termination for marginal safety and excessive outage for maintenance and repair. In view of the high capital cost of a reactor, it pays to stretch the life beyond the 40-year mark. Some of the problems to be addressed are (a) difficulty in finding spare parts, (b) corrosion and plugging in generators, (c) deterioration of electrical systems, (d) high radiation levels because of buildup of deposits, (e) corrosion of piping, and (f) radiation damage of reactor vessel welds from neutron bombardment. The last item is associated with the possibility of pressurized thermal shock (PTS), in which temperature changes in embrittled material result in vessel rupture. Location of fuel with low neutron production rate near the surface is a helpful solution.

License renewal. Extension of an operating license for 20 y can be granted by the NRC if several submissions by the operating company are provided. The procedure is described in NRC rules. Environmental matters are covered in 10CFR51, and license renewal is in 10CFR54 (see References). Regulatory Guide 1.188 tells how to apply, and a Standard Review Plan gives NRC's techniques. Special attention in the licensing must be given to the potential effects of aging of components and systems, with information on ways to mitigate the effects. The objective is to determine whether the plant can operate safely in the extended period. The license renewal process is outlined in an NRC Web site (see References). The applicant for a renewal license must submit an environmental report that analyzes the plant's impact during the continued operation. Use can be made of a Generic Environmental Impact Statement (GEIS, see References) prepared by NRC, with adaptation to fit the specific plant.

The first United States plants to seek license renewal were Calvert Cliffs, operated in Maryland by Baltimore Gas & Electric, and Oconee, operated in South Carolina by Duke Power. Subsequently, a number of plants initiated plans for license renewal.

24.4 EFFECT OF DEREGULATION AND RESTRUCTURING

The electrical generation industry faces problems related to access to its transmission lines. There are a growing number of nonutility producers of electricity that use wind, water, and cogeneration. Industrial consumers seeking the lowest cost electricity would like to buy power from such independent generators and use the existing utility-owned network. Users in the northern United States would like to import more power from Canada. The process of transferring large blocks of power around the grid is called "wheeling." Utilities are concerned about the effect of increased wheeling on system stability and reliability, on costs of new transmission lines, and on safety. The problem is not solely that of the utilities, because residential and commercial users may experience higher costs if the utilities lose large customers.

Various new approaches to energy management on the part of utilities have been required by public utility commissions. The broadest category is Integrated Resource Planning (IRP), which takes account of all aspects of energy, including environmental effects and social needs. Within it is Demand Side Management (DSM) that seeks to reduce usage rather than meeting customers' requirements. DSM emphasizes encouragement of conservation and avoidance of new large facilities by use of alternative energy sources. Related is Least Cost Planning (LCP), which requires the examination of all costs, including existing plants. This comes into play when a major equipment replacement such as steam generator is needed. Shutting down the plant might be more economical. In all these methods, the PUC played a more active role in decision making than previously.

The Energy Policy Act of 1992 (Section 23.8) will continue to have a significant effect on the electric utility industry. Some of its pertinent provisions are noted. For example, the Public Utility Holding Company Act of 1935 (PUHCA), which governed power production by utilities, was modified to allow greater competition among power producers, including a new category called "exempt wholesale generators" (EWGs), which are unregulated power producers. The objective was to let market forces play a greater role. Independent power producers (IPPs), those outside the utility structure, were encouraged to develop. The Federal Energy Regulatory Commission (FERC) was given greater power, especially to order transmission access, when it can be shown that it is in the public's interest (i.e., reliability is maintained and costs to users is reduced). The process of integrated resource planning (IRP) is required at the state level. The new law thus accelerates the process of utility industry restructuring that had been evolving since the energy crisis of the 1970s.

Nuclear power's position was enhanced by the Energy Act through streamlining of the reactor licensing process, the use of certification of standardized reactor designs, and the establishment of a government corporation for uranium enrichment. However, the effect of other features of the Act is uncertain. Mandatory efficiency standards were set for electrical equipment and the development of an electric automobile given greater support. Meeting efficiency goals clearly would tend to reduce the need for new electric power, whereas massive electrification of ground transportation would increase demand. From the regulatory standpoint, it will be easier to license new reactors, thus encouraging investors. On the other hand, new competition will increase the economic pressure on the utilities. They must cut costs but are required by NRC and INPO to maintain safety. The recurring question "How safe is safe enough?" needs to be addressed to the satisfaction of the industry, the regulator, and the public.

The Energy Act of 1992 requires that each state of the United States develop a plan for transition of electric generation by regulated monopoly to a free market. A variety of techniques to ensure equity among the various stakeholders have been developed. One of the key issues in the debates is how to handle "stranded costs." These costs to utilities result from the change itself and consist of several categories: (a) locked-in power purchase contracts with independent generators required by the Public Utilities Regulatory Policies Act (PURPA); (b) regulatory assets, which are programs for energy efficiency, low-income assistance, and deferred fuel costs, approved by regulatory bodies; (c) capital investment debt, incurred in the construction of nuclear power plants, normally to be paid off over many years by income from consumers; and (d) decommissioning funds required over and above those already accumulated. The question is, "Who should bear the burden of the stranded costs?" Users of electricity and their advocacy groups believe that the consumer should not have to pay for what is considered mismanagement on the part of utility executives and provide a bailout of the industry. They argue that investors in utility stocks and bonds must take their chances on loss just as with any other investment. Utilities on the other hand point out that there was a contract involving approved expenditures in exchange for reliable electricity and that decisions to build nuclear plants were fully supported by regulators. Those holding stocks or bonds obviously do not want the value of their investment to decline. The ideal solution of this problem is to devise a formula that gives each party a fair part of the burden, such that the transition can be effected smoothly and efficiently, with realization of the goal of reduced costs to consumers with assured reliability.

The subject of deregulation is very complex because of the many issues and variety of groups affected, as well as differing situations among the states, which are addressing the opportunities and problems. Several discussions of the subject from different vantage points are found in References.

24.5 ADVANCED REACTORS

Light water reactors of the Pressurized Water Reactor (PWR) and Boiling Water Reactor (BWR) type have performed very well over several decades. However, in the United States without any action being taken, a number of reactors would come to the end of their license period and be shut down. Many believe that it is in the best interests of society to continue the nuclear option as a part of an energy mix. To do so, nuclear power must be acceptable to the public, the utilities, the regulatory agencies, and the financial community. This implies the need for confidence in reactor safety and economy.

The United States nuclear power industry includes electric utilities that use reactors, equipment manufacturers and vendors, and service organizations. That industry is convinced that electricity from nuclear power will continue to be necessary to sustain economic growth. Leaders note that nuclear power does not contribute to pollution and potential global warming and helps provides energy security through the reduction of reliance on uncertain supplies of foreign oil. The industry believes that energy conservation and the use of renewable sources of energy are highly desirable but not sufficient for long-term needs, especially in light of a growing population and the demand for environmental protection.

Accordingly, a Strategic Plan for Building New Nuclear Power Plants (see References) was published with a final version dated 1998. The document serves to highlight the industry's commitment to encouraging new plant orders. The Plan identifies a number of "building blocks" for accomplishing goals. Among these are continued plant safety and reliability, stable licensing including NRC design certification, well-defined utility requirements, successful first-of-a-kind engineering, progress in disposal of high-level and low-level wastes, adequate fuel supply, enhanced government support, and improved public acceptance.

Crucial to the success of the mission are changes in the method of licensing of siting, construction, and operation by the NRC. The conduct of a single hearing for the license on the basis of standardization of designs will reduce the time required and eliminate much uncertainty. Experience gained in the more than 30 y of commercial reactor operation is to be applied to the design, operation, and maintenance of the new advanced reactors. Self-improvement initiatives through INPO will be continued.

The first major step in carrying out the Plan was the development of an Advanced Light Water Reactor Utility Requirements Document (see References). It provides policy statements about key features such as simplification of systems, margins of safety, attention to human factors, design for constructibility and maintainability, and favorable economics.

Two different concepts were specified: (a) A large output (1300 MWe) "evolutionary" design that benefits from current designs, and (b) A mid-size output (600 MWe) "passive" design that depends more on natural processes for safety instead of mechanical-electrical devices.

Numerical specifications include completion in 5 y, low worker radiation exposure (less than 100 mrems/y), refueling on a 24-month basis, and an ambitious 87% average availability over a 60-y design life.

A thorough analysis was made of the means by which standardization can be achieved in design, maintenance, and operation, along with the benefits that accrue:

(a) A reduction in construction time and costs comes from the use of common practices.
(b) Use of identical equipment in several plants favors both economy and safety.
(c) Standardized management, training, and operating procedures will lead to greater efficiency and productivity.

Three advanced reactor designs intended to meet the United States nuclear industry objectives were developed. A description of the principal candidates is given in the following.

The ABB Combustion Engineering System 80+ (see References) is an evolutionary 1300-MWe reactor that satisfies the Requirements Document. Its containment is spherical rather than the typical cylindrical, giving more working space. Its control system features the latest in electronics, including fiber optics, computers, and visual displays. Safety is enhanced by many features, including a gas turbine for emergency AC power. A combination of simplicity and economy of scale makes the cost of electricity competitive. Versions of the design have been built in Korea.

General Electric Co. has an Advanced Boiling Water Reactor (ABWR) design of 1300 MWe (see References). The ABWR circulates coolant by internal pumps. Passive safety features include containment cooling that uses natural convection. Analysis of the plant by probabilistic risk assessment (PRA) indicates negligible hazard to the public. The reactor design conforms to the Requirements Document and was reviewed by the NRC. Two ABWRs are being built in Japan and others are planned.

Westinghouse designed an advanced reactor with acronym AP600, with a lower power level of 600 MWe (see References). The principal design goals were simplicity and enhanced safety. The numbers of pipes, valves, pumps, and cables were greatly reduced in this design. The AP600 has a number of passive processes for safety, the use of gravity, convection, condensation, and evaporation. Examples are a large water storage reservoir for emergency cooling and another one for containment wall cooling.

Early reactor development in the period 1950–1965 was spearheaded by the federal government through the AEC. In the new advanced reactor program, the Department of Energy (DOE) is helping the endeavor, but there is now an imposing array of organizations cooperating to bring about the new generation

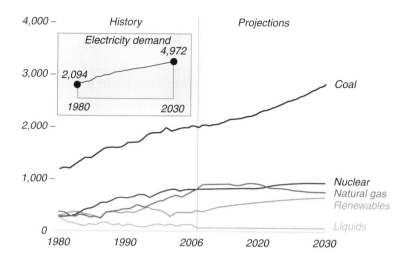

FIGURE 24.3 Energy generation by fuel, 1980–2030 (billion kWh). Source: Annual Energy Outlook 2008. DOE/EIA-0383(2008).

of reactors. In addition to overall guidance by the Nuclear Energy Institute, support is provided by the Edison Electric Institute (EEI), the Electric Power Research Institute (EPRI), and the Institute of Nuclear Power Operations (INPO). A review and forecast by the DOE of electrical generation in the period 1980–2030 is shown in Figure 24.3.

24.6 NUCLEAR POWER RENAISSANCE

The revival of interest in the United States in nuclear power in the first decade of the 21st century is characterized as a renaissance. The situation contrasts with the stagnant conditions of the previous two decades. A number of utilities have made application for reactor license renewals, and several have initiated planning for new nuclear plants, the first in more than 25 y.

Reasons for the new attitude toward nuclear are: (a) increased concern about the global warming consequences of carbon emissions from fossil fueled plants; (b) recognition of the vulnerability of the country to the uncertainty of foreign oil supply; (c) favorable economics compared with other energy sources; and (d) an anticipated greater demand for electricity but opinion that renewables are inadequate.

A consensus developed in the United States that it was time to expand nuclear power. The reactor concept most likely to be adopted in the nuclear renaissance is Westinghouse's Advanced Passive 1000 (AP1000). It is a system for which design preapproval has been issued by the NRC. That approval greatly simplifies the application by a nuclear company for a construction/operating license (COL).

The AP1000 has the same design goals and methods of achieving them as the AP600, discussed in Section 24.5. Again, passive processes such as gravity, convection, combustion, and evaporation are employed, and water reservoirs are provided. Changes from AP600 are limited to those required by the higher power. Modular construction is employed to minimize costs and time. The estimated core damage frequency is smaller than NRC requirements by a factor of 250. Some of the key parameters of the Westinghouse design (see References) are as follows:

> Net electrical/thermal output 1117 MWe/3400 MWt
> Number of 17 × 17 fuel assemblies 157
> Peak coolant temperature 321°C (610°F)
> Reactor vessel ID 399 cm (157 in)
> Operating cost 3.5 cents/kWh

Ownership of Westinghouse had gone through phases including purchase by British Nuclear Fuels (BNFL), which sold the company to the Japanese firm Toshiba.

Three ambitious programs were presented: the Energy Act of 2005 by Congress; Nuclear Power 2010 by the DOE; and Vision 2020 by the nuclear industry. The programs have in common a principal objective to make much greater use of nuclear processes. In the following, we describe the key features of the initiatives. A variety of activities, programs, and projects have been set into motion in preparation for an implementation of the renaissance. These are described briefly in the following.

> *Energy Policy Act of 2005.* First is the Energy Loan Guarantee Fund, which is a sort of insurance that nuclear and other sources of energy can use. The guarantee is up to 80% of the cost of the project and a long-term repayment is allowed. Second is a tax credit of 1.8 cents per kWh produced on 6,000 MW of new capacity for a period of 8 y. For example, if there is an allocation of 750 MW to a 1,000 MW plant, it can claim $(0.75)(1.8) = 1.35$ cents per kWh, with certain limitations. Third is the renewal for 20 y of the Price-Anderson Act, which expired in 2003. It provides insurance coverage in case of nuclear accident. Plants are required to purchase $300 million of private insurance as primary coverage. Then they must pay a fee of $95.8 million per reactor. With more than 100 United States reactors, the total is more than $10 billion. Fourth is standby support for new plants if there is delay caused by the NRC or from litigation. For the first two new plants, 100% coverage is up to $500 million each, with 50% for the next four plants. Fifth is a requirement on NRC to take several actions in support of counterterrorism. Sixth is authorization of approximately $3 billion for nuclear R&D and hydrogen projects. Further details on the Act are found in the References.

Nuclear Power 2010. This program, announced by DOE in 2002, is an industry-government cooperative effort to find sites for new nuclear power plants, to develop advanced technologies, and to test the regulatory process. The objective is to see orders for new reactors by 2010.

The motivation for the program is the expectation that the demand for electricity will increase by 50% by the year 2030. Advantage is to be taken of the fact that uranium is relatively cheap and readily available in Canada and Australia; interest rates are still low; and there is a favorable image of nuclear facilities.

In response to the initiative, three consortia of utilities and vendors have arisen. One plans to use GE's Economic Simplified Boiling Water Reactor (ESBWR); another a PWR, Westinghouse's AP1000; and a third GE's ABWR.

New technologies may assist in approaching a "hydrogen economy" that would greatly relieve the demand for foreign oil for transportation.

Vision 2020. The nuclear industry as represented by the Nuclear Energy Institute proposes to add 50,000 MWe of nuclear power by the year 2020, plus 10,000 MWe by 2012 through increased efficiency of operation and productivity, and a startup of a mothballed reactor. These goals are aimed at contributing to the prevention of global warming because of emissions of greenhouse gases. The amounts are believed necessary to achieve a 30% nonpolluting component of United States electric power generation as hydropower decreases. Immediate action is recommended for a demonstration project to produce hydrogen (and oxygen) from off-peak electricity. Nuclear reactors are needed as well to provide the heat for purification of contaminated water and desalination of seawater to relieve shortages in the Western United States and Florida. Vision 2020 calls also for expansion of uses of radioisotopes in medicine and food safety.

Future nuclear reactors. The growth in world population in the 21st century and the expectations of developing countries will require unusual demands for energy. Nuclear reactors can provide part of that energy economically, safely, and without environmental effects.

Looking toward further improvements in reactor technology, the United States DOE initiated a study of new nuclear designs titled Generation IV. Contributing to the study is the Generation IV International Forum (GIF), a consortium of 10 countries[†] and the European Union. The long-term goals are well described in a 2002 report titled, "Technology Roadmap for Generation IV Nuclear Energy Systems," (see References). The goals are: (a) Sustainable Nuclear Energy, which

---

[†]Argentina, Brazil, Canada, France, Japan, Republic of Korea, South Africa, Switzerland, United Kingdom, and United States.

focuses on waste management and resource utilization; (b) Competitive Nuclear Energy, which seeks low cost electricity and other products such as hydrogen; (c) Safe and Reliable Systems, which implies both prevention and responses to accidents; (d) Proliferation Resistance and Physical Protection, which means control of materials and prevention of terrorist action. Of a large group of reactor systems, six were identified as prospects for research and development, as follows:

1. Gas-cooled Fast Reactor (GFR). Helium cooled, fast neutron spectrum, closed fuel cycle for actinide burnup.
2. Lead-cooled Fast Reactor (LFR). Lead or Pb-Bi coolant, fast neutron spectrum, closed fuel cycle, metal fuel, long core life.
3. Molten Salt Reactor (MSR). Circulating coolant of molten Na-Zr-U fluorides that allows actinide feeds, graphite moderator, thermal spectrum.
4. Sodium-cooled Fast Reactor (SFR). Na coolant, fast neutron spectrum, closed pyrometallurgical fuel cycle, MOX fuel.
5. Supercritical-water-cooled Reactor (SCWR). Fast or thermal spectrum, operation above the critical point for water (22 MPa, 374°C), high thermal efficiency.
6. Very High Temperature Reactor (VHTR). Thermal spectrum, once-through cycle, either HTGR or pebble bed-type, high efficiency, useful for process heat.

Global Nuclear Energy Partnership. In 2006 an initiative with multiple benefits called Global Nuclear Energy Partnership (GNEP) was proposed by President Bush. It seeks to involve a number of countries in R&D needed to achieve an ideal nuclear system. As described in a set of Web sites, the program has the following seven elements.

Expansion of United States nuclear power. GNEP is the next step in the efforts to stimulate new nuclear plant construction. This element builds on the earlier initiatives of the Energy Policy Act, Nuclear Power 2010, Standby Support, and Early Site Permits.

Proliferation-resistant recycling. New chemical separation processes make plutonium less accessible to rogue nations or terrorists. As a substitute for the PUREX process, which creates a pure plutonium product, the UREX+ process will include other actinides—neptunium, americium, and curium. The mixture of elements will be recycled to burn out wastes and obtain maximum energy.

Minimization of wastes. Separated fission products would have a volume far less than that of spent fuel. The heat source and activity of wastes would be reduced to the extent that the Yucca Mountain repository would be adequate to handle both past and future needs.

Advanced burner reactors. New fast neutron reactors are to be developed to destroy transuranics by fission and transmutation. A small test reactor would be followed by a full-sized reactor, and ultimately by a fleet of advanced burners.

Reliable fuel services. Cooperation is sought between "fuel supplier nations" and "user nations" to reduce incentives for enrichment and reprocessing. Suppliers will ensure fuel to be used and returned for recycling and take care of residual wastes.

Small-scale reactors. Reactors that are suitable for developing nations are to be designed and tested. They would be simple, safe, long-lived, and proliferation-resistant. They would be able to provide heat applications such as district heating and desalination.

Nuclear safeguards. Full control and accounting for all nuclear materials is visualized. In the reactor development program, cooperation with International Atomic Energy Agency (IAEA) will enhance safeguards and deter proliferation of weapons.

The GNEP program is planned to start with studies of feasibility and economics, and be implemented well into the 21st century. Some observers believe that the GNEP will be unsuccessful, in part for lack of committed funding, and there is an opinion that the project distracts and detracts from the United States nuclear power renaissance (see articles by Kadak and by Wald in References).

24.7 SUMMARY

Half the cost of electric power is for generation. Electricity from plants that use coal or nuclear fuel is comparable in cost, with a tradeoff between capital costs and fuel costs. Costs of construction of nuclear plants and the time to complete them in the United States were exorbitant for several reasons. There have been no orders for new nuclear plants since 1978. The nuclear industry has several opportunities for improvements including license extension but is faced with the challenges of electricity restructuring. Several advanced reactor concepts are being promoted to preserve the nuclear option. A nuclear renaissance involves building additional power plants and carrying out research on new reactor concepts.

24.8 EXERCISES

24.1 Many different energy units are found in the literature. Some of the useful equivalences are:

$1 \text{ eV} = 1.602 \times 10^{-24} \text{ J}$

$1 \text{ cal} = 4.185 \text{ J}$

$1 \text{ Btu} = 1055 \text{ J}$

$1 \text{ bbl (oil)} = 5.8 \; 10^6 \text{ Btu}$

$1 \text{ quad} = 10^{15} \text{ Btu}$

$1 \text{ Q} = 10^{18} \text{ Btu}$

$1 \text{ exajoule (EJ)} = 10^{18} \text{ J}.$

(a) Find out how many barrels of oil per day it takes to yield 1 GW of heat power.

(b) Show that the quad and the EJ are almost the same.

(c) How many quads and Q correspond to the world annual energy consumption of approximately 300 EJ?

(d) How many disintegrations of nuclei yielding 1 MeV would be needed to produce 1 EJ?

24.2 Find the yearly savings of oil by use of uranium in a nuclear reactor, with rated power 1000 MWe, efficiency 0.33, and capacity factor 0.8. Note that the burning of one barrel of oil per day corresponds to 71 kW of heat power (see Exercise 24.1). At $75 a barrel, how much is the annual dollar savings of oil?

24.9 REFERENCES

Electricity Sources DOE/EIA
http://www.eia.doe.gov/cneaf/electricity/epa/epat1p1.html
Data on generation by type of producer.

The 1970's Energy Crisis
http://cr.middlebury.edu/es/altenergylife/70's.htm
Effects of the oil embargo.

Alvin M. Weinberg, *Continuing the Nuclear Dialogue*, American Nuclear Society, La Grange Park, IL, 1985. Essays spanning the period 1946–1985, selected and with introductory comments by Russell M. Ball.

Karl O. Ott and Bernard I. Spinrad, *Nuclear Energy: A Sensible Alternative*, Plenum Press, New York, 1985. Titles of sections: Energy and Society, Economics of Nuclear Power, Recycling and Proliferation, Risk Assessment, and Special Nuclear Issues Past and Present.

Philip C. Jackson, Jr., *Introduction to Artificial Intelligence*, 2nd Ed., Dover Publications, New York, 1985. A popular update of a classic work.

John A. Bernard and Takashi Washio, *Expert Systems Applications Within the Nuclear Industry*, American Nuclear Society, La Grange Park, IL, 1989.

Instrumentation and Control
http://portfolio.epri.com
Select 2009 Research Offerings/Nuclear/Instrumentation and Control.

Richard C. Dorf and Robert H. Bishop, *Modern Control Systems*, Addison-Wesley, Menlo Park, CA, 1998. Feedback control theory with practical examples. A chapter titled Digital Control Systems. Two nuclear problems.

Regulatory Guides on Digital Computer Software
http://www.nrc.gov/reading-rm/doc-collections/reg-guides
Select Power Reactors to access Regulatory Guides 1.168–1.173.

Digital Instrumentation and Control Systems in Nuclear Power Plants
http://www.nap.edu
Search on the topic above.

License Renewal Process
http://www.nrc.gov/reactors/operating/licensing/renewal.html
Overview, Process, Regulations, Guidance, and Public Involvement.

License Renewal Generic Environmental Impact Statement
http://www.nrc.gov/reading-rm/doc-collections/nuregs/staff/sr1437
Available to utilities to reference in applications. From NRC.

Electricity: The Road Toward Restructuring
http://www.cnie.org/NLE/CRS/abstract.cfm?NLEid=67
Congressional Research Service Report by Abel and Parker.

Energy Policy Act of 2005
http://www.nei.org
Search: Energy Policy Act of 2005; Select Highlights, etc. (from Nuclear Energy Institute).

The Global Nuclear Energy Partnership
http://www.gnep.energy.gov
Select from Implementing Elements.

Matthew Wald, "The Best Nuclear Option," *Technology Review*, July/August 2006, p. 59.

Advanced Reactors
http://www.uic.com.au/nip16.htm
A briefing paper from Australia. Table of reactor data, features, and status.

Advanced Reactors
http://www.nucleartourist.com
Links to the principal designs. By Joseph Gonyeau.

Howard J. Bruschi, "The Westinghouse AP1000—Final Design Approved," *Nuclear News*,
    November 2004, pages 30–35

"A Technology Roadmap for Generation IV Nuclear Energy Systems," DOE, December 2002.
    Idaho National Laboratory
http://nuclear.inl.gov/gen4
Click on A Technology Roadmap.

New Commercial Reactor Designs
http://www.eia.doe.gov/cneaf/nuclear/page/analysis/nucenviss2.html
From DOE/EIA.

ABB Combustion Engineering System 80+
http://www.nuc.berkeley.edu/designs/sys80/sys80.html
University of California Nuclear Engineering student report.

AP600
http://www.ap600.westinghousenuclear.com
Attractive and informative Web site.

Timothy J. Brennan, Editor, *A Shock to the System: Restructuring America's Electricity*,
    Resources for the Future, Washington, DC, 1996.

Nuclear Renaissance
http://www.eurekalert.org/features/doe/2003-12/danl-nr031804.php
AAAS/DOE.

Annual Energy Outlook 2008 With Projections for 2030
http://www.eia.doe.gov/oiaf/aeo
Report DOE/EIA-0383(2008) on all forms of energy.

International Nuclear Power

ALTHOUGH THE United States spearheaded research and development of nuclear power, its use in other parts of the world has expanded greatly. There are two reasons: (a) many countries do not have natural energy sources of coal and oil, and (b) some countries such as France and Japan have state-owned or strongly state-supported nuclear power systems. On the other hand, the distribution of the use of nuclear power throughout the world is quite uneven. We will now look at the global power situation and examine trends for large geographic or socioeconomic units—Western Europe, the Far East, the former U.S.S.R., and developing countries.

We will not be able to give a full and detailed account of the complex and changing energy situation abroad because of the many countries and organizations involved. Rather, we will concentrate on the status and trends of nuclear power plants. It turns out that the best information on international nuclear power is found on the Internet. For data on the countries themselves consult the CIA World Factbook and World Information. For many of the nuclear facilities see The Virtual Nuclear Tourist. Maps of locations of many nuclear plants can be found from the International Nuclear Safety Center. For a complete list of nuclear organizations around the world, use the IAEA's INIS. The Department of Energy (DOE) Information Administration provides a great deal of data on politics, economics, energy, and nuclear power.

In the decade of the 1990s, three major economic movements tended to dominate the nuclear electric industry of the world. These were privatization, globalization, and deregulation. In privatization, facilities are sold by a government to private industry to raise cash and potentially achieve better performance. Globalization involves the purchase of companies across national borders, for example Entergy of New Orleans buying London Electricity and then selling it to Electricité de France. Consolidation into multinational companies also is popular. Deregulation (also see Section 24.4) opens up new options for sources of supply of electrical energy to customers. The preceding three processes tend to make the nuclear power situation abroad more fluid and complex. In Europe, it is the European Union that is the driving force behind deregulation, rather than by states as in the United States. Reviews and analyses are found in References.

At the social/political level, the strength of opposition to all aspects of nuclear energy has a decided influence on actions of politicians. The resultant conflict between government and industry in some countries results in stagnation or the demise of nuclear power.

25.1 REACTOR DISTRIBUTION

A review of the status of nuclear power in countries around the world is provided in Table 25.1, which shows the number of reactors and the megawatts of power for those in operation and under construction for all nations that are committed to nuclear power. The damaged TMI-2 and Chernobyl reactors have been omitted from the table, but it includes some whose future is uncertain. The table should be treated as a snapshot of a status subject to change.

Table 25.1 World Nuclear Power as of December 31, 2007. (Source: *Nuclear News*, American Nuclear Society, March 2008)

Country	Units Operating		Total	
	No.	MWe	No.	MWe
Argentina	2	935	3	1,627
Armenia	1	376	1	376
Belgium	7	5,801	7	5,801
Brazil	2	1,901	3	3,176
Bulgaria	2	1,906	4	3,906
Canada	22	15,164	22	15,164
China	11	8,694	31	29,514
Czech Republic	6	3,472	6	3,472
Finland	4	2,656	5	4,256
France	59	63,363	60	64,963
Germany	17	20,429	17	20,429
Hungary	4	1,799	4	1,799
India	17	3,732	23	6,672
Iran	0	0	1	915

(*continued*)

Table 25.1 World Nuclear Power as of December 31, 2007. (Source: *Nuclear News*, American Nuclear Society, March 2008)—*cont* . . .

Country	Units Operating		Total	
	No.	MWe	No.	MWe
Japan	55	47,589	58	50,074
Lithuania	1	1,185	1	1,185
Mexico	2	1,360	2	1,360
Netherlands	1	485	1	485
Pakistan	2	425	3	725
Romania	2	1,412	5	3,272
Russia	31	21,743	38	26,328
Slovakia	5	2,034	7	2,844
Slovenia	1	666	1	666
South Africa	2	1,800	2	1,800
South Korea	20	16,810	26	23,610
Spain	8	7,439	8	7,439
Sweden	10	8,916	10	8,916
Switzerland	5	3,220	5	3,220
Taiwan	6	4,884	8	7,484
Ukraine	15	13,107	18	15,957
United Kingdom	19	10,982	19	10,982
United States	104	102,056	105	103,233
Non-U.S.	339	274,285	399	328,417
Total	**443**	**376,341**	**504**	**431,650**

Several observations can be made about the table. The United States has approximately one fourth of the reactors of the world. France, with its population approximately a fifth that of the United States, has by far the largest per capita use of nuclear power. When construction is complete, France will produce nearly half as much nuclear electricity as the United States. Japan has a growing nuclear power system, third in the world after the United States and France. Korea continues to add power plants. Except for a small program in South Africa, the continent of Africa is not represented; except for Brazil and Argentina, countries in South America have no power reactors. The developing nations of those regions may adopt nuclear power in the future. The People's Republic of China, despite its vast population, is just getting started on a power program.

Another perspective of the world's nuclear activities is provided in Figure 25.1, giving the percentage of the various countries' electricity that is supplied by nuclear power. The distributions of Table 25.1 and Figure 25.1 tend to reflect the status of technological development, with variations dependent on available natural resources and public acceptance. Finally, we note that two thirds of the more than 100 countries of the globe do not have any plans for reactors.

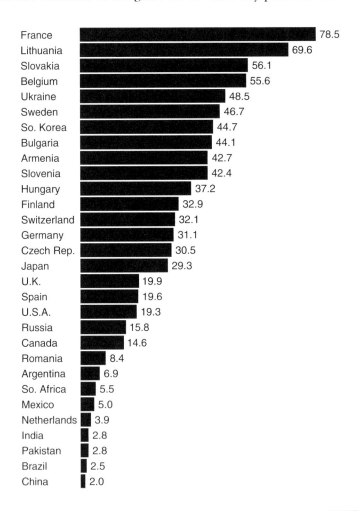

Country	Percentage
France	78.5
Lithuania	69.6
Slovakia	56.1
Belgium	55.6
Ukraine	48.5
Sweden	46.7
So. Korea	44.7
Bulgaria	44.1
Armenia	42.7
Slovenia	42.4
Hungary	37.2
Finland	32.9
Switzerland	32.1
Germany	31.1
Czech Rep.	30.5
Japan	29.3
U.K.	19.9
Spain	19.6
U.S.A.	19.3
Russia	15.8
Canada	14.6
Romania	8.4
Argentina	6.9
So. Africa	5.5
Mexico	5.0
Netherlands	3.9
India	2.8
Pakistan	2.8
Brazil	2.5
China	2.0

FIGURE 25.1 Nuclear share of electricity generation 2008. Numbers are percentages. (Source: IAEA).

25.2 **WESTERN EUROPE**

A transition has occurred in power generation in Western Europe. More electricity there comes from nuclear power than from any other power source.

The leading user of nuclear power in Europe is France. Its nuclear situation is dominated by the fact that the country does not have gas, oil, or coal but does have some uranium. Power is supplied by one company, Electricité de France (EdF), which is making a profit and reducing its debt despite a very large growth in facilities. All support for the French power system is provided by two companies Areva (formerly Framatome) for reactor design and construction and Cogema for fuel supply and waste management. The Ecole Polytechnique provides the education of all of the operators and managers, and thus the common training is transferable between units. Safety in reactor operation is thus enhanced. Because reactors are standardized and the system is state owned, France is able to avoid the licensing and construction problems of the United States. It only requires 6 y to build a nuclear power station. Over the years, there was little opposition to nuclear power in France, in part because the state had provided attractive amenities to local communities while emphasizing the necessity of the power source for the nation's economy. EdF sells low-cost electricity to other countries, including the United Kingdom, by use of a cable under the English Channel. Lacking fossil fuel resources, France has focused on energy security through the production and use of nuclear power. Almost 80% of its electricity comes from its 59 reactors. The growth in nuclear power has leveled, but public support remains high. The text of a PBS television program (see References) discusses why the French like nuclear power.

The fast breeder reactor Superphenix had operated successfully until sodium leaks caused it to be shut down, and attempts to revive the program failed. The lower power Phenix serves as a research vehicle.

A thorough account of the French nuclear power program is given by the Australian Uranium Information Centre (see References). In 2006 a decision was made to build a 1670-MWe European Power Reactor (EPR) and to design a sodium-cooled fast reactor, both with 2012 as the goal.

With the unification of Germany in 1990, the nuclear power program of the former East Germany was suspended in the interests of safety. Operation of the remaining plants was very successful, with high-capacity factors. Some electricity was available for export. One third of Germany's electricity comes from nuclear power in 17 plants. There is considerable ambivalence about the future, with generally favorable support by the public. However, there has been active opposition by antinuclear political parties. Utilities would like to extend operations, whereas elements of government would like them phased out. Among expressed concerns were the purported danger of transportation of fuel cycle materials and the nuclear waste disposal problem.

It is not clear what would substitute for the current nuclear fraction of the electricity supply. Higher costs and blackouts are likely.

In Belgium, the utility Electrabel provides generation and transmission, as well as several other public services. A large fraction (56%) of the country's electricity is provided by seven nuclear plants, some of which use MOX fuel. A phaseout is required by present law.

Sweden had a strong nuclear R&D program, leading to 12 reactors that produced approximately half of the country's electricity. Two reactors were shut down after a phaseout policy. After the TMI accident, a public referendum called for continuing operation of the plants. However, a decision was made not to expand nuclear power, and pressure caused one reactor to be closed. Power needs were to be made up by importation. In light of concerns about economics and global warming, the reactors may be allowed to continue operating. The situation is quite complex.

Finland has two Soviet Pressurized Water Reactors (PWRs) and two Swedish PWRs, all of which operate at very high capacity factors. The country has developed an effective underground nuclear waste storage system. Nuclear power is favored for economic and environmental reasons. Areva is building in Finland the first European Pressurized Reactor (EPR), with a start date around 2010.

Switzerland enjoys very high reactor capacity factors, with an average of approximately 85%. The Beznau plant is designed for district heating, providing 70 MWt heat power to 5,000 households in six towns within a 13-km radius. Increased demand has been met by increasing the power rating of existing reactors and through import of more electricity from France. Some Swiss favor nuclear plants because they reduce coal imports and are free of pollution. In 2003 voters rejected a proposal to phase out nuclear power, but there remains some opposition. A new energy policy calls for additional plants.

Spain is seeking to improve its economic condition through growth in electricity use. It has made good progress despite turmoil in the Basque region, where terrorist action forced suspension of construction on two reactors. The country has excellent facilities for production of nuclear equipment. There is a possibility that Spain's eight reactors will be phased out, but it is not clear what will replace the needed power. Some consideration is given to wind power.

Portugal is expected to continue to rely on natural gas and does not intend to develop nuclear facilities.

The United Kingdom has a long history of the use of gas-cooled nuclear reactors for commercial electricity. For years, the cooperation of a state agency and a commercial organization worked well. In 1990, the nuclear industry was privatized, with British Energy buying facilities. The Sizewell B PWR was put into operation in 1996. It features a highly modern computer management system. A few of the older Magnox reactors are being phased out after several decades of operation. Britain maintains reprocessing facilities, serving Japan. There is favorable government support for new nuclear capacity to avoid carbon emissions and to provide energy security.

Italy adopted nuclear power at an early date, but after a public vote in 1987 mothballed or closed all of its reactors. The country relies on oil-fired plants and imported electricity. No nuclear generation is likely in the future.

The Netherlands produces only a small part of its electricity from a reactor. After planning a shutdown, a reversal of policy calls for long-term extension.

Austria had completed one power reactor but never operated it because of a national referendum. The frustrated reactor operators decided to dismantle and sell the plant components.

Although few reactors will be coming on line in Europe, there is considerable interest in the European Pressurized Reactor (EPR) being built in Finland.

25.3 EASTERN EUROPE AND THE CIS

In the late 1980s, the former Soviet Union was embarked on a nuclear power expansion program aimed at increasing electricity approximately 10% per year, with a long-range goal of approximately 100,000 MWe. It was expected that the use of centralized factories making standardized designs and the use of specialist teams would permit construction times of less than 5 y.

With the advent of the Chernobyl accident and related adverse public reaction and the economic stresses associated with the political changes in Eastern Europe, the planned program lost momentum and did not recover until around 2000. Several PWRs are being planned, are under construction, or going into operation. Russia has a firm commitment to expand nuclear power. The breakup of the Soviet Union in 1991 and the creation of the Commonwealth of Independent States (CIS) resulted in a new national distribution of reactors. There are reactors in Russia, Ukraine, and Lithuania. Several countries formerly allied with the Soviet Union were dependent on it for designs and technical assistance. Those that still have reactors are Armenia, Czech Republic, Slovakia, Hungary, Romania, and Bulgaria.

Russia has in operation several graphite moderated light water-cooled reactors. An equal amount of power comes from VVERs, and several newer reactors are coming on line. One fast breeder BN-600 reactor, Beloyarskiy, is in operation.

Ukraine has severe economic problems, including difficulty in buying fuel for its reactors and the loss of skilled workers. Fossil and nuclear plants each provide nearly half of its electricity. All of the reactors are VVERs, with the RBMKs phased out. Ukraine continues to anticipate a growth in nuclear power.

Lithuania's two reactors at Ignalina were of the RBMK type, like Chernobyl, and the European Union made closure of one of them a condition for acceptance of Lithuania. A replacement is planned.

Kazakhstan's reactor at Aktau was shut down after many years of operation. It was unique in that it was a fast breeder reactor and used its waste heat for desalination of water. Kazakhstan has extensive uranium deposits and plans to build new reactors.

Armenia's one reactor was shut down because of concerns about earthquakes but was started again because of an energy crisis. The European Union wants the plant closed for safety reasons.

The Czech Republic has six modern VVERs.

Poland started building a reactor in the 1980s but stopped after Chernobyl. The country is planning for a nuclear plant in the distant future.

Hungary has a four-unit Russian plant that provides a large fraction of the country's electric power, but has no plans for additional reactors.

Romania's one pressurized heavy water reactor built by Canada is operating, with four others partially completed.

Bulgaria originally operated six Soviet-supplied PWRs, but they were criticized for poor safety. The European Union required shutdown of some as condition for Bulgaria's acceptance. In consequence, meeting of Kyoto standards will be problematic.

Slovakia has several Russian VVER-440 reactors in operation, providing approximately half of the country's electricity, and other reactors are under construction. Despite improvements in safety, the European Union required shutdowns as a condition for acceptance.

Slovenia, formerly part of Yugoslavia, has one American PWR that also serves Croatia. A second plant is being considered.

Concern has been expressed that the political and economic situation in Eastern Europe would result in more reactor accidents. It is generally conceded that many reactors of Eastern Europe and the Commonwealth of Independent States are not as safe as West European or American reactors and that in many cases maintenance and operating practices are not as rigorous. The countries have been urged to shut down some of the older reactors in the interests of safety but have resisted on the grounds that the need for electric power is crucial. The United States and other nations are providing technical advice and financial assistance to the countries of the former U.S.S.R. Organizations in the United States include the Department of Energy, the Nuclear Regulatory Commission, and the Nuclear Energy Institute. The countries of the European Union and the World Association of Nuclear Operators are also involved.

One justification for helping is the principle that "a reactor failure anywhere is a failure everywhere," reflecting extreme public sensitivity to reactor incidents. Another basis is enunciated by nuclear power commentator Simon Rippon,[†] "Help must be given to improve reactors in Eastern Europe and the CIS for one reason only: because the continued operation of most of these plants is vital to the economic survival of the countries concerned. The greater threat to the world at large is not of another Chernobyl, but the chaos that could ensue if the new Eastern economies fail."

[†]In *Nuclear News*, August 1992.

25.4 **THE FAR EAST**

The principal user of nuclear power in the Far East is Japan. Government and industry have been committed to a successful nuclear program. Starting with a nuclear capacity of 33 GWe in 1991, Japan had hoped to reach 50 GWe by 2000 and 72.5 GWe by 2010. These goals will not quite be reached, even though reactor construction times are low, slightly over 4 y. The operation of existing PWRs and boiling water reactors (BWRs) has been highly efficient, a result of the Japanese work ethic, mutual company-employee trust, and attention to detail. Japan's national goal of becoming essentially energy-independent is to be met by use of facilities for enrichment, fabrication, reprocessing, and waste disposal. Reprocessing is justified on grounds of assuring a stable fuel supply rather than on economics. In recovering plutonium and burning it in light water reactors (LWRs) or preferably fast breeders, Japan avoids large stockpiles of plutonium. Research is in progress on advanced PWRs, a breeder reactor, a high-temperature gas-cooled reactor (see References), and a tokamak fusion system. Several nuclear accidents in Japan have dampened enthusiasm for nuclear power expansion. A sodium leak occurred in the fast breeder MONJU, and there was a fire and explosion in a reprocessing plant. In 1999 a criticality accident happened when operators put too much enriched uranium in a vessel. Fears of contamination of the vicinity were unfounded, but one worker died from radiation exposure. Nevertheless, in the long run, concerns in Japan about gaseous emissions from fossil plants may outweigh concerns involving radioactivity.

South Korea has achieved very large growth in productivity over recent decades. Because it must import all of its oil and gas, it is expanding its nuclear power program. Four of the reactors are Canadian Deuterium Uranium (CANDU), the rest PWRs from Westinghouse, ABB-CE, and Framatome. One reactor was designed and built with Korean technology. Several additional plants are planned before 2015, two of which are third-generation System 80+.

North Korea has experienced very severe economic problems, and relations with South Korea remain tense. A small reactor was purportedly used to produce weapons plutonium. It may be shut down in exchange for assistance from other countries.

Taiwan, being an island, has no electrical power connections to other countries and for its rapid transition from agriculture to industry it has been highly dependent on imported oil. More than half of Taiwan's electricity comes from nuclear plants. Four LWR plants are operating, and General Electric is supplying two advanced boiling water reactors.

China's situation is different from that of many countries of the world. It has a tremendous need for electric power, its per capita consumption being a few percent of that of the United States. China's principal energy source is coal, creating serious environmental problems. A major hydroelectric dam giving 18.2 GWe is under construction. Expansion of nuclear power with the help of foreign firms is underway, but the added power will be minimal in terms of the large population and energy demand. The 300 MWe Qinshan-1 PWR is of indigenous design and construction. Others were provided by Canada, France, and Russia (VVERs). Several more are under construction or proposed.

Indonesia is planning an ambitious nuclear power program to start construction around 2010.

A Philippine power reactor was mothballed in 1987 and is unlikely to ever operate.

25.5 OTHER COUNTRIES

Nuclear programs of selected countries of several continents are reviewed briefly.

India has one BWR and several pressurized heavy water reactors of approximately 200-MWe capacity, with others under construction. The government nuclear program (see References) plans for 20,000 MWe by 2020, including two Russian VVER-1000s. India's fast reactor experimental facility is fueled by Pu-U carbide with a thorium blanket, intended to test the use of the large indigenous reserves of thorium. A number of coal, gas, and hydro plants are planned.

Pakistan has very low per capita electricity consumption but has a growing demand. This will likely be met by hydroelectric power and coal. A small 300-MWe PWR supplied by China augments the older 125-MWe plant. Plans call for a large increase in nuclear power within the next few decades.

Turkey is seeking to expand its electric power, mainly by use of imported natural gas and by developing new hydro power. Several nuclear plants are planned as well, with cooperation with Iran on energy matters.

Iran is working with Russia to complete a power reactor and seeks to enrich uranium with centrifuges. Other countries object, fearing the Iran has nuclear weapons ambitions, and use sanctions to try to stop enrichment.

In South Africa an aggressive program of electrification is designed to improve living conditions. It will be mainly based on coal. One nuclear station has two PWR reactors from Framatome. A demonstration pebble bed modular reactor is planned.

The heavy-water moderated reactors of Canada operated very successfully for many years. The CANDU uses natural or very slightly enriched uranium in pressure tubes that permit refueling during operation. Very high-capacity factors are thus possible. Canada has established a heavy water industry and uses uranium mined within the country. The government corporation Atomic Energy Canada Ltd. (AECL) provided heavy water reactors for Korea. Several of the CANDU reactors were shut down because of management and technical problems. Restarting was of high priority.

Mexico has one nuclear plant at Laguna Verde on the Gulf of Mexico, with two General Electric BWRs.

In Cuba, two Russian reactors of 408 MWe have been planned and partly constructed. The United States objects on safety grounds. Completion is problematic.

Brazil's nuclear electricity from two reactors provides approximately 4% of the main source, hydroelectric. One reactor was supplied by Westinghouse and the other by Germany's Kraftwerk Union. The country is recovering from a late 1990s financial crisis, which cut back on plans for eight reactors.

Argentina has ample oil, natural gas, and hydro potential. It has two pressurized heavy water reactors with a third halted for lack of funds.

The foregoing sections indicate that the rate at which nuclear power is being adopted varies greatly throughout the world, because each country has a unique situation. In some countries public opinion is a dominant factor; in others limited capital; in still others, especially developing countries, a lack of technological base. For several Latin American countries, large national debts are limiting. Despite problems, the amount of nuclear power abroad continues to grow slowly. Table 25.2 shows the number of reactors and their power for the sum of those in

Table 25.2 Reactors in Operation or Under Construction, for Ends of Years 1978-2006 (From Issues of *Nuclear News*, American Nuclear Society)

Year	U.S.		Non-U.S.	
	No.	MWe	No.	MWe
1978	195	189,604	328	215,364
1979	189	182,015	341	223,753
1980	172	163,549	361	244,910
1981	166	157,654	363	244,422
1982	147	135,534	374	257,609
1983	139	128,507	389	275,003
1984	129	119,006	399	285,991
1985	129	118,962	407	293,919
1986	127	116,989	426	311,475
1987	126	116,939	438	320,231
1988	125	114,461	435	319,870
1989	119	109,012	427	311,450
1990	119	109,184	406	302,744

(continued)

Table 25.2 Reactors in Operation or Under Construction, for Ends of Years 1978-2006 (From Issues of *Nuclear News*, American Nuclear Society)—*cont* . . .

Year	U.S.		Non-U.S.	
	No.	MWe	No.	MWe
1991	119	109,307	392	296,919
1992	116	107,573	388	296,360
1993	116	107,906	390	298,352
1994	115	106,517	375	285,023
1995	113	104,453	382	290,304
1996	112	104,062	384	293,425
1997	108	101,582	387	298,021
1998	107	101,382	387	298,531
1999	107	101,633	389	301,541
2000	107	101,686	383	298,806
2001	107	102,678	383	299,292
2002	107	102,637	387	303,499
2003	107	103,971	385	309,777
2004	107	104,063	385	307,035
2005	105	102,466	384	307,634
2006	105	102,691	386	314,087

operation and under construction in two categories: United States and non-United States. The total world power is almost the same at the end and beginning of the period. For 1978 it was 404,968 MWe; for 2007 it was 431,650. The decline in United States reactors from an initial share of 47% to a final share of 25% is closely matched by the rise in reactors abroad. The shift tends to parallel the decline in United States leadership in several areas of technology.

25.6 SUMMARY

The need for power and the lack of fuel resources in many countries has prompted the adoption of nuclear reactors for electric power. As of the end of 2007 there were 104 operating reactors in the United States and 339 abroad.

The leading countries, in decreasing order of operating nuclear power, are the United States, France, Japan, Germany, and Russia. The decline in United States reactors planned and under construction over the years is balanced by the rise in non-United States reactors.

25.7 REFERENCES

Uranium Information Centre (Australia)
http://www.uic.com.au
Select Briefing Papers, country name.

International Nuclear Safety Center
http://www.insc.anl.gov
Select Maps

International Nuclear Power Plants
http://www.radwaste.org/power.htm
From organization Radwaste.org (Disregard "The web page cannot be found").

IAEA: Power Reactor Information System (PRIS)
http://www.iaea.org/programmes/a2
Select country name, plant.

World Nuclear Association
http://www.world-nuclear.org
Click on Information Papers, select country name.

World List of Nuclear Power Plants
Nuclear News (annual March issue)

Energy Information Administration
Google: "EIA World Nuclear Reactors."

Dick Kovan, "ENC 2007: Europe Shows the Way," *Nuclear News*, September 2007, p. 34
Status and prospects of nuclear power in 16 European countries.

Privatization and the Globalization of Energy Markets
http://www.eia.doe.gov/emeu/pgem/contents.html
Discusses trends in many countries.

Electricity Reform Abroad and United States Investment
http://www.eia.doe.gov/emeu/pgem/electric
DOE report explaining processes and emphasizing privatization experience in Argentina, Australia, and the United Kingdom.

The World Factbook 2008
https://www.cia.gov/library/publications/the-world-factbook
Information about all countries—geography, people, government, economy, communications, transportation, and transnational issues. By Central Intelligence Agency.

World Information
http://www.infoplease.com/a0107262
Facts and histories of countries.

International Nuclear Information System (INIS)
http://www.iaea.org/inisnkm
Select Links. List of nuclear organizations. From International Atomic Energy Agency.

The Virtual Nuclear Tourist
http://www.nucleartourist.com
Large amount of information about domestic and foreign reactors.

European Nuclear Forums
http://www.foratom.org
A trade association for the nuclear energy industry in Europe.

Nuclear Energy Agency
http://www.nea.fr
Nuclear arm of the Organisation for Economic Co-operation and Development (OECD), based in
    Paris, with 28 member countries, mainly in Europe, North America, and the Far East.

Energy Technology Data Exchange
http://www.etde.org
Select ETDEWEB for a large collection of documents. Membership is by registration. From
    International Energy Agency.

Nuclear Power Corporation of India, Ltd.
http://www.npcil.nic.in
Performance data for all plants.

Nuclear Energy in Sweden
http://www.uic.com.au/nip39.htm
A thoughtful paper from Australia's Uranium Information Centre.

NEI Source Book on Soviet-Designed Nuclear Power Plants, 5th Ed.
http://www.insc.anl.gov/neisb/neisb5
Download chapters or full document in pdf (1.4 MB, 362 pages). Program histories and situa-
    tions; advantages and deficiencies of designs; actions being taken to upgrade.

International Energy Outlook
http://www.eia.doe.gov/oiaf/ieo
Select World Energy Demand and Economic Outlook or Electricity.

The International Status of Nuclear Power
http://www.uic.com.au/nip07.htm
Briefing paper with a table of reactors (No. and MWe) with electricity generation and uranium
    required. From Australia.

Office of Nuclear Affairs of the French Embassy in Washington, DC
http://ambafrance-US.org/spip.php?article949
General information on nuclear power in France.

Nuclear Reaction. Why the French Like Nuclear Energy
http://www.pbs.org/wgbh/pages/frontline/shows/reaction/readings/french.html
(Or Google "French nuclear power").

Nuclear Power in France (February 2007)
http://www.uic.com.au/nip28.htm

La Hague Reprocessing Plant.
Google Areva La Hague

Electricité de France (EDF)
http://www.edf.fr/html/fr/index.html
Operator of all nuclear plants in France. Select English version.

Areva
http://www.areva.com
About the world's leading nuclear manufacturer.

World Nuclear Association
http://www.world-nuclear.org
Select Information Papers/Germany.

Annual Energy Outlook 2008
http://www.eia.doe.gov/oiaf/aeo
Estimate of trends to 2030 in all forms of energy, provided by DOE's Energy Information
    Administration.

International Energy Annual (IEA)
http://www.eia.doe.gov/iea
Select World Energy Overview.

Nuclear Engineering International
http://www.neimagazine.com
Dedicated to nuclear matters around the world.

Nuclear Explosions

THE PRIMARY purpose of this book is to describe the peaceful and beneficial applications of nuclear energy. To attempt a discussion of the military uses is risky because of the emotional nature of the subject and the impossibility of doing justice to the complex problems involved. To neglect the subject, however, would be misleading, as if we wished to suggest that nuclear energy is entirely benign. Thus, we will review some important facts and ideas about nuclear explosions and their uses, with three objectives:

(a) Distinguish between nuclear power and nuclear weapons.
(b) Identify the technical aspects and strategic issues involved in the military use of nuclear processes.
(c) Indicate the continued need for control of nuclear materials.

We will describe nuclear explosions, nuclear weapons proliferation and safeguards, disarmament, and the options for disposal of weapons material.

26.1 NUCLEAR POWER VERSUS NUCLEAR WEAPONS

In the minds of many people there is no distinction between reactors and bombs, resulting in an inordinate fear of nuclear power. They also believe that the development of commercial nuclear power in countries abroad will lead to their achievement of nuclear weapons capability. As a consequence of these opinions, they favor dismantling the domestic nuclear industry and prohibiting United States commercial participation abroad.

Recalling some World War II history will help clarify the situation. The first nuclear reactor, built by Enrico Fermi's team in 1942, was intended to verify that a self-sustaining chain reaction was possible and also to test a device that might generate plutonium for a powerful weapon. The experiment served as a basis for the construction of plutonium production reactors at Hanford, Washington. These supplied material for the first atom bomb test at Alamogordo, New Mexico, and later for the bomb dropped at Nagasaki. The reactors used generated heat but no electric power and were designed to favor the production of plutonium-239. More recently, plutonium for weapons was produced by reactors at the Savannah River Plant in South Carolina.

Isotope separation production facilities at Oak Ridge during World War II yielded uranium enriched to approximately 90% U-235. The material was fabricated into the bomb used at Hiroshima. Subsequently, separation facilities have been used to give the 3% to 4% fuel for light water power reactors. Such fuel can be made critical when formed into rods and moderated properly with water, but it cannot be used for construction of a nuclear weapon. If the fuel is inadequately cooled while in a reactor, fission heat can cause cladding damage and, under the worst conditions, fuel melting. The resultant chemical reaction with water bears no resemblance to a nuclear explosion. Therefore it can be stated positively that a reactor cannot explode like a nuclear bomb.

The spent fuel in a reactor contains a great deal of U-238, some U-235, Pu-239, Pu-240, and Pu-241, along with fission products. If this "reactor grade" plutonium is chemically separated and made into a weapon, the presence of neutrons from spontaneous fission of Pu-240 will cause premature detonation and an inefficient explosion. For this reason spent fuel is a poor source of bombs. A much more likely avenue to obtain "weapons-grade" plutonium is the dedicated research reactor, with low levels of neutron exposure to prevent Pu-240 buildup. Another favorable means is a specially designed separation method to obtain nearly pure U-235. Neither of these approaches involves nuclear power reactors used for commercial electricity.

26.2 NUCLEAR EXPLOSIVES

Security of information on the detailed construction of modern nuclear weapons has been maintained, and only a qualitative description is available to the public. We will draw on unclassified sources (see References) for the following discussion of the earliest versions.

First, we note that two types of devices have been used: (a) the fission explosive ("atom bomb") that uses plutonium or highly enriched uranium and (b) the fusion or thermonuclear explosive ("hydrogen bomb"). The reactions described in earlier chapters are involved. Next, it is possible to create an explosive fission chain reaction by two different procedures—either by the "gun" technique or by "implosion." Figure 26.1 is a simplified sketch of the gun system, in which a plug of highly enriched uranium is fired into a hollowed-out cylinder of uranium to produce a supercritical mass. A natural U "tamper" holds the combined materials together momentarily. This atom bomb was given the name Little Boy. The gun technique was not feasible for a weapon that uses plutonium. Spontaneous fission of Pu-240 would release neutrons that would trigger a premature ineffective explosion. Figure 26.2 is a sketch of the alternative, the implosion method, in which chemical high explosives in the form of lenses compress a plutonium

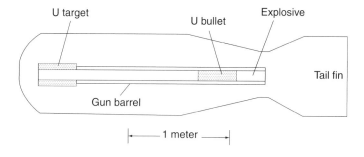

FIGURE 26.1 Uranium fission nuclear weapon, gun type Little Boy.

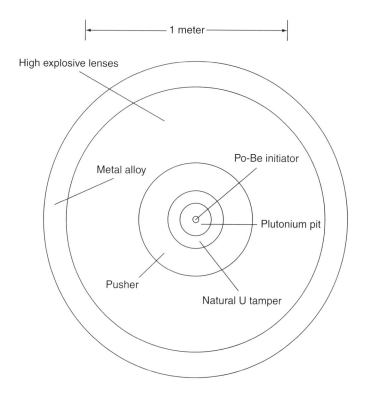

FIGURE 26.2 Plutonium fission nuclear weapon, implosion type Fat Man.

metal sphere to supercriticality. A uranium tamper is also used. This weapon was called Fat Man.

In either of these devices, an initial supply of neutrons is required. One possibility is the polonium-beryllium source, with the (α,n) reaction, analogous to Rutherford's experiment (Section 4.1). The excess reactivity of the supercritical masses causes a rapid increase in power and the accumulated energy blows the material apart, a process labeled "disassembly." In the case of implosion, when the fissile material is compressed, there is an increase in ratio of surface to volume that results in larger neutron leakage but a decrease in mean free path that reduces leakage. The latter effect dominates, giving a net positive increase of multiplication. Additional details on nuclear weapons can be found in several Web sites, for example, Atomic Archive and The High Energy Weapons Archive (see References).

According to report ANL-5800 (see References) an unreflected spherical plutonium assembly has a critical mass of approximately 16 kg, whereas that of a highly enriched (93.5%) uranium sphere is approximately 49 kg. By adding a 1-inch layer of natural uranium as reflector, the critical masses drop to 10 kg and 31 kg, respectively. The critical mass of uranium with full reflector varies rapidly with the U-235 enrichment, as shown in Table 26.1. It is noted that the total mass of a device composed of less than 10% U-235 is impractically large for a weapon.

An appreciation of the effect on critical mass of an implosion that increases uranium density can be gained by the study of Computer Exercise 26.A.

Details of the compact and versatile modern thermonuclear weapons are not available, but we can describe the processes involved in the first hydrogen bomb explosion, the Ivy/Mike shot in the South Pacific in 1952. It included heavy hydrogen as fusion fuel, involving the two reactions also to be used in fusion reactors,

$$\ _{1}^{2}\text{H} + \ _{1}^{2}\text{H} \rightarrow \ _{1}^{3}\text{H} + \ _{1}^{1}\text{H} \ \text{ or } \ _{2}^{3}\text{He} + \ _{0}^{1}\text{n}$$

$$\ _{1}^{2}\text{H} + \ _{1}^{3}\text{H} \rightarrow \ _{2}^{4}\text{He} + \ _{0}^{1}\text{n}.$$

The following description is an abbreviation of that found in the book *Dark Sun* (see References). As sketched in Figure 26.3, the unit called "Sausage" was a hollow steel cylinder 20 ft long and 6 ft 8 in in diameter. The cavity was lined with lead. At one end of the cavity was a "primary" sphere of plutonium and enriched uranium that would be caused to fission by implosion. In the middle of the cavity was a cylindrical container of liquid deuterium, much like a large thermos bottle. Along its axis was a stick of Pu called the "sparkplug,"

Table 26.1 Critical Masses of U-235 and U Versus Enrichment

% U-235	U-235 (kg)	U (kg)
100	15	15
50	25	50
20	50	250
10	130	1300

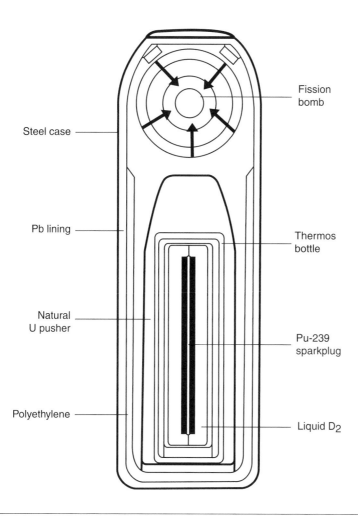

Steel case

Pb lining

Natural
U pusher

Polyethylene

Fission
bomb

Thermos
bottle

Pu-239
sparkplug

Liquid D$_2$

FIGURE 26.3 Thermonuclear weapon. (Adapted from Rhodes, see References.)

which served as a "secondary" fission source. The deuterium container was surrounded by a natural U "pusher." Finally, the inside of the casing was lined with
polyethylene.

The sequence of events was as follows. An electrical discharge to detonators
set off the high-explosive shell of the primary. A uranium tamper and shell vaporized and compressed the central plutonium ball while setting off a Po-Be source
inside, releasing neutrons. X-rays from the resulting supercritical fireball heated
the polyethylene to a plasma that reradiated X-rays to heat the U pusher. Neutrons
and energetic alpha particles were released in the heated deuterium and fission
took place in the sparkplug. Some tritium was formed, which contributed to
the fusion reaction. Additional energy and radiation came from fast neutron

fission in the uranium-238 in the tamper. The resultant explosion created a crater 200 feet deep and a mile across.

In later weapon versions, the fusion component was composed of lithium deuteride. Neutrons from fission interact with the lithium-6 according to the equation

$$^6_3\text{Li} + ^1_0\text{n} \rightarrow ^3_1\text{H} + ^4_2\text{He} + 4.8 \text{ MeV}.$$

The tritium produced allows for the D-T reaction to occur. Other thermonuclear devices used tritium as the principal explosive material.

Nuclear explosives release their energy in several ways. First is the blast effect, in which a shock wave moves outward in air, water, or rock, depending on where the event occurs. Second is the thermal radiation from the heated surrounding material, at temperatures typically 6000°C. Finally, there is the nuclear radiation, consisting mainly of neutrons and gamma rays. The percentages of the energy that go into these three modes are, respectively, 50, 35, and 15. There is a great deal of radioactive fallout contamination from fission products, in addition to the X-rays, gamma rays, and neutrons.

The energy yield of a weapon is measured in equivalent tons of chemical explosive. By convention, 1 ton of TNT corresponds to 10^9 calories of energy. The first atom bomb had a strength of 20,000 tons. A smaller device of 3 kilotons was exploded underground in the Gnome test. A large cavity was created, as shown in Figure 26.4. The Ivy/Mike explosion gave 10.4 megatons. Tests of

FIGURE 26.4 Underground cavity created by the *Gnome* fusion explosion, part of the Plowshare program. Note the person standing near the center. (Courtesy of Lawrence Livermore National Laboratory, United States Department of Energy.)

50-megaton devices have been reported. The energy of explosion is released in a very short time, of the order of a microsecond.

The radiation effect of a nuclear explosion is extremely severe at distances up to a few kilometers. Table 26.2 shows the distances at which neutron dose of 500 rems is received for different yields.

Special designs of devices have been mentioned in the literature. Included are "radiological weapons" intended to disperse hazardous radioactive materials such as Co-60 and Cs-137. Another is the "neutron bomb," a small thermonuclear warhead for missiles. Exploded at heights of approximately 2 km above the Earth, it would have little blast effect but would provide lethal neutron doses. In 1978 the United States canceled neutron bomb development.

By special arrangements of material in the fusion bomb, certain types of radiation can be accentuated and directed toward a chosen target. Examples of third-generation nuclear weapons could yield large quantities of lethal gamma rays or electromagnetic pulses (EMP) that disrupt solid-state electronic circuits. More detailed diagrams and descriptions of fission and fusion bombs are found in the book by Hansen (see References).

A consequence of a major exchange of nuclear missiles near the Earth's surface would be an increase in the particles suspended in air. Part would be dust created by the blast; part would be smoke from fires in forests and other combustibles ignited by the heat. As a result, the amount of sunlight reaching the ground would be reduced, causing cooling of the atmosphere. The situation has been called "nuclear winter" by some investigators, who predict serious modification of the climate, with a reduction in agricultural production. Such an effect occurred in the early 1800s as the result of the eruption of a volcano. The subject of atmospheric cooling has been studied a great deal, but there is disagreement among scientists as to the magnitude of the effect. The original theory was criticized for failure to take proper account of self-correcting processes, including increased precipitation that would tend to dispel dust and smoke.

Table 26.2 Distance-Yield Relation for Nuclear Explosion

Yield (tons)	Radius (meters)
1	120
2	450
10,000	1050
1,000,000	2000

26.3 THE PREVENTION OF NUCLEAR WAR

The nuclear arms race between the United States and the U.S.S.R. that began after World War II was stimulated by mutual suspicion and fear and by technological advances in nuclear weapons. Each of the superpowers sought to match and to exceed the other's military capability.

As of 1945 the United States clearly had nuclear weapons superiority, but by 1949 the U.S.S.R. had developed its own atom bomb. After considerable controversy, the United States undertook to develop the hydrogen bomb (Super bomb, or "Super") by use of thermonuclear fusion and by 1952 had restored the advantage. By 1953 the Soviets had again caught up. In the ensuing years each country produced very large numbers of nuclear weapons. If deployed by both sides in an all-out war, with both military and civilian targets, hundreds of millions of people would die.

The policy adopted by the two powers to prevent such a tragedy was deterrence, which means that each country maintains sufficient strength to retaliate and ruin the country that might start a nuclear war. The resultant stalemate is given the term "mutual assured destruction" (MAD). This "balance of terror" could be maintained unless one country develops an excessive number of very accurate missiles and chooses to make a first strike that disables all retaliatory capability.

The methods by which nuclear warheads can be delivered are: (a) carried by bombers, such as the United States B-52; (b) intercontinental ballistic missiles (ICBMs) launched from land bases; and (c) missiles launched from submarines such as the *Poseidon* and *Trident*, which are later versions of the first nuclear submarine, *Nautilus*.

The ICBM is propelled by rocket, but experiences free flight under the force of gravity in the upper atmosphere. The nuclear warhead is carried by a reentry vehicle. The ICBM may carry several warheads (MIRV, multiple independently targetable reentry vehicles), each with a different destination.

An alternative is the cruise missile, an unmanned jet aircraft. It can hug the ground, guided by observations along the way and by comparison with built-in maps, and maintain altitude by computer control (see References).

There are two uses of nuclear weapons. One is tactical, whereby limited and specific military targets are bombed. The other is strategic, involving large-scale bombing of both cities and industrial sites, with intent both to destroy and to demoralize. Most people fear that any tactical use would escalate into strategic use.

Thousands of nuclear warheads have been available to the superpowers for many years, with the number of megatons equivalent TNT per weapon ranging from 0.02 to 20. The area that could be destroyed by all these weapons is approximately 750,000 square kilometers, disrupting each country's functions such as manufacturing, transportation, food production, and health care. A civil defense program would reduce the hazard but is viewed by some as tending to invite attack.

The international aspect of nuclear weapons first appeared in World War II when the Allies believed that Germany was well on its way to producing an atomic bomb. The use of two weapons by the United States to destroy the cities of Hiroshima and Nagasaki alerted the world to the consequences of nuclear warfare. Many years have been devoted to seeking bilateral or international agreements or treaties that seek to reduce the potential hazard to mankind. The increase in fallout from nuclear weapons testing prompted the Limited Test Ban Treaty of 1963. It forbade nuclear tests in the air, water, or space, and the United States and the Soviet Union thereafter conducted all testing underground. However, this treaty did not control the expansion in nuclear arms.

In 1968 an international treaty was developed at Geneva with the title Non-Proliferation of Nuclear Weapons (NPT). The treaty is somewhat controversial in that it distinguishes states (nations) that have nuclear weapons (NWS) and those that do not (NNWS). The main articles of the treaty require that each of the latter would agree (a) to refrain from acquiring nuclear weapons or from producing them, and (b) to accept safeguards set by the International Atomic Energy Agency (IAEA) based in Vienna. The treaty involves an intimate relationship between technology and politics on a global scale and a degree of cooperation hitherto not realized. There are certain ambiguities in the treaty. No mention is made of military uses of nuclear processes as in submarine propulsion or of the use of nuclear explosives for engineering projects. Penalties to be imposed for noncompliance are not specified, and finally the authority of the IAEA is not clear. The treaty has been signed by five NWS (United States, Russia, Great Britain, France, and China) and 180 NNWS. In 1995 the NPT was extended indefinitely. India was a signatory as NNWS but proceeded to develop and test a nuclear weapon.

The nuclear weapons states (NWS) can withhold information and facilities from the nonnuclear weapons states (NNWS) and thus slow or deter proliferation. To do so, however, implies a lack of trust of the potential recipient. The NNWS can easily cite examples to show how unreliable the NWS are.

We have already discussed in Section 22.5 the attempt by President Carter to prevent proliferation. By banning reprocessing in the United States he had hoped to discourage its use abroad. It is continuing United States policy to prohibit the sale to foreign countries of sensitive equipment and materials, those believed to be adaptable for construction of nuclear weapons. If the policy is extended to the transfer of legitimate nuclear power technology, however, such policies can be counterproductive for several reasons. International relations suffer, and the United States loses any influence it might have on nuclear programs. Perceived inequity may strengthen a country's determination to achieve weapons capability and to seek alternative alliances that further that goal.

Negotiations began in 1967 on Strategic Arms Limitation Talks (SALT) and an accord was signed in 1972. SALT I led to a ceiling on strategic nuclear weapons and thus tended to achieve equality in strength. However, it said nothing about continued improvements in missiles. It restricted the deployment of Antiballistic Missile (ABM) defense systems. Each nation was allowed to defend its capital and one other location.

The SALT II agreement between leaders of the two nations in 1979 dealt with detailed limits on types of launchers and missiles, including the MIRV type. It placed emphasis on preserving the ability of both sides to verify compliance. The treaty was never ratified by the United States Congress, and talks were not resumed.

In 1983, a program of detection and interception of nuclear missiles was initiated by President Ronald Reagan. This research and development effort was called Strategic Defense Initiative (SDI) but soon became known popularly as "Star Wars" because of its space implications. In this multibillion dollar project, various devices were proposed and studied, including earth satellite weapons platforms, X-ray laser beams, small tactical nuclear bombs, and "smart pebbles," small high-speed objects that could destroy incoming missiles. The SDI program was controversial for technical reasons having to do with feasibility and political reasons related to the wisdom of mounting the program. Some believe, however, that it had a favorable influence on the achievement of an end to the Cold War.

Negotiations continued over the years, leading to the Intermediate-Range Nuclear Force Treaty (INF) of 1988, in which a number of missiles were destroyed in the United States and the U.S.S.R., with inspection teams from the other country functioning smoothly. Two new sets of accords called Strategic Arms Reduction Talks (START) were developed in the Reagan-Bush era. The use of the word "reduction" instead of "limitation" is significant. By 2001 the START I treaty reduced the number of nuclear warheads from 10,000 to 6,000 in each country. The START II, which dealt with intercontinental missiles and MIRVs, called for the reduction of strategic warheads by the end of 2007 to 3,000 to 3,500. Further progress in disarmament was made in 2003 with the ratification by Russia and the United States of the Strategic Offensive Reduction Treaty (SORT), which calls for a level of 1,700 to 2,200 warheads by the end of 2012.

With progress in arms reduction, the Star Wars program became less relevant and was greatly scaled down. The elimination of thousands of warheads was an important step in terms of world safety, but there still remain enough weapons for mutual destruction. The breakup of the Soviet Union left ICBMs in the independent states of Ukraine, Belarus, and Kazakhstan, but agreement was reached to transfer the weapons to Russia. Internal economic, political, and ethnic tensions make control difficult. Concern has been expressed also that weapons scientists and engineers of the former U.S.S.R. may be induced for economic reasons to emigrate to nations seeking nuclear capability (e.g., Iran and North Korea).

United States legislation called Cooperative Threat Reduction (CTR) provided funding and expertise to help Russia control nuclear weapons. Also known as the Nunn-Lugar-Domenici Act after the senators who authored it, the program was designed to account for weapons, to secure them, and to arrange for disposition. Of special importance was provision for the employment elsewhere of thousands of former weapons experts. Accomplishments include deactivation of some 6,900 warheads and the elimination of 290 tons of enriched uranium. According to Turpen (see References), there is still much to do.

Another byproduct of the international political changes is the purchase by the United States from Russia of highly enriched uranium (HEU) from dismantled nuclear weapons to be converted by blending into low-enriched uranium (LEU) for use in power reactors. The program, called "megatons to megawatts," calls for the transfer of 10,000 kg/y of HEU for 5 y and 30,000 kg/y for 15 y, giving a total of 500 metric tons. The LEU at 4.5% enrichment is sent by the Russian firm TENEX and received by United States Enrichment Corporation. As of the end of 2007 a total of 12,885 warheads had been eliminated. The program also includes downblending of part of the United States stockpile of HEU. Several virtues accrue: financial benefit to Russia, diversion of weapons grade material to peaceful purposes, and relief from the necessity by the United States to expand isotope separation capability. Computer Exercise 26.B considers the arithmetic of the process by which HEU is diluted into reactor grade uranium and investigates cost aspects of a United States purchase from Russia.

A large quantity of the United States weapons high-enriched uranium (160 tonnes) is slated to be diverted to naval ship reactors. Another 20 tonnes is to be blended with natural uranium to a concentration appropriate to commercial power reactors. An additional 20 tonnes will be held for space and research reactor fueling. All of these actions under the Department of Energy (DOE) National Nuclear Safety Administration (NNSA) will render the HEU less accessible to terrorists.

An agreement between the United States and Russia was reached in 2006 for each to convert 34 metric tons of excess weapons-grade plutonium into unusable forms. The United States is constructing a mixed oxide (MOX) fuel fabrication facility to produce fuel for power reactors.

26.4 NONPROLIFERATION AND SAFEGUARDS

We now discuss proliferation of nuclear weapons and the search for means to prevent it. Reducing the spread of nuclear materials has recently become more important as the result of increases in political instability and acts of violence throughout the world.

To prevent proliferation we can visualize a great variety of technical modifications of the way nuclear materials are handled, but it is certain that a country that is determined to have a weapon can do so. We also can visualize the establishment of many political institutions such as treaties, agreements, central facilities, and inspection systems, but each of these is subject to circumvention or abrogation. It must be concluded that nonproliferation measures can merely reduce the chance of incident.

We now turn to the matter of employment of nuclear materials by organizations with revolutionary or criminal intent. One can define a spectrum of such, starting with a large well-organized political unit that seeks to overthrow the existing system. To use a weapon for destruction might alienate people from their cause, but a threat to do so might bring about some of the changes they demand. Others include terrorists groups, criminals, and psychopaths who may have little to lose and thus are more apt to use a weapon. Fortunately, such organizations tend to have fewer financial and technical resources.

Notwithstanding difficulties in preventing proliferation, it is widely held that strong efforts should be made to reduce the risk of nuclear explosions. We thus consider what means are available in Table 26.3, a schematic outline.

Protection against diversion of nuclear materials involves many analogs to protection against the crimes of embezzlement, robbery, and hijacking. Consider first the extraction of small amounts of fissile material such as enriched uranium or plutonium by a subverted employee in a nuclear facility. The maintenance of accurate records is a preventive measure. One identifies a material balance in selected process steps (e.g., a spent-fuel dissolver tank or a storage area). To an initial inventory the input is added and the output subtracted. The difference between this result and the final inventory is the material unaccounted for (MUF). Any significant value of MUF prompts an investigation. Ideally, the system of accountability would keep track of all materials at all times, but such detail is probably impossible. Inspection of the consistency of records and reports is coupled with independent measurements on materials present.

Restricting the number of persons who have access to the material and careful selection for good character and reliability is a common practice. Similarly, limiting the number of people who have access to records is desirable. It is easy to see how falsification of records can cover up a diversion of plutonium. A discrepancy of only 10 kg of plutonium would allow for material for one weapon to be

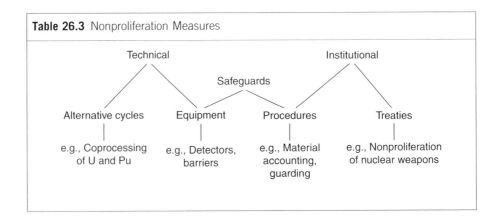

Table 26.3 Nonproliferation Measures

diverted. Various personnel identification techniques are available such as picture badges, access passwords, signatures, fingerprints, and voiceprints.

Protection against intruders can be achieved by the usual devices such as ample lighting of areas, use of a guard force, burglar alarms, TV monitoring, and barriers to access. More exotic schemes to delay, immobilize, or repel attackers have been considered, including dispersal of certain gases that reduce efficiency or of smoke to reduce visibility, and the use of disorienting lights or unbearable sound levels.

Illegal motion of nuclear materials can be revealed by the detection of characteristic radiation, in rough analogy to metal detection at airports. A gamma-ray emitter is easy to find, of course. The presence of fissile materials can be detected by observing delayed neutrons resulting from brief neutron irradiation.

In the transportation of strategic nuclear materials, armored cars or trucks are used, along with escorts or convoys. Automatic disabling of vehicles in the event of hijacking is a possibility.

26.5 IAEA INSPECTIONS

Shortly after the Nonproliferation Treaty of 1968 was signed, the IAEA set up a worldwide safeguards system. It applied to all source materials (uranium and thorium) and special fissionable materials (plutonium and uranium-233). The primary purpose of IAEA inspections has been to detect the diversion of significant quantities of nuclear material from peaceful to military purposes. Over the years since 1970, a large number of nondestructive portable instruments have been developed to carry out the surveillance. Gamma ray and neutron detectors are used to determine the enrichment of uranium and the content of plutonium in spent fuel.

The role of IAEA was highlighted in the 1991 investigation of the nuclear weapons program of Iraq under Saddam Hussein, under the auspices of the United Nations Security Council. Large amounts of uranium had been imported from other countries without being reported. Orders were placed abroad for equipment that could have a dual purpose. As revealed by IAEA inspectors, such equipment was channeled into the construction of modern versions of electromagnetic uranium isotope separators (Section 9.1), to centrifuges, and to reactors and reprocessing equipment for plutonium weapons production. In support of the field investigations after the end of the Gulf War, laboratory studies at the IAEA's laboratories in Austria were conducted. Samples taken by inspectors were found to contain as high as 6% enrichment in U-235. A particle spectral measurement confirmed the presence of polonium-210, which is a component of an initiator for an implosion-type nuclear weapon. Much of Iraq's nuclear capability was destroyed in the Gulf War and afterward in response to sanctions by the United Nations. The assumption by the United States that Iraq still had weapons of mass destruction (i.e., nuclear capability) led to the invasion of 2003.

On a long-range basis, the IAEA is concerned about the possibility that a repository for spent fuel, intended to isolate the waste from the biosphere, may in fact become a "plutonium mine" in some future era when energy shortages become acute and fissile materials become a very valuable commodity.

Various other countries are known or suspected of having or have at one time had nuclear weapon programs. Prominent among the lists given by Jones and McDonough and by Morrison and Tsipsis (see References) are Israel, India, Pakistan, North Korea, Iran, Iraq, and South Africa.

The Comprehensive Test Ban Treaty (CTBT) of 1996 seeks international agreement not to "carry out any nuclear weapon test explosion or any other nuclear explosion." This means underground tests, as well as those in the air, on water, or in space, even those for peaceful purposes. Excluded are explosions by inertial fusion devices or the destruction of any terrorist weapon. The treaty also calls for a system of monitoring and inspection to verify compliance. Ratification is required by the 44 countries that either helped draft the document or have power reactors or research reactors. Many countries have signed, but not all have ratified. The treaty seems to have public support in the United States, but some claim that acceptance by the United States would hamstring defense, while some nations would violate the treaty. In 1999 the United States Senate narrowly rejected the CTBT. For details on the treaty, see References.

26.6 PRODUCTION OF TRITIUM

Over the many years of the Cold War, the DOE and its predecessors had maintained a stockpile of weapons material, especially tritium and plutonium. The isotope H-3, tritium, as one of the ingredients of the hydrogen bomb, was produced in heavy water reactors at the Savannah River Plant in South Carolina. Because of safety concerns, the reactors were shut down. A program of refurbishing the old reactors was undertaken, and as supporting capacity to produce a continuing supply of 12.3 y tritium, a development program called "New Production Reactor" was started. Two types of reactors were designed—a heavy water reactor and a high temperature gas-cooled reactor. With the reduction in international tension, the United States determined that tritium supplies would be adequate for two decades and suspended design of the new reactors. The amount of tritium needed depends on the level of weapons capability that is maintained. In a scenario with a smaller number of warheads, recycling of tritium from dismantled weapons would provide an adequate source. Subsequently, however, it was decided that an alternative supply was needed to maintain the stockpile, because tritium has a half-life of 12.3 y, corresponding to a loss of 5.5% a year. Consequently, DOE sponsored two studies of production techniques either with a power reactor or with a particle accelerator.

Research was performed at Los Alamos on the alternative source of tritium, in the program called Accelerator Production of Tritium (APT). It involved use of a

1,000-ft linear accelerator to bring protons to 1 GeV. These bombarded a tungsten target to yield spallation neutrons, which would be absorbed in helium-3 to give tritium. Because the APT program was not financed after 2004, the accelerator was adapted for basic physics research. A decision was made in favor of the reactor route.

The production in a conventional reactor by neutron bombardment involves burnable poison rods. These are auxiliary to the main control, containing an isotope of large thermal neutron cross section such as boron-10, which burns out quickly and allows a larger initial fuel loading. It was proposed by DOE to replace the boron rods with an appropriate number containing lithium-6. These target rods would consist of concentric cylinders of zircaloy, lithium aluminide, and stainless steel. Absorption of a neutron in Li-6, with thermal cross section 940 barns, yields tritium and an alpha particle. Tests at the Tennessee Valley Authority reactors indicated that production of tritium would be adequate and that the reactor would operate safely. In 2003 TVA's Watts Bar reactor started producing the first United States tritium in more than 15 y. Irradiated rods were shipped in 2005 to the Savannah River Site for storage and extraction in a new facility.

26.7 MANAGEMENT OF WEAPONS URANIUM AND PLUTONIUM

During the Cold War both the United States and the U.S.S.R accumulated large amounts of highly enriched uranium and weapons-grade plutonium. A program of dismantlement is under way as part of the START treaties. An excess of these materials over that needed for continued nuclear deterrence will be disposed of in some way. It has been estimated that there is a total of 100 tonnes of Pu and 200 tonnes of U, roughly in equal amounts in the two countries. The enriched uranium can be readily diluted with natural uranium to produce a low-enrichment fuel, helping meet the demand of current and future power reactors. The plutonium is not as easy to handle, because there are no Pu isotopes to serve as diluent. Thus the stockpiles of Pu are vulnerable to diversion to nations or groups who might use, or threaten to use, the material to gain their ends.

The plutonium of principal concern is in pure form in contrast with that present in spent fuel. The latter would require special equipment to extract the Pu, and the product would be less suitable for a weapon because of the presence of Pu-240.

Plutonium is far from being "the most dangerous substance known to man," as claimed by some, but it is highly radiotoxic and requires special precautions in all handling. Use of Pu increases the chance of radioactive contamination as was experienced at various DOE sites, especially at Rocky Flats, Colorado.

There are several possibilities for managing the plutonium. Some believe that it should be stored in anticipation of a need for its energy values some time during the 21st century. One could visualize a storage facility like Fort Knox where gold and silver are secured. Storage over a long period would require protection against chemical attack and accidental criticality, as well as from theft.

A National Academy of Sciences (NAS) panel composed of prominent knowledgeable people identified three principal options: (a) vitrification of the Pu with a highly radioactive contaminant to deter diversion and processing. This would result in glass logs that could be treated as spent fuel and put in an underground repository. Future mining of the Pu would be very unattractive. (b) Blending the plutonium as the oxide with a suitable amount of uranium oxide to form MOX that could serve as fuel for power reactors. This would eliminate the plutonium and have the advantage of a beneficial use. The disadvantage is the cost of processing and fabrication, which is significantly higher than that for uranium because of the hazard of ingesting the radioactive material. This approach requires the development of a suitable fuel fabrication plant. Several countries, notably France, England, Belgium, and Japan, are in a position to prepare and use the MOX, whereas the United States has little experience or inclination to use it, having abandoned the option of reprocessing spent fuel. (c) To place the Pu in a deep drilled hole in the ground. Although this is feasible, there is no strong support for the idea. The NAS panel also examined the option of the use of an accelerator-driven subcritical system to burn the plutonium but concluded that there were too many uncertainties, including the possible need for reprocessing. The NAS recommended carrying along options (a) and (b) in parallel, a strategy that was adopted by the DOE. The surplus weapons plutonium is stored at several DOE sites, with the bulk at Savannah River and Hanford. Most is in the form of "pits," the spherical weapons cores, but there are some 12-ft-long rods at Hanford. Eventually, all will be stored at one location to improve security and reduce storage costs.

For disposal of Pu by immobilization and burial, a criterion called the "spent fuel standard" is applied (i.e., the Pu should be as inaccessible for weapons use as that in spent fuel from commercial reactors).

It is expected that the burning option would consist of a once-through fuel cycle. To use up the 50 tonnes of excess Pu in a reasonable period would require relatively few commercial reactors. It is straightforward arithmetic to determine the combination of time and number of reactors to perform the task. See Exercise 26.3.

Whatever method of disposal is finally adopted, meticulous procedures and records must be maintained and special rigorous precautions taken to prevent the material from getting into unscrupulous hands. The NAS report urged that agreements be reached between the United States and Russia, and mechanisms established through the IAEA that would assure that each nation fulfilled its commitments. This would reduce mutual concerns that one party might retrieve Pu and rearm nuclear weapons.

When one realizes the enormous damage that nuclear explosions can create, it is clear that all possible steps must be taken to prevent them from occurring. In addition to continued efforts to reduce the stockpile of armaments, to secure workable treaties, and to use technology to provide protection, there is an urgent need to eliminate all the unfavorable conditions—social, economic, and cultural—that prompt conflict in the world.

26.8 SUMMARY

Although spent fuel from power reactors contains plutonium, it is not the same as a nuclear weapon. The original atom bombs used U-235 and Pu, but the much more powerful modern weapons are based on the fusion of hydrogen isotopes. Intercontinental ballistic missiles from land and missiles from submarines make up the bulk of the arsenals of the United States and the former U.S.S.R. Continual efforts are made to prevent further proliferation of nuclear weapons. It is imperative that nuclear explosions be avoided.

26.9 EXERCISES

26.1 The critical mass of a uranium-235 metal assembly varies inversely with the density of the system. If the critical mass of a sphere at normal density 18.5 g/cm^3 is 50 kg, how much reduction in radius by compression is needed to make a 40 kg assembly go critical?

26.2 A proposal is advanced to explode fusion weapons deep underground, to pipe to the surface the heat from the cavity produced, and to generate and distribute electricity. If no energy were lost, how frequently would a 100-kiloton device have to be fired to obtain 3,000 MW of thermal power? Alternately, how many weapons per year would be consumed?

26.3 Find out how many commercial reactors would be needed to consume 50 tonnes of Pu in 30 y, assuming the following data: reactor power, 1,000 MWe; efficiency, 0.33; capacity factor, 0.75; 60 assemblies removed and new ones installed per year, 3-year irradiation to fuel burnup 30,000 MWd/tonne, fuel weight per assembly 1000 lb, one third of new fuel containing MOX at 2.5% Pu. NOTE: there are two ways to solve the problem.

26.4 Assuming an annual need for 4% U-235 fuel of 50 tonnes per power reactor, how many reactor-years of operation can be achieved with 20 tonnes of 90% U-235, when blended with 0.7% natural U?

Computer Exercises

26.A The implosion of a mass of fissionable material can be studied by use of the computer program FASTR, introduced in Chapter 13. It is a neutron multigroup method for calculating criticality in a pure U-235 metal assembly.

(a) Calculate the critical size and mass for several values of the uranium density, including higher densities than normal as would be achieved by implosion of a nuclear warhead. Suggested values of the parameter UN (line 2310) besides 0.048 are 0.036 and 0.060.

(b) From the results of (a) above, deduce a good value of x in a formula for critical mass as a function of metal density of the form

$$M = M_0(\rho/\rho_0)^x$$

where M_0 is the critical mass at ordinary density ρ_0.

26.B Arrangements are made for the purchase by the United States of Russian uranium at enrichment 94 w/o to be blended with natural U to create 3 w/o fuel for power reactors. With computer program ENRICH3 (Chapter 9), estimate a fair price to pay per kilogram of HEU if the blending is done (a) in Russia, or (b) in the United States. If importation amounts to 10 tonnes/y for 5 y followed by 30 tonnes/y for 15 y, which would take about half the stockpile, what is the total worth in each case? What additional information would be useful to arrive at a proper figure?

26.10 REFERENCES

Atomic Archive
http://www.atomicarchive.com
Includes the history of the Manhattan Project.

Jeremy Bernstein, *Nuclear Weapons: What You Need to Know*, Cambridge University Press, New York, 2007. By an accomplished science writer.

Cynthia C. Kelly, Editor, *The Manhattan Project: The Birth of the Atomic Bomb in the Words of its Creators, Eyewitnesses, and Historians*, Blackdog & Leventhal, New York, 2007. An excellent selection of documents. See review in *Nuclear News*, June 2008.

Nuclear Timeline
http://www.gsinstitute.org/dpe/timeline.html
Political history pre-1940 to 2000s. Global Security Institute.

Megatons to Megawatts
http://www.cdi.org/friendlyversion/printversion.cfm?documentID=2210
Detail about processes and politics. From Center for Defense Information.

Richard L. Garwin and Georges Charpak, *Megawatts and Megatons: A Turning Point in the Nuclear Age*, Random House, New York, 2001.

Megatons to Megawatts
http://www.usec.com/v2001_02/HTML/megatons.asp
Transfer of HEU from Russia to the United States.

Strategic Offensive Reductions Treaty (SORT)
http://www.armscontrol.ru/Start/sort.htm

Status, comments, expert opinions.

Robert Serber, *The Los Alamos Primer*, University of California Press, Berkeley, 1992. A set of five lectures on the early (1943) state of knowledge on the possibility of an atom bomb, with extensive explanatory annotations. Included are history, nuclear physics, and chain reactions.

Reactor Physics Constants, ANL-5800, 2nd Ed., United States Atomic Energy Commission, Washington, DC, 1963. Data on critical masses in Section 7.2.1.

Richard Rhodes, *The Making of the Atomic Bomb*, Simon & Schuster, New York, 1986. Thorough, readable, and authoritative, this is regarded as the best book on the Manhattan Project.

Richard Rhodes, *Dark Sun: The Making of the Hydrogen Bomb*, Simon & Schuster, New York, 1995. Highly readable political and technical aspects. Role of espionage. Diagrams of H-bomb explosion on p. 506.

Chuck Hansen, *The Swords of Armageddon*
http://www.coldwar.com
History of post-1945 United States weapons development. CD-ROM, Chucklea Publications.

Chuck Hansen, *United States Nuclear Weapons: The Secret History*, Aerofax, Arlington, TX, 1988. Very detailed descriptions with many photos of all weapons and tests. Diagram of H-bomb explosion on p. 22.

The Nuclear Weapon Archive
http://nuclearweaponarchive.org
In FAQ by Carey Sublette select 8.0 The First Nuclear Weapons.

WMD Nuclear Resources
http://www.fas.org/nuke
Weapons of mass destruction. From Federation of American Scientists.

Thomas B. Cochran, William M. Arkin, and Milton M. Hoenig, *Nuclear Weapons Databook,* Vol. I, *U.S. Nuclear Forces and Capability*, Ballinger Publishing Co., Cambridge, MA, 1984. Facts and figures on weapons and delivery systems.

Samuel Glasstone and Philip J. Dolan, Compilers and Editors, *The Effects of Nuclear Weapons*, 3rd Ed., United States Department of Defense and United States Department of Energy, U.S. Government Printing Office, Washington, DC, 1977.

M. S. Yadav, *Nuclear Weapons and Explosions: Environmental Impacts and Other Effects*, SDS New Delhi, 2007.

Elizabeth Turpen, "The Human Dimension is Key to Controlling Proliferation of WMD," *APS News*, April 2007, p. 8. The article discusses the activation of weapons facilities to peaceful production.

Senator Pete Domenici, *A Brighter Tomorrow: Fulfilling the Promise of Nuclear Energy*, Rowman & Littlefield, New York, 2004.

Susan Thaul and Heather O'Maonaigh, Editors, *Potential Radiation Exposure in Military Operations*, National Academy Press, Washington, DC, 1999. Information and recommendations. Also read at http://www.nap.edu/catalog.php?record_id =9454

The Effects on the Atmosphere of a Major Nuclear Exchange, Committee on the Atmospheric Effects of Nuclear Explosions, National Academy Press, Washington, DC, 1985. Also read at http://www.nap.edu/catalog.php?record_id =540

Management and Disposition of Excess Weapons Plutonium: Reactor-Related Options, National Academy Press, Washington, DC, 1995. Describes options for Pu use and disposal with recommendation that two be carried along in parallel. A limited-edition Executive

Summary was also published in 1994. A Web site on the Summary is also available at http://www.nap.edu/catalog.php?record_id =4754

K. K. S. Pillay, "Plutonium: Requiem or Reprieve," *Radwaste Magazine*, American Nuclear Society, Vol. 4, No. 3, January 1996, pp. 59-65.

James E. Nolan and Albert D. Wheelon, "Third World Ballistic Missiles," *Scientific American*, August 1990, p. 34.

Kosta Tsipis, "Cruise Missiles," *Scientific American*, Vol. 236, No. 2, February 1977, p. 20.

Dietrich Schroeer, *Science, Technology, and the Nuclear Arms Race*, John Wiley & Sons, New York, 1984.

Lawrence Scheinman, *The Nonproliferation Role of the International Atomic Energy Agency: A Critical Assessment* , Resources for the Future, Washington, DC, 1985. Explains, notes problems, and suggests initiatives.

Comprehensive Test Ban Treaty
http://www.ctbto.org/fileadmin/content/treaty/treatytext.tt.html
Full text of treaty.

Rodney W. Jones, et al., *Tracking Nuclear Proliferation: A Guide in Maps and Charts, 1998*, Carnegie Endowment for International Peace, Washington, DC, 1998.

Stephen Schwartz, Ed., *Atomic Audit: The Costs and Consequences of U.S. Nuclear Weapons Since 1940*, Brookings Institute Press, Washington, DC, 1998. Highly critical of secrecy, carelessness, and waste of money.

Philip Morrison and Kosta Tsipis, *Reason Enough To Hope: America and the World of the Twenty-first Century*, MIT Press, Cambridge, 1998. Recommends reduction of military forces and emphasis on Common Security.

Lawrence Livermore National Laboratory
http://www.llnl.gov
Select Site Map/Mission & Programs/Nuclear Weapons Stockpile Stewardship.

MILNET: Nuclear Weapons
http://www.milnet.com/milnet/nuclear.htm
Essays on several topics.

The Future

27

"We should all be concerned about the future because we will have to spend the rest of our lives there" (Charles F. Kettering).

THE UNITED States has a serious energy problem. Viewpoints as to its solution from three different constituencies are discussed by Jane C. S. Long (see References). The first, environmentalists, are deeply concerned about climate change caused by greenhouse gases. Their solution is achievement of greater energy efficiency and developing more renewable energy. The second, the energy security constituency, is concerned about reliance on uncertain foreign oil supplies. It favors expansion of ethanol production and making liquid fuel from coal. The third, the economic vitality group, is concerned with high prices that could depress the United States economy. It would increase domestic oil exploration and production.

From the viewpoint of supporters of nuclear energy, the construction of a number of new nuclear power plants to generate electricity and hydrogen for transportation would go a long way to improving the energy situation. However, there are broader questions that require answers.

What should be the role of nuclear power in the United States in the more distant future compared with other energy sources?

What will be the use of nuclear energy sources on a global and long-term basis?

What will be the ultimate energy source for mankind after fossil fuels are gone?

The future can be viewed in several ways. The first is acceptance, as by a fatalist, who has no expectation either of understanding or of control. The second is prediction, on the basis of belief and intuition. The third is idealization, as by a utopian, who imagines what would be desirable. The fourth is analysis, as by a scientist, of historical trends, the forces that are operative, and the probable effect of exercising each of the options available. Some combination of these views may be the answer, including the realization that the future always will bring surprises. Nevertheless, if the human species is to survive and prosper, we must believe we have some control of our destiny and take positive action to achieve a better world.

473

The oil crisis of 1973 involving an artificial shortage alerted the world to the importance of energy. A number of studies were published. Some of these are still relevant; others are very much out of date. Only a few will be cited here in References. In subsequent years, prices of oil declined, the oil supply was adequate, and natural gas became abundant, so public concern relaxed, and few updates of the studies were made. Other reasons for reservations about the literature of the 1970s and 1980s can be cited.

1. Principal emphasis has been on the situation in the United States or in developed countries, with less attention to developing countries.

2. Various investigators come to quite different conclusions even if they use the same data, depending on their degree of pessimism or optimism. At one extreme is the book *The Limits to Growth* and later works (called "doomsday" studies) and the upbeat energy reports of the Hudson Institute (see References).

3. Extrapolations of data can be very wrong, as evidenced by predictions on the growth of nuclear power made in the 1960s.

4. Sharp differences exist between writers' opinions on the future role of nuclear power. For example, Worldwatch Institute dismisses it at the outset, whereas the United States. National Academies view it as a desirable option.

5. Analyses may be irrelevant if they do not take account of social and political realities in addition to technical and economic factors.

In the next section, we identify some of the factors that need to be considered in planning for the energy future.

27.1 DIMENSIONS

Many aspects of the world energy problem of the future affect nuclear's role. We can view them as dimensions because each has more than one possibility.

The first is the time span of interest, including the past, present, immediate future (say the next 10 y), a period extending well into the 21st century, and the indefinite future (thousands or millions of years). Useful markers are the times oil and coal supplies become scarce.

The second is location. Countries throughout the world all have different resources and needs. Geographic regions within the United States also have different perspectives.

The third is the status of national economic and industrial development. Highly industrialized countries are in sharp contrast with underdeveloped countries, and there are gradations in between the extremes. Within any nation there are differences among the needs and aspirations for energy of the rich, of the middle class, and of the poor.

The fourth is the political structure of a country as it relates to energy. Examples are the free enterprise system of the United States, the state-controlled electricity production of France, the centrally planned economy of the People's Republic of China, and the transitional economy of the Commonwealth of Independent States.

The fifth is the current nuclear weapons capability, the potential for acquiring it, the desire to do so, or the disavowal of interest.

The sixth is the classification of resources available or sought: exhaustible or renewable; and fossil, solar, or nuclear.

The seventh is the total cost to acquire resources and to construct and operate equipment to exploit them.

The eighth is the form of nuclear power that will be of possible interest: converters, advanced converters, breeders, actinide burners, accelerators, and several types of fusion devices, along with the level of feasibility or practicality of each.

The ninth is the relationship between the effect of a given technology on social and ethical constraints such as public health and safety and the condition of the environment.

The tenth is the philosophical base of people as individuals or groups, with several contrasting attitudes: a view of man as a part of nature versus man as central; preference for simple lifestyle vs. desire to participate in a "high-tech" world; pessimism versus optimism about future possibilities; and abhorrence versus acceptance of nuclear. In addition, cultural and religious factors, national pride, and traditional relationships between neighboring nations are important.

27.2 WORLD ENERGY USE

The use of energy from the distant past to the present has changed dramatically. Primitive man burned wood to cook and keep warm. For most of the past several thousands of years of recorded history, the only other sources of energy were the muscles of men and animals, wind for sails and windmills, and water power. The Industrial Revolution of the 1800s brought in the use of coal for steam engines and locomotives. Electric power from hydroelectric and coal-burning plants is an innovation of the late 1800s. Oil and natural gas became major sources of energy only in the 20th century. Nuclear energy has been available for only approximately 50 y.

To think about the future, as a minimum it is necessary to understand the present. Data on energy production and use are available from the United States Department of Energy (DOE). Table 27.1 gives world consumption by geographic region. Of special note is the disparity in per capita consumption. Because productivity, personal income, and standard of living tend to follow energy consumption, the implications of these numbers for the human condition in much of the world is evident. The data on ratio of consumption and production confirm our knowledge that the Middle East is a major energy supplier through petroleum and shows that Western Europe and the Far East are quite dependent on imported energy.

Table 27.1 World Primary Energy, 2005

Region	Consumption (10^{15} Btu)	Per capita (10^6 Btu)	Consumption/ Production
Africa	14.4	16	0.42
North America	121.9	280	1.23
Central & South America	23.4	52	0.83
Europe	86.3	146	1.76
Eurasia	45.8	160	0.67
Middle East	22.9	125	0.35
Asia & Oceania	148.1	41	1.29
World	462.8	72	1.01

Source: International Energy Annual *(see References).*

A breakdown of the electrical production according to primary energy source by region is given in Table 27.2. We see that there is essentially no nuclear power in Africa or South America. Of that in the Far East, most is in Japan and Korea.

Predictions have been made on world future energy consumption patterns. Figure 27.1 is taken from a DOE report. It shows that nuclear power continues to increase. The projection may not take enough account of eventual acceptance of the virtue of nuclear power in avoiding gaseous emissions. Also, the curve for liquids does not take account of the large increase in oil prices. Finally, the rapid growth in renewables may be optimistic.

Table 27.2 World Total Net Electricity Generation (2005, BkWh)

Region	Thermal	Hydro	Nuclear	Other[*]	Total
Africa	430	89	12	2	533
North America	3238	658	880	119	4895
Central & So. America	253	613	16	26	908
Europe	1838	540	957	160	3495
Middle East	582	21	0	neg.	603
Asia & Oceania	4270	735	524	59	5588
World	11455	2900	2626	370	17351

[*]geothermal, solar, wind, wood, and waste.

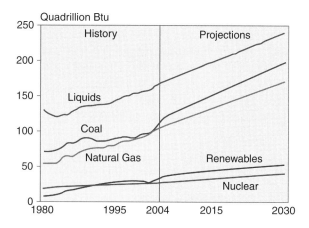

FIGURE 27.1 World energy consumption 1980–2030. Source: International Energy Outlook 2007.

The most recent data on world population are shown in Table 27.3. Note that Latin America includes the Caribbean. The fertility rate is defined as the number of children per woman. It is seen to be highest in underdeveloped regions. The trend of population in the future depends crucially on that parameter, as shown in Figure 27.2. The three growth projections involve fertility rates that vary with country and with time. The "high" case leads to a world population of almost 11 billion by the year 2050. The population in developed countries is expected to become flat.

Table 27.3 World Population Data, 1998

	Millions of Inhabitants	**Fertility Rate**	**Life Expectancy**
Africa	761	5.4	51
North America	301	2.0	76
Latin America	507	2.8	69
Asia	3363	2.7	65
Near East	166	4.4	69
Europe	798	1.5	72
Oceania	30	2.4	72
World	5927	2.9	63

Source: World Population Profile: 1998 (no later report), *United States Bureau of the Census.*

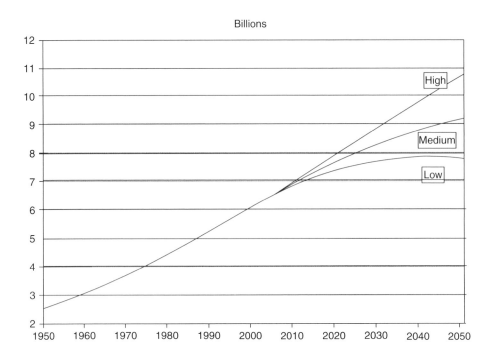

Billions

FIGURE 27.2 Past and projected world population growth, 1950–2050, with three fertility scanarieos. Source: World Population Prospects, 2006 Revision.

27.3 NUCLEAR ENERGY AND SUSTAINABLE DEVELOPMENT

Throughout history there has been little concern for the environment or human welfare. European countries systematically extracted valuable resources from Mexico, South America, and Africa, destroying cultures on the way. In the expansion to the west in the United States, vast forests were cleared to provide farmland. The passenger pigeon became extinct, and the bison nearly so. Slavery flourished in the United States until 1865. Only after European countries lost their colonies after world wars did African nations and India gain autonomy.

The environmental movement of the 1960s was stimulated by the book *Silent Spring* by Rachel Carson. That overuse of resources could be harmful was revealed by Garrett Hardin's essay "The Tragedy of the Commons." As early as 1798, Malthus had predicted that exponential growth of population would exceed linear growth of food supply, leading to widespread famine. The idea was revived by use of sophisticated computer models by Meadows, et al. in *The Limits to Growth* (see References), which predicted the collapse of civilization under various pressures associated with continued growth.

Finally, in the 1970s and 1980s the United Nations sponsored several international conferences on global problems and potential solutions. Out of these came the concept of "sustainable development." The phrase gained great popularity among many organizations that were concerned with the state of the world. The original definition of the term was "... meets the needs of the present without compromising the ability of future generations to meet their own needs." As noted by Reid (see References), the phrase can be interpreted to support business-as-usual or to require drastic cutbacks. However, it generally implies conserving physical and biological resources, improving energy efficiency, and avoiding pollution, while enhancing living conditions of people in developing countries. Ideally, all goals can be met. The subject is broad in that it involves the interaction of many governments, cultures, and economic situations. Several conferences have been held under United Nations auspices to highlight the issues, obtain agreements, and map out strategies. One prominent conference was the Earth Summit held in Rio de Janeiro in 1992, which included Agenda 21, a list of 2000 suggestions for action. A follow-up appraisal of results was made in 1997. Progress since is monitored by a "watch" organization.

Johannesburg hosted another Earth Summit in 2000. There is a wealth of Web sites on the subject. The objectives of sustainable development are furthered by nongovernmental organizations (NGOs). Unfortunately, implementation of goals have been frustrated by wars, the HIV/AIDS epidemic, drought and famine, and disease. One might be pessimistic and question whether there is any hope of achieving the desired improvements in light of failure over half a century. Or one might be optimistic that the concept can bring all parties together in a concerted effort and ultimate breakthrough.

A potential cure for a runaway population and continued misery is improved economic conditions. However, the gap between conditions in rich and poor countries persists, and no improvement is in sight. The problem has become more complex by the concerns about the environment related to the destruction of the rain forest in Central and South America. There are no easy solutions, but a few principles seem reasonable. Protection of the environment is vital, but it should not thwart the hopes of people in underdeveloped countries for a better life. It is obvious that simple sharing of the wealth would result in uniform mediocrity. The alternative is increased assistance by the developed countries, in the form of capital investment and technological transfer. This must be done recognizing the principle that the people of the country being helped should lead the program to improve.

There was a time in the past when international cooperation and assistance was considered to be highly desirable. The post-World War II Marshall Plan brought Germany and Japan back to a high level of productivity and prosperity. The Peace Corps effected improvements in many countries. The Atoms for Peace program of President Eisenhower in the 1950s provided nuclear information and assistance to dozens of countries, forming the basis of the international nuclear industry. The trend in recent years has been in the opposite direction, with emphasis on

United States industrial competitiveness and United States leadership in world politics. It is quite possible that greater stability in the world would result from efforts to find more ways to cooperate—through partnerships of commercial organizations, bilateral national agreements, and arrangements developed under United Nations auspices.

One would expect that a philosophy that embraces human rights and supports justice would be implemented by major efforts to help less fortunate people around the world. But even if the motivation were only enlightened self-interest, helping bring up standards of living should open new markets for goods and services and avoid the problem of competing products on the basis of cheap labor.

Success in effecting improvements depends on the means by which help is provided. An issue to resolve is whether to help developing countries shape an overall economic and social plan that includes energy management or to advise how energy should be handled in the country's own plans.

Technology can be introduced in two ways: (a) supplying devices that are appropriate to the receiving country's urgent needs and that are compatible with existing skills to operate and maintain equipment; or (b) supplying equipment, training, and supervision of sophisticated technology that will bring the country quickly to industrial status. Arguments for and against each approach can easily be found. It is possible that both should be followed to provide immediate relief and further the country's hopes for independence.

Advanced countries have applied restrictions to the transfer of nuclear technology to some developing nations in an attempt to prevent the achievement of nuclear weapons capability. Third-World countries resent such exclusion from the opportunity for nuclear power.

One major objective of sustainable development is the improvement of human health in developing countries. If nuclear medicine for diagnosis and treatment were expanded universally, it could make a great difference to the health of people such as those in Africa. For countries that cannot afford to import coal, oil, and natural gas, the introduction of nuclear power for widespread supply of electricity could facilitate pollution-free industrial and commercial development while enhancing human comfort. Nuclear plants can be built to use the waste heat for desalination of seawater, providing safe water for human consumption. For such to be implemented, a reactor type is needed that avoids the high capital cost of conventional light water reactors, requires little maintenance, and is passively safe.

27.4 GREENHOUSE EFFECT AND GLOBAL CLIMATE CHANGE

The greenhouse effect is one of the processes by which the Earth is warmed. Sunlight of short wavelength can readily pass through water vapor and gases such as carbon dioxide in the atmosphere. Energy is absorbed by the Earth's surface, which emits long wavelength infrared radiation that is stopped by the vapor and gases. The effect accounts for an increase in natural temperature of approximately $30°C$. Figure 27.3 shows the energy flows for the effect.

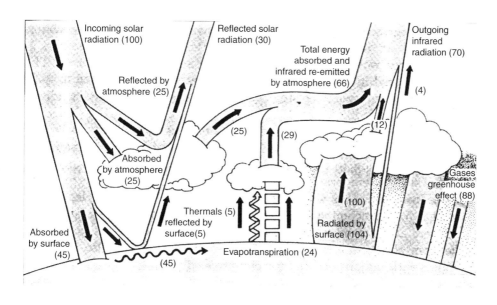

FIGURE 27.3 Earth's radiation energy balance in relation to the greenhouse effect. Numbers are percentages of incoming sunlight. (From Schneider, see References.)

There is good evidence that the carbon dioxide content of the air has increased from a preindustrial level of 200 ppm to a current value of approximately 350 ppm. Less certain is the amount of temperature change over that period because of natural fluctuations related to sun activity, volcanic dust, and shifting ocean currents.

Greenhouse gases are the collection of natural and manmade substances including water (H_2O), carbon dioxide (CO_2), methane (CH_4), nitrous oxide (N_2O), and fluorochlorocarbons. Each of these has been increasing in concert with industrialization and increased biomass burning. Estimates have been made of a possible increase of 3°C to 8°C in global temperature by the middle of the 21st century if action is not taken. Consequences of such global warming that have been proposed are more severe weather including droughts, storms, and floods; higher incidence of tropical disease; and a melting of ice near the poles that would cause a rise in ocean level that would inundate coastal cities.

International concern led to the Kyoto Protocol of December 1997, which calls for a reduction in carbon emission by all countries, with different percentages for each. The United States would be expected to reduce 7% from 1990 levels. Many nations have signed the treaty, but few have ratified it.

Nuclear power was deliberately excluded from the Kyoto protocol by representatives of European Green Parties. This omission was criticized in a report of the Nuclear Energy Agency (see Reference). That report notes that current nuclear electric plants worldwide are reducing CO_2 emissions by approximately 17%.

Scientists from more than 100 countries have contributed to a study and evaluation of recent trends in observations, research, and modeling results related to global warming. Their work is for the Intergovernmental Panel on Climate Change (IPCC), which is sponsored by the World Meteorological Organization (WMO) and the United Nations Environment Programme (UNEP). The mission of IPCC is to assess all information on causes, impacts, and options for adaptation and mitigation. A relatively brief but comprehensive review of the situation and future is found in a Web site (see References).

That there is a serious potential world problem is highlighted in a multipart report on the science of climate change issued in 2007 by a committee of the IPCC. A key finding was the conviction that the source of global warming is human activities. Estimates are made of the magnitude of temperature increases and the amount of rise in sea levels that would force migration of millions of people in low-lying countries. Predictions are made on the loss of arctic ice and increases in heat waves and tropical storms. Impacts are described for various parts of the world in several categories: water, ecosystems, food, coasts, and health. Adaptations and mitigations are suggested. The most obvious solution is the reduction of CO_2 emissions. Increases in time by developing countries are expected, whereas developed countries resist limitations. Improvements in efficiency of energy use, especially in vehicles, are desired. Capture and sequestration of carbon is under consideration. Nuclear power as an alternate source of electricity is mentioned in the report but not emphasized (see References).

The subject of global warming is controversial for several reasons. Some believe the potential consequences are so severe that it is urgent to take immediate action. Waiting for additional confirmation through research is considered to be too late. Others are concerned about the worldwide economic disruption that might be caused by drastic reduction of energy production to meet the Kyoto goals. Opposition to action has been expressed in the United States about the low limits on emission by developing countries. The United States Congress has refused to ratify the Protocol on the grounds that it is unfair and if implemented could seriously affect the economy. From a scientific viewpoint, some believe that there is no real correlation between CO_2 increase and global temperature, that the modeling of trends is yet inadequate in that it does not take proper account of the role of clouds or the absorption of carbon in the ocean, and that the computer models have not been able to reproduce past history correctly.

Singer (see References) recommends a program of adaptation, if necessary, noting that if there were global warming, it could result in more evaporation of water from the oceans and ice accumulation in Greenland and Antarctica. He favors research on the sequestering of carbon by fertilization of the ocean to enhance the population of phytoplankton, a side effect of which would be an increased supply of food fish.

The nuclear industry calls attention to the fact that nuclear reactors provide electric power without any emission of carbon dioxide or other greenhouse gases. This gives a rationale for the continued operation of nuclear power plants, for extending life through relicensing, and for building new plants. Country data are presented by Australia on the weights of annual carbon release currently avoided by use of nuclear power throughout the world (see References). The total is 2600 million tons.

A number of reports, books, and Web pages provide ample reading material on the subject (see References).

27.5 PERSPECTIVES

Let us examine the role of nuclear energy in the global sense over centuries by developing a qualitative but logical scenario of the future. Any analysis of world energy requires several ingredients—an objective, certain assumptions, a model, necessary constraints, input data, performance criteria, and output information.

A primary assumption is that fossil fuels will ultimately become excessively expensive: oil within a few decades and coal within a century or so. Thus the objective of a meaningful analysis must be to effect a smooth transition from the present dependence on fossil fuels to a stable condition that uses resources that are essentially inexhaustible or are renewable.

One constraint vital to the analysis is that a minimum first level of sufficient energy must be available to provide mankind's needs for food, shelter, clothing, protection, and health. This status corresponds to an agrarian life that uses locally available resources, little travel, and no luxuries. A desirable second level is an energy that will provide a quality of life that provides transportation, conveniences, comforts, leisure, entertainment, and opportunities for creative and cultural pursuits. This situation corresponds to an abundant life of Americans living in the suburbs and working in a city, amply supplied with material goods and services. It is mandatory that the first level is assured and that the second level is sought for all people of the world. This goal implies existing differences between conditions in developed and developing nations should be eliminated to the best of our ability.

Conservation provides a means for effectively increasing the supply of energy. Experience has shown that great savings in fuel in developed countries have resulted from changes in lifestyle and improvements in technology. Examples that work are the use of lower room temperatures in winter, shifts to smaller automobiles with more efficient gasoline consumption, increased building insulation, energy-efficient home appliances and industrial motors, and electronically controlled manufacturing. Unfortunately, the move to larger vehicles in the United States was in the wrong direction. There remains considerable potential for additional saving, which has many benefits—conservation of resources, reduced emission of pollutants, and enhanced industrial competitiveness. Finally, there is limited applicability of the concept to underdeveloped countries, where more energy use is needed, not less.

Protection of the environment and of the health and safety of the public will continue to serve as constraints on the deployment of energy technologies. The environmental movement has emphasized the damage being done to the ecology of the rain forest in the interests of development; the harm to the atmosphere, waters, and land from industrial wastes; and the loss of habitats of endangered or valuable species of wildlife. Air pollution caused by emissions from vehicles and coal-fired power plants poses a problem in cities. Less well known is the release of radioactivity from coal plants, in amounts greater than those released from nuclear plants in normal operation. Although a core meltdown followed by failure of containment in a nuclear plant would result in many casualties, the probability of such a severe accident is extremely low. In contrast, there are frequent deaths resulting from coal mining or offshore oil drilling. There is an unknown amount of life-shortening associated with lung problems aggravated by emissions from burning coal and oil.

Eventually, people will appreciate the fact that no technology is entirely risk free. Even the production of materials for solar energy collectors and their installation results in fatalities.

The use of electric power is growing faster than total energy because of its cleanliness and convenience. It is wasteful to use electrical power for low-grade heat that could be provided by other fuels. However, it is likely that the growth will be even more rapid in the future as computer-controlled robot manufacturing is adopted worldwide.

The needs for transportation in developed countries absorb a large fraction of the world's energy supply, largely in the form of liquid fuel. Petroleum serves as a starting point also for the production of useful materials such as plastics. To stretch the finite supply and give more time to develop alternatives, several conservation measures are required. Examples are more efficient vehicles and expansion of public transportation. Later, as oil becomes scarce, it will be necessary to obtain needed hydrocarbons by liquefying coal. This need suggests that coal should be conserved. Rather than expanding coal-fired electrical production, nuclear reactors could be built.

To counter rising costs of gasoline, the "plug-in" automobile is likely to become very popular. The demand for electrical power would increase dramatically and enhance the role of nuclear power.

Nuclear energy itself may very well follow a sequential pattern of implementation. Converter reactors, with their heat energy coming from the burning of uranium-235, are inefficient users of uranium because enrichment is required and spent fuel is disposed of. Breeder reactors, in contrast, have the potential of making use of most of the uranium, thus increasing the effective supply by a large factor. Sources of lower uranium content can be exploited, including very low-grade ores and the dissolved uranium in seawater, because almost all of the contained energy is recovered. To maintain an ample supply of uranium, storage of spent fuel accumulated from converter reactor operations should be considered instead of permanent disposal by burial as a waste. Conventional arguments that reprocessing is uneconomical are not as important when reprocessing is needed as a step in the planned deployment of breeder reactors. Costs for storage of spent fuel should be examined in terms of the value of uranium in a later era in which oil and coal are very expensive to secure. Eventually, fusion that uses deuterium and tritium as fuel may be practical, and fusion reactors would supplant fission reactors as the latter's useful lives end.

Because of the chemical value in the long term of natural gas, oil, and coal, burning them to heat homes and other buildings seems entirely wasteful. Electricity from nuclear sources is preferable. Resistance heating involves use of a high-quality energy for a low-grade process, and it would be preferable to use heat pumps, which use electricity efficiently for heating purposes. As an alternative, it may be desirable to recover the waste heat from nuclear power plants for district heating. To make such a coupling feasible, excellent insulation would be required for the long pipes from condenser to buildings, or the always-safe power plant would be built in close proximity to large housing developments.

Solar energy has considerable potential as a supplement to other heat and electricity sources for homes and commercial buildings, especially where sunlight is abundant. Direct energy solar boilers or arrays of photovoltaic cells are promising sources of auxiliary central-station power to be used in parallel with nuclear systems that augment the supply at night. The large variations in output from solar devices can also be partially compensated for by thermal storage systems, flywheels, pumped water storage, and compressed gas.

The ultimate system for the world is visualized as a mixture of solar and nuclear systems, with distribution dependent on climate and latitude. Breeder reactors or fusion reactors would tend to be located near large centers of manufacturing, whereas the smaller solar units would tend to be distributed in outlying areas. Solar power would be very appropriate for pumping water or desalination of seawater to reclaim desert areas of the world. Other sources, such as hydroelectric, geothermal, and wind, would also be used where those resources exist.

A conclusion that seems inevitable is that every source of energy imaginable should be used in its appropriate niche in the scheme of things. The availability of a variety of sources minimizes the disruption of life in the event of a transportation strike or international incident. Indeed, the availability of several sources that can be substituted for one another has the effect of reducing the possibility of conflict between nations. Included in the mix is extensive use of conservation measures. A corollary is that there should be a great deal of recycling of products. The reclamation of useful materials such as paper, metals, and glass would be paralleled by treating hazardous chemical wastes to generate burnable gas or application as fuel for the production of electricity.

Another conclusion from the preceding scenario is that a great deal of research and development remains to be done to effect a successful transition. Resources of energy and materials are never completely used up; they merely become harder to acquire, and eventually the cost becomes prohibitive. The effects of the status of the world of various assumptions and actions related to energy can be examined by application of program FUTURE in Computer Exercise 27.A.

The oil embargo of 1973, in which limits were placed on shipments from producing countries to consuming countries, had a sobering effect on the world. It
prompted a flurry of activity aimed at expanding the use of alternative energy
sources such as solar, wind, biomass, and oil shale, along with conservation. Easing of the energy crisis reduced the pressure to find substitutes, and as oil prices
fell, automobile travel increased. Use of energy in general is dominated by current
economics. If prices are high, energy is used sparingly; if prices are low, it is used
freely without concern for the future. Ultimately, however, when the resource
becomes more and more scarce and expensive, its use must be curtailed to such
an extent that social benefits are reduced. If no new sources are found, or if no
renewable sources are available, the quality of existence regresses and man is
brought back to a primitive condition. The use of fossil fuels over the long term
is dramatically portrayed by the graph of Figure 27.4 presented by Hubbert
approximately 50 y ago but no less meaningful today.

Future civilizations will be astounded at the careless way the irreplaceable
resources of oil, natural gas, and coal were wasted by burning, rather than used
for the production of durable and recyclable goods.

International tensions are already high because of the uneven distribution of
energy supplies. The condition of the world as supplies become very scarce is difficult to imagine. To achieve a long-term solution of the energy problem, money
and effort must be devoted to energy research and development that will yield
benefits decades and centuries into the future. Individual consumers cannot contribute except to conserve, which merely postpones the problem slightly. One
cannot expect the producers of energy to initiate major R&D projects that do
not have immediate profits.

The cost of carrying out expanded research and development programs would
be very great indeed and difficult to justify in terms of immediate tangible products. But that R&D must be carried out while the world is still prosperous,

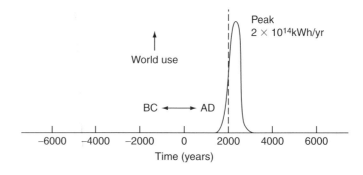

FIGURE 27.4 The epoch of fossil fuels. (Adapted from *Energy Resources: A Report to The
Committee on Natural Resources,* M. King Hubbert, National Academy of Sciences – National
Research Council, Washington, DC, 1962.)

not when it is destitute as a result of resource exhaustion. The world must take the enlightened view of a prudent person who does not leave his future to chance but invests carefully to survive in later years. In energy terms the world is already approaching its old age.

27.6 DESALINATION

Throughout the world there are regions that badly need fresh water for drinking and industrial use. Ironically, many are located next to the sea. Removal of salt by application of nuclear heat is a promising solution. Experience with nuclear desalination has been gained in more than 100 reactor-years in Kazakhstan by use of a fast reactor and in Japan by use of light water reactors.

Two modes of reactor application are (a) heat only and (b) electricity and waste heat. The first of these is simpler in that less equipment and maintenance are needed. The second has the benefit of providing a source of electric power.

In either mode, the contribution to the desalination process is the steam from a heat exchanger. There are two general technologies in which the heat can be used by desalination equipment. The first is distillation, in which the water evaporates on contact with a steam-heated surface and is separated from the salt. There are two versions of this technology: multistage flash distillation (MSF), and multi-effect distillation (MED). The second is reverse osmosis (RO), in which a porous membrane with a pressure difference separates water from salt. Two subsets of this approach have been tested. To protect the membrane considerable pretreatment is required.

A key parameter of performance of a system is the water volume per unit of heat power. The maximum value, assuming no losses of heat, is calculated from conservation of energy. One watt of heat energy corresponds to 86,400 joules per day. To bring water from 20°C to 100°C requires 80 cal/g and vaporization takes 539 cal/g. The total is $(619)(4.185) = 2590$ J. Thus one watt gives $86,400/2,590 = 33.4$ g/d. For a 1,000-MWe nuclear reactor with 2,000 MW waste heat, the maximum daily yield is $(2 \times 10^9)(33.4) = 6.68 \times 10^{10}$ g/d or 6.68×10^4 m^3/d. According to the IAEA approximately 23 million m^3/d of desalted water is produced in the world by 12,500 plants, an average of 1,840 m^3/d. To double the global water production would require a minimum of $2.3 \times 10^7/(6.68 \times 10^4) = 344$ reactors.

The International Atomic Energy Agency (IAEA) is actively promoting the concept of nuclear desalination in several countries. It has published a guidebook to aid member states in making decisions and implementing projects (see References). The IAEA has developed a PC-based computer code DEEP to analyze the economics of an installation. It provides descriptions of the concepts MSF, MED, and RO. A comprehensive article in the magazine *Desalination* by Megahed gives the history and future possibilities for nuclear desalination, along with a description of activities in Canada, China, Egypt, India, Korea, and Russia (see References).

27.7 RECYCLING AND BREEDING

The word "cycle" has come to mean the mode of management of nuclear fuels. There are several possibilities: once-through, the method currently used commercially in the United States, leading to used fuel being stored or buried; recycle, in which spent fuel is reprocessed and some of the products reused; closed cycle, with all materials retained in a system, involving nuclear and chemical treatment and burning out many undesired radioisotopes; and breeding, where as much or more fissile material is generated as was supplied initially.

The equipment required for each mode of operation is distinctly different; there are different consequences in terms of safety and security and various associated costs of operation. In the following, we address the topic of breeding, which has had a checkered history as discussed in Chapter 13.

In the post-World War II period 1945–1970, when reactor R&D was underway, it was believed that the reserves of uranium were limited, especially in terms of an expected growth in nuclear power. Thus it was thought necessary to use the U-238 by conversion into Pu-239 by means of reprocessing. The growth did not occur and there turned out to be much more uranium available. Reprocessing was stopped by President Carter and reinstated by President Reagan, but in the meanwhile, industry concluded that the process was too expensive.

Wilson (see References) gives a cogent discussion of reprocessing and breeding, stating that it may be 50 to 100 y before they are needed because uranium ore costs become excessive. However, he proposes an alternative environmental reason for continuing to pursue reprocessing (viz., objection to underground disposal of high-level waste). The breeding cycle that requires reprocessing would reduce the volume of wastes and eliminate many of the long-lived radioisotopes that determine repository character.

One can visualize another important reason for continuing programs of research, development, testing, and deployment of the nuclear, physical, and chemical aspects of recycling and reprocessing. It is the need to maintain an information base that involves data, methodology, and people with technical knowledge and skills. The latter aspect is especially significant in light of the continued retirement of nuclear scientists, engineers, and technicians who have accumulated vast experience over their professional careers. It is noteworthy in this connection that the United States has a habit of abandoning projects that later are found to be of value. With these ideas in mind, we examine the activities identified as promising for the future.

In the Generation IV program two of the concepts are the lead-cooled fast reactor (LFR) and the sodium-cooled fast reactor (SFR), as noted in Chapter 24. A variant of both that features recycling and breeding is the revival of the Integral Fast Reactor (IFR) in the Advanced Fast Reactor (AFR) of Argonne National Laboratory (see References). At the time of its cancellation the IFR was judged to be highly successful. It had the promise of full use of uranium instead of the low percents of light water reactors. The claim was made that such a reactor system has the potential of making nuclear power essentially inexhaustible and of satisfying humanity's long-term energy needs.

As described in Section 13.4, it was a closed cycle with sodium coolant and metal fuel, with pyrometallurgical chemical processing. Because of the high levels of radioactivity in the fuel, diversion and proliferation were virtually impossible. The system had the potential of burning the isotopes that dominate waste repository performance—plutonium, neptunium, americium, and curium. Wastes remaining would have a relatively short half-life and be of small volume, making a second repository unnecessary. Depending on the mode of operation, IFR/AFR-type systems could process surplus weapons plutonium, existing spent fuel, depleted uranium accumulated from isotope separation, and eventually natural uranium.

A powerful case for R&D on a fast reactor is found in an article by Hannum, Marsh, and Stanford (see References). There, a comparison is made among three cycles—once-through, plutonium recycle, and full recycle. The virtues of recycling are thoroughly discussed in testimony to Congress of Philip Finck of Argonne National Laboratory, and the need for the AFR as an extension of IFR is fully described in an article by Chang. An analysis of nuclear fuel reserves and use is provided by an article by Lightfoot, et al. Finally, a 2005 position statement of the American Nuclear Society is titled, "Fast Reactor Technology: A Path to Long-Term Energy Sustainability."

27.8 THE HYDROGEN ECONOMY

The potential connection between nuclear power and hydrogen as a fuel was discussed as early as 1973 in an article in *Science* (see References). There, hydrogen was viewed as an alternative to electricity that is storable and portable. Recently, there has been a growing interest in the use of hydrogen gas instead of oil, natural gas, or coal. This is prompted by several concerns: (a) the uncertain supply of foreign oil and natural gas; (b) the health and environmental impact of polluting combustion gases; and (c) the potential for global warming from the increasing emission of carbon dioxide. A virtue of H_2 is that it burns with only water as a product.

Hydrogen is useful in the enhancement of low-grade sources of oil, but its greatest application would be in fuel cells, where chemical reactions yield electricity. To avoid the problem of excessive weight of containers to withstand high pressure, hydrogen could be held as a metal hydride, with density of H_2 comparable to that as a liquid. Gas is released on heating to approximately 300°C. Studies are in progress on the storage of hydrogen on special surfaces or nanostructures.

Of major importance is the means by which hydrogen is generated. At present it is obtained by treatment of natural gas. It could be derived from coal as well. Both sources give rise to undesired products such as CO_2. If the carbon dioxide could be successfully sequestered, this problem would be eliminated.

There are two ways in which nuclear reactors could provide the hydrogen as an energy carrier (not as a source in itself). The obvious technique is electrolysis of water, which uses the electricity from a nuclear power plant. Alternately, the heat from a high-temperature gas-cooled reactor is sufficient to initiate thermochemical reactions that have higher efficiency. Of many possible reactions, the leading candidate is the following set. The temperature at which they take place is indicated.

$$2H_2SO_4 \rightarrow 2SO_2 + 2H_2O + O_2 (800°C)$$

$$2HI \rightarrow I_2 + H_2 (450°C)$$

$$I_2 + SO_2 + 2H_2O \rightarrow 2HI + H_2SO_4 (120°C)$$

Because the sulfur and iodine are recycled the net reaction is

$$2H_2O \rightarrow 2H_2 + O_2.$$

The energy required is merely the heat of combustion of hydrogen.

Fission of uranium in the reactor as the primary source of energy does not yield carbon dioxide or other gases and provides a more continuous and reliable supply of electricity than wind or solar power. There is an obvious need for an infrastructure for large-scale commercial application of hydrogen fuel, especially for vehicle transportation.

There are two ways hydrogen can be used for transportation—by burning in an internal combustion engine and by serving as a source for fuel cells. A fuel cell is an energy conversion device that produces electricity by the reaction of hydrogen and oxygen to produce water, the reverse of electrolysis. A set of fuel cells can power electric motors that drive an automobile or truck. The most likely type to be used is the polymer exchange membrane fuel cell (PEMFC). Its components are anode, cathode, membrane, and catalyst.

To be successful technically, onboard storage of hydrogen sufficient for a 300-mile auto drive is required. Tanks with compressed gas at very low temperature would probably be heavy, expensive, and complicated to use. Research is underway on alternatives. One possibility is an ultra-thin metal alloy film, in which there are unusual quantum effects. Lightweight materials like magnesium are promising candidates. Hydrides and carbon nanostructures are being considered as well. To be successful commercially, it would be necessary to provide refueling capability at ordinary gas stations.

Further information on what is now called the hydrogen economy is available in the literature and on the Web (see References). An article in the magazine *Physics Today* can be downloaded. It highlights the basic research needed to achieve a breakthrough. A committee of the National Research Council developed a 256-page book that can be purchased, but its Executive Summary is on the Web. A description of the hydrogen program of the DOE is also found on the Web. Three articles in *Nuclear News* provide technical details and calculations on the nuclear power requirements.

27.9 **NEXT GENERATION NUCLEAR PLANT**

The nuclear power field in the United States has come full circle in the choice of reactor type. Graphite served as moderator for reactors at Chicago, Hanford, Oak Ridge, and Brookhaven. Except for reactors at Peach Bottom and Fort St. Vrain, light water reactors have dominated since. But for the future, graphite-moderated reactors seem very promising.

Two concepts have recently been studied. Both make use of coated particles as fuel, with uranium oxide and layers of silicon carbide and carbon to retain fission products. Both use helium gas as coolant and operate at high temperatures.

The first is the Pebble Bed Modular Reactor (PBMR), initiated by the South African company Eskom. Research is underway at MIT and INL. In the PBMR the coated particles are held in spheres of approximately 3 inches diameter. The reactor core would contain some 450,000 of such spheres, which would flow through the reactor vessel and be irradiated. Information on the reactor is found in the PBMR Web site (see References).

The second is the Gas Turbine Modular Helium Reactor (GT-MHR), designed by General Atomics. The core consists of hexagonal graphite blocks of 36 cm across flats. The prisms are pierced by holes that contain 1.25 cm diameter rods of coated particles mixed with a carbon binder and have holes for the helium coolant. A complete description of the reactor is found in a *Nuclear News* article (see References).

The Energy Policy Act of 2005 called for the establishment of the Next-Generation Nuclear Plant (NGNP). It specified that it should be located at the Idaho National Laboratory and be capable of producing hydrogen. An Independent Technology Review Group (ITRG) was assigned the task of selecting the most promising reactor concept and assessing the requirements for R&D leading to a prototype reactor. The type selected out of an initial group of six was the Very High Temperature Reactor (VHTR), which is said to satisfy all requirements on safety, economics, proliferation resistance, waste reduction, and fuel use. Its form could be either the prismatic block or the pebble bed, with the principal investigation into the performance of the coated particle fuel elements at the 900°C to 950°C helium temperatures needed for hydrogen production. ITRG noted several needs: (a) to develop a high-temperature H_2 facility; (b) to address the role of an intermediate heat exchanger to isolate the reactor from the hydrogen unit; (c) to determine the proper dynamic coupling of the two components; (d) to achieve successful fabrication of fuel kernels; and (e) to develop a high-performance helium turbine. Figure 27.5 shows the coupled reactor and hydrogen unit.

Some of the preliminary design features are as follows:

Coolant inlet/outlet temperatures 640/1,000°C
Thermal power 600 MW
Efficiency >50%
Helium flow rate 320 kg/s
Average power density 6 to 10 MWt/m^3

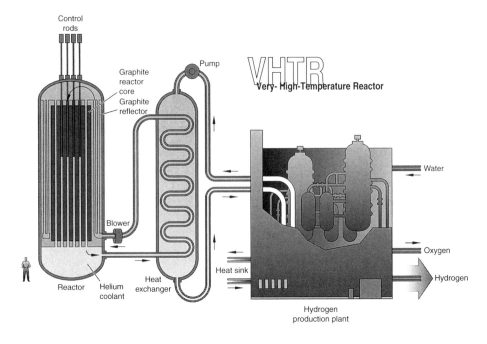

FIGURE 27.5 Very high temperature reactor with hydrogen processing unit. (Courtesy of DOE.)

Additional details are found in a DOE "roadmap" and in an assessment of features and uncertainties by the ITRG (see References).

We can estimate the impact of reactors producing hydrogen gas for use as fuel for transportation in place of gasoline. Consider one reactor of power 600 MWt with 50% efficiency of H_2 generation. As noted in Section 27.8, the sulfur-iodine process causes the dissociation of water into component gases. By use of the heat of combustion of 142 MJ/kg with hydrogen density 0.09 kg/m³, we find the production rate is 2.03 million cubic meters per day. If that hydrogen is burned in place of gasoline with a heat of combustion of 45 MJ/kg, the equivalent volume of gasoline is 2.06×10^5 gallons per day. The dollar value then depends on the price of gasoline. Computer Exercise 27.B provides details of the preceding calculation.

Completion of tests of fuel integrity for the VHTR is expected around the year 2020. If the tests are successful, a number of very high temperature reactors could be in operation for hydrogen generation by the middle of the 21st century. In addition to the saving of foreign oil, the reactors would contribute significantly to a reduction in emissions of greenhouse gases and help alleviate global climate change.

27.10 SUMMARY

The energy future of the world is not clear, because both optimistic and pessimistic predictions have been made. The population growth of the world remains excessive, with growth rates in underdeveloped countries being highest. Nuclear power may play an important role in achieving sustainable development and in easing international energy tensions. Research and development are seen as key ingredients in the quest for adequate energy. Goals for the future are recycling and breeding, desalination with nuclear power, and the production of hydrogen for transportation both to help prevent global warming and to reduce the need for imported oil. The Very High Temperature Reactor is regarded as the best candidate for the next generation nuclear system.

27.11 EXERCISES

27.1 The volume of the oceans of the Earth is 1.37×10^{18} m^3, according to *Academic American Encyclopedia*, Vol. 14, p. 326. If the deuterium content of the hydrogen in the water is 1 part in 6700, how many kilograms of deuterium are there? With the heat available from deuterium, 5.72×10^{14} J/kg (see Exercise 14.4) and assuming a constant world annual energy consumption of approximately 300 EJ, how long would the deuterium last?

27.2 A plan is advanced to bring the standard of living of all countries of the world up to those of North America by the year 2050. A requisite would be a significant increase in the per capita supply of energy to other countries besides the United States and Canada. Assume that the Medium Growth Case of Figure 27.2 is applicable, resulting in a growth from 6 billion to 9 billion people. By what factor would world energy production have to increase? If the electricity fraction remained constant, how many additional 1000-MWe reactors operating at 75% capacity factor (or equivalent coal-fired power plants) would be needed to meet the demand?

Computer Exercises

27.A Computer program FUTURE considers global regions and levels of development, mixes of source technology, energy efficiency, resource limitations, population, pollution, and other factors. Some of the methods and data of Goldemberg (see References) are used. Explore the menus, make choices or insert numbers, and observe responses.

27.B Computer program VHTR calculates the number of gallons of gasoline equivalent in combustion energy to that in the daily production of hydrogen by a Very High Temperature Reactor. Study the program and make changes in input.

27.12 **REFERENCES**

Jane C. S. Long, "A Blind Man's Guide to Energy Policy," *Issues in Science and Technology*, Winter 2008, p.51.

International Energy Annual
http://www.eia.doe.gov/iea
Comprehensive data from Energy Information Administration, United States Department of Energy, on production and consumption by type, country and region, and year.

International Energy Outlook 2008
http://www.eia.doe.gov/oiaf/ieo
1990 through 2030.

Palmer Cosslett Putnam, *Energy in the Future*, Van Nostrand, New York, 1953. A classic early study of plausible world demands for energy over the subsequent 50 to 100 years, sponsored by the U.S. Atomic Energy Commission. Includes a large amount of data and makes projections that were reasonable at the time. Written before the development of commercial nuclear power and before the environmental movement got under way, the book is quite out of date but is worth reading for the thoughtful analysis.

Resources and Man, National Academy of Sciences–National Research Council, W. H. Freeman, San Francisco, 1969. Especially Chapter 8, "Energy Resources," by M. King Hubbert. A sobering study of the future that has been cited frequently.

Donnella H. Meadows, Dennis L. Meadows, Jorgen Randers, and William W. Behrens III, *The Limits to Growth: A Report for The Club of Rome's Project on the Predicament of Mankind*, Universe Books, New York, 1972. An attempt to predict the future. Conclusion: The world situation is very serious. Influential in its time.

Herman Kahn, William Brown, and Leon Martel, *The Next 200 Years: A Scenario for America and the World*, William Morrow, New York, 1976. Contrast of pessimism and optimism.

Thomas Hoffmann and Brian Johnson, *The World Energy Triangle: A Strategy for Cooperation*, Ballinger, Cambridge, 1981. A thoughtful investigation of the energy needs of the Third World and assessment of ways developed countries can help to their own benefit. Sponsored by the International Institute for Environment and Development.

Alvin M. Weinberg, *Continuing the Nuclear Dialogue: Selected Essays*, American Nuclear Society, La Grange Park, IL, 1985 (selected and with introductory comments by Russell M. Ball). Dr. Weinberg was a pioneer and philosopher in the nuclear field. Writings span 1946–1985.

Jose Goldemberg, *Energy, Environment and Development*, Earthscan Publications, London, 1996. Facts, figures, and analyses of energy planning, taking account initially of broad societal goals.

Nuclear Power: Technical and Institutional Options for the Future, National Academy Press, Washington, DC, 1992. A study by a committee of the National Research Council of ways to preserve the nuclear fission option. An analysis of nuclear power's status, obstacles, and alternatives.

An Appropriate Role for Nuclear Power in Meeting Global Energy Needs
http://www.acus.org
Select Publications/Energy Environment/Nuclear Power. From the Atlantic Council of the United States.

Richard Rhodes, *Nuclear Renewal: Common Sense About Energy*, Penguin Books, New York, 1993. Assessment of risks; programs of France and Japan.

Alan E. Waltar, *America the Powerless: Facing Our Nuclear Energy Dilemma*, Cogito Books, Madison, WI, 1995. Poses issues and states realities.

Albert B. Reynolds, *Bluebells and Nuclear Energy*, Cogito Books, Madison, WI, 1996. Includes a discussion of new reactor designs.

Max W. Carbon, *Nuclear Power: Villain or Victim? Our Most Misunderstood Source of Electricity*, Pebble Beach Publishers, Madison, WI, 1997.

Alan M. Herbst and George W. Hopley, *Nuclear Energy Now*, John Wiley & Sons, Hoboken, NJ, 2007. Subtitle: Why the Time Has Come for the World's Most Misunderstood Energy Source.

Gwyneth Cravens, *Power to Save the World*, Alfred A. Knopf, New York, 2007. Subtitle: The Truth About Nuclear Energy. A former skeptic embraces nuclear power.

David Bodansky, *Nuclear Energy*, 2nd Ed., Springer/AIP Press, New York, 2003. Subtitle: Principles, Practices, and Prospects.

David Reid, *Sustainable Development: An Introductory Guide*, Earthscan Publishers, London, 1995. A thoughtful and informative book that describes the issues candidly.

David Pearce, *Economics and Environment: Essays on Ecological Economics and Sustainable Development*, Edward Elgar, Cheltenham, UK, 1998. A collection of papers by a distinguished author.

Earth Summit in Rio de Janeiro, 1992
http://www.un.org/geninfo/bp/enviro.html
Agenda 21 adopted.

Earth Summit+5
http://www.un.org/esa/earthsummit
Special Session of the General Assembly to review and appraise the implementation of Agenda 21.

Earth Summit 2002
http://earthsummit2002.org
Johannesburg (Rio + 10).

State of the World 2008: Innovations for a Sustainable Economy
http://www.worldwatch.org
Excerpts from book. From Worldwatch Institute.

Nuclear Energy and the Kyoto Protocol
www.nea.fr/html/ndd/reports/2002/nea3808.html

Intergovernmental Panel on Climate Change (IPCC)
http://www.ipcc.ch
Select Full Report or Summary for Policymakers on one or more of
these: "The AR4 Synthesis Report," "The Physical Science Basis," "Impacts, Adaptation and Vulnerability," "Mitigation of Climate Change."

Al Gore, *An Inconvenient Truth: The Crisis of Global Warming*, Viking, New York, 2007.

Congressional Research Service Reports
http://ncse.online.org/NLE/CRS
Search topic Climate Change.

Donald J. Wuebbles and Jae Edmonds, *Primer on Greenhouse Gases*, Lewis Publishers, Chelsea, MI, 1991. An update of a Department of Energy report tabulating gases and their sources with data on effects.

S. H. Schneider, "Introduction to Climate Modeling" in *Climate System Modeling*, K. E. Trenbeth, Ed., Cambridge University Press, New York, 1992.

S. Fred Singer, *Hot Talk Cold Science: Global Warming's Unfinished Business*, The Independent Institute, Oakland, CA, 1998.

S. Fred Singer, *Global Warming's Unfinished Debate*, Independent Institute, Oakland, CA, 1999.

Global Warming
http://www.uic.com.au/nip.htm
Select Climate Change/Global Warming-Science. CO_2 avoided with nuclear generation.

Congressional Research Service Reports
http://ncse.online.org/NLE/CRS
Search topic Climate Change.

Design Concepts of Nuclear Desalination
http://www.iaea.org/nucleardesalination
Select report from Virtual Office and download.

Mohamed M. Megahed, "An Overview of Nuclear Desalination: History and Challenges," *Nuclear Desalination*, Vol. 1, 2003, p. 2–18. Inaugural issue of magazine with all articles on the topic.

William H. Hannum, Gerald E. Marsh, and George S. Stanford, "Smarter Use of Nuclear Waste," *Scientific American*, December 2005
http://www.nationalcenter.org/NuclearFastReactorsSA1205.pdf

Advanced Fast Reactor (AFR)
http://www.ne.anl.gov/research/ardt/afr/index.html

Fast Neutron Reactors
http://www.uic.com.au/nip98.htm
Comprehensive essay.

Yoon I. Chang, "Advanced Fast Reactor: A Next-Generation Nuclear Energy Concept," *Forum on Physics & Society*, April 2002.
http://units.aps.org/units/fps/newsletters/2002/april/a1ap02.cfm

Phillip J. Finck, Hearing June 16, 2005 on Nuclear Fuel Reprocessing Before the House Committee on Science, Energy Subcommittee
http://www.anl.gov/Media_Center/News/2005/testimony050616.html

Pebble Bed Modular Reactor
https://www.pbmr.com
Use "Read more" features for details of the program.

M. P. LaBar, A. S. Shenoy, W. A. Simon, and E. M. Campbell, "The Gas Turbine-Modular Helium Reactor," *Nuclear News*, La Grange Park, IL, October 2003.

W. E. Winsche, K. C. Hoffman, and F. J. Salzano, "Hydrogen: Its Future Role in the Nation's Energy Economy," *Science*, June 1973, p. 1325.

George W. Crabtree, Mildred S. Dresselhaus, and Michelle V. Buchanan, "The Hydrogen Economy," *Physics Today*, December 2004, p. 39. Available on line at www.aip.org/pt/vol-57/iss-12/p39.html or Google with title and magazine.

Committee on Alternatives and Strategies for Future Hydrogen Production and Use, National Research Council, *The Hydrogen Economy: Opportunities, Costs, Barriers, and R&D Needs*, National Academies Press, Washington, DC, 2004.
http://www.nap.edu/catalog.php?record_id=10922.
Select Read Full Text.

Charles W. Forsberg and K. L. Peddicord, "Hydrogen Production as a Major Nuclear Energy Application," *Nuclear News*, September 2001, p.41.

Charles W. Forsberg, Paul L. Pickard, and Per Peterson, "The Advanced High Temperature Reactor for Production of Hydrogen or Electricity," *Nuclear News*, February 2003, p. 30.

Charles W. Forsberg, "What is the Initial Market for Hydrogen from Nuclear Energy?" *Nuclear News*, January 2005, p. 24.

DOE Hydrogen Program Home Page
www.hydrogen.energy.gov
Links to subjects of production, delivery, storage, manufacturing, etc.

"A Technology Roadmap for Generation IV Nuclear Energy Systems."
http://gif.inel.gov/roadmap/pdfs/gen_iv_roadmap.pdf
DOE report 2002, 91 pages.

Independent Technology Review Group (Idaho Nuclear Laboratory), "Design Features and Technology Uncertainties for the Next Generation Nuclear Plant," 2004.
http://nuclear.inl.gov/deliverables/docs/itrg-report-rev-2008-26-04.pdf

Appendix

CONVERSION FACTORS

To convert from numbers given in the British or other system of units to numbers in SI units, multiply by the factors in the following table, adapted from *ASTM Standard for Metric Practice*, 2nd Ed., American Society for Testing and Materials, Philadelphia, PA, 1989. For example, multiply the energy of thermal neutrons of 0.0253 eV by 1.602×10^{-19} to obtain the energy as 4.053×10^{-21} J. Note that the conversion factors are rounded off to four significant figures.

Original System	SI	Factor
Atmosphere	Pascal (Pa)	1.013×10^5
Barn	Square meter (m²)	1.000×10^{-28}
Barrel (42 gal for petroleum)	Cubic meter (m³)	1.590×10^{-1}
British thermal unit, Btu	Joule (J)	1.055×10^3
Thermal conductivity (Btu/h-ft)	W/m-°C	1.73
Calorie (cal)	Joule (J)	4.185
Centimeter of mercury	Pascal (Pa)	1.333×10^3
Centipoise	Pascal-second (Pa-s)	1.000×10^{-3}
Curie (Ci)	Disintegrations per second (dps)	3.700×10^{10}
Day (d)	Second (s)	8.640×10^4
Degree (angle)	Radian	1.745×10^{-2}
Degree Fahrenheit (°F)	Degree Celsius (°C)	$°C = \frac{5}{9}(°F - 32)$
Electron-volt (eV)	Joule (J)	1.602×10^{-19}
Foot (ft)	Meter (m)	3.048×10^{-1}
Square foot (ft²)	Square meter (m²)	9.290×10^{-2}
Cubic foot (ft³)	Cubic meter (m³)	2.832×10^{-2}
Cubic foot per minute (ft³/min)	Cubic meter per second (m³/s)	4.719×10^{-4}
Gallon (gal) U.S. liquid	Cubic meter (m³)	3.785×10^{-3}
Gauss	Tesla (T)	1.000×10^{-4}
Horsepower (hp) (550 ft-lb/s)	Watt (W)	7.457×10^2
Inch (in.)	Meter (m)	2.540×10^{-2}
Square inch (in.²)	Square meter (m²)	6.452×10^{-4}
Cubic inch (in.²)	Cubic meter (m³)	1.639×10^{-5}
Kilowatt hour (kWh)	Joule (J)	3.600×10^6
Kilogram-force (kgf)	Newton (N)	9.807
Liter (l)	Cubic meter (m³)	1.000×10^{-3}
Micron (μ)	Meter (m)	1.000×10^{-6}
Mile (mi)	Meter (m)	1.609×10^3
Miles per hour (mi/h)	Meters per second (m/s)	4.470×10^{-1}
Square mile (mi²)	Square meter (m²)	2.590×10^6
Pound (lb)	Kilogram (kg)	4.536×10^{-1}

(continued)

Original System	SI	Factor
Pound force per square inch (psi)	Pascal (Pa)	6.895×10^3
Rad	Gray (Gy)	1.000×10^{-2}
Roentgen (r)	Coulomb per kilogram (C/kg)	2.580×10^{-4}
Ton (short, 2000 lb)	Kilogram (kg)	9.072×10^2
Watt-hour (W-h)	Joule (J)	3.600×10^3
Year (y)	Second (s)	3.156×10^7

ATOMIC AND NUCLEAR DATA

(a) *Atomic Weights of the Elements 2007*([†]) (based on mass of carbon-12 as exactly 12). For some elements that have no stable nuclide, the mass number of the isotope with longest half-life is listed.

Atomic Number	Name	Symbol	Atomic Weight
1	Hydrogen	H	1.00794
2	Helium	He	4.002602
3	Lithium	Li	6.941
4	Beryllium	Be	9.012182
5	Boron	B	10.811
6	Carbon	C	12.0107
7	Nitrogen	N	14.0067
8	Oxygen	O	15.9994
9	Fluorine	F	18.9984032
10	Neon	Ne	20.1797
11	Sodium	Na	22.9897693
12	Magnesium	Mg	24.3050
13	Aluminum	Al	26.9815386
14	Silicon	Si	28.0855
15	Phosphorus	P	30.973762
16	Sulfur	S	32.065
17	Chlorine	Cl	35.453
18	Argon	Ar	39.948
19	Potassium	K	39.0983
20	Calcium	Ca	40.078
21	Scandium	Sc	44.955912
22	Titanium	Ti	47.867
23	Vanadium	V	50.9415
24	Chromium	Cr	51.9961
25	Manganese	Mn	54.938045
26	Iron	Fe	55.845

(*continued*)

Atomic Number	Name	Symbol	Atomic Weight
27	Cobalt	Co	58.933195
28	Nickel	Ni	58.6934
29	Copper	Cu	63.546
30	Zinc	Zn	65.38
31	Gallium	Ga	69.723
32	Germanium	Ge	72.64
33	Arsenic	As	74.92160
34	Selenium	Se	78.96
35	Bromine	Br	79.904
36	Krypton	Kr	83.798
37	Rubidium	Rb	85.4678
38	Strontium	Sr	87.62
39	Yttrium	Y	88.90585
40	Zirconium	Zr	91.224
41	Niobium	Nb	92.90638
42	Molybdenum	Mo	95.96
43	Technetium	Tc	(97.9072)
44	Ruthenium	Ru	101.07
45	Rhodium	Rh	102.90550
46	Palladium	Pd	106.42
47	Silver	Ag	107.8682
48	Cadmium	Cd	112.411
49	Indium	In	114.818
50	Tin	Sn	118.710
51	Antimony	Sb	121.760
52	Tellurium	Te	127.60
53	Iodine	I	126.90447
54	Xenon	Xe	131.293
55	Cesium	Cs	132.905452
56	Barium	Ba	137.327
57	Lanthanum	La	138.90547
58	Cerium	Ce	140.116
59	Praseodymium	Pr	140.90765
60	Neodymium	Nd	144.242
61	Promethium	Pm	(144.9127)
62	Samarium	Sm	150.36
63	Europium	Eu	151.964
64	Gadolinium	Gd	157.25
65	Terbium	Tb	158.92535
66	Dysprosium	Dy	162.500
67	Holmium	Ho	164.93032
68	Erbium	Er	167.259
69	Thulium	Tm	168.93421
70	Ytterbium	Yb	173.054

(continued)

Atomic Number	Name	Symbol	Atomic Weight
71	Lutetium	Lu	174.9668
72	Hafnium	Hf	178.49
73	Tantalum	Ta	180.94788
74	Tungsten	W	183.84
75	Rhenium	Re	186.207
76	Osmium	Os	190.23
77	Iridium	Ir	192.217
78	Platinum	Pt	195.084
79	Gold	Au	196.966569
80	Mercury	Hg	200.59
81	Thallium	Tl	204.3833
82	Lead	Pb	207.2
83	Bismuth	Bi	208.98040
84	Polonium	Po	(209)
85	Astatine	At	(210)
86	Radon	Rn	(222)
87	Francium	Fr	(223)
88	Radium	Ra	(226)
89	Actinium	Ac	(227)
90	Thorium	Th	232.03806
91	Protactinium	Pa	231.03588
92	Uranium	U	238.02891
93	Neptunium	Np	(237)
94	Plutonium	Pu	(244)
95	Americium	Am	(243)
96	Curium	Cm	(247)
97	Berkelium	Bk	(247)
98	Californium	Cf	(251)
99	Einsteinium	Es	(252)
100	Fermium	Fm	(257)
101	Mendelevium	Md	(258)
102	Nobelium	No	(259)
103	Lawrencium	Lr	(262)
104	Rutherfordium	Rf	(267)
105	Dubnium	Db	(268)
106	Seaborgium	Sg	(271)
107	Bohrium	Bh	(272)
108	Hassium	Hs	(270)
109	Meitnerium	Mt	(276)
110	Darmstadtium	Ds	(281)
111	Roentgenium	Rg	(280)
112	Ununbium	Uub	(285)
113	Ununtrium	Uut	(284)
114	Ununquadium	Uuq	(289)
115	Ununpentium	Uup	(288)

(*continued*)

Atomic Number	Name	Symbol	Atomic Weight
116	Ununhexium	Uuh	(293)
117	Ununseptium	Uus	
118	Ununoctium	Uno	(294)

[†]*Commission on Atomic Weights and Isotopic Abundances of the International Union of Pure and Applied Chemistry (IUPAC) http://www.chem.qmul.ac.uk/iupac/AtWt/.*

(b) Selected Atomic Masses (rounded to six decimals)

Electron	0.000549
Proton	1.007276
Neutron	1.008665
$^1_1$H	1.007825
$^2_1$H	2.014102
$^3_1$H	3.016049
$^3_2$He	3.016029
$^4_2$He	4.002603
$^6_3$Li	6.015123
$^7_3$Li	7.016005
$^9_4$Be	9.012182
$^{10}_5$B	10.012937
$^{11}_5$B	11.009305
$^{12}_6$C	12.000000
$^{14}_6$C	14.003242
$^{13}_7$N	13.005739
$^{14}_7$N	14.003074
$^{16}_8$O	15.994915
$^{17}_8$O	16.999132
$^{92}_{37}$Rb	91.919729
$^{140}_{55}$Cs	139.917282
$^{234}_{92}$U	234.040952
$^{235}_{92}$U	235.043930
$^{236}_{92}$U	236.045562
$^{238}_{92}$U	238.050788
$^{239}_{94}$Pu	239.052163
$^{240}_{94}$Pu	240.053814

Reference: G. Audi, A. H. Wapstra, and C. Thibault, "The Ame2003 atomic mass evaluation (II)," Nuclear Physics A729, p. 337 (2003). Complete data on Internet at http://csnwww.in2p3.fr/amdc/web/masseval.html. Note that conversion factor used is 931.49386 MeV/amu rather than the CODATA figure below.

(c) Values of Fundamental Physical Constants

Speed of light, c	299792458 m/s
Elementary charge, e	$1.602176487 \times 10^{-19}$ C
Electron-volt, eV	$1.602176487 \times 10^{-19}$ J
Planck constant, h	$6.62606896 \times 10^{-34}$ J-s
Avogadro constant, N_A	$6.02214179 \times 10^{23}$/mol
Boltzmann constant, k	$1.3806504 \times 10^{-23}$ J/K
Electron rest mass, m_e	$9.10938215 \times 10^{-31}$ kg or 0.510998910 MeV
Proton rest mass, m_p	$1.672621637 \times 10^{-27}$ kg
Neutron rest mass, m_n	$1.674927211 \times 10^{-27}$ kg
Atomic mass unit, u	$1.660538782 \times 10^{-27}$ kg or 931.494028 MeV
Magnetic constant, μ_0	$4\pi \times 10^{-7} = 12.566370\ldots \times 10^{-7}$
Electric constant, ε_0	$1/\mu_0 c^2 = 8.854187\ldots \times 10^{-12}$

Reference: Peter J. Mohr, Barry N. Taylor, and David B. Newell, "The Fundamental Physical Constants," Physics Today, July 2007, p. 52. Also on the Web site of National Institute of Science and Technology (NIST), http://physics.nist.gov/constants. Download in pdf.

ANSWERS TO EXERCISES

1.1.	2400 J
1.2.	20°C, 260°C, −459°F, 1832°F
1.3.	2.25×10^4 J
1.4.	512 m/s
1.5.	149 kW, 596 kWh
1.6.	2×10^{20}/s
1.7.	(a) (proof), (b) 2.22×10^{-9} g
1.8.	3.04×10^{-11} J
1.9.	3.38×10^{-28} kg
1.10.	3.51×10^{-8} J
1.11.	8.67×10^{-4}
1.12.	(proof)
1.13.	(a) (proof), (b) 0.140, 0.417, 0.866
1.14.	(a) 6.16×10^4 Btu/lb, (b) 1.43×10^5 J/g, (c) 3.0 eV
2.1.	0.0828×10^{24}/cm^3
2.2.	1.59×10^{-8} cm, 1.70×10^{-23} cm^3
2.3.	2200 m/s
2.4.	(proof)
2.5.	2.1 eV
2.6.	3.26×10^{15}/s
2.7.	−1.5 eV, 4.77×10^{-8} cm, 12.0 eV, 2.9×10^{15}/s
2.8.	(sketch)
2.9.	(proof)

(continued)

ANSWERS TO EXERCISES—Cont'd

2.10. 8.7×10^{-13} cm, 2.4×10^{-24} cm^2

2.11. 1.35×10^{-13}

2.12.   28.3 MeV

2.13.   1783 MeV

2.14. 1.46×10^{17} kg/m^3, 1.89×10^4 kg/m^3, 0.99×10^{13} kg/m^3

2.15.   (proof)

2.16. 1.21×10^{14}, 2.48×10^{-4} cm

3.1. 7.26×10^{-10} /s, 2.18×10^{10} Bq, 0.589 Ci

3.2. 3.66×10^{10}/s vs. 3.7×10^{10}/s

3.3.   1.65 μg

3.4. 3.21×10^{14}/s, 8.68×10^3 Ci, 1.06×10^{14}/s, 2.86×10^3 Ci

3.5.   (diagram)

3.6.   (graph)

3.7. 2.47×10^{20}, 1.74×10^{-17}/s, 4.30×10^3 dps and Bq, 0.116 μCi

3.8.   (a) (graph), (b) 1.82 h, argon-41

3.9. 1.61×10^3/y, radium

4.1.   (proof)

4.2. $^{14}_{6}$C, $^{10}_{5}$B

4.3.   1.19 MeV

4.4.   4.78 MeV

4.5. 3.95×10^{-30} kg, 3.55×10^5 m/s, 1.3×10^{-3} MeV

4.6. 2.04×10^7 m/s, 1.38×10^{-12} J or 8.67 MeV

4.7.   1.20 MeV

4.8.   1.46/cm, 0.68 cm

4.9. 1.70×10^7 m/s, 4.12×10^4/cm^3

4.10. 6×10^{13}/cm^2 – s, 0.02/cm, 1.2×10^{12}/cm^3 – s

4.11.   0.207, 0.074, 88, 0.4 cm

4.12.   (a) (proof), (b) 382 barns

4.13. 2.74×10^{12}/cm^3 – s, 1.49×10^{12}/cm^3 – s

4.14. 4.56×10^{-7}, 0.456

4.15.   0.1852 b. vs. 0.19 b. from Table 4.1

4.16.   0.504/cm, 0.099 cm, 4.9%

4.17.   1.9%

4.18.   0.0795 cm^{-1}, 0.328 cm^{-1}, 0.305 cm, 1.02 cm

4.19. $v_1 \cong -u_1$, $v_2 \cong 0$, particle rebounds with original speed

4.20.   (proof), $\alpha = 0.98333$, $\xi = 0.00838$

4.21.   (proof)

5.1.   0.0233, 42.8

5.2. 1.45×10^{21}/s, 2.07×10^{-13} m

5.3.   (a) 0.245 MeV, (b) (proof), (c) $E' = E_0 /2$

5.4.   0.62 MeV

5.5. $\cong 0.001$ cm

5.6. 0.033×10^{24}/cm^3, 0.46/cm, 1.51 cm

5.7.   0.289 cm

(continued)

ANSWERS TO EXERCISES—Cont'd

5.8.	0.39 cm, 1.92×10^{-5} C/cm^3, 6.15×10^{-4} J/g
5.9.	0.00218 MeV, 2206
5.10.	8.16×10^{-14} J, 5.91×10^9 K
6.1.	6.53 MeV
6.2.	$^{100}_{38}$Sr
6.3.	(a) 66.4 MeV, 99.6 MeV, (b) 140, 93, (c) 0.96×10^7 m/s, 1.44×10^7 m/s
6.4.	168.5 MeV
6.5.	2.50
6.6.	1.0%, 99%
6.7.	0.00812 g/day
6.8.	8.09×10^6, 5.89×10^6, 5.18×10^6
7.1.	0.0265 amu, 24.7 MeV
7.2.	(proof)
7.3.	0.453 kg, 13,590 kg/day
7.4.	(a) 3.10×10^6 m/s, (b) 1.92×10^{17}/cm^3
7.5.	9.3×10^5 K
7.6.	2.73×10^5 eV
8.1.	0.114 V
8.2.	2.5×10^6/s
8.3.	1.31×10^{-7} s
8.4.	(proof)
8.5.	0.183 Wb/m^2
8.6.	(a) 1.96 MeV, (b) = 299791633 m/s, (c) 1.33 tesla
8.7.	214.7 amu, 0.99999
8.8.	746 mA, 373 MW
8.9.	(proof), 5.2×10^{-11}
8.10.	20.958 μs, 9 picoseconds
8.11.	(a) $\Delta E(\text{keV}) \cong [88.46/R(\text{m})][E(\text{GeV})]^4$, (b) 8.8×10^{-14}, (c) $p \cong ae^2/(6\pi\varepsilon_0 c^3)$
9.1.	(proof)
9.2.	1.0030
9.3.	0.0304, 0.0314
9.4.	9.59×10^4 kg, 7.66×10^4
9.5.	0.71%
9.6.	195 kg/d
9.7.	488
9.8.	(proof)
9.9.	1.0030
9.10.	0.422 kg/d, 0.578 kg/d
9.11.	238.028915; 99.283621, 0.710971, 0.005408
9.12.	at 3 w/o $12.2M, at 5 w/o $23.8M, gain $6.0M
9.13.	(a) 639 mA, (b) 32 kW, (c) no
9.14.	(a) 2.5×10^6 m/s^2, 2.55×10^5, (b) 1.147
10.1.	1.19×10^{21}/cm^3, 1/45
10.2.	0.0165

(*continued*)

ANSWERS TO EXERCISES—Cont'd

10.3.	6.0×10^5
10.4.	0.30
10.5.	10
10.6.	(a) n = 1: P(0) = 0.5, P(1) = 0.5; n = 2: P(0) = 0.25, P(1) = 0.50, P(2) = 0.25; n = 3: P(0) = 0.125, P(1) = 0.375, P(2) = 0.375, P(3) = 0.125, (b) throw once: P(0) = 5/6, P(1) = 1/6, throw twice: P(0) = 25/36, P(1) = 10/36, P(2) = 1/36
10.7.	$n = 1(\bar{x} = 1/6)$: P(0) = 0.846, P(1) = 0.141; $n = 2(\bar{x} = 1/3)$: P(0) = 0.717, P(1) = 0.239, P(2) = 0.040
10.8.	740 cps, 4.44×10^4; 211; 4.4×10^{-8}
10.9.	(a) proof, (b) 0.2907, (c) 0.2624
10.10.	(derivation)
11.1.	2.21
11.2.	$2.04 \times 10^{10}/cm^2 - s$
11.3.	1.171, 1.033, 0.032
11.4.	1.850, 1.178, 2.206
11.5.	2.049; yes
11.6.	8.64×10^6; 89,700 kg, 2691 kg; $40.4M
11.7.	(a) $28.8\ m^3$, 1.51 m, (b) 0.318
11.8.	(a) (proof), (b) 0.037
11.9.	$156\ ft^3$
11.10.	1.43 min
11.11.	0.0345
11.12.	3.58 cm, $0.0987\ cm^{-2}$, 0.441
11.13.	(plot)
11.14.	(a) 1.84 w/o, (b) 2.23 w/o
12.1.	(proof)
12.2.	(discussion)
12.3.	$150\ W/cm^2$; $3\ W/cm^2$-°C
12.4.	315°C
12.5.	30°F
12.6.	1830 MW, 1350 MW, 26%
12.7.	664 kg/s, 2.6%
12.8.	$8.09 \times 10^6\ m^2$, $8.26 \times 10^5\ J/m^2$-h
12.9.	20.5 million gal/day
12.10.	(proof)
12.11.	(proof), 0.76
13.1.	1.7, 0.7
13.2.	0.986
13.3.	(discussion)
13.4.	2.61, 0.20
13.5.	6300 kg; 3877 d or 10.6 y
14.1.	0.1 mm, 0.65 cm
14.2.	(proof)
14.3.	(proof)
14.4.	5.72×10^{14} J/kg, 0.116; $500/kg, 0.003 mills/kWh

(*continued*)

ANSWERS TO EXERCISES—Cont'd

14.5. (a), (b) proofs, (c) 2.56×10^{-13} m, (d) 3.57×10^{-13} m, (e) 202 times, 177 times

15.1. (sample)

15.2. (activity)

16.1. 6.25×10^{10}, 2.3×10^{-9}

16.2. 200

16.3. 1.67 mrads, 3.34 mrads, 6.7×10^{-4}

16.4. 0.8%

16.5. 400 mrems, 4 mSv

16.6. 1/3

17.1. Fe-59

17.2. $^{6}_{3}\text{Li} + ^{1}_{0}\text{n} \rightarrow ^{3}_{1}\text{H} + ^{4}_{2}\text{He}$,
       $^{3}_{1}\text{H} + ^{16}_{8}\text{O} \rightarrow ^{18}_{9}\text{F} + ^{1}_{0}\text{n}$

17.3. 0.63 mm

17.4. 3.0 s

17.5. 3.15×10^{8} y

17.6. 2378 y ago

17.7. 5.9×10^{-5}

17.8. $N_{Rb}/N_{Sr} = 1/[\exp(-\lambda t) - 1]$

17.9. (discussion)

17.10. (discussion)

17.11. 11.97 d

17.12. 2.64 y

17.13. 4.66 μg, 0.00764 cm

17.14. Ir-192, Co-60, Cs-137

17.15. 0.0871; 0.1%

17.16. 2.4×10^{-4}

17.17. 14

18.1. 5 mCi

18.2. 241 rads

18.3. 89.8 kg

18.4. 19,500 Ci, 289 W

18.5. $3.46 \times 10^{13}/\text{cm}^2-\text{s}$

19.1. 0.0157, 2.40; 7.7×10^{-4} s; 63.8 s

19.2. 30.2 s

19.3. (a) 3.9×10^{-8} s, (b) 2.56×10^{-8} s

19.4. 40°C

19.5. 0.90 s

19.6. 0.0068, 0.0046, 0.0034, 0.0021

19.7. −0.0208

19.8. 0.6 or 60%

19.9. 117, 138, 150, 152, 153; yes

19.10. 0.0195

19.11. 9.2%

19.12. (proof)

(*continued*)

ANSWERS TO EXERCISES—Cont'd

19.13.   29,680 MWd/tonne
19.14. $B(3) = (3/2)B_1$, $B(4) = (8/5)B_1$
20.1.   359 μg
20.2.   7.81 km/s; 22,284 mi, 35,855 km
20.3.   96%
21.1.   1560 mrems/y, 371 mrems/y
21.2.   54.6 μCi
21.3. 5×10^{-4} μCi/cm^3
21.4.   Boron
21.5.   382/cm^2–s
21.6.   3, 10, 0.6, 0.3, 20
21.7.   (in μCi/ml): 3.15×10^{-7}, 2.98×10^{-7}, 3.34×10^{-7}
21.8.   7.60 d, 94.6 d, 69.6 d
21.9. 3.35×10^{-6} μCi/g
22.1.   25,400 (49%); 95,900 (85%)
22.2.   PWR 65,500 MWe, BWR 32,500 MWe, Total 98,000 MWe; waste 4,496 m^3
22.3.   0-10 d I-131; 10 d-114 d Ce-141; 114 d-4.25 y Ce-144; 4.25 y-100 y Cs-137
22.4.   98.7%
22.5. 1.05×10^{13} cm^3/s, 2.22×10^{10} ft^3/min
22.6.   0.113 MW, 30.9 d
22.7.   (a) Percents: 4.6, 56.6, 34.6, 4.3, (b) 987 kg, (c) 1205 kg, (d) 82%
22.8.   33.219; 967 y, 1003 y, 801,000 y
22.9.   (a) proof, (b) 1.216×10^7/cm^2–s
24.1.   14,120; 284 quads, 0.284 Q; 6.24×10^{30}
24.2.   12.5 million barrels, $937.5 million
26.1.   0.57 cm
26.2.   Every 1.61 d, 227/y
26.3.   Minimum of 8
26.4.   10.4 y
27.1. 2.29×10^{16} kg, 43.6 billion y
27.2.   6.63, 1142

COMMENTS

Contact. The author welcomes suggestions about the text, References, Exercises, Answers, Solutions, Computer Exercises, and Computer Programs. Please contact as follows:

Dr. Raymond L. Murray
Nuclear Engineering Department
Box 7909 North Carolina State University
Raleigh, NC 27695

murray@eos.ncsu.edu
Tel. (919)848-7235
Cell (919)280-6821

25

References. The Internet has become a primary source of information, with Google as the main search engine. It produces an enormous number of Web sites, requiring considerable effort to select those of reliability and high quality. Unfortunately, Google does not indicate the date of posting. It is tempting to use Wikipedia, but not wise to rely on the correctness of its entries.

Exercises. These problems at the ends of chapters generally can be answered by use of a hand-held calculator.

Answers. Numerical results from chapter Exercises are listed in this Appendix. They are given to only about three significant figures.

Computer programs in Qbasic, Excel, and MATLAB for the solution of Computer Exercises in the text can be found at http://elsevierdirect.com/companions/9780123705471.

Instructor support materials including Solutions to Exercises and PowerPoint slides are available by registering at http://textbooks.elsevier.com.

Computer Programs. A complete set that uses Qbasic and Excel is supplemented by newer routines being developed with MATLAB. For comments and questions about the latter contact:

Dr. Randy J. Jost   rjost@engineering.usu.edu
Utah State University   Tel. (435)-797-0789
Department of Electrical & Computer Engineering
Department of Physics
4120 Old Main Hill
Logan, UT 84322-4120

ORIGINAL COMPUTER PROGRAMS

Following is a list of titles of the Qbasic and Excel* programs, the Computer Exercise number in which they are used, and a brief indication of function. For comments or questions, contact the author at murray@eos.ncsu.edu.

Title	Computer Exercise	Function
ALBERT	1.A	Relativistic properties of particles
ELEMENTS	2.A	Symbols, A, and Z for elements
BINDING	2.B	Semi-empirical mass formula for B and M
DECAY1*	3.A	Radioactive decay: activity
DECAY	3.B	Decay equations, calculations, graph
GROWTH	3.C	Constant generation growth of nuclide
RADIOGEN	3.D	Parent-daughter radioactivity
MOVENEUT	4.A	Displays moving particle
CURRENT	4.A	Displays stream of particles

(continued)

Title	Computer Exercise	Function
CAPTURE	4.A	Displays capture of neutron by nucleus
HEADON	4.A	Displays direct elastic collision
RANDY	4.B	Prints out set of random numbers
RANDY1	4.B	Average value of random numbers
RANDY2	4.B	Statistics for random numbers
ABSCAT	4.C	Compares absorption and scattering
SCATTER	4.D	Displays general elastic collision
ENERGY	4.E	Variation in number of collisions
ALBERT	4.F	Relation of neutron v, T, and E
COMPTON	5.A-C	Photon-electron scattering collision
FISSION	6.A	Displays fission process with fragments
YIELD	6.B	Fission yields for long-lived nuclides
REACT2	7.B	Q-values for fusion reactions
REACT3	7.C	Survey of potential fusion reactions
ALBERT	8.A-B	High velocity particles in accelerators
ENRICH3	9.A-B	Material flows in isotope separator
STAT	10.A-C	Binomial, Poisson, Gauss distributions
EXPOIS	10.D	Simulates counting data
CRITICAL	11.A	Critical conditions U and Pu assemblies
MPDQ92	11.B	Criticality with space dependence
XETR*	11.C	Xenon-135 reactivity transient
SLOWINGS	11.D	Scattering, absorption, and leakage
ASSEMBLY	11.E	Displays PWR fuel assembly
BWRASEM	11.F	Displays BWR fuel assemblies and rod
CONDUCT	12.A	Integral of thermal conductivity
TEMPLOT	12.B	Temperature distribution in fuel pin
BREED	13.A	Breeder reactor with early data
BREEDGE	13.A	Breeder reactor with newer data
FASTR	13.B	Fast reactor criticality, Hansen-Roach
FUSION	14.A	Fusion parameters and functions
MAXWELL	14.A	Calculates and plots a distribution
VELOCITY	14.A	Four characteristic speeds
DEBYE	14.A	Debye length of fusion plasma
IMPACT	14.A	Parameters for ionic collision
RADIUS	14.A	Cyclotron radius and other quantities
MEANPATH	14.A	Mean free path of charged particles
TRANSIT	14.A	Fusion plasma parameters
CROSECT	14.A	Fusion cross section and reactivity
RADIOGEN	16.A	Radon activity in closed room
RADOSE	16.B	Conversion of dose, radioactivity units
RADIOGEN	17.A	Mo-Tc radionuclide generator
PREDPREY*	18.A	Predator-prey simulation
ERADIC	18.B	Application of sterile male technique
OGRE*	19.A	One-delayed-group reactor transient

(continued)

Title	Computer Exercise	Function
KINETICS	19.B	Time-dependent behavior of reactor
RTF*	19.C	Reactor transient with feedback
COREFUEL	19.D	Display of core fuel arrangements
RUBBLE	19.E	Sketch of TMI-2 damaged core
CIRCLE6	19.F	Fuel rod array in Chernobyl reactor
SQRCIR6	19.F	Holes in graphite core of Chernobyl
CORODS	19.F	Control rod arrangement, Chernobyl
ORBIT1	20.A	Trajectory of spacecraft from Earth
PLANETS	20.B	Display of motion of Earth and Mars
PLANETS1	20.B	Phase difference of Earth and Mars
ALBERT	20.C	Mass increase of space ship
EXPOSO	21.A	Gamma attenuation, buildup factors
NEUTSHLD	21.B	Fast neutron shielding by water
CLUSTER	21.C	Statistical demonstration of clustering
EXPOSO	21.D	Array of sources in an irradiator
FUELPOOL	22.A	Displays water pool with spent fuel
WASTPULS	22.B	Displays motion of waste pulse
WTT	22.C	Dispersion in waste transport
ACTIVE	22.D	Displays activation product
LLWES	22.E	Expert system, waste classification
FASTR	26.A	Critical mass as it depends on density
ENRICH3	26.B	Blending Russian HEU material
FUTURE	27.A	Global energy analysis
VHTR	27.B	Very high temperature reactor

Index

A

ABB Combustion Engineering System80+, 428
Absolute zero, 7
Absorption cross section for boron (graph), 51
Absorption cross sections (table), 50
Absorption of neutrons, 43, 180
Abundance ratio in centrifuge, 117
Abundance ratio in isotope separation, 112
Acceleration in Newton's law, 3
Acceleration of gravity, 4
Accelerator Production of Tritium (APT), 378, 466–467
Accelerator Transmutation of Wastes (ATW), 378
Accelerators, 85, 93-103, 135, 258, 276, 333, 382, 467
Accident at Chernobyl. *See* Chernobyl accident
Accident at Three Mile Island. *See* Three Mile Island accident
ACRS (Advisory Committee on Reactor Safeguards), 397
Actinides, 155
Activation, 363, 379, 387
Activation products in reactor coolant (table), 379
Activity, 33, 35-36, 348, 351, 364, 397
Activity (definition), 33
Advanced Fast Reactor, 187, 488
Advanced Light Water Reactor Utility Requirements Document, 427
Advanced Liquid Metal Reactor (ALMR), 187
Advanced reactors, 427–429
Advisory Committee on Reactor Safeguards (ACRS), 397
AEC (Atomic Energy Commission), 395–396, 420
Age of minerals, 252
Age of neutrons (definition), 54
Age of the earth, 252
Age to thermal of neutrons (table), 54
Agreement State, 399
Aircraft attack, 313
Aircraft carriers, 324
Aircraft nuclear propulsion, 325
Alamogordo, 221
ALARA (As low as is reasonably achievable), 237, 347, 348, 350
ALI (Annual limit of intake), 353–354
Allied General Nuclear Services (AGNS), 369, 372
ALMR (Advanced Liquid Metal Reactor), 187
Alpha particle (definition), 29
Alpha particle range, 70

Alpha particles as projectiles, 64
Alpha particles in radioactive chain, 34
ALSEP (Apollo Lunar Surface Experimental Package), 328
Alsos Mission, 219
American Association of Mechanical Engineers (ASME), 407
American National Standards Institute (ANSI), 407, 409
American Nuclear Society (ANS), 406, 407, 409
American Physical Society (APS), 308
American Public Power Association (APPA), 406, 407
American Society for Testing and Materials (ASTM), 407
Americium-241, 258, 261
Amplifier in counter circuit, 130
Anger camera, 250
Angular frequency in accelerator, 97
Angular speed of ions, 96, 97
Annihilation of electron and positron, 69, 129, 250
Annual radiation exposure, 233
Anode of counter, 126
ANP (Aircraft Nuclear Propulsion), 325
ANS (American Nuclear Society), 406, 407, 409
ANS publications, 406
ANSI (American National Standards Institute), 407, 409
Answers to Exercises, 504–509
Antarctic ice cap waste disposal, 373
Antibody, 251
Antigen, 251
Antimatter, 32
Antimony-124 in neutron source, 258
Antiparticle, 32, 101
Antiprotons, 100, 101
AP1000, 429–431
Apollo-12 moon mission, 328
APPA (American Public Power Association), 407, 409
Approvals for irradiation of food (table), 274
APS (American Physical Society), 308
APT (Accelerator Production of Tritium), 378, 466–467
Arab oil embargo, 223
Areva, 441, 442
Argentina, 440, 447
Armenia, 443
Art work authentication by NAA, 256

513

Artificial heart, 331
Artificial intelligence (AI), 423
Artificial radioactivity, 218
ASLB (Atomic Safety and Licensing Board), 397
ASME (American Association of Mechanical Engineers), 407
Assurance of radiation safety, 398
Assurance of reactor safety, 293–300
Asteroid, 253, 330, 331, 333
ASTM (American Society for Testing and Materials), 407
Astronomical studies by NAA, 256
Atom and light, 16–20
Atom bomb, 453, 454, 458, 460
Atomic and nuclear data tables, 500–503
Atomic and nuclear structure (diagram), 22
Atomic Energy Act of 1946, 395, 409
Atomic Energy Act of 1954, 396
Atomic Energy Commission (AEC), 395–396, 420
Atomic mass unit (AMU), 23, 24
Atomic masses, 25
Atomic masses (table), 503
Atomic number (definition), 15
Atomic number of hydrogen, 21–22
Atomic Safety and Licensing Board (ASLB), 397
Atomic structure, hydrogen isotopes (diagram), 21
Atomic theory, 15–16
Atomic vapor laser isotope separation (AVLIS), 118
Atomic weight (definition), 15
Atomic weights (table), 500–503
Atoms and nuclei, 15–25
Atoms-for-Peace program, 479
Attenuation factor in shielding, 344
Attenuation of gamma rays in matter, 70
Attenuation of particles, 48–49
ATW (Accelerator Transmutation of Wastes), 378
Australian Uranium Information Centre, 441
Austria, 443
Automobile energy transformations, 4
Average change in logarithm of energy, 54
Average cosine of scattering angle, 57
Average delay of neutrons in fission, 291
AVLIS, 118
Avogadro's number, 23, 79
Avogadro's number (definition), 15

B

Babcock and Wilcox Co., 325
Backfill, 373
Background radiation, 35
Balance equations, 182
Bare core, 141

Barn (unit), 48
Barnwell disposal site, 383, 385
Barrier in gaseous diffusion separator, 111
Basalt, 375
Base load, 153
Bathtub effect in waste disposal, 382
Bayes theorem, 305
Beamlet, 205
Beatty disposal site, 383
Becker, 218
Becquerel (scientist), 217
Becquerel (unit), 33
BEIR V, 237
BEIR VII, 236
Belgium, 442
BELLE, 236
Below regulatory-concern wastes (BRC), 364
Berson, 250
Beryllium-9, 258, 261
Beta particle (definition), 29
Beta particle energies, 64
Beta particles in radioactive chain, 34
Beta particles, ionization by, 64
Betatron, 98
Big Bang, 69, 72, 101
Binding energy, 23–25
Binomial distribution, 132
Bioaccumulation factor, 351
Biological effects of radiation, 229–239
Biological half-life, 348, 353
Biological research, 381
Blanket of breeder reactor, 184
BN-600 reactor (table), 184
BNCT (boron neutron capture therapy), 271
Bohr theory of hydrogen atom, 17–18, 218
Boiling point, 7
Boiling water reactor (BWR), 150, 168–169, 295, 427–428, 445
Boltzmann's constant (definition), 5
Boltzmann's constant in gas theory, 16
Boolean algebra, 304–305
Bootstrap current in fusion, 208
Boric acid, 152
Boron
    absorption cross section (graph), 51
    in fusion reaction, 206
    neutron counter, 127
    in reactor control, 152–154
Boron neutron capture therapy (BNCT), 271
Boron trifluoride, 127
Boron-10 in neutron counter, 127
Bosons, 101

Bothe, 218
BR (breeding ratio), 181
Brachytherapy, 269
Bragg formula, 281
Brazil, 440, 447
BRC (Below regulatory-concern wastes), 364
Breakeven in fusion, 198
Breeder reactor, 179-191
Breeder reactor in the future, 363
Breeder reactor, use of depleted U, 114
Breeding, 75, 179-181
Breeding blanket, 184, 196
Breeding gain (BG), 181
Breeding ratio (BR), 181
Breit, Gregory, 218
Bremsstrahlung, 64, 86
British units, 6
Brookhaven National Laboratory, 99-102, 283
Buckling (definition), 145
Building spray system, 301
Buildup factor for radiation, 346-347
Bulgaria, 443-444
Burn front in fusion, 203
Burning of fuels, 5
Burning of uranium, 9
Burnup (definition), 298
Burnup calculation, 298
Bush, President George H. W., 331
Bush, President George W., 331
BWR (boiling water reactor), 150, 168-169, 295, 427-428, 445
Byproduct material, 395

C

Cadmium in neutron detector, 129
Californium-252
    neutrons for treatment, 247
    properties, 237
    spontaneous fission, 69
Calutron isotope separator, 110, 221
Canada, 425, 446
Cancer risk, 390
Cancer treatment by radiation, 269-271
CANDU (Canadian deuterium uranium) reactor, 150, 446
Canisters of high-level waste, 372-373
Capacitor as accelerator, 94
Capacity factor (CF), 121, 155, 389, 423, 434, 441, 442, 446, 469, 493
Capital costs of nuclear plants, 417-418
Capture gamma ray, 42
Carbon cycle in the sun, 83

Carbon dating, 252-253
Carbon diffusion length, 55
Carbon dioxide as greenhouse gas, 480-481
Carbon-14, 32, 252
Carter, President Jimmy, 307
Cascade of isotope separator, 112-113
Cask design tests, 366-367
Cassini spacecraft, 330
Catalytic exchange, 119
Cathode of counter, 126
CDF (Collider Detector Fermilab), 100
Cell, biological, 229
Celsius temperature scale, 5
Centrifugal force, 116
Centrifuge isotope separation, 118
Cerenkov counter, 135
CERN, 99, 101
Cesium-137, 31
    as radiation source, 351
    for radiation gauge, 259
    for radiation processing, 272
    for radiography, 258
Chadwick, 218
Chain of radioactive decays, 33-34
Chain reaction, 76, 78, 141-156
Challenger accident, 331, 332
Charge of electron, 22
Charged-particle beams in fusion, 197
Chemical shim, 294
Chemical symbol, 21
Chernobyl accident
    description, 309-311
    effect on public opinion, 422
    psychological effects, 311
    radiation exposure, 310-311
    radioactivity release, 309
Chernobyl building (diagram), 310
Chernobyl-4 reactor features, 150
China, 440
China Syndrome, 307
CHP (combined heat and power), 173
Chromosomes, 229-230
Classification of reactors, 148-152
Clinch River Breeder Reactor Project (CRBRP), 183
Cobalt gamma ray treatment, 254
Cobalt-60
    for radiation processing, 272
    gamma rays for sterilization, 273
    in radiography, 257
    production, 42
    properties, 33, 34
Cockroft-Walton machine, 95, 100

Code of Federal Regulations Title 10, Energy
  environmental rules, 383, 397
    key sections, 398
    low-level waste, 390
    radiation protection, 340
    radionuclide emissions, 397
Cogema, 441
Cogeneration, 173
Coherent, 20
Cold fusion, 207
Collider, 99-101
Collider Detector Fermilab (CDF), 100
Collisions of particles, 45-46
Collisions, number of, 46
Color *vs.* temperature, 17
Columbia University, 221
Combined heat and power (CHP), 173
Commercial LLW disposal sites, 386-387
Committed effective dose equivalent, 340, 352-353
Commonwealth of Independent States (CIS),
      443-444
Compaction of wastes, 379
Compound nucleus, 43, 75
Comprehensive Test Ban Treaty (CTBT), 466
Compton effect, 66-69
Computer assistance, 423
Computer assistance in reactor operations, 423
Computer programs, 39, 423-424
Condenser for steam, 168-171
Conditional probability, 305
Conduction of heat, 149, 161-162
Conductivity of semiconductor, 131
Conductivity, thermal, 162-163
Confinement time in fusion, 197, 208
Consequences (definition), 305
Conservation
  benefits, 483
  of charge, 68
  of energy, 4, 163
  of energy and momentum, 43-44
  of mass-energy, 43, 75
  of momentum, 44
Construction costs for nuclear units (table), 418
Construction times for nuclear power plants
      (graph), 419
Consumption of U-235, 153-155
Consumption of uranium, 80
Consumption of water and fish (table), 351
Containment, 300-301
Contamination, 374, 379, 385, 387
Control absorber, 154
Control rod reactivity worth (graph), 296

Control rods, 42-43, 152, 295-297
Controlled fusion, 84
Convection of heat, 161, 165
Conversion factors for units (table), 499-492
Conversion of nuclear fuels, 78, 115
Conversion ratio (CR), 180-181
Converter reactors, 484
Coolant, 149-150, 152, 163-168
Cooling tower, 171-172
Cooperative Threat Reduction (CTR), 462
Core flooding tank, 300
Core of reactor, 152, 298, 491
Cosmic radiation, 229, 239
Cosmic rays, 35
Cosmotron, 99, 100
Cost of nuclear fuel, 153
Costs of electrical generation, 423, 433
Costs, stranded, 426
Coulomb force, 64, 85
Coulomb's relation, 44-45
Counter, 126
Counterterrorism, 135, 430
Counting rate, 35
CR (conversion ratio), 180-181
CRBRP (Clinch River Breeder Reactor Project), 183
Crick, 248
Crime investigation by NAA, 256
Critical condition, 145, 290
Critical experiment, 143, 290, 293
Critical heat flux (CHF), 165
Critical mass, 141
  of plutonium, 456
  of uranium, 456
Critical organ, 340, 348
Criticality, 141, 153, 290, 367
Criticality accidents, 293, 445
Crookes, 217
Crop mutations, 277-278
Cross section, 45-48
  for D-D reaction (graph), 85
  of boron-10, 127
  of elemental boron (graph), 51
  of natural uranium (graph), 50
Cross-linking of polymers, 279
CT (Computerized Axial Tomography), 250
CT scan (photo), 251
CTBT (Comprehensive Test Ban Treaty), 466
Cuba, 447
Curie (unit), 33
Curies (scientists), 33, 217
Curium-242, 258
Current density, 46-48, 56

Cycle time, neutron, 291-293
Cyclotron, 97-98, 218, 249-250
Czech Republic, 444

D

DAC (Derived air concentration), 353-354
Dark energy, 101
Dark matter, 101, 328
Dating of artifacts, 252-253
D-D reaction, 84, 85, 87, 195, 207-209
Dead Sea Scrolls, 252
Dead time of counter, 127
Debye length, 201, 212
Decay constant, 32, 33, 35, 231, 247, 248, 317, 348
Decay heat, 301, 307, 365, 389, 405
Decay law, 30-33
Decommissioning, 386-387
  costs, 387
  options, 386
DECON, 386
Decontamination, 239, 381, 386, 388
Dees of cyclotron, 97, 98
Defense sites, 385-386
Defense wastes, 363-365
Delayed neutrons, 76, 291-293, 313
Delta-k, 143
Density, 15, 16
Density measurement, 260, 261
Department of Energy (DOE), 399-400
Department of Homeland Security, 136, 245,
      404, 412
Departure from nucleate boiling ratio (DNBR), 165
Depleted uranium
  as resource, 186-187
  for breeder reactors, 363
  from isotope separation, 114
Deregulation, 425-427, 437
Desalination, 173, 431, 433, 443, 480, 485, 487
Design basis threat, 313, 409
Detection of thermal neutrons, 127
Detectors, 125-145
  demands on, 125
  for reactor safety, 289-293
  in half-life measurement, 35-36
  operating modes, 127
Deuterium, 21
  abundance, 85
  in fusion, 85
  separation of, 119
Deuterons, 95
Diagnosis of disease by NAA, 256
Differential equations, 34, 38, 159, 182, 291, 317, 318

Diffusion coefficient, 56-58, 158
Diffusion length, 55, 58, 145
Diffusion lengths (table), 55
Digital control of nuclear plants, 424
Diphenyl coolant, 222
DIPS (Dynamic Radioisotope Power System),
      330, 331
Dirty bomb, 135, 136, 239, 240, 243
Disassembly, 456
Discriminator, 133, 134
Disintegration, 29, 36
Disintegrations per second, 33
Dissociation, 24
  of molecules, 70
  of nucleus, 24
Distillation in desalination, 487
District heating, 173, 433, 442, 485
DNA, 230, 236
  fingerprinting, 248
  radiation damage of, 234
DNBR (Departure from nucleate boiling ratio), 165
DOE (Department of Energy), 246, 277,
      364, 377
DOE Energy Information Administration, 418
Doping of semiconductors 279-280
Doppler broadening, 293
Dose (definition), 231
  calculation, 341-342
  calculations for nuclear power station, 344
  effective, 341
  equivalent, 240
  from TMI accident, 307
  limits, 234
  limits for public, 309
  rate, 233
  types, 341
Dosimeters, 130, 347
Double-strand breaks of DNA, 234
Doubling time (DT), 181
Downblending, 463
Drift velocity of ions, 126
Drip shield, 376
Driver in fusion, 203
D-T generator, 135
D-T reaction, 195-197, 201, 202, 206, 208, 211
Dysprosium-165 for treatment, 270

E

Earth Summit, 479
Earth's radiation energy balance (diagram), 481
ECCS (emergency core cooling system), 300-302
Ecole Polytechnique, 441

Economics
    definition, 415
    of nuclear energy, 404
EdF (Electricité de France), 441
Edison Electric Institute (EEI), 405, 429
Effect of distance on radiation intensity, 341
Effective dose, 340, 348, 352-354
Effective generation time, 291
Effective half-life, 231, 248, 348, 353, 357
Effective mass of gamma ray, 44
Effective multiplication factor, 147, 289-290
Efficiency, 169, 170
Einstein, 9, 217, 219
Einstein's equation, 99
Einstein's mass-energy relation, 11, 23
Einstein's theory, 8, 9, 67, 101
Eisenhower, President Dwight, 400
Elastic collisions, 51, 59
Electric and magnetic forces, 93-94, 197
Electric charge, 17
Electric energy (table), 417
Electric field, 100
Electric Power Research Institute (EPRI), 313,
    405, 429
Electric propulsion in space, 332
Electric vehicle, 408
Electrical energy, 4
Electrical power cost, 415-418
Electricité de France (EdF), 441
Electrochemical conversion, 167
Electrolysis, 119, 490
Electromagnet, 97, 98
Electromagnetic energy, 7
Electromagnetic fields (EMF), 239
Electromagnetic process, 110
Electromagnetic pulses (EMP), 459
Electromagnetic separation of uranium isotopes,
    109-110
Electron, 9
    as projectile, 63-64
    as beta particle, 63
    charge, 6
    orbits, 19
    symbol, 23
Electron volt (unit), 6
Electron-hole pair, 131, 136
Electrostatic force
    in binding energy, 23-25
    in fission process, 73-74
    in fusion process, 119, 196
Electrostatic generator, 95
Electroweak force, 101

Elements, 15
Emergency core cooling system (ECCS) 300-302
Emergency plan, 305
EMF (electromagnetic fields), 239
EMP (electromagnetic pulses), 459
Energy, 3-10
    average, 5, 48
    economics, 415-433
    effect of adding (Figure), 8
    electromagnetic, 7
    flow (diagram), 415
    from fission, 76
    from fission (table), 77
    from fusion, 195, 196
    levels in hydrogen atom (diagram), 19
    nuclear *vs.* chemical, 9
    of combustion, 9
    of photon, 66, 67
    radiant, 7-8
    research, 486
    transformation, 4
    world needs, 9-10
    yield from fusion, 195
    yield of nuclear weapon, 458
Energy flow (diagram), 415
Energy Information Administration of DOE, 418
Energy Loan Guarantee Fund, 430
Energy Policy Act of 1992, 408, 409, 425
Energy Policy Act of 2005, 408, 409, 430, 491
Enriched uranium management, 445
Enrichment, 112, 115, 117
Enterprise, aircraft carrier, 324
ENTOMB, 386
Envirocare, 385
Environmental effect of heated water, 167-169
Environmental movement, 223
Environmental protection, 427
Environmental Protection Agency (EPA), 376, 396-397
Environmental Radiological Assessment, 350-352
Environmental restoration, 385-386
EPA limits on radon, 349-350
EPRI (Electric Power Research Institute), 313, 405
Equivalence of matter and energy, 8-9
Error, 132
Escape velocity, 327
European Pressurized Water Reactor (EPR), 441-443
Evaporation of neutrons from nucleus, 102
Evaporation of wastes, 380
Event trees, 303
Evolutionary reactor design, 426
Excitation, 63-64
Excitation energy, 73

Excited state, 73, 130
Exclusion area, 302
Experimental Breeder Reactor, 182, 186
Expert system, 384
Explosive flaw detection by neutron
        radiography, 259
Exponential, 48, 70
Exponential decay, 30
Exponential growth, 289

F
Faraday's law of induction, 98
Fast breeder reactor, 182–186
Fast Flux Test Facility (FFTF), 183
Fast neutrons, 51, 73–74
    detection of, 129
Fast reactor, 145, 147, 180
Fat Man nuclear weapon (diagram), 455
Fault trees, 304–305
Federal Emergency Management Agency (FEMA),
        306, 397
Federal Energy Regulatory Commission (FERC), 425
Feed of isotope separator, 109
Feedback, 293
FEL (free electron laser), 102
FERC (Federal Energy Regulatory
        Commission), 425
Fermi I reactor, 183
Fermi, Enrico
    quotation, 245
    reactor experiment, 218
Fermilab, 100
Fertile, 79
Fertile materials, 79, 179
FFTF (Fast Flux Test Facility), 183
Fick's law of diffusion, 56
Film badges, 347
Film boiling, 165
Filters, 380
Final Safety Analysis Report (FSAR), 299
Finland, 442
Fissile, 75, 78, 79
Fission chamber, 128
Fission explosive, 454
Fission fragments, 73, 76, 102, 125, 128
Fission process, 73–74
    energy release, 9
    role of binding energy, 23–25
Fission products (graph), 77
    decay gammas, 76
    discovery of, 218
    hazard, 289
    heat, 300
    volume, 361
    weight, 361
Fission reaction equation, 78
Fissionable, 75
Flibe, 202, 204
Flow half-life, 248
Flow rate analysis, 247
Fluorine-18, 250
Flux, neutron, 46, 128
Food irradiation, 273
Food spoilage, 284
Force-on-force exercises, 313
Forces, 3
Fossil fuel exhaustion, 473
Fossil fuel plants, 153, 170, 222
Fossil fuels, 473
Fractional distillation, 119
Framatome, 441, 445, 446
France, 437
Free electron laser (FEL), 102
Frequency
    light, 7
    of incidents, 293
    photons, 17
Frisch, 218
FSAR (Final Safety Analysis Report), 299
Fuel, 149
    assembly, 366
    consumption, 148, 153
    cost, 153
    element, 152
    fabrication controls, 297
    isotopic composition, 313
    rod, 152
    rod consolidation, 366
    storage pool, 365
    zones, 299
Fuel cells, 409, 489
Fuel cycles, 153, 179, 362
Fundamental physical constants (table), 504
Fusion, 83
    energy release, 85
    future use, 372
    inertial confinement, 197
    magnetic confinement, 197
    pellets, 198
    power density, 198
    reaction rates (graph), 196
    reactions, 195–197
    reactors, 195–211
Future, views of, 473

G

Galileo spacecraft, 330
Gamma rays (definition), 7
  cobalt-60, 34
  cross sections (graph), 68
  flux, 341–315
  radiography, 258
  reaction product, 39
  scattering on electron, 66–70
  shielding, 345–346
  spectroscopy, 253
Gap, 162
Gas centrifuge, 116–118
Gas counters, 126–127
Gas law 16
Gas turbine measurements by neutron
    radiography, 259
Gas Turbine Modular Helium Reactor
    (GT-MHR), 491
Gaseous core reactor, 325–326
Gaseous diffusion, 110–115
Gases, 15
Gate of fault tree diagram, 304
Gaussian distribution (graph), 132
Geiger-Mueller (GM), 127, 260, 347
General Atomics, 206
General Electric, 187
General Electric Advanced Boiling Water
    Reactor, 428
Generation IV International Forum (GIF), 431
Generation IV nuclear designs, 431
Generator, electrical, 331
Genes, 230
Genetic effects of radiation, 231
Geologic media, 375
Geological applications of NAA, 257
Germanium, 131
Germany, 438
Global climate change, 480–483
Global Nuclear Power Partnership (GNEP), 453
Global warming, 481
Globalization, 437
Glow curve of phosphor (graph), 130
GM (Geiger-Mueller) counter, 127, 260, 347
Godiva, 141–142
Google, 510
GPHS (General Purpose Heat Source), 330
Granite, 375
Gravity, 3, 4
Gray (unit), 231
Greater-than-Class-C wastes, 384
Greek characters, 7

Greenhouse effect, 480–483
Greenhouse gases, 481
Groves, General Leslie, 219
Gulf War, 324, 465
Gun weapon technique, 454

H

Hadron calorimeter, 135
Hahn, 218
Half-life (definition), 30
  biological, 348
  carbon-14, 252
  effective, 357
  measurement of, 35–36
  values, 36
Half-thickness (definition), 70
Half-thickness for gamma rays, 70
Hanford, 220
Hastelloy, 376
Head-on collision, 51
Health physics, 350
Health Physics Society (HPS), 407
Heat, 4
  exchanger (diagram), 167
  flow, 162
  generation, 161–167
  of combustion, 11
  of vaporization, 7
  transfer coefficient, 163
  transmission, methods of, 161
Heavy charged particles, 64–66
Heavy water diffusion length, 55
Heisenberg, Werner, 219
Heisenberg's uncertainty principle, 20
Helium, 15
Helium detector, 127
Helium-3 in fusion, 84, 206
Helix, 93, 199
Heterogeneous, 149, 184
HEU (highly-enriched uranium), 115
Higgs boson, 101
High pressure injection system, 305
High temperature gas-cooled reactor (HTGR),
    150, 156
High-level waste (HLW), 363
High-level waste disposal, 372
High-pressure injection system, 305
High-voltage machines, 95, 96
Highly enriched uranium (HEU), 115
Hiroshima, 221
History of nuclear energy, 217–225
HLW (high-level waste), 363

HLW nuclides, 353
Hohlraum, 204
Homeland Security Act of 2002, 404
Homogeneous, 149
Homogeneous aqueous reactor, 222
Hormesis, 236
HPS (Health Physics Society), 407
HTGR (high temperature gas-cooled reactor),
    150, 445, 466, 490
Human Genome Project, 248, 265
Hungary, 443
Huygens probe, 330
Hybrid reactor, 206
Hybridization, 248
Hydroelectric, 4
Hydrogen
    atom (diagram), 18
    atom, Bohr theory, 18
    atomic weight, 15
    bomb, 83, 197, 203
    heavy, 6
    isotopes, 21
    number density, 47
    propellant for rocket, 326
    radius of electron motion, 47
    scattering by, 47
Hydrogen economy, 431, 489, 490
Hydrogen gas, 17, 489, 492

I

IAEA (International Atomic Energy Agency)
    and Chernobyl, 300
    functions, 400–401
    inspections, 278
    safeguards role, 399, 433
ICBM (intercontinental ballistic missile), 460
Ice, 7
Icebreakers, 325
ICRP (International Commission on Radiological
    Protection), 234
IDCOR (Industry Degraded Core Rulemaking), 308
IEEE (Institute of Electric and Electronics
    Engineers), 406, 407
IFR (Integral Fast Reactor), 186–187
Ignition temperature, 87, 195, 197, 211
Implantation of radionuclides, 269
Implosion weapon technique 454–456
Inadvertent intruder, 384
Incineration, 379
Independent Technology Review Group (ITRG), 491
India, 446
Indium, 128

Individual Plant Examination (IPE), 309
Indonesia, 446
Induction accelerator (betatron), 98
Industry Degraded Core Rulemaking (IDCOR), 308
Inelastic scattering, 47, 280
Inelastic scattering research, 280
Inertial confinement fusion (ICF), 197
    Lawson criterion, 197
    machines, 203–206
    progress (graph), 209
Infinite multiplication factor, 147
Ingestion dose conversion factors (table), 352
Innovation in Nuclear Infrastructure and
        Engineering (INIE), 407
INPO (Institute of Nuclear Power Operations), 401–404
INSC (International Nuclear Safety Center), 437
Insect control, 278–279
Institute of Electric and Electronics Engineers
        (IEEE), 406, 407
Institute of Nuclear Power Operations (INPO),
        401–404
Institutional and industrial wastes (table), 382
Instrumentation system, 296
Intact decommissioning, 387
Integral Fast Reactor (IFR) 167–168
Intercontinental ballistic missile (ICBM), 460
Intergovernmental Panel on Climate Change
        (IPCC), 482
Intermediate heat exchanger, 184, 185, 491
Internal radiation exposure 321–322
International Atomic Energy Agency. *See* IAEA
        (International Atomic Energy Agency)
International Commission on Radiological
        Protection (ICRP), 234
International nuclear power, 437–449
International Nuclear Safety Center (INSC), 437
International Thermonuclear Experimental
        Reactor (ITER), 210, 211
Internet, 437
Interstate compacts (map), 383
Inverse square spreading, 343, 345
Invert, 376
Iodine radioisotopes, 249
Iodine-125 for treatment, 269
Iodine-131, 270
Ion engine for space propulsion, 333
Ion propulsion in space, 325
Ion-exchange, 381
Ionization
    Bohr theory, 64
    by electrons, 63–64
    in biological tissue, 231

Ionization chamber, 126
Ions, 19
IPE (Individual Plant Examination), 309
Iraq, 110
Iraq weapons program, 465
Iridium-192, 258
Iridium-192 for treatment, 269
Irradiation facility (diagram), 273
Isotopes (definition), 21
 information from, 277
 production and consumption, 179
 radioactive, 246
 separation methods, 111
 separators, 109
 stable, 110
Italy, 183, 218, 442
ITER (International Thermonuclear Experimental
 Reactor), 210, 211
Ivy/Mike shot, 456

J

Japan, 430-431, 445
JET (Joint European Torus), 202, 209
Joint Committee on Atomic Energy (JCAE), 396
Joint European Torus (JET), 202, 209
Joliot, 218
Jupiter, 328-330

K

K-capture, 32
Kazakhstan, 443
Kelvin, 5
Kemeny Commission, 307, 401
Kinetic energy (definition), 4
 atoms, 5
 electron in electric field, 98-99
 fission fragments, 77
 hydroelectric, 4
 relativistic, 44
Kinetic theory of gases, 16
Kink in plasma, 201
Krypton-85, 342, 354
Kuiper Belt, 328, 330
Kyoto Protocol, 481

L

Lady Godiva, 141-142
Large Hadron Collider, 101
Laser (definition), 20
 applications, 21
 beams, 20-21
 beams in fusion, 197

fusion, 204-205
 isotope separation, 118-119
 materials, 19
Laser-fusion reactor (diagram), 205
Latent heat, 7
Lawrence, Ernest O., 97, 218, 221
Laws, regulations, and organizations, 395-409
Lawson criterion, 197-198, 208
Lead (Pb), 29
Lead-206, 34
Leakage of neutrons from reactor, 141, 145, 456
LET (linear energy transfer), 66, 230
LEU (low-enrichment uranium), 115, 463
License renewal, 424-425
Light, Bohr theory, 18
Light, speed of, 7, 44
Light, visible, 7
Limited Test Ban Treaty of 1963, 461
Limits of exposure, 234-238
Linac, 96-97
Linear accelerator, 96-97
Linear attenuation coefficient, 344
Linear energy transfer (LET), 66
Linear no threshold (LNT), 236
Linear-quadratic model, 235-236
Liquid level measurement, 381
Liquid metal fast breeder reactor (LMFBR), 150,
 184-185
Liquid sodium, 183, 185
Lithium, 196, 197, 202
Lithium deuteride, 458
Lithium iodide, 129
Lithuania, 443
Little Boy nuclear weapon (diagram), 455
LLRWPA (Low-Level Radioactive Waste Policy
 Act), 383
LLW (low-level waste), 364, 381-383
LLW disposal concepts, 385, 387
LLW sources, 366
LMFBR (liquid metal fast breeder reactor), 150,
 183, 185
LOCA (loss of coolant accident), 300-302
Logarithmic energy change, 64
Logging, 261
Loop fast breeder reactor, 184-185
Los Alamos, 141, 221, 327, 466
Loss of coolant accident (LOCA), 300-302
Louisiana Energy Services, 118, 123
Love Canal, 224
Low-enrichment uranium (LEU), 115, 463
Low-level radioactive waste disposal
 facility, 237

Low-Level Radioactive Waste Policy Act (LLRWPA), 383
Low-Level Radioactive Waste Policy Amendments Act of 1985, 383
Low-level waste (LLW), 364
   disposal sites, 386-387
   management, 378-383
   streams (table), 382
Low-pressure injection system, 301
Luckey, 236
Lunar landing, 325, 327

M
Macroscopic cross section, 46-48, 344
MAD (mutual assured destruction), 460
Magnetic confinement, 86
Magnetic confinement fusion (MCF), 175
   Lawson criterion, 197
   machines, 199-203
   progress (graph), 209
Magnetic field, 93-95, 97-99, 104, 109, 199-201
Magnetic induction, 94
Magnetic Resonance Imaging (MRI), 250
Main Ring of Tevatron, 100
Maintenance, 154, 206, 225, 298, 309, 347, 386, 398, 401, 402, 405, 423, 424, 427, 428, 444, 464, 480, 487
Maintenance Rule, 398
Manhattan Project, 220, 221
Mars mission, 331-332
Mass, 3, 8, 9
Mass attenuation coefficients (table), 345
Mass flow rate of coolant, 163
Mass number, 20-21
Mass of electron, 8
Mass ratio, proton to electron, 17
Mass spectrograph, 109-110
Mass, rest, 8
Mass-energy, 43, 68
Masses of atoms and nuclei, 23
Material balances, 113
Material unaccounted for (MUF), 464
Materials Testing Reactor, 222
Materials, radiation effect on, 246
Matter, state of, 86
Maxey Flats disposal site, 382
Maximum permissible concentration (MPC), 348-349
Maxwell's gas theory, 16
Maxwell's speed distribution (formula), 26
Maxwellian distribution 16 (graph), 55

McMillan, E. M., 99
MDS Nordion, 272, 273
Mean free path, 48, 144
Mean free path for gamma rays, 70
Mean free path, transport, 56
Mean life, 33, 291
Meat irradiation, 275
Medical diagnosis by X-rays, 257
Medical imaging, 249-250
Medical imaging by immunoassay, 250-251
Medical supply sterilization, 276-277
Medical treatment, 269-271
Megatons to Megawatts, 112, 463
Megawatt-days per metric ton, 155
Meitner, 218
Melting point of sodium, 183
Merchant ship, 325
Mercury detection in the environment, 256
Methane in counter, 129
Metric system of units, 6
Metric ton (tonne), 155
MeV (unit), 6
Mexico, 446
Microscopic cross section, 46
Microshells for ICF, 204
Microwaves, 201
Mill tailings, 349, 361, 364
Millirem (unit), 232, 233
Mirror machine, 200
Mixed oxide (MOX), 184, 363, 370, 463
Mixed wastes, 364
Mobility, 126
Moderator (definition), 54
   effect on neutron energy distribution, 145-146
   materials, 149
   properties, 54, 55
   purpose, 78
   reactor, 119
Moderator temperature effect, 293
Molecular biology research, 282-283
Molecular structure studies, 283-284
Molten Salt Reactor Experiment (MSRE), 187
Molybdenum-99, 249
Momentum, 43
Monitored Retrieval Storage (MRS), 377
MONJU reactor, 183, 445
Monoclonal antibodies (MAbs), 270-271
Moon mission, 238, 333
Most probable speed, 16, 47
MOX (mixed oxide), 184, 188, 370, 468
MPC (maximum permissible concentration), 348-349

MRI (Magnetic Resonance Imaging), 250
MRS (Monitored Retrieval Storage), 377
MSRE (Molten Salt Reactor Experiment), 187
MUF (Material unaccounted for), 464
Multibarrier, 373
Multichannel analyzer, 134
Multiplication factor, 142–147
    fuel consumption, 293
    reactor safety, 293–300
Multiplication of neutrons, 142
Muons, 207
Murphy's law, 224
Mutsu, merchant ship, 325
Mutual assured destruction (MAD), 460

N

NAA (neutron activation analysis), 253–257
Nagasaki, 221
NaK coolant, 326
NARUC (National Association of Regulatory Utility
    Commissioners), 407
NAS (National Academy of Sciences), 408, 468
National Academy for Nuclear Training, 402
National Academy of Sciences (NAS), 408, 468
National Association of Regulatory Utility
    Commissioners (NARUC), 407
National Council on Radiation Protection and
    Measurements (NCRP), 234, 236, 237
National Environmental Policy Act of 1969
    (NEPA), 396
National Ignition Facility (NIF), 205, 208
National laboratories, 396
National Rural Electric Cooperative Association
    (NRECA), 406, 407
National Synchrotron Light Source, 102
Natural radiation, 233, 238, 239
Natural radioactivity, 34
Natural reactor, 156
Natural uranium, 16, 112–114, 187
Nautilus, nuclear submarine, 323–324
Naval propulsion, 323–325
NCRP (National Council on Radiation Protection
    and Measurements), 234, 236, 237, 340
Negative feedback in reactor, 293
Negatron, 32
NEI (Nuclear Energy Institute), 405–406, 407
NEPA (National Environmental Policy Act of
    1969), 396
NEPA (Nuclear Energy for the Propulsion of
    Aircraft), 325
Neptune, 330
Neptunium-239, 181, 370

NERVA (Nuclear Engine for Rocket Vehicle
    Application), 327, 332
Net current density, 56
Netherlands, 443
Neural networks, 424
Neutrino (definition), 29
    detectors, 135
    fission source, 76
Neutron (definition), 21, 22
    absorption, 75–76
    activation, 387
    capture, 42, 44, 196
    chain reactions, 141–156
    cross section measurements, 280
    cross sections, 49–51
    cycle (diagram), 146
    cycle time, 293
    decay, 32
    detectors, 127–129
    diffraction, 281
    energy from fission, 78
    energy groups, 182
    energy loss, 54
    flux, 48, 147, 154, 155, 296–297
    from fission, 74
    from fusion, 195–196
    gas, 47
    interaction with matter, 281
    interferometry, 281
    migration, 51–56
    number, 22
    number density, 127
    per absorption in U-235, 78, 144
    per fission, 78, 343
    population growth, 289–293
    radiation damage, 70
    radiography, 258–259
    reactions, 70
    scattering in carbon, 53
    shielding, 344
    source, 143, 258–259, 290
    symbol for, 23
    transmutation doping (NPD), 280
    wave length, 282
Neutron activation analysis (NAA), 253–257
Neutron-free fusion reactions, 206
Nevada Test Site, 375
Newton's law, 3
Next Generation Nuclear Power Plant
    Project, 409
NIF (National Ignition Facility), 205, 208

NIST (National Institute of Science and Technology), 13, 66, 70, 72, 360, 504
Nitrogen-13, 250
NNOG (Nuclear Non-Operating Owners Group), 407
NNWS (Nonnuclear weapons states), 461
Non-leakage factor, 145
Non-Proliferation of Nuclear Weapons (NPT), 461
Nonnuclear weapons states (NNWS), 461
Nonproliferation and safeguards, 463–465
North Carolina State College, 407
North Korea, 445
Nova fusion machine, 205
NPD (Neutron transmutation doping), 280
N-P junction, 131
NPT (Non-Proliferation of Nuclear Weapons), 461
NRC (Nuclear Regulatory Commission), 236, 237, 300, 313, 340, 350, 352, 364, 376, 377, 387, 397–399, 409, 421, 424, 427, 430
NRC policies and practices, 383
NRECA (National Rural Electric Cooperative Association), 406–407
Nuclear controversy, 223–225
Nuclear emulsion, 135
Nuclear Energy Institute (NEI), 405–407
Nuclear Energy Research Initiative (NERI), 407
Nuclear energy, benefits of, 221, 224
Nuclear explosions, 453–469
Nuclear fission, 9
Nuclear force, 24, 85
Nuclear forensics, 136
Nuclear fuel cycle, 361–363
Nuclear fuel cycles (diagram), 362
Nuclear Fuel Services (NFS), 369
Nuclear gauges, 261
Nuclear heat energy, 161–173
Nuclear Incident Response Team (NIRT), 404
Nuclear journals, 406
Nuclear medicine, 249, 269
Nuclear Non-Operating Owners Group (NNOG), 407
Nuclear physics, 217–218
Nuclear power
    future, 473–493
    opinion of, 223–224
    plant dose limits, 237, 341, 354–355
    plant radioactive releases, 310–311, 349–351
    reactors, 148–156
Nuclear Power 2010 Partnership, 409
Nuclear power renaissance, 190, 407, 429–433
Nuclear power *vs.* nuclear weapons, 453–454
Nuclear processes, 41–56
Nuclear propulsion, 323–333
Nuclear reactions, 41–43

Nuclear reactor, 47, 48
Nuclear Regulatory Commission (NRC), 234, 236–237, 299–300, 306, 308, 309, 313, 350, 364, 376–377, 386, 387, 390, 397–399, 409, 421, 424–426, 429–430
Nuclear rocket, 326, 327, 332
Nuclear security, 312–313
Nuclear share of electricity generation (chart), 440
Nuclear standards, 406
Nuclear structure, 21–23
Nuclear Waste Fund, 377
Nuclear Waste Policy Act of 1982 (NWPA), 365, 399
Nuclear weapons, descriptions 453–454
    descriptions, 453–454
    development, 219–221
    energy yield, 458
    radiation effect, 459
Nuclear weapons states (NWS), 461
Nuclear winter, 459
Nucleate boiling, 165
Nuclei, number of, 30
Nucleon, 21, 23–25
Nucleus of atom, 19
Nucleus of biological cell, 229
Number density, 47, 127
Number density in fusion, 208
Number density of neutrons, 127
Number of neutrons per absorption, 144, 179
Number of neutrons per fission, 343
Number prefixes (table), 6
NWPA (Nuclear Waste Policy Act of 1982), 365–366
NWS (Nuclear weapons states), 461

O

O&M (operation and maintenance), 417, 423
Oak Ridge, 102, 110, 187, 221, 386, 454
Occupational exposure dose limits, 237
OCRWM (Office of Civilian Radioactive Waste Management), 375, 399
Office of Nuclear Material Safety and Safeguards, 399
Oil crisis of 1973, 474
Oklo phenomenon, 156
OMEGA inertial confinement fusion machine, 205
Once-through fuel cycle, 468
OPEC (Organization of Petroleum Exporting Countries), 188
Operating limits of reactors, 297–298
Operation and maintenance (O&M) costs, 417, 423
Oppenheimer, J. Robert, 221
Opposition to nuclear power, 421, 441

Orbits, electron, 18, 212
Organ and tissue weighting factors (table), 354
Organic coolant, 222
Outage management, 155, 160
Oxygen, 5, 7, 20, 22, 83, 271, 490

P

Paducah, KY, 112
Pair annihilation, 69
Pair production, 129
Pakistan (table), 190, 439
Palladium-103 for treatment, 269
Parent-daughter relation
    fuel production and consumption, 181–182, 186
    radioactive chain, 33–34
    radionuclide generator, 248, 264
Particle accelerators, 93–107
Particle attenuation, 48–49
Particle Beam Fusion Accelerator (PBFA), 205
Particles, 4, 5, 7, 23, 32, 41, 44, 65, 99, 104, 197, 230, 257, 332, 380
Parts per million (ppm), 154
Passive reactor design, 428–429
Pathogen reduction by irradiation, 277
PBFA (Particle Beam Fusion Accelerator), 205
Pebble Bed Modular Reactor (PBMR), 446, 491
Pellets of nuclear fuel, 151–152, 222
Perfect gas law, 16
Period of reactor, 291, 292, 295
Periodic table, 15
Pesticide investigations by NAA, 256
PET (Positron Emission Tomography), 228–229, 250–251
PET scan (photo), 251
Petroleum processing using NAA, 255
PGNAA (Prompt gamma neutron activation analysis), 254, 255
Phenix, 183, 441
Philippines, 446
Philosophy of safety, 311–312
Phosphor in detector, 129, 130
Phosphorus-32, 246, 248, 270
Photoelectric effect, 67, 68, 130
Photomultiplier, 129, 130, 137, 250
Photon, 7, 18, 20, 45, 66
Physics research, 104, 136, 467
Physiological effects of radiation, 231
Pile, 220
Pinch effect, 199, 205
Pioneer Anomaly, 328
Pioneer spacecraft, 328
Planck's constant, 17

Plasma, 3, 86–87
Plutonium
    disposal options, 445
    for weapons, 454
    from reprocessing, 385
    management, 467–469
    production, 220, 453, 454
Plutonium-238, 328
Plutonium-239
    as fissile isotope, 79
    effect on reactor operation, 152
    in breeder reactor, 180–183
    production equations, 84
Plutonium-240, 75, 155, 181, 372, 454, 467
Plutonium-241, 34, 155, 180, 374
Plutonium-242, 155, 181
Poisson distribution, 132, 133
Polonium-210, 34, 382, 465
Population growth (graph), 478
Portugal, 442
Positive void coefficient, 310
Positron, 32, 43, 69, 250
Positron emission tomography (PET), 250
Positron-electron pair, 32
Pot fast breeder reactor, 183, 184
Potassium-40, 32, 272
Potassium-Argon dating, 253
Potential energy, 3, 4, 24, 116
Power, 5
Power coefficient, 318
Power loss rate in fusion, 198
Power of reactor, 147
Power, thermal, 149, 163
PRA (probabilistic risk assessment), 303–305, 398
Predictions of energy and population, 446, 477–478
Preliminary Safety Analysis Report (PSAR), 299
Preservation of artistic objects, 279
Pressure, 15
Pressure in plasma, 79
Pressure vessel, 151, 152
Pressurized water reactor (PWR), 150, 151, 153, 295, 406, 446
Pressurizer, 168, 306
Prevention of nuclear war, 426–429, 460–463
Price-Anderson Act, 306
Privatization, 437
Probabilistic Risk Assessment (PRA), 303, 305, 309
Probability, 147, 235, 305
Product of isotope separator, 109
Projectiles, 42, 46
Projections (graphs), 429
Proliferation resistance, 432, 491

Prometheus project, 333
Prompt gamma neutron activation analysis
        (PGNAA), 254-257
Prompt gamma rays, 73, 76
Prompt neutrons, 76, 291
Proportional counter, 115, 127
Protactinium-233, 79
Protection against weapons diversion, 464
Protection from radiation exposure, 339
Protection of the environment, 479, 483
Protective measures, 339-341
Proton, 21, 32, 44
Proton recoil counter, 127-128
Prudence review, 397-398
PSAR (Preliminary Safety Analysis Report), 299
Public concerns about food irradiation, 275-276
Public Utilities Commissions (PUCs), 422
Public Utility Holding Company Act of 1935
        (PUHCA), 425
Pulse, 115, 122, 123
Pulse height analysis, 133-134
Purex process, 370
Purposes of reactors, 148
PWR (pressurized water reactor), 149, 150, 222,
        294, 325, 365, 427, 442
PWR system flow diagram, 168

Q

Quality assurance (QA), 300, 398
Quality control (QC), 300
Quality factor (table), 232
Quantum mechanics, 20
Quantum numbers, 18
Quark, 101
Quench, 127, 306

R

Racks, fuel storage, 365
Rad (unit), 232
Radiant energy, 7
Radiation, 7
Radiation and materials, 63-72
Radiation chemistry, 279
Radiation damage, 70, 201, 279, 280, 297, 424
Radiation detectors, 125-136
Radiation dosage, sources of, 238-239
Radiation dose in nuclear power plants (table), 238
Radiation dose units, 231-232
Radiation doses for foods (table), 275
Radiation effect of nuclear weapon, 459
Radiation effects research, 234, 236
Radiation Effects Research Foundation (RERF), 235, 242

Radiation exposure trend (diagram), 403
Radiation exposure, average U.S., 233
Radiation fluxes (table), 342
Radiation from cell phones, 239
Radiation gauges, 259-261
Radiation losses in plasma, 87
Radiation preservation of food, 271-276
Radiation protection, 339-355
Radiation shielding, 332
Radiation sickness, 231
Radiation therapy, 271-272
Radiation treatment of sewage sludge, 277
Radiation treatment of wood, 279
Radiation, heat, 149
Radiative capture, 73
Radii of nuclei, 23
Radio-frequency, 97, 99
Radioactive capsule ("seed"), 269
Radioactive chains, 33-34
Radioactive decay, 29
Radioactive isotopes, 245, 246
Radioactive isotopes (table), 31
Radioactive probe, 248
Radioactive releases from nuclear power plant, 340, 344
Radioactive waste disposal, 361-387
Radioactive waste inventories (table), 382
Radioactive wastes in fusion, 207
Radioactive wastes, definition, 339
Radioactivity, 27-34
Radioactivity in sodium coolant, 222
Radiography, 257-259
Radioimmunoassay, 250-251
Radioisotope heater units (RHUs), 328, 330
Radioisotope shortage, 246
Radioisotope thermoelectric generator (RTG),
        327-333
Radioisotopes, 245
Radiological protection analysis, 340
Radiological weapons, 459
Radionuclide generator, 248
Radionuclides in waste, 345
Radionuclides used in therapy (table), 249
Radiopharmaceuticals (table), 248-249
Radium hazard, 234
Radium-226
    and definition of curie, 33
    as source of radon, 349
    in uranium series, 34, 361
Radius of gyration, 93
Radius of hydrogen atom, 23
Radius of nucleus (formula), 23
Radon, 233

Radon problem, 349-350
Radon-222, 349, 361
Radura, 274
Random motion, 47
Random numbers, 47
Range, 66
Range of beta particles (formula), 64, 71
RBMK reactor, 309, 310
Reaction rate (definition), 46
    absorption, 48
    detector, 119, 127
    fusion (graph), 196
    production of Pu-239, 79
    production of U-233, 75, 79
    theory, 44-48
Reactivity, 143, 157, 159, 292, 295-296
Reactivity of control rods, 294-296
Reactor
    first, 148
    fuel zones, 299
    life extension, 387
    materials (table), 149
    natural, 156
    operation, 152-156
    orders, 421
    oversight process, 398
    power, 79, 147
    R&D, 327
    safety, 289-314
    types, 148-152
Reactor fuel inspection by neutron radiography, 259
Reactor Safety Study, 303
Reactors under construction and planned
        (table), 447
Reagan, President Ronald, 462
Recycle, 361-362
Recycling of plutonium, 179
Red Book, 188
Reflector, 145, 184
Regulations, 340, 396-399
Regulatory Guides, 398
Relation of DAC and ALI, 354
Relativistic Heavy Ion Collider (RHIC), 101
Relativity
    Compton effect, 66
    kinetic energy, 9, 44
    mass increase, 9
    mass-energy relation, 24
    speed in accelerator, 93-94
Relief valve, 306
Rem (unit), 232-233
Removal cross section, 344

Renaissance, nuclear, 190, 407, 429-433
Repository, 373-378
Reprocessing
    future use, 468
    history, 369
    merits of, 372-373
    operations, 370-371
Reprocessing, commercial, 186, 372
Reproduction factor, 147
Reproduction factor (table), 180
Reproduction, biological, 230-231
Research and development projects, 396, 430,
        486-489
Resonance absorption, 145
Resonance capture, 180, 181
Resonance cross section, 128
Resonance escape probability, 145
Rest energy, 9, 66, 99
Rest mass, 8, 43, 98
Restructuring, 425-426
RHU (Radioactive Heating Units), 328, 330
Ribosome, 282-283
Richland, 383
Rickover, H. G., 323
Risk, 303
Risk-Informed Performance-Based regulation,
        398-399
Robotics, 423
Robots for space missions, 332
Rocket equation, 327
Rocket, nuclear, 326, 327, 332
Rod-drop technique, 296
Roentgen (scientist), 217
Romania, 443
Room temperature, 5, 47, 49, 55, 60
Roosevelt, President Franklin D., 219
Rover project, 327
RSH (Radiation, Science, and Health), 236
RTG (Radioisotope thermoelectric generator),
        327-328, 333
RTG for SNAP-27 (table), 329
Runaway electrons, 201
Russia, 443
Rutherford, 16, 218
Rutherford's experiment, 41

S
Sachs, 219
Safety features of reactors, 300
Safety goals of NRC, 311
Safety philosophy, 311
SAFSTOR, 386

Salmonella, 272
Salt, 369
SALT (Strategic Arms Limitation Talks), 461
Sarcophagus of Chernobyl-4, 311
Saturn, 329-330
Sausage in plasma, 201
Savannah, merchant ship, 325
Scattering cross section, 49
Scattering cross section for hydrogen (graph), 53
Scattering length, 281
Scattering of electrons, 66
Scintillation counter, 129-130
Scintillation detection system (diagram), 129
Screwworm fly, 278, 285
SDI (Strategic Defense Initiative), 462
Seaborg, Glenn, 219
Second law of thermodynamics, 169
Secular equilibrium, 34
Seed germination study by neutron
       radiography, 258
SEI (Space Exploration Initiative), 331
Semiconductor, 130, 131, 269
Semilog plot, 35
Separation factor in gaseous diffusion, 117
Separation factors in centrifuge, 117
Separation of deuterium, 119
Separation of U-235, 111-122, 363
Separative work units, 114
September 11, 2001, 136, 245, 312, 404
Sheffield waste disposal site, 382
Shell model of atoms, 19
Shipping casks, 367-369
Shippingport, 222
Shock wave, 203
Shroud of Turin, 252
SI units, 6, 13
Sievert (unit), 232, 233
Silex process, 118
Silicon, 131
Simulation, 423
Single-channel analyzer, 134
SIT (sterile insect technique), 278-279
Sizes of atoms and nuclei, 23
Slovakia, 444
Slovenia, 444
Smyth Report, 122, 139, 226
SNAP (Systems for Auxiliary Nuclear Power), 326
SNS (Spallation Neutron Source), 102
Sodium as coolant, 183
Sodium iodide, 129, 249
Sodium-22, 32
Sodium-24, 38, 247

Sodium-graphite reactor, 222
Soil moisture measurement, 261
Sojourner, 330
Solar energy, 485
Solar radiation, 238
Solid-state detector, 130-132
Solid-state detector (diagram), 131
Soluble poison, 152
Solvent extraction (diagram), 370
Somatic effects of radiation, 231
Source terms, 308
Sources of radioisotopes, 258
South Africa, 446
South Korea, 445
Space isotopic power, 327-331
Space reactors, 325-327
Space shuttle, 224
Spain, 442
Spallation, 102-103, 467
Spallation neutron research, 102, 104, 378, 467
Spallation Neutron Source (SNS), 102
Sparkplug of hydrogen bomb, 456
Special relativity, 8, 11, 28
Specific heat, 5, 16, 163
Specific inventory, 181
SPECT (Single Photon Emission Computer
       Tomography), 250
Speed, 3, 4
   composition, 366, 367
   molecular, 17
   most probable, 16
   of falling water, 4
   of gas particles, 16
   of light, 7-8
   of particles, 8, 41
   of particles in betatron, 98
Spoilage of food, 272
Spontaneous fission, 75
Stable isotopes, 246
Stage of isotope separator, 111
Standard deviation, 133
Standard Model, 101
Star Wars, 462
START (Strategic Arms Reduction Talks), 462, 467
Statistics of counting, 132-133
Steam generation, 167-169
Steam generator, 167-169
Stellarator, 200
Sterile insect technique (SIT), 278-279
Sterilization of medical supplies, 276-277
Stewardship, 386
Stonehenge dating, 252

Stopping power, 66
Storage pool, 365
Storage racks, 365
Stranded costs, 426
Strassman, 218
Strategic Arms Limitation Talks (SALT), 461
Strategic Arms Reduction Talks (START), 462, 467
Strategic Defense Initiative (SDI), 462
Strategic Offensive Reduction Treaty (SORT), 462
Strategic Plan for Building New Nuclear Power Plants, 427
Stream of particles, 46
Stripping reaction, 102
Strontium-90, 35, 259, 356
Subcritical, 141, 156
Subcritical multiplication factor, 290
Submarine, nuclear, 323–324
Supercollider, 101
Supercompactors, 379
Superconducting magnets, 100
Supercritical, 141, 153
Superphenix, 183, 441
Survey meters, 347
Sustainable development, 478–480
Sweden, 442
Switzerland, 442
SWU, 114
Symbols for isotopes, 22
Synchrotron, 99–102
Synchrotron radiation, 102, 105
Synchrotron X-rays, 283–284
Systeme Internationale, 6
Szilard, 219

T

Tactical weapons, 460
Tails of isotope separation, 118
Taiwan, 445
Tamper, 454
Tanks, underground, 363
Technetium-99m, 248–249
Technical Specifications, 299
Temperature, 5
    coefficient, 293
    conversions, 10
    difference in fuel, 148
    effect on neutron multiplication, 293
    gas, 16
    plasma, 86
Terraforming of Mars, 333
Terrorism, 125, 239, 397, 404, 409

Terrorists, 463–464
Tevatron, 100
Textile manufacturing using NAA, 255
TFTR (Tokamak Fusion Test Reactor), 202
Thermal efficiency, 169, 173
Thermal energy, 5–6, 73
Thermal neutron absorption cross sections (table), 50
Thermal neutron flux, 280
Thermal neutrons, 80
Thermal neutrons, detection of, 127
Thermal pollution, 171
Thermal reactor, 145, 180, 181
Thermal utilization, 147
Thermodynamics, 169
Thermodynamics, laws of, 4
Thermoelectric conversion, 326
Thermoluminescence, 252
Thermoluminescent dosimeter (TLD), 130
Thermonuclear explosions, 197, 454
Thermonuclear reactions, 86–87, 203
Thermonuclear weapon (diagram), 457
Thickness gauges, 259
Thorium-232, 34, 79, 179
Three Mile Island accident
    causes, 307
    effect on nuclear power industry, 422
    recovery, 307
Three Mile Island Unit 2 (TMI), 318
Threshold, 236
Thyroid gland, 231, 348
Time per revolution in accelerator, 97
TMI (Three Mile Island Unit 2), 318
Tokamak, 200
Tokamak Fusion Test Reactor (TFTR), 202
Tokamak-60 (Japan), 203
Torus, 200
Tracers, 248
Transmutation, 41–43
Transmutation doping of semiconductors, 279
Transport mean free path, 56, 58
Transportation, 366
Transportation of wastes, 387
Transportation regulations, 366
Transuranic wastes (TRU), 364, 386, 388
Trident strategic missiles, 324
Tritium
    in fusion, 42, 457
    in reactor waste, 382
    isotope of hydrogen, 32
Tritium
    production, 42, 466–467

TRU (Transuranic wastes), 364, 386, 388
TRUPAC II, 386
Tsetse fly, 278
Tuff, 375
Turbine, 167–168, 173
Turkey, 446
Two-phase flow, 167

U

U.S. energy problem, 473
U.S. Enrichment Corporation (USEC), 112, 115, 118, 408
Ukraine, 443
Ulysses spacecraft, 330
Uncollided flux, 345
Underground waste tanks, 363
Unique radiolytic products (URP), 272
United Kingdom, 441
University of Chicago, 220
Unplanned scrams trend (diagram), 403
Uranium
    atomic weight, 15
    composition, 366
    cross section (graph), 52
    decay, 29
    demand and resources (table), 189
    deposits in U.S., 188
    economics, 426
    energy content, 190
    hexafluoride, 112, 362
    in explosive, 454
    isotope half-lives, 156
    macroscopic cross section, 46–48
    ore, 361
    oxide, 151
    reserves of U.S. (table), 190
    resources (table), 179, 187–191
    separation by laser, 118–119
Uranium-233
    fissile isotope, 75
    fuel for breeder, 181
    reaction equations to produce, 76
Uranium-234 content in uranium, 114
Uranium-235
    absorption cross section, 47
    consumption, 153–154
    critical mass (table), 456
    enrichment, 298
    fission, 76, 77
    half-life, 31
    separation, 118–119
    weapon material, 219

Uranium-236
    from radiative capture, 78–79, 144
    in fission, 73, 74
Uranium-238
    consumption, 181–182
    fertile isotope, 79
    half-life, 31, 34
    precursor of radon, 349
Uranium-lead dating, 263
USEC (U.S. Enrichment Corporation), 112, 115, 118
Useful radiation effects, 269–284

V

Valence shell, 19
Van de Graaff accelerator, 95, 96, 99
Vapor shield in fusion machine, 202
Vaporization, 7
Veksler, V. I., 99
Velocity of light, 10
Venn diagrams, 305
Very High Temperature Reactor (VHTR), 432, 491, 492
Viability Assessment document, 375
Virtual Nuclear Tourist, 437
Virtual reality, 424
Vision 2020, 430, 431
Vitrification of plutonium, 468
Voltage multiplier, 95
Volume flow rate of coolant, 164
Volume of nucleus, 17
Voyager spacecraft, 329

W

WANO (World Association of Nuclear Operators), 404, 444
Ward Valley, 385
Waste classification, 363–364
Waste heat, 169–173
Waste Isolation Pilot Plant (WIPP), 386, 388
Waste of isotope separator, 112, 113
Waste power, 170
Waste radionuclides, 390
Waste volume, 379
Water, 54
    diffusion length, 55
    moderator, 149–150, 165
    molecular number density, 121
    neutron mean free path, 293
Watson, 248
Watt (unit), 5
Wave length of particle, 281

Wave mechanics, 281
Wavelength, 7, 17
Weapons
  plutonium, 188, 453–457
  thermonuclear, 456–460
  uranium, 454–457
Weapons-grade plutonium, 454
Weight percent (w/o), 113, 114
Weighting factors, organ and tissue (table), 354
West Valley, 369
Westinghouse, 168, 222, 323, 407, 420, 428, 430,
      445, 447
Westinghouse AP600, 428
Wheeling of electric power, 425
Wigner, Eugene, 218, 219
Wikipedia, 12, 101, 510
WIPP (Waste Isolation Pilot Plant), 386, 388
Wood treatment by radiation, 284
Work (definition), 3
World annual electricity production (table), 476
World Association of Nuclear Operators (WANO),
      404, 444
World energy consumption trends (graph), 477
World energy problem, 474
World energy use, 475–478
World nuclear power (table), 438
World population data (table), 477
World population growth (graph), 478
World primary energy (table), 476

World Total Net Electricity Generation
      (table), 476
World War II, 219
WWW Virtual Library, 12, 106, 160

X
X-rays
  fluorescence spectrometry, 257
  from K-capture, 32
  hazard, 229
  in hydrogen bomb, 454
  medical diagnosis, 257
  origin, 7, 63
  radiography, 257–259
  treatment, 269
  use in inertial confinement fusion, 197
Xenon-135, 155

Y
Yalow, 250
Yellowcake, 361
Yield of fission products (graph), 77
Yield of neutrons from spallation, 102
Yucca Mountain, 375–378

Z
Zircaloy, 149
Zirconium, 58